Statistical Methods for Engineers and Scientists

STATISTICS: Textbooks and Monographs

A SERIES EDITED BY

D. B. OWEN, Coordinating Editor

Department of Statistics
Southern Methodist University
Dallas, Texas

Vol. 1: The Generalized Jacknife Statistic, *H. L. Gray and W. R. Schucany*
Vol. 2: Multivariate Analysis, *Anant M. Kshirsagar*
Vol. 3: Statistics and Society, *Walter T. Federer*
Vol. 4: Multivariate Analysis: A Selected and Abstracted Bibliography, 1957-1972, *Kocherlakota Subrahmaniam and Kathleen Subrahmaniam* (out of print)
Vol. 5: Design of Experiments: A Realistic Approach, *Virgil L. Anderson and Robert A. McLean*
Vol. 6: Statistical and Mathematical Aspects of Pollution Problems, *John W. Pratt*
Vol. 7: Introduction to Probability and Statistics (in two parts), Part I: Probability; Part II: Statistics, *Narayan C. Giri*
Vol. 8: Statistical Theory of the Analysis of Experimental Designs, *J. Ogawa*
Vol. 9: Statistical Techniques in Simulation (in two parts), *Jack P. C. Kleijnen*
Vol. 10: Data Quality Control and Editing, *Joseph I. Naus* (out of print)
Vol. 11: Cost of Living Index Numbers: Practice, Precision, and Theory, *Kali S. Banerjee*
Vol. 12: Weighing Designs: For Chemistry, Medicine, Economics, Operations Research, Statistics, *Kali S. Banerjee*
Vol. 13: The Search for Oil: Some Statistical Methods and Techniques, *edited by D. B. Owen*
Vol. 14: Sample Size Choice: Charts for Experiments with Linear Models, *Robert E. Odeh and Martin Fox*
Vol. 15: Statistical Methods for Engineers and Scientists, *Robert M. Bethea, Benjamin S. Duran, and Thomas L. Boullion*
Vol. 16: Statistical Quality Control Methods, *Irving W. Burr*
Vol. 17: On the History of Statistics and Probability, *edited by D. B. Owen*
Vol. 18: Econometrics, *Peter Schmidt*
Vol. 19: Sufficient Statistics: Selected Contributions, *Vasant S. Huzurbazar (edited by Anant M. Kshirsagar)*
Vol. 20: Handbook of Statistical Distributions, *Jagdish K. Patel, C. H. Kapadia, and D. B. Owen*
Vol. 21: Case Studies in Sample Design, *A. C. Rosander*
Vol. 22: Pocket Book of Statistical Tables, *compiled by R. E. Odeh, D. B. Owen, Z. W. Birnbaum, and L. Fisher*
Vol. 23: The Information in Contingency Tables, *D. V. Gokhale and Solomon Kullback*
Vol. 24: Statistical Analysis of Reliability and Life-Testing Models: Theory and Methods, *Lee J. Bain*
Vol. 25: Elementary Statistical Quality Control, *Irving W. Burr*
Vol. 26: An Introduction to Probability and Statistics Using BASIC, *Richard A. Groeneveld*
Vol. 27: Basic Applied Statistics, *B. L. Raktoe and J. J. Hubert*
Vol. 28: A Primer in Probability, *Kathleen Subrahmaniam*
Vol. 29: Random Processes: A First Look, *R. Syski*

Vol. 30: Regression Methods: A Tool for Data Analysis, *Rudolf J. Freund and Paul D. Minton*
Vol. 31: Randomization Tests, *Eugene S. Edgington*
Vol. 32: Tables for Normal Tolerance Limits, Sampling Plans, and Screening, *Robert E. Odeh and D. B. Owen*
Vol. 33: Statistical Computing, *William J. Kennedy, Jr. and James E. Gentle*
Vol. 34: Regression Analysis and Its Application: A Data-Oriented Approach, *Richard F. Gunst and Robert L. Mason*
Vol. 35: Scientific Strategies to Save Your Life, *I. D. J. Bross*
Vol. 36: Statistics in the Pharmaceutical Industry, *edited by C. Ralph Buncher and Jia-Yeong Tsay*
Vol. 37: Sampling from a Finite Population, *J. Hajek*
Vol. 38: Statistical Modeling Techniques, *S. S. Shapiro*
Vol. 39: Statistical Theory and Inference in Research, *T. A. Bancroft and C.-P. Han*
Vol. 40: Handbook of the Normal Distribution, *Jagdish K. Patel and Campbell B. Read*
Vol. 41: Recent Advances in Regression Methods, *Hrishikesh D. Vinod and Aman Ullah*
Vol. 42: Acceptance Sampling in Quality Control, *Edward G. Schilling*
Vol. 43: The Randomized Clinical Trial and Therapeutic Decisions, *edited by Niels Tygstrup, John M. Lachin, and Erik Juhl*
Vol. 44: Regression Analysis of Survival Data in Cancer Chemotherapy, *Walter H. Carter, Jr., Galen L. Wampler, and Donald M. Stablein*
Vol. 45: A Course in Linear Models, *Anant M. Kshirsagar*
Vol. 46: Clinical Trials: Issues and Approaches, *edited by Stanley H. Shapiro and Thomas H. Louis*
Vol. 47: Statistical Analysis of DNA Sequence Data, *edited by B. S. Weir*
Vol. 48: Nonlinear Regression Modeling: A Unified Practical Approach, *David A. Ratkowsky*
Vol. 49: Attribute Sampling Plans, Tables of Tests and Confidence Limits for Proportions, *Robert E. Odeh and D. B. Owen*
Vol. 50: Experimental Design, Statistical Models, and Genetic Statistics, *edited by Klaus Hinkelmann*
Vol. 51: Statistical Methods for Cancer Studies, *edited by Richard G. Cornell*
Vol. 52: Practical Statistical Sampling for Auditors, *Arthur J. Wilburn*
Vol. 53: Statistical Signal Processing, *edited by Edward J. Wegman and James G. Smith*
Vol. 54: Self-Organizing Methods in Modeling: GMDH Type Algorithms, *edited by Stanley J. Farlow*
Vol. 55: Applied Factorial and Fractional Designs, *Robert A. McLean and Virgil L. Anderson*
Vol. 56: Design of Experiments: Ranking and Selection, *edited by Thomas J. Santner and Ajit C. Tamhane*
Vol. 57: Statistical Methods for Engineers and Scientists. Second Edition, Revised and Expanded, *Robert M. Bethea, Benjamin S. Duran, and Thomas L. Boullion*
Vol. 58: Ensemble Modeling: Inference from Small-Scale Properties to Large-Scale Systems, *Alan E. Gelfand and Crayton C. Walker*

OTHER VOLUMES IN PREPARATION

Statistical Methods for Engineers and Scientists

Second Edition, Revised and Expanded

Robert M. Bethea

Chemical Engineering Department
Texas Tech University
Lubbock, Texas

Benjamin S. Duran

Department of Mathematics
Texas Tech University
Lubbock, Texas

Thomas L. Boullion

Department of Mathematics and Statistics
University of Southwestern Louisiana
Lafayette, Louisiana

MARCEL DEKKER, INC. New York and Basel

Library of Congress Cataloging in Publication Data

Bethea, Robert M.
Statistical methods for engineers and scientists.

(Statistics, textbooks and monographs ; v. 57)
Includes bibliographical references and index.
1. Mathematical statistics. 2. Probabilities.
I. Duran, Benjamin S., [date]. II. Boullion, Thomas L. III. Title. IV. Series.
QA276.B425 1984 519.5 84-7774
ISBN 0-8247-7227-X

MARCEL DEKKER, INC.
270 Madison Avenue, New York, New York 10016

Current printing (last digit):
10 9 8 7 6 5 4 3 3 2 1

PRINTED IN THE UNITED STATES OF AMERICA

In gratitude and affection for their patience, understanding, and assistance during the preparation of our text, we dedicate this book to Nancy Jo, Dolores, and Nancy.

PREFACE TO THE SECOND EDITION

We have used the first edition of this book as a text many times. Based on this experience and the comments made by our students, colleagues, and others, we have revised and updated (in terms of computer usage) the text to make it a better instrument for classroom instruction or self-study. Our suggestions for course arrangements remain as stated in the Preface to the First Edition. This text is also useful for professional engineers and scientists especially because of its emphasis on regression, experimental design, analysis of variance, and computerized data reduction and interpretation.

The major changes include an approximately forty percent increase in the number of homework problems, incorporation of the Statistical Analysis System (SAS*) programs throughout, addition of several topics, and a reorganization of several chapters.

Chapters 1-7 remain essentially the same as in the first edition except for the inclusion of more homework problems and blending in the use of the SAS System in several worked examples. The chi-square goodness-of-fit test has been added in Chapter 7.

The major reorganization has taken place in Chapters 8, 9, and 11. Some material, including the three-way analysis of variance, has been moved from Chapter 8 to Chapter 11. A section dealing with multiple comparisons has been added in Chapter 8. Several changes have been made in Chapter 9, including an extension of the nonlinear regression section. Several topics have been

---

*SAS is the registered trademark of SAS Institute Inc., Cary, North Carolina 27511-8000.

added in Chapter 11, including analysis of covariance and nested designs. Numerous examples, worked by the use of the SAS System programs, which are included, have been woven into these chapters. In addition to the major revisions, slight changes have been made throughout the text to improve readability and presentation.

We wish to thank those persons, societies, publishers, and organizations listed in the Preface to the First Edition who have graciously extended their permissions to reproduce and use data for examples and homework problems in this edition of our text. To that list, we wish to add Imperial Chemical Industries PLC; Butterworth's; John Wiley & Sons, Inc.; and the National Academy of Sciences. All of the above courtesies are acknowledged in the appropriate locations in the text. We are grateful to the Literary Executor of the late Sir Ronald A. Fisher, F.R.S. to Dr. Frank Yates, F.R.S., and to Longman Group Ltd., London for permission to reprint Table XXIII from their book Statistical Tables for Biological, Agricultural and Medical Research (6th Edition, 1974).

We are grateful to SAS Institute Inc. for permission to use and reproduce the SAS programs and other material which are their property. We appreciate the comments we have received on the first edition from many colleagues, reviewers, and numerous students. We have incorporated many of those suggestions in this revision. We thank those students and others who have allowed us to use data from research and laboratory experiments for examples and problems. We especially wish to thank Sue Willis for typing all the revisions and the editorial staff who have been so helpful.

It goes without saying that without the patience and understanding of our families, this revision would never have been completed.

Robert M. Bethea
Benjamin S. Duran
Thomas L. Boullion

PREFACE TO THE FIRST EDITION

This book is intended as a basic introductory text in applied statistical techniques for undergraduate students in engineering and the physical sciences. Theoretical developments and mathematical treatment of the principles involved are included as needed for understanding of the validity of the techniques presented. More intensive treatment of these subjects should be obtained, if desired, from the many theoretical statistics texts available.

The material in this text can be arranged for either two three-credit quarter courses or one three-credit semester course at the option of the instructor. For the former case, the material would be covered *in toto* with additional time allowed throughout for applications in the disciplines of the students involved. The most logical place for division is at the end of Chapter 7 on statistical inference. For the latter case the material in the first four chapters is presented in three weeks. Chapter 5 is allotted one week; Chapters 6 through 8, two weeks each; Chapters 9 and 10 are covered in four weeks total; and Chapter 11, three weeks. Instructors desiring only one three-credit quarter course in methodology would probably use only Chapters 6 through 9 provided the students have been introduced to basics.

Although many of the example problems are oriented toward chemical engineering and chemistry, by no means is this text limited to those areas. The examples merely illustrate general statistical principles as applicable to all fields, but particularly engineering and the related physical sciences. The problems at the end of each chapter are graded in difficulty for the convenience of the instructor. They are arranged to logically follow the outline of material presented in the text.

Our deepest appreciation is extended to our good friends and former departmental chairmen, Arnold J. Gully and Patrick L. Odell, for their encouragement, constructive criticism of all parts of this text, and above all for allowing us adequate time for the preparation of the manuscript. Appreciation is also extended to James M. Davenport for his assistance in the tabulation of the F-distribution.

We are indebted to the many students and associates who have contributed to this text by way of suggestions for improvements, problem statements, experimental data, and consultation. We are grateful to Dr. H. T. David and Dr. D. V. Huntsberger of the Department of Statistics at Iowa State University and Dr. F. B. Cady now of the Statistics Department of the University of Kentucky for permission to use as illustrative examples and/or homework problems some of the problems they assigned one of us (RMB) in their respective classes or on preliminary examinations. We are indebted to the Literary Executor of the late Sir Ronald A. Fisher, F.R.S., to Dr. Frank Yates, F.R.S., and to Oliver and Boyd, Ltd., Edinburgh, for permission to reprint Table XXIII from their book, _Statistical Tables for Biological, Agricultural and Medical Research_.

In addition to these, our thanks go to The Iowa State University Press, Addison-Wesley Publishing Co., Prentice-Hall, Inc., and Gulf Publishing Co. for permission to reprint data for use in examples and problems. Thanks are also due to the editors of the American Institute of Chemical Engineers, the Committee of Editors of the American Chemical Society, the American Society for Quality Control, the American Society for Testing and

Materials, and the American Society of Mechanical Engineers for permission to reprint data extracted from articles in their respective publications. All these courtesies are specifically acknowledged at the appropriate location in the text.

Our thanks go to Diane Frazier for typing the original manuscript and to Lillian Lavender, Peggy Boyd, Jill Perry, Elizabeth Pigg, Kathy Radenz, and Lillian Bonner who patiently and carefully retyped all the revisions, problems, etc. Without their help, this text would still be in our notes and on several boxes of dictaphone belts.

Robert M. Bethea
Benjamin S. Duran
Thomas L. Boullion

CONTENTS

Preface to the Second Edition v

Preface to the First Edition vii

List of Worked Examples xvii

List of Tables xxiii

Chapter 1. INTRODUCTION 1

Chapter 2. PROBABILITY 6

2.1 Introduction 6

2.2 Definition of Probability 6

2.3 Possible Ways for Events to Occur 11

2.3.1 Permutations 11

2.3.2 Combinations 13

2.4 Probability Computation Rules 14

2.5 A Posteriori Probability 21

Problems 24

Chapter 3. DISTRIBUTIONS 31

3.1 Introduction 31

3.2 Definitions 31

3.2.1 Discrete Distributions 33

3.2.2 Continuous Distributions 35

3.2.3 Experimental Distributions 36

3.3 Theoretical Distributions 50

3.3.1 Bionomial Distribution 55

3.3.2 Poisson and Exponential Distributions 59

3.3.3 Negative Binomial Distribution 67

3.3.4 Hypergeometric Distribution 68

3.3.5 Weibull Distribution 71
3.3.6 Gamma Distribution 73
3.3.7 Normal Distribution 73
3.4 Other Theoretical Distributions 81
3.4.1 Chi-Square Distribution 81
3.4.2 The "Student's" t-Distribution 82
3.4.3 F-Distribution 82
Problems 83

Chapter 4. DESCRIPTIVE STATISTICS 102
4.1 Introduction 102
4.2 Measures of Location 103
4.3 Measures of Variability 105
Problems 110

Chapter 5. EXPECTED VALUES AND MOMENTS 111
5.1 Introduction 111
5.2 Discrete Distributions 113
5.3 Continuous Distributions 115
5.4 Joint Distributions and Independence of Random Variables 116
5.5 Moments 121
5.6 Examples 122
Problems 125

Chapter 6. STATISTICAL INFERENCE: ESTIMATION 127
6.1 Introduction 127
6.2 Statistical Estimation 127
6.3 Point Estimates 128
6.4 Interval Estimates 132
6.5 Chi-Square Distributions 134
6.6 The t-Distribution 136
6.7 The F-Distribution 138
6.8 Estimation of the Mean 140
6.9 Comparison of Two Means 146

6.10 Estimation Involving Paired Observations 153
6.11 The Variance 157
6.12 Estimation of a Variance 157
6.13 Comparison of Two Variances 159
6.14 Estimation of a Proportion P 162
6.15 Comparison of Two Proportions 167
Problems 169

Chapter 7. STATISTICAL INFERENCE: HYPOTHESIS TESTING 179
7.1 Introduction 179
7.2 Types of Errors 180
7.3 Testing of Hypotheses 181
7.4 One-Tailed and Two-Tailed Tests 182
7.5 Tests Concerning the Mean 186
7.6 Tests on the Difference of Two Means 193
7.7 Paired t-Test 202
7.8 Testing a Proportion P 205
7.9 Testing the Difference of Two Proportions 206
7.10 Tests Concerning the Variance 208
7.11 Testing the Equality of Variances 210
7.12 Other χ^2 Tests 213
7.13 Contingency Tests 217
Problems 220

Chapter 8. ANALYSIS OF VARIANCE 241
8.1 Introduction 241
8.2 General Linear Model 243
8.3 One-Way Analysis of Variance 245
8.3.1 Pooled Variance Estimates 245
8.3.2 Variance of Group Means 247
8.3.3 Model for One-Way Analysis of Variance 249
8.4 Two-Way Analysis of Variance 259
8.4.1 Model for Two-Way Analysis of Variance 259
8.5 Confidence Intervals and Tests of Hypotheses 266

8.6 Multiple Comparisons Among Treatment Means 271
8.7 Bartlett's Test for Equality of Variances 278
Problems 282
References 299

Chapter 9. REGRESSION ANALYSIS 301
9.1 Introduction 301
9.2 Simple Linear Regression 301
9.2.1 Interval Estimation in Simple Linear Regression 312
9.2.2 Hypothesis Testing in Simple Linear Regression 317
9.2.3 Inverse Prediction in Simple Linear Regression 319
9.2.4 Analysis of Variance in Simple Linear Regression 321
9.2.5 Lack of Fit 328
9.2.6 Regression Through a Point 329
9.3 Regression Using Matrices 330
9.4 Multiple Linear Regression 332
9.5 Polynomial Regression 343
9.6 Nonlinear Regression 351
9.7 Correlation Analysis 363
9.7.1 Correlation in Simple Linear Regression 364
9.7.2 Correlation in Multiple Linear Regression 368
9.8 Stepwise Regression 373
9.9 Testing Equality of Slopes 388
9.10 Transformation of Data in Regression Analysis 392
9.10.1 Propagation of Error 392
9.10.2 On Transforming the Data 394
Problems 394
References 421

Chapter 10. ORTHOGONAL POLYNOMIALS IN POLYNOMIAL REGRESSION 424
10.1 Introduction 424
10.1.1 Orthogonal Model 425
10.2 Fitting a Quadratic by Orthogonal Polynomials 433
10.3 Tests of Significance 434
Problems 442
References 444

Chapter 11. EXPERIMENTAL DESIGN 445
11.1 Introduction 445
11.2 Sources of Error 462
11.3 Completely Randomized Designs 462
11.3.1 Analysis of Variance 463
11.3.2 Subsampling in a Completely Randomized Design 468
11.3.3 Interaction 473
11.3.4 Model for Two-Way Analysis of Variance 473
11.3.5 Model for the Three-Way Analysis of Variance 481
11.3.6 Four-Way Analysis of Variance 488
11.3.7 Nested Designs 493
11.4 Randomized Complete Block Design 499
11.4.1 Analysis of Variance, RCB 500
11.4.2 Missing Data, RCB 504
11.4.3 Paired Observations, RCB 507
11.4.4 Subsampling in a Randomized Complete Block Design 508
11.5 Latin Square Designs 522
11.5.1 Analysis of Variance for the Latin Square 524

11.5.2 Missing Data, LS 529
11.6 Graeco-Latin Square 533
11.7 Factorial Experiments 534
11.7.1 Main Effects 535
11.7.2 Confounding 537
11.8 Other Designs 541
11.8.1 Split-Plot Designs 541
11.8.2 Incomplete Block Designs 545
11.8.3 Box-Wilson Composite Rotatable Design 549
11.9 Design Efficiency 553
11.10 Analysis of Covariance 555
Problems 562
References 584

Appendix A: Matrix Algebra 585

Appendix B: Introduction to SAS 593

Appendix C: Tables of Statistical Functions 600
Table I. Binomial Cumulative Distribution 600
Table II. Poisson Cumulative Distribution 609
Table III. Standard Normal Cumulative Distribution 615
Table IV. Cumulative t-Distribution 623
Table V. Cumulative Chi-Square (χ^2) Distribution 625
Table VI. Cumulative F-Distribution 630

Appendix D: Answers to Selected Problems 655

Index 670

LIST OF WORKED EXAMPLES

Example No.	Illustration of	
2.1	Use of permutations	12
2.2	Use of permutations	13
2.3	Use of combinations	13
2.4	Formulas for probabilities of compound events	15
2.5	Formulas for probabilities of compound events	17
2.6	Formulas for conditional probability	18
2.7	Formulas for conditional probability	19
2.8	Binomial distribution	21
2.9	Bayes formula	22
2.10	Bayes formula	23
3.1	Experimental distribution: computer solution	46
3.2	Binomial distribution	57
3.3	Binomial distribution	58
3.4	Binomial distribution	58
3.5	Binomial distribution	58
3.6	Binomial distribution	59
3.7	Poisson distribution	62
3.8	Poisson distribution: approximation to the binomial	62
3.9	Negative binomial distribution	68
3.10	Hypergeometric distribution	68
3.11	Hypergeometric distribution	70
3.12	Use of normal distribution tables	77
3.13	Use of normal distribution tables	78

Example No.	Illustration of	
3.14	Use of normal distribution tables	78
3.15	Use of normal distribution tables	79
3.16	Use of normal distribution tables	80
4.1	Sample means	104
4.2	Measures of location and variation	108
4.3	Measures of location and variation	109
4.4	Measures of location and variation: computer solution	110
5.1	Population mean and variance	122
5.2	Probability density function	122
5.3	Population mean and variance	123
5.4	Population mean and variance of binomial distribution	123
5.5	Population mean and variance of binomial distribution	125
6.1	Confidence interval on μ; σ^2 known	142
6.2	Confidence interval on μ; σ^2 unknown	144
6.3	Confidence interval on μ; σ^2 unknown	145
6.4	Confidence interval on $\mu_1 - \mu_2$; σ_1^2 and σ_2^2 known	148
6.5	Confidence interval on $\mu_1 - \mu_2$; σ_1^2 and σ_2^2 are unknown but equal	151
6.6	Confidence interval on $\mu_1 - \mu_2$; paired observations	154
6.7	Confidence interval on $\mu_1 - \mu_2$; paired observations	156
6.8	Confidence interval on σ^2	158
6.9	Confidence interval on σ_1^2/σ_2^2	161
6.10	Confidence interval on a proportion P	164
6.11	Confidence interval on a proportion P	166
6.12	Confidence interval on a proportion P	166
6.13	Confidence interval on $P_1 - P_2$	168

Example No.	Illustration of	
7.1	Test of hypothesis $\mu = \mu_0$	185
7.2	Test of hypothesis $\mu = \mu_0$	189
7.3	Test of hypothesis $\mu \geq \mu_0$	191
7.4	Test of hypothesis $\mu \geq \mu_0$	192
7.5	Test of hypothesis $\mu_1 = \mu_2$	195
7.6	Test of hypothesis $\mu_1 = \mu_2$	196
7.7	Test of hypothesis $\overline{\mu} = \overline{\mu}_1$	198
7.8	Computerized data reduction: $\mu_1 = \mu_2$	201
7.9	Test of hypothesis $\mu_D \leq 0$ $(\mu_1 - \mu_2 \leq 0)$	203
7.10	Computerized data reduction: paired comparisons	204
7.11	Test of hypothesis $P \leq P_0$	205
7.12	Test of hypothesis $P_1 = P_2$	207
7.13	Test of hypothesis $\sigma^2 \geq \sigma_0^2$	209
7.14	Test of hypothesis $\sigma_1^2 = \sigma_2^2$	211
7.15	Chi-square test that $P = P_0$	213
7.16	Chi-square: test for normality	215
7.17	Chi-square test: contingency table	217
8.1	One-way AOV	253
8.2	One-way AOV: computer solution, equal observations	255
8.3	One-way AOV: computer solution, unequal observations	257
8.4	Two-way AOV	262
8.5	Two-way AOV: computer solution	264
8.6	Confidence interval for contrasts	268
8.7	Confidence interval for contrasts	270
8.8	Multiple confidence intervals: Scheffé	272
8.9	Multiple comparisons: Scheffé and Duncan's methods: computerized solution	274
8.10	Bartlett's test of homogeneity of variances	280

Example No.	Illustration of	
9.1	Simple linear regression	308
9.2	Simple linear regression and related confidence intervals	315
9.3	Hypothesis testing in simple linear regression	317
9.4	Confidence interval for the true value of X	320
9.5	Simple linear regression: computer solution	324
9.6	Regression through the origin	329
9.7	Multiple linear regression	335
9.8	Multiple linear regression: computer solution	339
9.9	Polynomial regression	345
9.10	Quadratic regression: computer solution	346
9.11	Nonlinear regression	355
9.12	Nonlinear regression by log-log transformation: computer solution	357
9.13	Nonlinear regression	362
9.14	Correlation in simple linear regression	365
9.15	Test of hypothesis $\rho \geq \rho_0$	368
9.16	Multiple correlation index	369
9.17	Curve fitting	370
9.18	Empirical modeling by computer: stepwise regression	374
9.19	Empirical model selection by R^2: computer solution	382
9.20	Comparison of slopes	390
10.1	Orthogonal polynomials in regression: cubic	436
10.2	Orthogonal polynomials in regression: quadratic	439
11.1	Grouping of experimental units	448
11.2	Grouping of experimental units	448

Example No.	Illustration of	
11.3	Power of a test	454
11.4	Determination of number of replicates	455
11.5	Determination of number of replicates	455
11.6	Analysis of completely randomized design	465
11.7	Subsampling in completely randomized design: computer solution	470
11.8	Two-way AOV with interaction: computer solution	477
11.9	Two-way AOV with interaction: computer solution	480
11.10	Three-way AOV	484
11.11	Analysis of completely randomized design with factorial treatment combinations	488
11.12	Analysis of nested design	495
11.13	Analysis of RCB	501
11.14	Analysis of RCB with missing data	505
11.15	Analysis of RCB with subsampling: orthogonal polynomial and computer solutions	508
11.16	Analysis of RCB design: subsampling	519
11.17	Analysis of a Latin square design	522
11.18	Analysis of a Latin square design	525
11.19	Analysis of a Latin square design: missing data	530
11.20	Analysis of a factorial experiment	539
11.21	Analysis of a split-plot design	542
11.22	Analysis of balanced incomplete block design	546
11.23	Box-Wilson type design	549
11.24	Analysis of covariance: computer solution	559

LIST OF TABLES

Table No.	Table Illustrating	
2.1	Number of Heads per Group in Coin-Tossing Experiment	9
3.1	Experimental Distribution of Die Outcome Data	37
3.2	Ring and Ball Melting Point Data	38
3.3	Rank Order for Melting Point Data	40
3.4	Grouped Frequencies of Melting Point Data with 0.5°C Class Length	42
3.5	Grouped Frequencies of Melting Point Data with 1°C Class Length	44
7.1	Testing μ When σ^2 Is Known	189
7.2	Testing μ When σ^2 Is Unknown	190
7.3	Testing $\mu_1 - \mu_2$ When σ_1^2 and σ_2^2 Are Known	194
7.4	Testing $\mu_1 - \mu_2$ When σ_1^2 and σ_2^2 Are Unknown but Equal	195
7.5	Testing $\mu_D = 0$ When Observations Are Paired	203
7.6	Testing a Proportion P	205
7.7	Testing $P_1 = P_2$	207
7.8	Testing the Variance σ^2	209
7.9	Testing $\sigma_1^2 = \sigma_2^2$	211
8.1	Data Array	242
8.2	One-Way Analysis of Variance	253
8.3	Two-Way Analysis of Variance	262
9.1	Test Procedures in Simple Linear Regression	318
9.2	Analysis of Variance for Simple Linear Regression	322

Table No.	Table Illustrating	
9.3	Analaysis of Variance for Multiple Linear Regression	335
10.1	ξ_i' Values	429
11.1	Replications Required for a Given Probability of Obtaining a Significant Result for One-Class Data	451
11.2	Replications Required for a Given Probability of Obtaining a Significant Result for Two-Class Data	456
11.3	Analysis of Variance for Completely Randomized Design with Equal Numbers of Observations per Treatment	464
11.4	Analysis of Variance for Completely Randomized Design with Unequal Numbers of Observations per Treatment	468
11.5	Analysis of Variance for Subsampling in a Completely Randomized Design (Equal Subclass Numbers)	469
11.6	Analysis of Variance for Factorial Treatment Combinations in Completely Randomized Design, Model I	474
11.7	Analysis of Variance for Factorial Treatment Combinations in a Completely Randomized Design, Model II	475
11.8	F-Ratios for Hypothesis Testing in Completely Randomized Designs with Factorial Treatment Combinations	476
11.9	Three-Way Analysis of Variance	482
11.10	Analysis of Variance of Randomized Complete Block Design with One Observation per Experimental Unit	500

Table No.	Table Illustrating	
11.11	Analysis of Variance for Randomized Complete Block Design with Subsampling in the Experimental Units (Model I)	509
11.12	Analysis of Variance for Cubic Regression in Randomized Complete Block Experiments	515
11.13	Analysis of Variance for an $m \times m$ Latin Square with One Observation per Experimental Unit	526
11.14	Results for Testing the Effect of Water Pressure, Air Flow, and Nozzle Operation on Efficiency	531
11.15	Latin Square for the Effect of Water Pressure, Air Flow, and Nozzle Operation on Efficiency	532
11.16	Analysis of Variance for Scrubber Tests	532
11.17	Analysis of Variance for Split Plots in Randomized Complete Block Experiment of r Replications	544
11.18	Analysis of Variance for Split Plots in Latin Square Experiments with r Replications	545
11.19	Results for the Analysis of the Effect of Air Flow Rate and Water Pressure on Efficiency at High Inlet Particulate Concentration Levels	547
11.20	Balanced Incomplete Block Design for the Effect of Air Flow and Water Pressure on the Nozzle Taps on Efficiency	548
11.21	Analysis of Variance for the Second Series of Air Pollution Control Tests	549
11.22	Results of Covariance Analysis in the One-Way AOV	557

1
INTRODUCTION

The subject matter of the field of statistics is often defined as the scientific study of techniques for collecting, analyzing, and drawing conclusions from data. The engineer and scientist use statistics as a tool, which, when correctly applied, is of enormous assistance in the study of the laws of physical science. This is why an introduction to statistical methods is useful to students who are preparing themselves for careers in these areas. There are no statistical procedures which are applicable only to specific fields of study. Instead, there are general statistical procedures which are applicable to any branch of knowledge in which observations are made. Statistical procedures now constitute an important part of all branches of science. Procedures which have been developed for use in one field are inevitably adapted for use in a number of other fields. However, some procedures are more frequently used in one group of related disciplines than in another. In this text, we concentrate on those procedures which are most widely used by engineers and scientists.

The scientific discipline of statistics has been used for describing, summarizing, and drawing conclusions from data. We, as scientists and engineers, have made enormous use of this most important and highly versatile discipline.

We use statistics principally to aid us in four ways. The first of these is to assist us in designing experiments and surveys. We desire that our experiments yield adequate answers to the questions which prompted their performance. The experiments should also be efficient ones. That is, they should provide the desired answers with maximum precision and accuracy with a minimum expenditure of time and other resources.

Another way in which the engineer and scientist use statistics is in describing and summarizing experimental data. This is properly termed descriptive statistics. The individual's first concern with a body of data is whether it is to be considered as all possible data or as part of a larger body. It is extremely important to make this distinction, since failure to do so often results in loose thinking and erroneous results being obtained. To be perfectly clear about this distinction, we define a population as the whole set or collection of similar values, attributes, etc., which are characteristic of a given experiment. Measurements or characteristics which are not constant, but show variability upon repeating the experiment under the same conditions, are called values of random variables. Thus, a population could be defined as the set of all possible values of a random variable. These values need not all be different nor finite in number.

A sample is a part of a population. In some situations, a sample may include the whole population. Usually, we desire to use sample information to make an inference about the population. For this reason, it is particularly important to define the population under consideration and to obtain a representative sample from the population. To obtain a representative sample, we embody in the rules for drawing the sample items the principle of randomness.

Randomly selected samples guarantee us that the mathematical laws of probability will hold. Thus, we are able to make statements about the population from which a sample is taken and give a quantitative measure of chance that these statements are true for the population.

A third use made of statistics is in testing hypotheses. You would not have performed an experiment or collected a set of data about some phenomenon without some definite purpose in mind. Your purpose may have been simply to satisfy your curiosity. It may have been an attempt to predict the behavior of a process or group of individuals. In any event, you will have had some idea, feeling, or belief that a population possessed a particular attribute. This, we say, is an hypothesis concerning the population. This is where statistics becomes invaluable. Through its proper use, we can plan experiments to evaluate any of these hypotheses. This is done by determining whether or not the results are reasonable and likely valid or are probably due strictly to random variation.

Thus far, the uses of statistics described above have been qualitative in nature. In many cases it is of equal if not greater importance to obtain quantitative relationships between variables. This is the last of the principal ways in which we use statistics. You have probably done this quite often in the past, many times subconsciously. Perhaps some of your most conscious efforts at quantification in this regard started with analytic geometry or the calculus. There you were confronted with the problem of describing a line, curve, surface, or other shape which itself defined one or more variables or occurrences in terms of others. Your most simple effort would have been to draw a line through a collection of data points and from it determine the equation relating the independent and dependent variables. You undoubtedly would have drawn that line so as to minimize the variation of data points around it. But the line you drew and the line drawn by someone else to describe the same data will almost invariably be different. We therefore say that the estimates obtained from these "eyeball" curve fits are empirical and quite subjective. A better method would have been for you both to use some consistent manner of determining the probable relation, if any, between the variables involved. Statistics here has provided us with just such procedures. The most common is the method of least squares.

In this text we will begin by studying the basic concepts of simple and compound probabilities of events occurring. From this you should obtain a feeling for the role played by chance in the outcome of investigations. These concepts are then applied to discrete and continuous distributions. You should then learn how to recognize the type of distribution involved so that you can handle sample data effectively.

After studying the ways in which data are distributed and the probabilities affecting those distributions, we then turn to ways of describing data. These descriptions are given in terms of measures of location and variability. The chapter on descriptive statistics is followed by a chapter on expectations. In that portion of the text, we will delve into theory to a sufficient degree so that it will become obvious why statistics calculated from samples can be used to adequately describe populations.

We next direct our attention to the methods involved in estimation and testing hypotheses about populations by use of sample data. Estimation includes procedures for constructing confidence intervals for the population parameters. In hypothesis testing the necessary statistical techniques for evaluating the hypotheses are developed. As a continuation of these topics we will then study the techniques of analysis of variance. These procedures are used for comparing several things at a time or several estimates of the same thing whenever certain assumptions are satisfied. They can also be used to aid us in making qualitative judgments concerning our observations.

The statistical methods involved in estimating quantitative relations between variables is the subject of the chapter dealing with regression and correlation analysis. In regression analysis we try to find the best estimates of the relationships between population variables from sample data and to test the validity of those estimates by suitable procedures. The technique for regression analysis presented is the method of least squares. This procedure is described for single and multiple independent variables in both linear and nonlinear forms. Correlation analysis is concerned with measuring the degree or ability of a regression equation to describe the actual relationship involved. A separate chapter is devoted to regression analysis involving orthogonal polynomials.

The final chapter of this book is devoted to experimental design. This section begins with the criteria used in designing efficient and effective experiments and continues with discussions concerning implementation of the designs and interpretation of the results. You will find the techniques developed in hypothesis testing, analysis of variance, and regression analysis used extensively throughout this portion of the text.

2
PROBABILITY

2.1 INTRODUCTION

Statistical treatment of data and the inferences or conclusions derived therefrom are the result of experimentation which involves the element of chance. The problem then is to measure or assess the degree of uncertainty in drawing an inference from a statistical treatment of the data. The degree of uncertainty is quantified by utilizing the notion of probability. Thus a person who expects to use statistical methodology should have at least a passing acquaintance with elementary probability. The objective of this chapter is to present probability concepts considered essential in statistical methodology and in the interpretation of the results of statistical analyses.

2.2 DEFINITION OF PROBABILITY

Consider an investigation which may be characterized in part by the fact that repeated experimentation, under essentially the same conditions, is necessary for its execution. For instance, an engineer may wish to study the effect that surface roughness has on the endurance of a certain bearing material. The only way in which an investigator can obtain information about any such phenomenon is to perform an experiment. Each experiment terminates with an outcome which cannot be predicted with certainty prior to the performance of the experiment. Thus we say there is an element of chance or randomness in the experiment.

Suppose we have such an experiment. If all the possible outcomes of the experiment can be described prior to its performance, and if it can be repeated under the same conditions, then it will be referred to as a random experiment. The set of all possible outcomes will be referred to as the sample space. For instance, the toss of a coin, assuming that the toss may be repeated under the same conditions, is an example of a random experiment having as its sample space the set {H,T}, where H denotes heads and T denotes tails. We will take a set to be a collection of objects and use braces to enclose the elements of a set.

Since the idea of a set allows us to discuss probability expeditiously, we will define several set operations which will be used later on. The set of outcomes contained in either A or B (or both) will be denoted by $A \cup B$ and is called the union of A and B. The set of outcomes contained in both A and B is called the intersection of A and B and is denoted by $A \cap B$ (or AB). In the language of probability we say the events A and B are mutually exclusive if $A \cap B$ contains no outcomes, that is, $A \cap B$ is the empty set, denoted by ϕ. If every outcome in A is an outcome in B, then A is said to be a subset of B, denoted by $A \subset B$. An event consisting of a single outcome will be called a simple event.

Let S denote the set of all possible outcomes of a random experiment; that is, S is the sample space. The outcomes in S will on occasion be referred to as sample points. We desire to define a function P such that if E is a subset of S, hereafter called an event, then P(E) is the probability that the outcome of the random experiment is an element of E. Since we are only considering experiments which are such that all possible outcomes can be described prior to their performance and which can be repeated under the same conditions, we will take $P(E)$ to be that number about which the relative frequency of the occurrence of E tends to stabilize after a long series of repetitions of the experiment. For instance, in the coin tossing experiment we would take $P(H) = 1/2$ and $P(T) = 1/2$ since we would expect the relative frequency of each to stabilize around the same value after a large number of performances of the experiment. Table 2.1 illustrates the results of the coin-tossing experiment.

The number of heads is recorded for

(a) Every 20 tosses for 15 times
(b) Every 200 tosses for 15 times
(c) Every 2000 tosses for 15 times

From each of these groups a relative frequency is calculated.

Each value of the experimental relative frequency of occurrence of H is only approximately equal to the probability of occurrence of H. As the number of tosses is increased the relative frequency values approach the defined or expected value of 1/2. It should be noted that even after 30,000 tosses, the relative frequency (0.4998) is still only approximately equal to the expected value of 0.5.

We now summarize the relative frequency definition (or interpretation) of probability.

Definition 1 If a random experiment is repeated a large number of times, n, then the observed relative frequency of occurrence, n_E/n, of the event E will tend to stabilize at some unknown constant $P(E)$, which is called the probability of E or the theoretical relative frequency of E.

Another definition of probability, which preceded the frequency definition, is the so-called classical definition.

Definition 2 If a random experiment can result in n equally likely and mutually exclusive outcomes and if n_E of these outcomes possess attribute E, then the probability of E is the ratio n_E/n.

TABLE 2.1

Number of Heads Per Group in Coin-Tossing Experiment[a]

	Number of Tosses					
	20		200		2000	
Group	Frequency	Relative frequency	Frequency	Relative frequency	Frequency	Relative frequency
1	14	0.70	104	0.520	1010	0.5050
2	11	0.55	91	0.455	990	0.4950
3	13	0.65	99	0.495	1012	0.5060
4	7	0.35	96	0.480	986	0.4930
5	14	0.70	99	0.495	991	0.4955
6	10	0.50	108	0.540	988	0.4940
7	11	0.55	101	0.505	1004	0.5020
8	6	0.30	101	0.505	1002	0.5010
9	9	0.45	101	0.505	976	0.4880
10	9	0.45	110	0.550	1018	0.5090
11	9	0.45	108	0.540	1021	0.5105
12	6	0.30	103	0.515	1009	0.5045
13	6	0.30	98	0.490	1000	0.5000
14	10	0.50	101	0.505	998	0.4990
15	13	0.65	109	0.545	988	0.4940

[a]Data from R. Lowell Wine, STATISTICS FOR SCIENTISTS AND ENGINEERS, ©1964, p. 105. Adapted by permission of Prentice-Hall, Inc., Englewood Cliffs, N. J.

A certain amount of caution regarding the terms "equally likely" and "mutually exclusive" must be exercised in using the classical definition. That the outcomes are equally likely cannot be proved but must be assumed. This assumption is reasonable in some problems. The classical definition also requires n to be finite. The frequency definition is more useful than the classical definition in that one generally does not know the true probability structure associated with the sample space.

The frequency and classical definitions of probability both suggest some of the properties that we would want $P(E)$ to have. Since the relative frequency is never negative, we would want P to be a nonnegative function. If S is the whole sample space, then its relative frequency is always 1, and we would want $P(S) = 1$. Finally, if $E_1, E_2, \ldots$ are mutually exclusive subsets of S, that is, no two of these subsets have a point in common, then the relative frequency of the union of these sets is the sum of the relative frequencies of the sets, so we would want

$$P(E_1 \cup E_2 \cup E_3 \cup \ldots) = P(E_1) + P(E_2) + P(E_3) + \ldots .$$

The classical definition suggests the same properties although it is restricted to a finite number of outcomes.

Thus we are led to the following axiomatic definition of probability.

<u>Definition 3</u> Let S be a sample space. If P is a function defined on subsets (events) of S satisfying

(a) $P(E) \geq 0$, for every event E,

(b) $P(E_1 \cup E_2 \cup \ldots) = P(E_1) + P(E_2) + \ldots$, where the events E_i, $i = 1,2,\ldots$, are such that no two have a point in common,

(c) $P(S) = 1$,

then P is called a probability function. The number $P(E)$ will be called the probability that the outcome of the random experiment is an element of the set E, or simply the probability of the event E.

2.3 POSSIBLE WAYS FOR EVENTS TO OCCUR

Many problems in elementary probability involve usage of the classical definition of probability, that is, the sample space has a finite number of equally likely and mutually exclusive outcomes. In such cases, evaluating the probability of a given event E consists of determining or counting the number of outcomes, n_E, favorable to the event E and then computing $P(E) = n_E/n$. In some problems the determination of n_E can become quite tedious. Consequently it is advantageous to develop some enumeration procedures which will be less time consuming and more efficient than total enumeration procedures.

We now state a counting principle which will be useful in determining the number of ways in which events can occur.

Counting Rule If an operation can be performed in m_1 ways, and if for each of these a second operation can be performed in m_2 ways, then the two operations can be performed in m_1m_2 ways. In probability terms, if an event E consists of the occurrence of an event B, followed by the occurrence of an event C, then we can obtain the number of ways that E can occur by taking the product of the number of ways that B can occur times the number of ways that C can occur.

We now consider permutations and combinations, two ideas which will help in determining the number of ways in which certain events can occur.

2.3.1 Permutations

Given a set of n distinguishable objects, all ordered sets formed with r objects chosen from the n given objects in any manner are called permutations of the n objects taken r at a time. The number P(n,r) of such permutations is given in most algebra books to be

$$P(n,r) = n!/(n - r)! \quad . \tag{2.1}$$

In particular the number of ways of rearranging n objects is n!.

Actually the result of Eq. (2.1) follows from the counting rule of Section 2.3 since there are n ways of choosing the first object, n - 1 ways of choosing the second object, ... , n - (r - 1) ways of choosing the rth object. Consequently,

$$\begin{aligned} P(n,r) &= n(n - 1) \ldots (n - r + 1) \\ &= \frac{n(n - 1) \ldots (n - r + 1)(n - r)!}{(n - r)!} \\ &= n!/(n - r)! \end{aligned}$$

Example 2.1 A fellowship, a research assistantship, and a teaching assistantship are available in this department this semester. In how many different ways may they be awarded to the nine applicants?

$$n = 9, \; r = 3$$
$$\text{Number of ways} = P(9,3) = 9!/6! = 504.$$

Permutations can be extended quite simply to the case where k groups of objects are involved, provided that the objects within each group are alike and that they are different from the objects of all other groups. Suppose there is a total of N objects of which n_1 are alike, n_2 are alike, ... , n_k are alike. Then the number of permutations of these N objects taken N at a time is given by

$$P(N;n_1,n_2,n_3,\ldots,n_k) = N!/n_1!n_2!n_3!\ldots n_k! \quad . \tag{2.2}$$

Example 2.2 How many different seating arrangements are possible for a materials science class composed of 7 Ch.E.'s, 3 T.E.'s, 9 C.E.'s, and 1 Pet. E.?

$$N = 20,\ n_1 = 7,\ n_2 = 3,\ n_3 = 9,\ n_4 = 1$$

Number of arrangements = 20!/7!3!9!1! = 221,707,200.

2.3.2 Combinations

Suppose that we select r objects from among n objects without regard to the order of arrangement of the objects among themselves. Any such selection is called a combination of n objects taken r at a time or a combination of n objects of order r. The number of combinations of n objects taken r at a time, $\binom{n}{r}$, is given as

$$\binom{n}{r} = n!/(n - r)!r! \tag{2.3}$$

Example 2.3 In how many different ways can a pair of students choose two calculators from the six available machines?

$$n = 6,\ r = 2$$

$$\text{Number of ways} = \binom{6}{2} = 6!/4!2! = 15 \quad .$$

To tell whether a permutation or combination is involved, remember that the order of selection is important in permutations. The order of selection is immaterial for combinations. If we consider the letters u, v, and w, we have the following two-letter permutations: uv, uw, vw, vu, wu, and wv. There are, however, only three combinations of pairs of letters. We see that uv and vu, uw and wu, and vw and wv represent the pairs of permutations possible for each two-letter combination. From this example, one should realize that the number of permutations of n objects taken r at a time is equal to the number of combinations of these n objects taken r at a time multiplied by the number of permutations of the r objects taken r at a time which is r!. Thus,

$$P(n,r) = \binom{n}{r} r! \quad .$$

2.4 PROBABILITY COMPUTATION RULES

If the probabilities of simple events are known, then the probabilities of compound (i.e., nonsimple) and other related events may be computed in terms of the simple event probabilities. For clarity, certain terms, some of which have been previously defined, associated with such compound and related events are defined as follows (A and B are considered to be arbitrary events of an experiment):

1. If A is an event of an experiment, then the event in which A does not occur, denoted by $\overline{A}$, is called the complementary event of A, or simply the complement of A.
2. The event "A and B" is just $A \cap B$.
3. The event "A or B" or "$A \cup B$" is that event in which at least one of A or B occurs.
4. Two or more events, A, B, C, etc., are mutually exclusive if the occurrence of any one event precludes the occurrence of each of the others.
5. The probability of occurrence of event A, given that event B has occurred, is called the conditional probability of event A given B and is denoted $P(A|B)$.
6. A collection of events is exhaustive if the collection includes all possible outcomes of the experiment.
7. The event E_1 is independent of the event E_2 if the probability of occurrence of event E_1 is not affected by the prior occurrence of event E_2 and vice versa so that

$$P(E_1|E_2) = P(E_1) \text{ and } P(E_2|E_1) = P(E_2). \qquad (2.4)$$

Rule 1: Multiplication

If $E_1, E_2, E_3, \ldots, E_n$ are independent events having respective probabilities $P(E_1)$, $P(E_2)$, $P(E_3), \ldots, P(E_n)$, the probability of occurrence of E_1 and E_2 and E_3 and ... E_n is

$$
\begin{aligned}
P(E_1 \cap E_2 \cap E_3 \cap \ldots \cap E_n) &= P(E_1 E_2 E_3 \ldots E_n) \\
&= [P(E_1)]\,[P(E_2)][P(E_3)] \ldots [P(E_n)] \\
&= \prod_{i=1}^{n} P(E_i). \qquad (2.5)
\end{aligned}
$$

This is immediate from the definition of independence of two events given in 7 above.

Example 2.4 If color, solution viscosity, and percent ethoxyl are independent characteristics of ethyl cellulose resin and the probabilities of any given batch being off-grade with respect to these characteristics are respectively 0.03, 0.05, and 0.02, what is the probability of a batch being on-grade with respect to all three characteristics?

The event of the batch being on-grade is the complementary event of it being off-grade. Thus, the probabilities of a batch being on-grade with respect to the three characteristics are 0.97, 0.95, and 0.98 and the probability of it being on-grade with respect to all three is

$$P(E_1 \cap E_2 \cap E_3) = (0.97)(0.95)(0.98) = 0.9031.$$

Rule 2: Addition

(a) If E_1, E_2, and E_3 are events, the probability of at least one of these events is

$$P(E_1 \cup E_2 \cup E_3) = P(E_1) + P(E_2) + P(E_3) - P(E_1E_2) - P(E_1E_3) - P(E_2E_3) + P(E_1E_2E_3). \quad (2.6)$$

(b) If only two events E_1 and E_2 are considered,

$$P(E_1 \cup E_2) = P(E_1) + P(E_2) - P(E_1E_2) \quad . \quad (2.7)$$

While no proof of these relationships is given here, they are clarified by observation of the Venn diagrams in which areas represent probabilities. A diagram for two events is shown in Fig. 2.1 and that for three events is shown in Fig. 2.2 in which the circles represent the probabilities of single events. The areas included in the intersecting circles represent the probabilities of the indicated compound events.

Equation (2.6) may be generalized for more than three events by adding all odd membered probabilities and subtracting all even membered probabilities.

(c) If the events $E_1, E_2, \ldots, E_n$, are mutually exclusive, then the addition rule becomes

$$P(E_1 \cup E_2 \cup E_3 \cup \ldots \cup E_n) = \sum_{i=1}^{n} P(E_i). \quad (2.8)$$

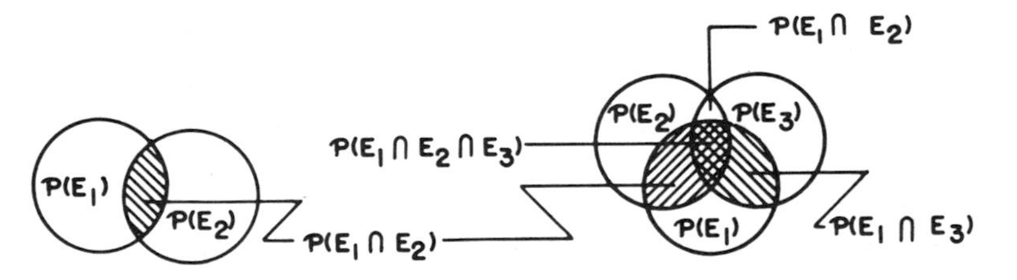

Fig. 2.1 Venn diagram for two events E_1 and E_2.

Fig. 2.2 Venn diagram for three events.

Example 2.5 From the data given in Example 2.4, calculate the probability of a given batch being off-grade with respect to any of the characteristics color, solution viscosity, and percent ethoxyl. The event E_1 is a batch not meeting the color, E_2 the solution, and E_3 the percent ethoxyl specification. Thus

$$P(E_1) = 0.03,\ P(E_2) = 0.05,\ \text{and}\ P(E_3) = 0.02,$$

$$\begin{aligned} P(E_1 \cup E_2 \cup E_3) &= 0.03 + 0.05 + 0.02 - (0.03)(0.05) \\ &\quad - (0.03)(0.02) - (0.05)(0.02) \\ &\quad + (0.03)(0.05)(0.02) \\ &= 0.0969 \quad . \end{aligned}$$

It will be noted that the event above is the complementary event of that in Example 2.4 and that the sum of the probabilities of the two events is 1.0.

Rule 3: Conditional Probability

The probability of the compound event $E_1 \cap E_2$ is equal to the probability of one event times the conditional probability of occurrence of the other event, given that the first event has occurred. That is,

$$\begin{aligned} P(E_1 \cap E_2) &= P(E_1)P(E_2|E_1) \\ &= P(E_2)P(E_1|E_2) \quad . \end{aligned} \tag{2.9}$$

The proof of Rule 3 is not given here: however, the rule can be clarified by making reference to Fig. 2.3 If event E_1 has occurred, then the probability that E_2 occurs is given by

$$P(E_2|E_1) = n_{E_1 \cap E_2}/n_{E_1}$$

where $n_{E_1 \cap E_2}$ and n_{E_1} denote the number of sample points in $E_1 \cap E_2$

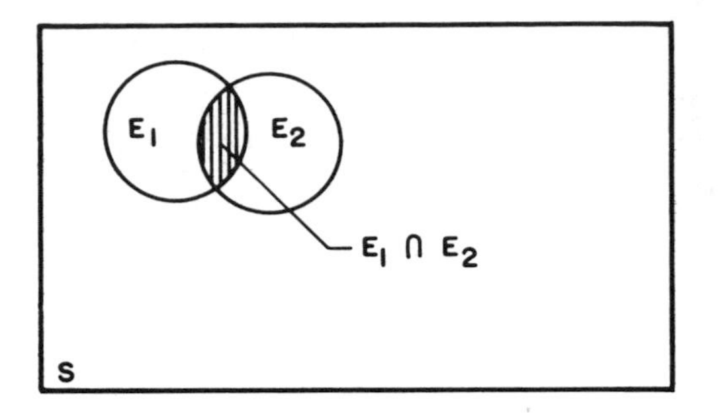

Fig. 2.3 Sample space S and Venn diagram for events E_1 and E_2.

and E_1, respectively. Thus

$$P(E_2 | E_1) = \frac{n_{E_1 \cap E_2}/n}{n_{E_1}/n} = \frac{P(E_1 \cap E_2)}{P(E_1)}$$

or

$$P(E_1 \cap E_2) = P(E_1)P(E_2 | E_1) \quad .$$

If the events E_1 and E_2 are independent, then there is no contingency of one on the other. Thus

$$P(E_2 | E_1) = P(E_2) \tag{2.10}$$

and

$$P(E_1 | E_2) = P(E_1) \tag{2.11}$$

so

$$P(E_1 \cap E_2) = P(E_1)P(E_2) \quad , \tag{2.12}$$

which is a proof of Rule 1 in the case of two events.

Example 2.6 In a batch manufacturing sequence, the material is tested after each processing step to determine its suitability for further processing. Batches not meeting test specifications are reprocessed. Past records indicate that the percentages of on-grade material produced by steps A and B, based on the input to the particular step, are 97.0 and 95.0, respectively. What is the probability of a batch passing the requirements for both steps A and B?

Since input to step B is contingent on satisfactory material being produced in step A, the 0.95 probability of success in the second step is a conditional probability. Thus

$$P(A \cap B) = P(A)P(B|A) = (.97)(.95) = 0.9215 \quad .$$

Example 2.7 The contents of two boxes used for storing hydrometers in the stockroom have become mixed. Box A, which is easier to reach and thus gone into by two-thirds of the students, contains 6 API hydrometers and 4 Baumé hydrometers. Box B contains 10 hydrometers of which 5 are Baumé and 5 are API hydrometers. If you remove one from each box, what is the probability that the two thus selected will be of the same type? They will be of the same type if they are both API hydrometers or if they are both Baumé hydrometers. Thus,

$$\begin{aligned} P(\text{2 alike}) &= P(\text{2 Baumé or 2 API}) = P(\text{2 Baumé}) \\ &\quad + P(\text{2 API}) = (0.4)(0.5) + (0.6)(0.5) \\ &= 0.5 \quad . \end{aligned}$$

What is the probability that a second hydrometer removed from one of the boxes will be of the same type as the one initially removed from that box? Note that since two-thirds of the students select box A, the probability of box A being selected is 2/3 and for box B it is 1/3. The desired event can occur by selecting a box and then selecting two hydrometers of the same type from it. Hence, the desired probability is

P(box A and 2 hydrometers alike or box B and 2 hydrometers alike)

$= P(\text{box A})\, P(2 \text{ hydrometers alike} | \text{box A}) + P(\text{box B})\, P(2 \text{ hydrometers alike} | \text{box B})$

$= 2/3\, P(2 \text{ Baumé or } 2 \text{ API} | \text{box A}) + 1/3\, P(2 \text{ Baumé or } 2 \text{ API} | \text{box B})$

$= 2/3\, [(0.4)3/9 + (0.6)5/9] + 1/3\, [(0.5)4/9 + (0.5)4/9]$

$= 62/135$.

To conclude this example, suppose that one of the seniors asked for a Baumé hydrometer before you started to sort the contents of the two boxes. What is the probability that if one hydrometer is selected from each box that you will get exactly 1 Baumé hydrometer?

P(exactly 1 Baumé) $= P$(Bé from A and API from B or API from A and Bé from B)

$= (0.6)(0.5) + (0.4)(0.5) = 0.5.$

Consider the situation where the probability of an event E_1 occurring in a single performance of an experiment is $p = P(E_1)$ and this probability remains the same for n independent trials of the experiment. If we desire the probability that the event E_1 will occur r times out of n trials of the experiment we reason as follows: The probability that E_1 occurs the first r times the experiment is performed is $p \ldots p = p^r$ and does not occur the next n - r times is $(1 - p) \ldots (1 - p) = (1 - p)^{n-r}$. Letting $q = 1 - p$ we see that the probability that E_1 occurs on each of the first r trials and does not occur on the last n - r trials is given by $p^r q^{n-r}$. Now, since we are only interested in E_1 occurring r times, we are not concerned about the order in which E_1 occurred and did not occur. Hence, since there are $\binom{n}{r}$ rearrangements of n objects of which r are alike and n - r are alike, yet different from the other r objects, we conclude that there are $\binom{n}{r}$ ways for us to get E_1 occurring r times and not occurring n-r times. Since these are all independent of each other, the probability of E_1 occurring r times and not occurring n-r times is $\binom{n}{r} p^r q^{n-r}$, where r is any integer between 0 and n.

Example 2.8 If n copper elbows are tossed into k bins in the store-room so that each elbow is equally likely to land in any of the bins (the sophomores often seem to return materials in this haphazard manner!), what is the probability that a particular bin, say bin A, will contain m of the n elbows?

$$P(\text{any elbow landing in bin A}) = 1/k$$

$$P(\text{m elbows landing in bin A}) = \binom{n}{m} (1/k)^m (1-1/k)^{n-m} \quad .$$

For this situation, let us assume that n = 12 elbows are involved, that only k = 4 bins are in use, and that we desire to obtain the probability that 3 elbows land in bin A.

$$\begin{aligned} P(\text{3 elbows land in bin A}) &= \binom{12}{3} (1/4)^3 (1-1/4)^{12-3} \\ &= (12!/9!3!)(1/4)^3(3/4)^9 \\ &= 220\ (1/64)(3/4)^9 \\ &= 0.2581 \quad . \end{aligned}$$

This rule finds frequent applications in quality control where batches of material are repeatedly sampled for defects.

2.5 A POSTERIORI PROBABILITY

In the preceding section, we assumed that the events would happen, that is, that the probabilities were known a priori and that they could be predicted or calculated directly. It is not uncommon, however, for the opposite situation to exist. In that case, we wish to determine the probability that a particular set of conditions or circumstances existed from the results already obtained. These a posteriori or "after the fact" probabilities can be determined from Bayes theorem which we now derive.

Suppose we have a two-stage experiment. Let $B_1, B_2, \ldots, B_n$ be the mutually exclusive and exhaustive events in the first stage of the experiment. In the second stage of the experiment, an event

E can occur only in conjunction with one or more of the events $B_1, B_2, \ldots, B_n$. Thus, we have

$$\begin{aligned} P(E) &= [P(E \cap B_1) \cup (E \cap B_2) \cup \ldots \cup (E \cap B_n)] \\ &= P(E \cap B_1) + P(E \cap B_2) + \ldots + P(E \cap B_n) \\ &= \sum_{i=1}^{n} P(E \cap B_i) \\ &= \sum_{i=1}^{n} P(B_i)P(E|B_i) \ . \end{aligned}$$

Given that the event E has occurred, if one is interested in obtaining the probability that event B_k occurred in the first stage, it may be calculated as follows:

$$\begin{aligned} P(B_k|E) &= \frac{P(B_k \cap E)}{P(E)} \\ &= \frac{P(B_k)P(E|B_k)}{\sum_{i=1}^{n} P(B_i)P(E|B_i)} \end{aligned} \qquad (2.13)$$

Formula (2.13) is known as Bayes formula.

Example 2.9 Consider the case of two bottles of reagents. One of these bottles, which we will call B_1, is filled with KOH pellets. The other bottle, which we will call B_2, is filled with an equal number of KOH and NaOH pellets. A pellet is drawn at random from one bottle and is found to be KOH. What is the probability that the pellet came from B_1? Let E be the event that the pellet drawn is KOH.

$$P(B_1) = P(B_2) = 1/2; \ P(E|B_1) = 1; \ P(E|B_2) = 1/2$$

$$P(B_1|E) = \frac{P(B_1)P(E|B_1)}{P(B_1)\,P(E|B_1) + P(B_2)P(E|B_2)}$$

$$P(B_1|E) = 1/2(1)/[1/2(1) + 1/2(1/2)] = 2/3$$

Example 2.10 In the production of orlon-cotton fabrics for consumer use, orlon spinning solution is forced through hundreds of tiny orifices in a spinneret which is mounted in a spinning machine. The filaments so produced are dried by a rising air stream, collated into fibers, and then combined into tow. Tow is a fairly ordered arrangement of parallel fibers. The next step is to process the tow in a crimping machine in which the fibers are uniformly stretched. Samples of the finished tow are then routinely inspected before baling.

Among the tests performed on each sample is one to measure stretch resistance. If the tow has been insufficiently stretched, fabrics made from it will very likely be deformable in use. If the tow has been stretched too much, there will be too many broken or weakened filaments to use in making top quality garments.

As an orlon production engineer your responsiblities include all aspects of filament and tow production starting with the arrival of the spinning solution at the spinning heads and ending with the tow-baling operation. The hourly tow report has just come in from the lab and indicates that the sample has poor stretch resistance. How can you quickly decide which mechanical operation is probably at fault?

The event E is the occurrence of poor stretch resistance. The probability of crimper malfunction is known to be $P(B_1) = 0.04$; for clogged spinnerets, $P(B_2) = 0.41$; and for ineffective filament collation, $P(B_3) = 0.12$.

From past experience we know poor stretch resistance occurs half the time when the crimper malfunctions. So $P(E|B_1) = 0.5$. Drawing in a like manner on previous observations, we have $P(E|B_2) = 0.1$ and $P(E|B_3) = 0.8$.

Let us utilize Bayes formula to calculate the equipment malfunction probabilities, $P(B_i|E)$ for the different mechanical operations.

$$P(B_1|E) = \frac{0.04(0.5)}{0.04(0.5) + 0.41(0.1) + 0.12(0.8)}$$

$$= \frac{0.02}{0.157} = 0.1274$$

$$P(B_2|E) = \frac{0.41(0.1)}{0.157} = 0.2611$$

$$P(B_3|E) = \frac{0.12(0.8)}{0.157} = 0.6115 \quad .$$

Thus, we conclude that we are most probably getting ineffective filament collation.

PROBLEMS

2.1 In how many ways could this Department's two \$500 and two \$250 freshman scholarships have been awarded to the 21 applicants last spring?

2.2 If you are taking three Departmental courses this semester out of the eleven offered, in how many ways could you have selected your course load in this curriculum?

2.3 In how many different ways can a student select an engineering major from the 7 engineering sciences on our campus and a minor from the remaining 8 sciences?

2.4 In how many ways can the 8 experiments in ChE 3351 be assigned to the 2 groups in each section if no group can do more than 1 experiment in the allotted time and equipment limitations are such that the groups cannot do the same experiment at the same time?

2.5 Our sophomore laboratory course in physical property measurement techniques has four groups in each section. In how many ways may the 10 experiments be arranged so that only one group is assigned the same experiment at any given lab period? There are 4 students in each group.

2.6 A single die is tossed. What is the probability that the outcome will be a one or a two?

2.7 The die is tossed again. For this toss, what is the probability that an even number or a 1 or 2 will be on the upper face?

2.8 This time, two dice are cast simultaneously. What is the probability that the sum of the numbers on their upper faces will be 7 or 8?

2.9 The same die is tossed twice. What is the probability of getting 7 by rolling, in order, a 2 followed by a 5? How does this compare with the probability of rolling 7 by any combination if two dice are thrown together?

2.10 A balanced die is fairly tossed. Let the event A denote the occurrence of an odd number, B the occurrence of an even number, C the occurrence of a number greater than 3, D the occurrence of a number less than 3, and E the occurrence of the number 3.

(a) Find $P(A)$, $P(B)$, $P(D)$.
(b) Which of the events are equally likely?
(c) Which of the events are mutually exclusive?
(d) Find the probability of "event C or event D".

2.11 Cut out the letters, one per small card, of the word ENGINEERING, place them in an empty ball mill, and tumble thoroughly for 5 minutes.

Open the mill and draw out 7 letters, one at a time. What is the probability of drawing, in the correct order, the letters of the word GINNING?

2.12 In your class of 12, the probabilities are that there will be 3 A's, 3 B's, 4 C's, 1 D, and 1 F at the end of the semester. Assuming that the grades are not curved in any way, what is the chance of your earning a B or a C? At least a C?

2.13 The results of the rigorous examination of a dozen automobile tires showed the following: one tire was perfect; three only had slight flaws in appearance; two had incompletely formed treads; one had a serious structural defect; and the rest had at least two of these defects. What is the probability that the next set of four tires you buy of this particular brand will be perfect? Will have at most only undesireable appearances? Will have less than two defects?

2.14 The storeroom has received 19 dozen disposable pipets and 1 dozen "class A" pipets (definitely not disposable). The new stockboy, not realizing that the pipets were of two different varieties, yesterday opened all 20 dozen packets and placed all the pipets in the same supply bin. What are the chances that you will receive 1, 2, or 5 "class A" pipets when you check out a dozen tomorrow?

2.15 The probability of our ancient mass spectrometer operating satisfactorily for a month is 0.4. The probability of the oscilloscope performing adequately during that same period is 0.9. What is the probability of both instruments needing a service call this month? Of one needing a service call?

2.16 The probability of a microsyringe being broken in our analytical instrumentation course is 1/8 for each student enrolled. What is the probability of survival of any particular syringe for the current semester? Thirty students are enrolled. They are arranged into 3 lab sections each having 2 groups of 5 students.

2.17 The chances of making a flawed pellet for the infrared examination of solid samples is 0.8. What is the probability of making 4 consecutive usable pellets in one 3-hour lab period? Approximately 15 minutes are needed to make each pellet.

2.18 A testing organization wished to rate a particular brand of table radios. Five radios are selected at random from stock and the brand is judged satisfactory if nothing is wrong with any of the five. (a) What is the probability that the brand will be rated as satisfactory if 10% of the radios are actually defective? (b) What is the probability of getting one or less defective from the sample if 10% of the stock radios are defective?

2.19 A company manufactures a lubricant which must pass on accelerated heat stability test. Past records indicate that the normal percentage of failures is 5%. During one month of operation 120 batches were produced and 11 were found defective from the heat stability standpoint. What is the probability that the manufacturing operation is not functioning normally?

2.20 A lot contains 1400 items. A sample of 200 items is taken and the lot accepted if no more than two defective items appear in the sample. If the lot actually contains 28 defectives what is the probability of its acceptance?

2.21 The following numbers are obtained from a mortality table based on 100,000 individuals.

Age	Deaths per 1000 during year
17	7.688
18	7.727
19	7.765
20	7.805
21	7.855

If these numbers are used to define probabilities of death for the corresponding age group and if A, B, and C denote individuals of ages 17, 19, and 21, respectively, calculate the probability that during the year: (a) A will die and B will live, (b) at least one of A, B, and C will die, and (c) A and B will both die.

2.22 When buying flasks for the storeroom, the clerk has the habit of checking 3 flasks out of each case of 24 of chipped lips, cracks, etc. If any defects are found in this sample of 3, the entire case is rejected. Assuming that each possible combination of 3 flasks had the same chance of being selected as the sample, what is the probability that a case of flasks containing 3 defective units will be accepted?

2.23 Box A contains 3 white and 4 black balls. Box B contains 2 white and 3 black balls. (a) If 2 balls are chosen, 1 from each box, what is the probability that they will be of the same color? (b) If 2 balls are chosen from Box A, what is the probability that both balls will be white? (c) If a box is selected (at random) and 2 balls drawn from it, what is the probability that they will be of the same color? All drawings are done without replacement.

2.24 Two balls are to be drawn, without replacement, from an urn containing 2 white, 3 black, and 4 green balls.

(a) What is the probability that both balls will be green?
(b) What is the probability that both balls will be of the same color?

2.25 A relatively new batch process is still producing 10% defective (off-grade) material.

(a) What is the probability of producing no off-grade material in the next 8 batches to be produced?
(b) What is the probability of more than one defective batch in the next 8 produced?
(c) What is the expected number of defective batches in 30, assuming each is independent?
(d) Would you say that zero off-grade batches in a successive series of 30 production run indicate a change in the process? What is the probability of being wrong in so stating?

2.26 A sample of 20 independent pieces is taken from a production line and none of the pieces is defective.

(a) What can be said about the percentage defective in the process?
(b) If the actual percentage defective in the process is $p = 2\%$, what is the probability of the above event?
(c) What is the probability if $p = 4\%$?

2.27 In the manufacture of a modified natural resin the three key product properties are melting point (A), color (B), and acid number (C), each of which is independent of the others. Marketing suggests sales specifications of minimum M.P. of $130^{\circ}C$, maximum color of 3 (arbitrary scale), and maximum acid number of 12. Analyses of available data indicate that the probabilities of material being off-grade with respect to each of these properties are $P(A) = 0.03$, $P(B) = 0.05$, and $P(C) = 0.02$. Assuming no process changes:

(a) What percentage of the resin produced will be off-grade?
(b) For those customers for which color is unimportant, what percentage off-grade is to be expected?
(c) If off-grade material is valued at 20¢/lb and first grade (on specification with respect to all properties) is priced at 50¢/lb, what reduction in price can be given to non-color-conscious customers?

2.28 In the manufacture of plastic objects by an injection molding process, it is found that on the average 5% are oversize as to length, 91% satisfy the length specification, and 4% are undersize as to length. With regard to thickness specification 2% are undersize, 96% are on specification and 2% are oversize. In a sample of 10 independent pieces, what is the probability of finding

(a) all good pieces
(b) exactly one piece oversize as to length
(c) eight pieces or less on specification with respect to both length and thickness?

3
DISTRIBUTIONS

3.1 INTRODUCTION

Most statistical methods useful to engineers and scientists deal with the collection, organization, analysis, and presentation of data. Such analyses of experimental data are used in making reasonable decisions which are at least partly based on the data.

Application of the methods of statistics requires an understanding of data and its characteristics, the most common of which are average and variation. These characteristics are reflected by distributions, either experimental (empirical) or theoretical. Most statistical methods are based on theoretical distributions which approximate the actual distributions. Thus the user of statistical methods must be familiar not only with the theoretical distributions but also with the distribution characteristics of the population(s) under consideration.

3.2 DEFINITIONS

Certain terms relative to distributions have special meanings in statistical usage. The more common of these terms are:

Population: A collection of objects which have at least one characteristic (attribute) in common.

Sample: A subset of a population.

Observation: A recording of information on some characteristic of an object.

Measurement: A numerical value indicating the extent, intensity, or size of a characteristic of an object.

Sample Space: The set of all possible outcomes of an experiment, that is, all possible results of a process that generates data.

Random Variable: A function from a sample space to the real line, that is, a function which assigns a real value to each outcome in a sample space.

Discrete Variable: A variable which can assume only isolated values, that is, values in a finite or countably infinite set.

Continuous Variable: A variable which can assume any value between two distinct numbers.

Cumulative Frequency: The sum of the frequencies of all values less than or equal to a particular value.

All of these terms will be used in the ensuing sections where their meanings will become increasingly clearer to the reader. However, since the terms population and sample are quite often confused with each other, a word of clarification is in line here. A population can be considered to be the set of all objects which have at least one characteristic in common. For example, the set of all engineers in the world. A sample is a subset of the population; for example, all the engineers living in Texas.

It should also be noted that a random variable is just a "rule" that associates a real number with each outcome in the sample space. Thus, in tossing two coins $S = \{(H,H), (H,T), (T,H), (T,T)\} = \{s_1, s_2, s_3, s_4\}$. Let X denote the number of heads so that $X(s_1) = 2$, $X(s_2) = X(s_3) = 1$, $X(s_4) = 0$. Upper case letters such as X,Y,Z, etc., will denote random variables and the corresponding lower case letters x, y, z, etc., will denote values of the random variables.

3.2.1 Discrete Distributions

Discrete distributions are used to describe count or enumeration data. Examples of such data are the number of attendees at the next Sigma Xi luncheon, the score of the next Super Bowl game, and the number of heads in n flips of a coin. All data of this type have one characteristic in common: the number of values in each case is finite. If one considers the random variables associated with the sample spaces of these examples, then there are measurable step changes between values of the random variables involved. Consequently, there is a specific, fixed probability associated with each value of the random variable. The function relating these probabilities to the values of the random variable is called the probability function.

Assume that the random variable X may take on any value of the finite, ordered, discrete set of values $S = \{x_1, x_2 . x_3, \ldots, x_n\}$, where n is finite or countably infinite. Let E be any subset of S, that is, let E be an event. The probability of E, $P(E)$, is a real, nonnegative number. In the case of a finite sample space S with equally likely x_i's, if f_i is the frequency of occurrence of x_i and N is the sum of the f_i then the probability function, also called the point probability function, $f(x_i)$, can be expressed as

$$P(X = x_i) = f(x_i) = f_i/N = f_i / \sum_{j=1}^{n} f_j \tag{3.1}$$

and will have the following properties:

(a) $f(x_i)$ is positive and real for each x_i,

(b) $\sum_{i=1}^{n} f(x_i) \equiv 1.$

(c) $P(E) = \Sigma f(x_i)$ where the sum is taken over all those x_i which are in E.

Any function $f(x_i) = P(X = x_i)$ satisfying properties (a), (b), and (c) is said to be a probability function. A random variable corresponding to a point probability function is called a discrete random variable.

For a die, the probability $f(x_i)$ for getting any particular value on a single toss is 1/6. A plot of the point probability function is then

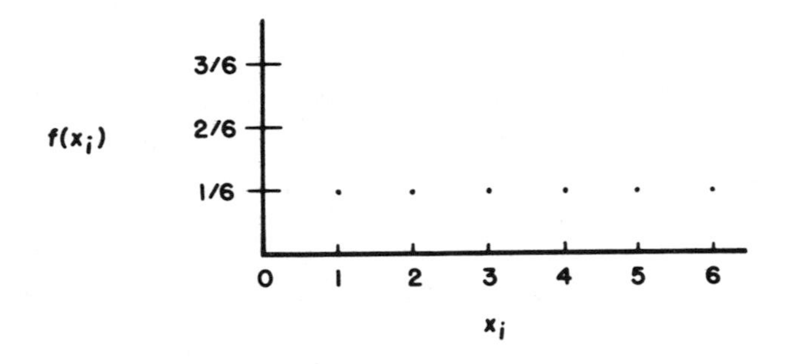

Corresponding to each probability function is a function, F(x), which is called the cumulative distribution function. F(x) can be obtained from the probability function according to the following definition. For $x_1 < x_2 < x_3 < \ldots < x_n$, F(x) is defined by

$$F(x) = P(X \leq x) = \begin{cases} 0 & \text{for } x < x_1 \\ \sum_{i=1}^{r} f(x_i) & \text{where } x_r \leq x < x_{r+1} \\ 1 & \text{for } x \geq x_n \end{cases} \tag{3.2}$$

The cumulative distribution function is a non-decreasing function with the following properties:

(a) $0 \leq F(x) \leq 1$,

(b) $\lim_{x \to -\infty} F(x) = 0$,

(c) $\lim_{x \to \infty} F(x) = 1$.

The cumulative distribution function F(x) corresponding to the probability function $f(x_i)$ illustrated above is graphed below. It should be noted that F(x) takes a jump at x = 1,2,...,6. This is indicated by a small tick on the graph. For instance F(2) = 2/6 and F(x) = 1/6 for $1 \leq x < 2$.

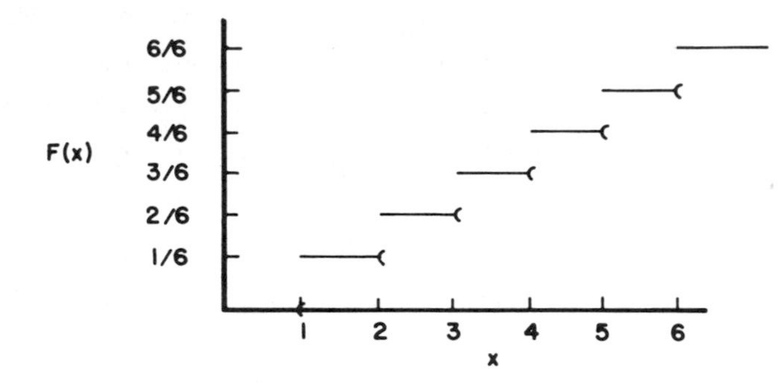

It follows from the definition of the cumulative distribution function that

$$P(x_1 < X \leq x_2) = F(x_2) - F(x_1) \quad . \tag{3.3}$$

In the next section we will consider an extension of the cumulative distribution function which is useful when the outcomes of an experiment are not countable.

3.2.2 Continuous Distributions

If there is a function f(t) such that

$$F(x) = \int_{-\infty}^{x} f(t)\, dt \tag{3.4}$$

then we say that F(x) is a continuous distribution function. The function f(t) is then referred to as a probability density function. From (3.4) we see that the probability density function can be obtained from the distribution function by taking a derivative, since

$$f(x) = \frac{d}{dx}\left[F(x)\right] \quad . \tag{3.5}$$

A random variable X such that $P(X \leq x_0) = F(x_0)$ with F as given in (3.4) is referred to as a continuous random variable. Continuous distribution functions of continuous random variables differ from those of discrete random variables in that the former do not show step-changes with changing values of x as do the latter. Indeed, the variable x cannot make step-changes but may assume any value within an interval. This was not allowed for discrete random variables. It should be noted that if X is a continuous random variable then $P(X = a) = 0$ for any real number a.

Some examples of variables with continuous distributions are flow rate, pressure drop, and temperature. The common factor in all these variables is that they cannot physically undergo abrupt step-changes but change continually with time.

3.2.3 Experimental Distributions

The distributions defined in Sections 3.2.1 and 3.2.2 are theoretical distributions which correspond to some populations. For example, $S = \{1,2,3,4,5,6\}$ is the population of values corresponding to the outcomes in rolling a die. The probability function $P(X = x_i) = \frac{1}{6}$ for $x_i = 1,2,3,4,5,6$ gives the "theoretical distribution" for this population. However, the die may not be fair so that the theoretical distribution is given by some other unknown probability function. One may, however, by rolling the die a large number of times, and using the frequency definition of probability in Chapter 2, arrive at an empirical probability function which is "close" to the theoretical distribution. For example, suppose that 1000 rolls of the die yielded the results summarized in Table 3.1. The resulting function $f(x_i)$, $x_i = 1,2,3,4,5,6$, in Table 3.1 satisfies all the conditions of a "theoretical" distribution but is, however, an "experimental" distribution.

TABLE 3.1

Experimental Distribution of Die Outcome Data

Outcome x_i	1	2	3	4	5	6
Frequency	142	200	58	328	122	150
Relative frequency $f(x_i)$	0.142	0.200	0.058	0.328	0.122	0.150

Before considering experimental distributions we define a few terms pertinent to such distributions.

Classes: Groups into which data are distributed.

Class boundary: That numerical value which divides two successive classes.

Class length: The numerical difference in the boundaries of a class.

Class mark: The mid-value of a class.

Class frequency: The frequency with which values of observations occur.

Relative frequency: The frequency expressed as a fraction of the total number of observations.

The methods commonly used in arriving at and displaying the distribution characteristics of data are illustrated using the raw data given in Table 3.2 which gives the melting points by Ring

TABLE 3.2

Ring and Ball Melting Point Data (Melting Point, °C)

108.4	108.2	108.9	106.2	107.1	108.5
105.5	107.2	107.4	107.7	108.4	107.6
107.4	107.0	109.2	107.1	108.8	109.7
111.5	109.4	107.2	107.6	108.8	109.2
110.9	108.8	106.2	108.5	108.4	108.9
107.1	108.3	108.5	107.3	108.3	107.7
109.1	107.3	107.0	108.3	106.9	106.9
108.7	108.4	108.2	107.6	108.6	110.0
108.9	107.7	106.5	106.2	109.4	109.9
108.1	108.4	108.5	108.0	109.3	107.8
108.8	106.3	108.3	108.3	108.3	108.4
108.4	109.0	109.1	109.0	108.2	109.1
108.2	106.9	108.3	107.6	107.1	108.3
109.2	108.0	109.1	107.1	107.4	109.6
108.3	108.6	106.0	106.8	108.2	106.3
106.6	106.4	106.8	107.4	109.1	107.4
107.3	107.7	108.9	109.2	108.4	108.4
107.9	108.9	108.6	107.6	108.5	106.3
106.7	107.6	106.1	107.3	106.8	107.1
109.1	109.3	106.3	107.7	107.9	108.2
106.2	107.7	108.8	107.8	108.6	107.3
109.2	107.6	109.5	107.9	107.5	108.3
109.2	108.3	108.1	106.8	108.8	108.1
109.5	108.4	108.3	108.1	108.1	107.8

and Ball method of batches of modified resin produced during a 6-month period. It should be noted that the values in Table 3.2 comprise a sample from the "conceptual" population of all batches that would be obtained henceforth.

In summarizing large quantities of raw data, grouping is commonly employed. For convenience, the values may be placed in rank order as in Table 3.3. In grouping data, a decision must be made as to how many classes should be used. The related question of class length must also be answered. Too many classes make the population characteristics difficult to picture; too few classes destroy the details of the picture. It is convenient, but not absolutely necessary, that all class lengths be equal. Generally 10 to 20 classes are adequate for even large sets of observations.

Some trial and error may be required in deciding on the class boundaries which result in the best picture of the data. The class length chosen should be a convenient one to work with. The classes should cover the range of the population with no empty intervals, however this is not always possible.

For the data under consideration, a class length of 0.5°C was selected, and since the range of the values is 105.5 to 111.5°C, thirteen classes are required. In order to assure that there is no doubt as to which class a value belongs, class boundaries are set at 105.05, 105.55, 106.05,...,111.55°C. Results of the data grouping are shown in Table 3.4

The data may be conveniently presented graphically in the form of frequency diagrams. The most commonly used diagram is the frequency histogram. The data of Table 3.4 are shown graphically in Fig. 3.1 in the form of a histogram.

In the construction of the histogram, class marks and boundaries are put on the horizontal axis and frequency and/or relative frequency on the vertical axis of a rectangular coordinate system.

TABLE 3.3

Rank Order for
Melting Point Data

Rank	Value	Rank	Value	Rank	Value
1	111.5	26	109.0	51	108.4
2	110.9	27	109.0	52	108.4
3	110.0	28	108.9	53	108.4
4	109.9	29	108.9	54	108.4
5	109.7	30	108.9	55	108.4
6	109.6	31	108.9	56	108.4
7	109.5	32	108.8	57	108.3
8	109.5	33	108.8	58	108.3
9	109.5	34	108.8	59	108.3
10	109.4	35	108.8	60	108.3
11	109.4	36	108.8	61	108.3
12	109.3	37	108.7	62	108.3
13	109.3	38	108.6	63	108.3
14	109.2	39	108.6	64	108.3
15	109.2	40	108.6	65	108.3
16	109.2	41	108.6	66	108.3
17	109.2	42	108.5	67	108.3
18	109.2	43	108.5	68	108.3
19	109.2	44	108.5	69	108.2
20	109.1	45	108.5	70	108.2
21	109.1	46	108.5	71	108.2
22	109.1	47	108.4	72	108.2
23	109.1	48	108.4	73	108.2
24	109.1	49	108.4	74	108.2
25	109.1	50	108.4	75	108.1

TABLE 3.3 (Continued)

Rank	Value	Rank	Value	Rank	Value
76	108.1	100	107.5	124	106.9
77	108.1	101	107.5	125	106.8
78	108.1	102	107.4	126	106.8
79	108.1	103	107.4	127	106.8
80	108.0	104	107.4	128	106.8
81	108.0	105	107.4	129	106.7
82	107.9	106	107.3	130	106.6
83	107.9	107	107.3	131	106.6
84	107.9	108	107.3	132	106.5
85	107.8	109	107.3	133	106.4
86	107.8	110	107.3	134	106.3
87	107.8	111	107.2	135	106.3
88	107.7	112	107.2	136	106.3
89	107.7	113	107.2	137	106.3
90	107.7	114	107.1	138	106.2
91	107.7	115	107.1	139	106.2
92	107.7	116	107.1	140	106.2
93	107.7	117	107.1	141	106.2
94	107.6	118	107.1	142	106.1
95	107.6	119	107.1	143	106.0
96	107.6	120	107.0	144	105.5
97	107.6	121	107.0		
98	107.6	122	106.9		
99	107.5	123	106.9		

TABLE 3.4

Grouped Frequencies of Melting Point Data with $0.5^{\circ}C$ Class Length

Class	Class mark	Class frequency	Cumulative frequency	Relative frequency	Cumulative relative frequency
105.05-105.55	105.3	1	1	0.00694	0.00694
105.55-106.05	105.8	1	2	0.00694	0.01388
106.05-106.55	106.3	11	13	0.07639	0.09028
106.55-107.05	106.8	12	25	0.08333	0.17361
107.05-107.55	107.3	21	46	0.14583	0.31944
107.55-108.05	107.8	19	65	0.13194	0.45137
108.05-108.55	108.3	38	103	0.26388	0.71528
108.55-109.05	108.8	16	119	0.11111	0.82639
109.05-109.55	109.3	19	138	0.13194	0.95833
109.55-110.05	109.8	4	142	0.02778	0.98611
110.05-110.55	110.3	0	142	0.00000	0.98611
110.55-111.05	110.8	1	143	0.00694	0.99306
111.05-111.55	111.3	1	144	0.00694	1.00000

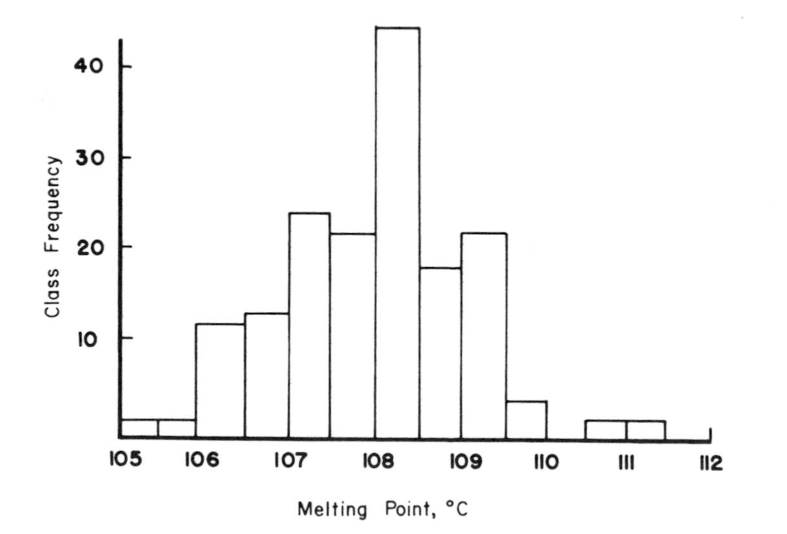

Fig. 3.1 Histogram of melting point data.

Thus each of the group frequencies is represented as a rectangle, the height of which is equal to or proportional to the corresponding group frequency.

The effect of class length on the distribution "picture" obtained is illustrated by a comparison of Figs. 3.1 and 3.2 in which the class length is changed from 0.5° to 1.0°C. By decreasing the number of groups, the irregularities in the histogram are eliminated. It is unreasonable, considering the nature of the operation on which the data are based, that the irregularities are significant. Hence it is believed that in this case a class length of 1.0°C gives a better picture of the data as a whole than does a 0.5°C length.

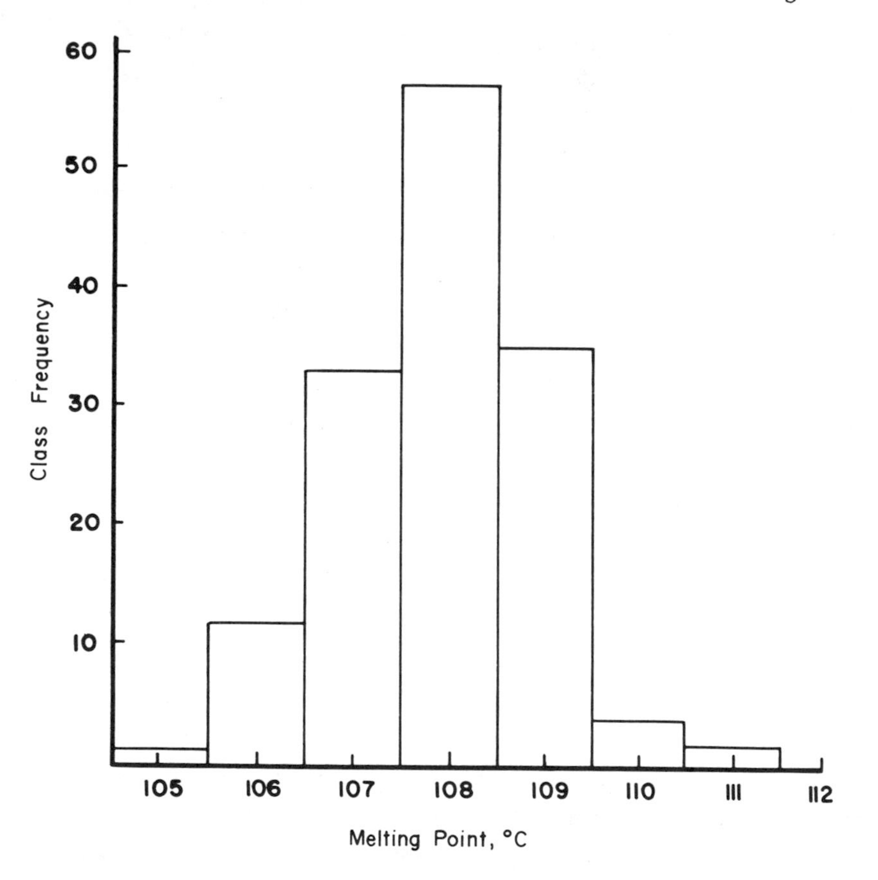

Fig. 3.2 Histogram of Table 3.5 data.

The grouped frequencies associated with the longer class length are given in Table 3.5

TABLE 3.5
Grouped Frequencies of Melting Point Data with $1^{\circ}C$ Class Length

Class	Class mark	Class frequency	Cumulative frequency	Relative frequency	Cumulative relative frequency
104.55-105.55	105.05	1	1	0.00694	0.00694
105.55-106.55	106.05	12	13	0.08333	0.09028
106.55-107.55	107.05	33	46	0.22917	0.31944
107.55-108.55	108.05	57	103	0.39583	0.71528
108.55-109.55	109.05	35	138	0.24306	0.95833
109.55-110.55	110.05	4	142	0.02778	0.98611
110.55-111.55	111.05	2	144	0.01388	1.00000

Grouped data may also be depicted by means of the frequency polygon or cumulative frequency polygon. The upper class boundary and cumulative frequency of the corresponding group locate points which are connected by straight lines to form the cumulative frequency polygon. The same data from Table 3.5 are shown in Fig. 3.3 which is a cumulative frequency polygon. Class mark and frequency of each group locate points which are connected by straight lines to form the frequency polygon. Since the cumulative frequency polygon shown in Fig. 3.3 depicts the total of frequencies having values less than a particular x_i value, the x_i value plotted is not the class mark (midpoint) but the upper class boundary.

Since the major purpose of frequency polygons and diagrams is to depict distributions, little is gained by drawing smooth curves through the vertices of the polygons to give frequency curves for predictive purposes. For such purposes it is preferable to fit an appropriate theoretical function to the relative frequency histogram and use this function for predictions.

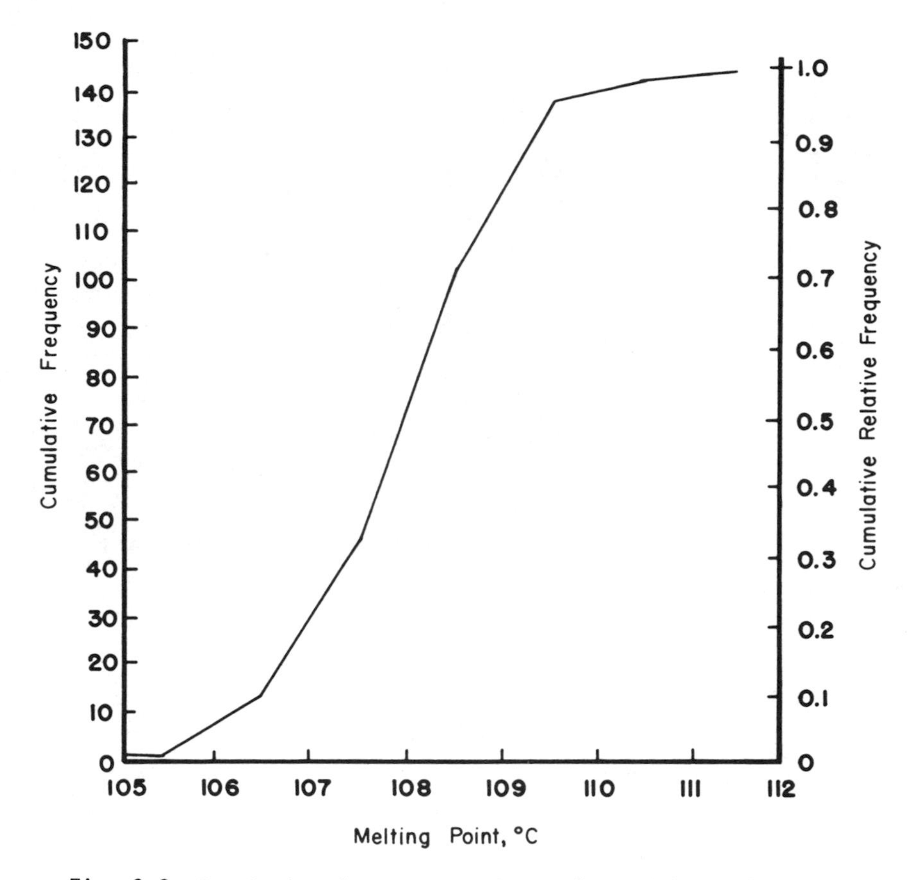

Fig. 3.3 Cumulative frequency polygon for melting point data.

In the handling of data representing observations of discrete random variables (count data), frequency is represented by a line on the diagram and the cumulative frequency polygon is a step polygon. These differences between continuous and discrete distributions arise from the fact that discrete variables can take only isolated values. The diagrams in Fig. 3.4 show the frequency histogram and cumulative frequency polygon for the die outcome data given in Table 3.1 obtained in rolling a die 1000 times.

Now that you have learned to handle experimental distributions, let us examine the use of commonly available computer programs for obtaining histograms and relative frequency poly-

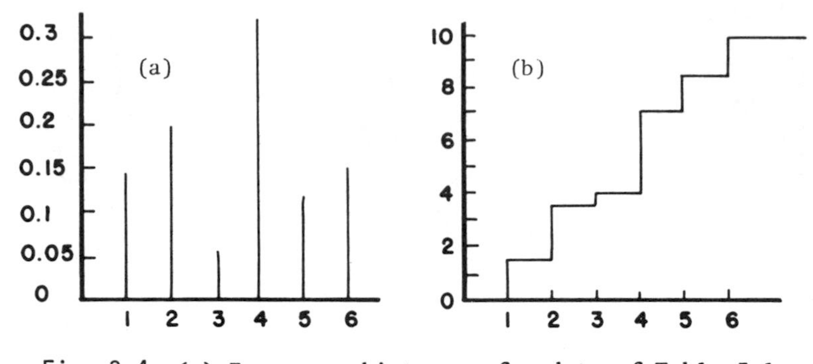

Fig. 3.4 (a) Frequency histogram for data of Table 3.1.
(b) Cumulative frequency polygon for data of Table 3.1.

gons. Two such commonly used packages are BMD® (trademark registered, UCLA) and SAS® (trademark registered, SAS Institute, Cary, NC). Although we feel that the former is better documented, the latter is, in our opinion, easier to use. We will include, from here on, appropriate example problems using SAS programs. Before you begin to read this example, study the material of Appendix B carefully so that you know how to set up your data into the required formats for executing SAS programs. We emphasize that the material there and in the SAS programs illustrated in this book is presented in abbreviated form. The serious student should obtain copies of all the SAS Institute publications to which we refer.

Example 3.1 An experiment entitled "Friction Losses in Valves and Fittings" is currently being performed in the unit operations laboratory. The experimental apparatus is equipped with a small orifice which is used to determine the flow rate through the system. For purposes of comparison, several values of the orifice coefficient which have been calculated by different lab groups are presented below. Prepare an appropriate frequency table, histogram, and frequency polygon from these data. Using

*SAS is the registered trademark of SAS Institute, Inc., Cary, N.C., 27511-8000.

the PROC PRINT, PROC FREQ, and PROC CHART programs (SAS Introductory Guide. SAS Institute Inc., Cary, NC: SAS Institute, Inc. 1978. 83 pp.), the data are printed (not shown again); arranged in ascending order with the relative frequency (percent column) and cumulative relative frequency (cum percent column); and displayed in histogram form (Fig. 3.5). As the job control language (JCL) and control cards vary with the computer, we will only show the programs and the results.

```
(put initial JCL cards/statements here)
DATA CO;
INPUT @@;
LABEL LABORATORY ORIFICE COEFFICIENT VALUES;
CARDS;
        0.626       0.670       0.500       0.648
        0.615       0.579       0.517       0.898
        0.625       0.627       0.495       0.883
        0.640       0.605       0.488       0.882
        0.623       0.628       0.526       0.964
        0.625       0.588       0.610       0.830
        0.608       0.559       0.630       0.654
        0.615       0.624       0.633       0.603
        0.633       0.615       0.610       0.621
        0.622       0.616       0.650       0.621
        0.624       0.579       0.640       0.584
        0.615       0.670       0.773       0.583
        0.616       0.517       0.641       0.632
PROC PRINT DATA = CO;
PROC FREQ DATA = CO;
TITLE 'RESULTS OF USING PROC FREQ';
PROC CHART DATA = CO;
VBAR VALUES/TYPE=FREQ NOSPACE MIDPOINTS= 0.4875 TO 1.005 BY 0.01;
TITLE 'RESULTS OF USING PROC CHART';
(put final JCL cards/statements here)
```

SAS

RESULTS OF USING PROC FREQ FOR CO DATA

CO	FREQUENCY	CUM FREQ	PERCENT	CUM PERCENT
0.488	1	1	1.923	1.923
0.495	1	2	1.923	3.846
0.5	1	3	1.923	5.769
0.517	2	5	3.846	9.615
0.526	1	6	1.923	11.538
0.559	1	7	1.923	13.462
0.579	2	9	3.846	17.308
0.583	1	10	1.923	19.231
0.584	1	11	1.923	21.154
0.588	1	12	1.923	23.077
0.603	1	13	1.923	25.000
0.605	1	14	1.923	26.923
0.608	1	15	1.923	28.846
0.61	1	16	1.923	30.769
0.615	4	20	7.692	38.462
0.616	2	22	3.846	42.308
0.621	2	24	3.846	46.154
0.622	1	25	1.923	48.077
0.623	1	26	1.923	50.000
0.624	2	28	3.846	53.846
0.625	2	30	3.846	57.692
0.626	1	31	1.923	59.615
0.627	1	32	1.923	61.538
0.628	1	33	1.923	63.462
0.63	1	34	1.923	65.385
0.632	1	35	1.923	67.308
0.633	2	37	3.846	71.154
0.64	3	40	5.769	76.923
0.641	1	41	1.923	78.846
0.648	1	42	1.923	80.769
0.65	1	43	1.923	82.692
0.654	1	44	1.923	84.615
0.67	2	46	3.846	88.462
0.773	1	47	1.923	90.385
0.83	1	48	1.923	92.308
0.882	1	49	1.923	94.231
0.883	1	50	1.923	96.154
0.898	1	51	1.923	98.077
0.964	1	52	1.923	100.000

```
                         SAS
              RESULTS OF USING PROC CHART
                 HISTOGRAM OF CO DATA
                 FREQUENCY BAR CHART

FREQUENCY
  10 +              *
     |              *
     |              *
     |              *
   9 +             **
     |             **
     |             **
     |             **
   8 +             **
     |             **
     |             **
     |             **
   7 +             **
     |             **
     |             **
     |             **
   6 +             ***
     |             ***
     |             ***
     |             ***
   5 +             ***
     |             ***
     |             ***
     |             ***
   4 +            ****
     |            ****
     |            ****
     |            ****
   3 +          * ****
     |          * ****
     |          * ****
     |          * ****
   2 + * *     ** ***** *
     | * *     ** ***** *
     | * *     ** ***** *
     | * *     ** ***** *
   1 +** **  * ** *******          *    *    ***      *
     |** **  * ** *******          *    *    ***      *
     |** **  * ** *******          *    *    ***      *
     |** **  * ** *******          *    *    ***      *
     -----------------------------------------------------
      0000000000000000000000000000000000000000000000000000
      ....................................................
      4455555555556666666666777777777788888888889999999999
      8901234567890123456789012345678901234567890123456789
      7777777777777777777777777777777777777777777777777777
      5555555555555555555555555555555555555555555555555555

                              CO
```

Fig. 3.5 Results of using PROC CHART: frequency bar chart for orifice coefficients.

Use either the output from PROC FREQ or the text-editing capabilities of your computer to save and enter the values of the orifice coefficient CO and the corresponding cumulative relative frequency CRF into a new SAS routine, PROC PLOT (SAS Introductory Guide). Having saved the output of PROC FREQ under the retrieval name ORIFICE, the procedure to generate the CRF graph is shown below. The CO data were in columns 13-18 and the CRF values in columns 57-63 of the output from PROC FREQ.

```
(put initial JCL cards/statements here)
DATA ORIFICE;
INPUT CO 13-18 CRF 57-63;
CARDS;
(type in CO and CRF values here or call them in from storage)
PROC PLOT DATA = ORIFICE;
PLOT CRF*CO;
TITLE 'DISTRIBUTION OF ORIFICE COEFFICIENT';
(put final JCL cards/statements here)
```

The result of this SAS procedure is the cumulative relative frequency polygon shown in Fig. 3.6.

3.3 THEORETICAL DISTRIBUTIONS

Frequency distributions are useful largely as a means of indicating which type theoretical distribution best describes the statistical properties of the population under consideration. For many scientific and engineering applications, the normal (or Gaussian) distribution, to be discussed later, is an appropriate model. However, many cases can be expected in which the frequency distribution is strongly skewed. Application of statistical methods based on the Gaussian distribution to such cases will lead to erroneous results and conclusions. Often careful study of the nature of the physical situa-

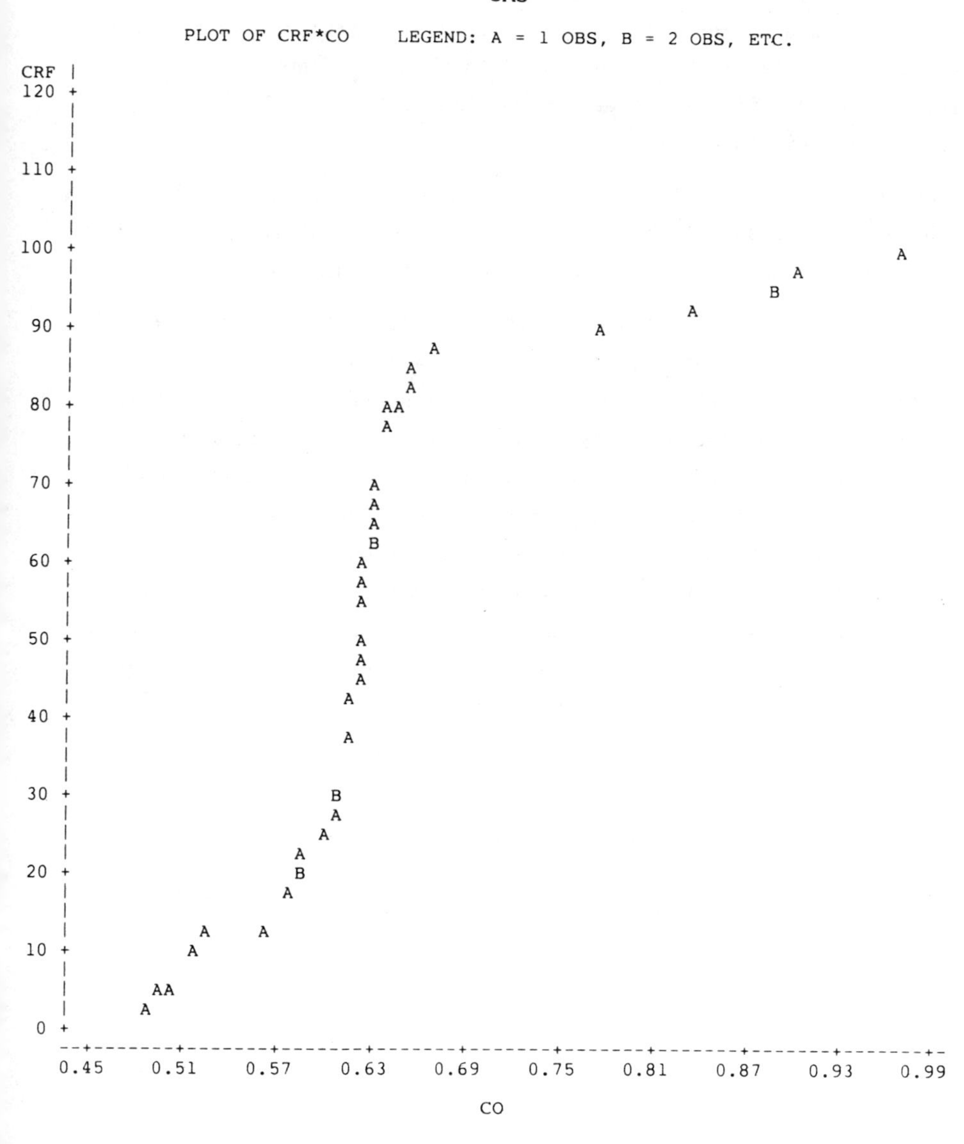

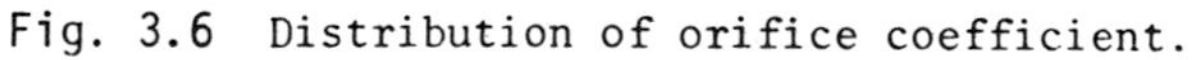
Fig. 3.6 Distribution of orifice coefficient.

tion from which the data are obtained will indicate the type distribution followed by the data. Before the statistical analysis of any experimental data is attempted, the data should be appropriately plotted. Often this will give a strong indication of the nature of the relationship involved.

The three most commonly used theoretical distributions are the Binomial and Poisson, both of which are discrete distributions, and the normal (or Gaussian) distribution which is continuous.

Before discussing some important theoretical distributions, we will define two characteristics associated with a distribution. In general, a characteristic associated with a distribution is called a parameter. A parameter is defined to be a numerical value associated with a theoretical distribution. The two parameters we consider here are the mean and variance, the former being a measure of location and the latter a measure of variation. Since the theoretical probability distribution gives a complete description of the corresponding random variable, we will call the two parameters the mean and variance of the random variable X or, equivalently, of the probability distribution. The parameters of a probability distribution, be it discrete or continuous, will be denoted by Greek letters, for example μ and σ.

Suppose a random variable X can take on the values of a discrete (finite) set $S = \{x_1, x_2, \ldots, x_n\}$ according to the probability function $f(x_i) = 1/n$, $i = 1,2,\ldots,n$. The mean, denoted by μ, of the discrete random variable X is defined by

$$\mu = \frac{\sum_{i=1}^{n} x_i}{n} = \sum_{i=1}^{n} x_i f(x_i), \tag{3.6}$$

since $f(x_i) = 1/n$. If X can assume the values of the set $S = \{x_1, x_2, \ldots, x_n\}$, which may be countably infinite, according to the probability function with values $f(x_1), f(x_2), \ldots, f(x_n)$, then the mean of X is likewise given by

$$\mu = \sum_{i=1}^{n} x_i f(x_i) \quad . \tag{3.7}$$

It should be noted the mean in (3.6) is the usual arithmetic average of the numbers $x_1, x_2, \ldots, x_n$. The mean in (3.7) is the usual weighted arithmetic average of $x_1, x_2, \ldots, x_n$. The probability function in each case is theoretical, which is not known in general. This prompts the usage of the term expectation or expected value. We say the mean μ is the expectation of X or the expected value of X, and denote this operation by

$$\mu = E(X) = \sum_{i=1}^{n} x_i f(x_i) \quad . \tag{3.8}$$

The second parameter of interest is the variance of X, denoted by σ^2. The variance of the discrete random variable X is the expected value of $(X-\mu)^2$ or

$$\sigma^2 = E\left[(X-\mu)^2\right] = \sum_{i=1}^{n} (x_i-\mu)^2 f(x_i) \quad . \tag{3.9}$$

The variance of X is simply the weighted average of $(x_1-\mu)^2$, $(x_2-\mu)^2, \ldots, (x_n-\mu)^2$, with respect to the weights $f(x_1), f(x_2), \ldots,$ $f(x_n)$.

Since a continuous variable may assume any value between two distinct numbers, the function describing a continuous distribution must be continuous and the geometric representation of the relationship between the probability density function f(x) and x must be a continuous curve. While measurement of a continuous variable may yield data which appear discrete because of the manner in which data are recorded, the data should be handled as a sample from a continuous distribution. For example, temperature is a continuous variable even though the thermometer readings are recorded only to the nearest degree, giving a series of recordings which appear discrete. The probability density function of a continuous random variable has the following properties:

(a) f(x), the probability density function is single valued and nonnegative, that is, $f(x) \geq 0$ for all x,

(b) $$\int_{-\infty}^{+\infty} f(x)\,dx = 1, \tag{3.10}$$

(c) $$\int_{x=a}^{x=b} f(x)\,dx = P(a < X \leq b). \tag{3.11}$$

Thus the probability of a value x falling between the values a and b is geometrically represented as the area under the probability density function between the limits a and b as shown in Fig. 3.7.

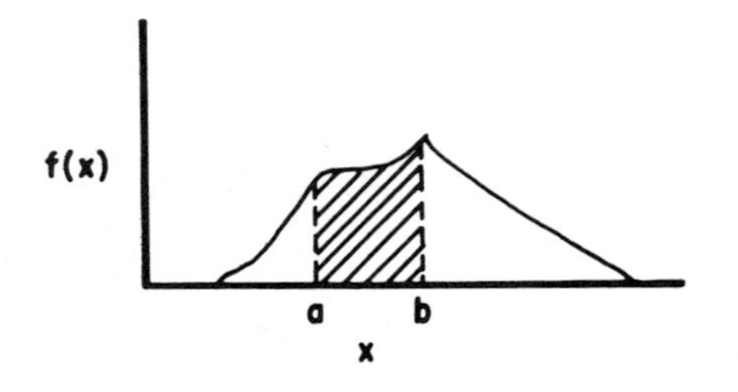

Fig. 3.7 Probability density function.

The mean μ and variance σ^2 of a continuous variable X are given by

$$\mu = E(X) = \int_{-\infty}^{\infty} xf(x)\,dx \tag{3.12}$$

and

$$\sigma^2 = E\left[(X-\mu^2\right] = \int_{-\infty}^{\infty} (x-\mu)^2 f(x)\,dx \quad . \tag{3.13}$$

The positive square root of σ^2 is called the standard deviation of X and is denoted by σ. The definitions (3.12) and (3.13) are suggested by similar considerations that lead to the definition of the Riemann integral.

The parameters μ and σ^2 are referred to as the population mean and variance, respectively. They are generally unknown and information about them, in the form of inferences, is obtained by considering a sample from the appropriate population. This leads to the sample analogues of μ and σ^2, which are called the sample mean $\overline{X}$ and sample variance S^2 or S_x^2. These ideas will be discussed in more detail in Chapter 4. However, we note in passing that $\overline{X}$ and S^2 are computed on the basis of the observations in the sample, which represents the corresponding population, according to (3.6) and (3.9) with $f(x_i) = \frac{1}{n}$ and $f(x_i) = \frac{1}{n-1}$, respectively. The reason for this will be pointed out later. The quantities $\overline{X}$ and S^2 are used in Section 3.3.7 which deals with the normal distribution.

3.3.1 Binomial Distribution

When an observation is placed in one or the other of two mutually exclusive categories, as is often the case in practice, a discrete distribution which is called the binomial or dichotomous distribution results. Although the variable is usually considered qualitative (yes or no, pass or fail, dead or alive), the distribution may be made quantitative by assigning the values 0 and 1 to the two categories. It makes no difference how these are assigned. The two categories are usually labeled success and failure and are assigned the values 1 and 0, respectively.

Let p denote the probability of success (the event occurring) and q the probability of failure (the event not occurring) in one trial of an experiment. Since these events are mutually exclusive and exhaustive, q = 1 - p. If the probability of a success is the same for each independent trial of the experiment the probability of exactly x successes in n trials is given by the expression

$$f(x) = \frac{n!}{x!(n-x)!} p^x q^{n-x}, \quad x = 0,1,2,\ldots,n \quad . \tag{3.14}$$

where $n!/x!(n-x)!$ is the number of combinations of n things taken x at a time. It is recognized that Eq. (3.14) is the general term of the familiar binomial expansion in which $x = 0,1,2,\ldots,n$, and the factor $n!/x!(n-x)!$ is the binomial coefficient of the xth term.

Table 1 of Appendix C contains values of the binomial cumulative distribution function

$$P(X \leq x) = \sum_{k=0}^{x} \binom{n}{k} p^k (1-p)^{n-k}, \quad x = 0,1,2,\ldots,n,$$

for selected values of n and p. By means of this table one can compute other probabilities such as $P(x_1 \leq X \leq x_2)$ since

$$P(x_1 \leq X \leq x_2) = P(X \leq x_2) - P(X \leq x_1 - 1)$$

$$= \sum_{k=0}^{x_2} \binom{n}{k} p^k (1-p)^{n-k} - \sum_{k=0}^{x_1 - 1} \binom{n}{k} p^k (1-p)^{n-k}$$

If we plot f(x) for a fixed number of observations, say n = 8, the shape of the resulting graphs change with p and q as shown in Fig. 3.8 where each f(x) is associated with a fixed value of n. The first of these (a) is markedly skewed, the second (b) only slightly skewed, and the third (c) is symmetric. As we will see, as n becomes large while the value of p stays fixed, the binomial distribution corresponding to (3.14) approaches the normal distribution and then reverts to skewness as p increases.

The mean and variance of the binomial distribution are given by

$$\mu = np = E(X) \tag{3.15}$$

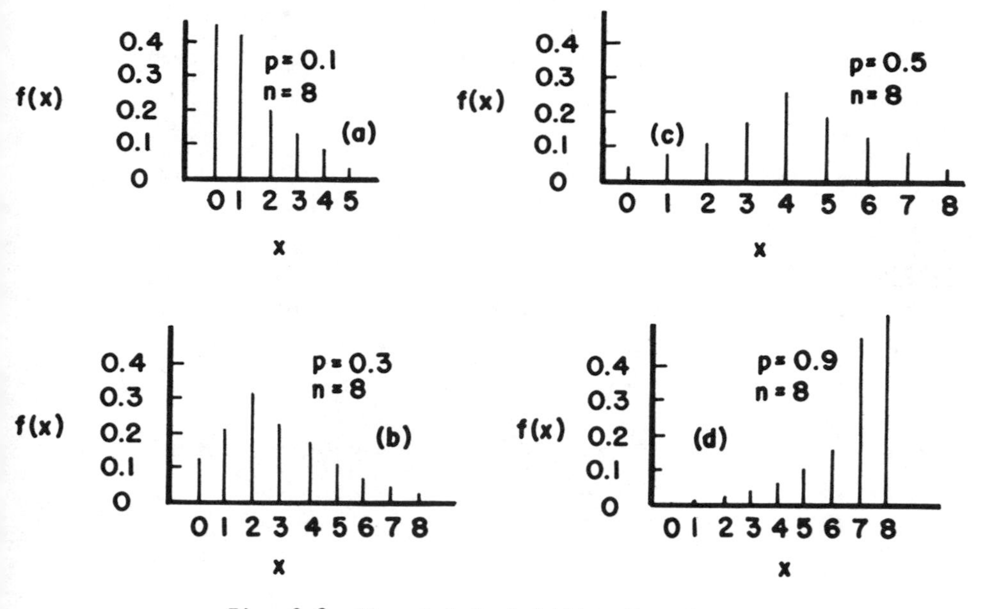

Fig. 3.8 Binomial Probability Functions.

and

$$\sigma^2 = npq = E\left[(X-\mu)^2\right] \quad . \tag{3.16}$$

Example 3.2 The probability that a compression ring fitting will fail to seal properly is 0.1. What is the expected number of faulty rings and their variance if we have a sample of 200 rings? Assuming that we have a binomial distribution, we have $\mu = np = 0.1(200) = 20$ and $\sigma^2 = npq = 200(0.1)(0.9) = 18$.

The special case of the binomial distribution when $n = 1$ is commonly called the point binomial or Bernoulli distribution and is completely defined by one parameter p (or $q = 1 - p$). Obviously, for the point binomial $\mu = p$ and $\sigma^2 = pq$.

Specific binomial probabilities, such as the probability that there will be exactly b successes in n trials, may be calculated from Table I by considering

$$P(X = b) = P(X \leq b) - P(X \leq b - 1). \qquad (3.17)$$

Example 3.3 Past history indicates that the probability of a given batch of chemical being on-grade is 0.90. Let X denote the number of off-grade batches. What is the probability of getting exactly one off-grade batch in 10 runs?

From Table I, with $n = 10$ and $p = 1 - 0.90 = 0.10$, we obtain

$$P(X=1) = P(X \leq 1) - P(X=0)$$
$$= 0.736 - 0.349$$
$$= 0.387.$$

Example 3.4 For the case of the preceding example, what is the probability of getting more than two off-grade batches in 10 runs? From Table I we obtain

$$P(X > 2) = 1 - P(X \leq 2)$$
$$= 1 - 0.930$$
$$= 0.070.$$

Example 3.5 The probability that an entering freshman at one of the Ivy League schools will graduate is 0.4. Of any group of six entering freshmen, what are the probabilities that (a) one will graduate and (b) at least one will graduate? Let X denote the number that will graduate. From Table I we get

(a) $P(X=1) = P(X \leq 1) - P(X=0)$

$= 0.233 - 0.047$

$= 0.186$

(b) $P(X \geq 1) = 1 - P(X=0)$

$= 1 - 0.047$

$= 0.953.$

Example 3.6 Twenty percent of the items produced in a certain stamping process are defective. Here $p = 0.2$ since we associate success with obtaining a defective. Let X denote the number of defectives in a randomly chosen sample of 8.

(a) What is the probability that 0, 1, or 2 items in the randomly chosen sample will be defective: From Table 1 we obtain

$$P(X = 0,1, \text{ or } 2) = P(X \leq 2) = 0.797.$$

(b) What is the probability of obtaining at least two defective items?

$$P(X \geq 2) = 1 - P(X \leq 1) = 1 - 0.503 = 0.497.$$

(c) What is the probability of obtaining at most 2 defective stampings?

$$P(X \leq 2) = 0.797.$$

3.3.2 Poisson and Exponential Distributions

The relation between the Poisson and binomial distributions is quite simple. To begin with let $\lambda = np$ where n is the number of observations (occurrences) and p is the probability of success. Thus $p = \lambda/n$ and $q = 1-\lambda/n$. The binomial distribution can then be written as

$$p(x) = \binom{n}{x}\left(\frac{\lambda}{n}\right)^x \left(1 - \frac{\lambda}{n}\right)^{n-x} \tag{3.18}$$

If n approaches infinity and p approaches zero in such a way that np is constant, then the binomial distribution approaches the Poisson distribution with probability function f(x) given by

$$f(x) = \lambda^x e^{-\lambda}/x!, \quad x = 0,1,2,\ldots \tag{3.19}$$

f(x) approximates the probability of x successes and e is the base of the natural logarithm, 2.71828. The cumulative Poisson distribution function F(x) is given by

$$F(x) = \sum_{k=0}^{x} e^{-\lambda}\lambda^k/k! \quad , \tag{3.20}$$

where the parameter λ is the mean and also the variance of the Poisson. Small values of λ yield skewed probability functions as shown in Fig. 3.9. As in the case of the binomial distribution the Poisson approaches the normal distribution as λ increases. The Poisson distribution function F(x) given by (3.20) is given in Table II of Appendix C for selected values of x and λ.

From the way the Poisson distribution was introduced as a limiting case of the binomial, we see that for small p and large n it could be used as a relatively good approximation to the binomial distribution. In general, if $p \leq 0.1$ and $n \geq 20$ the Poisson distribution can be used to approximate the binomial without introducing any significant errors. This will be illustrated in the examples below.

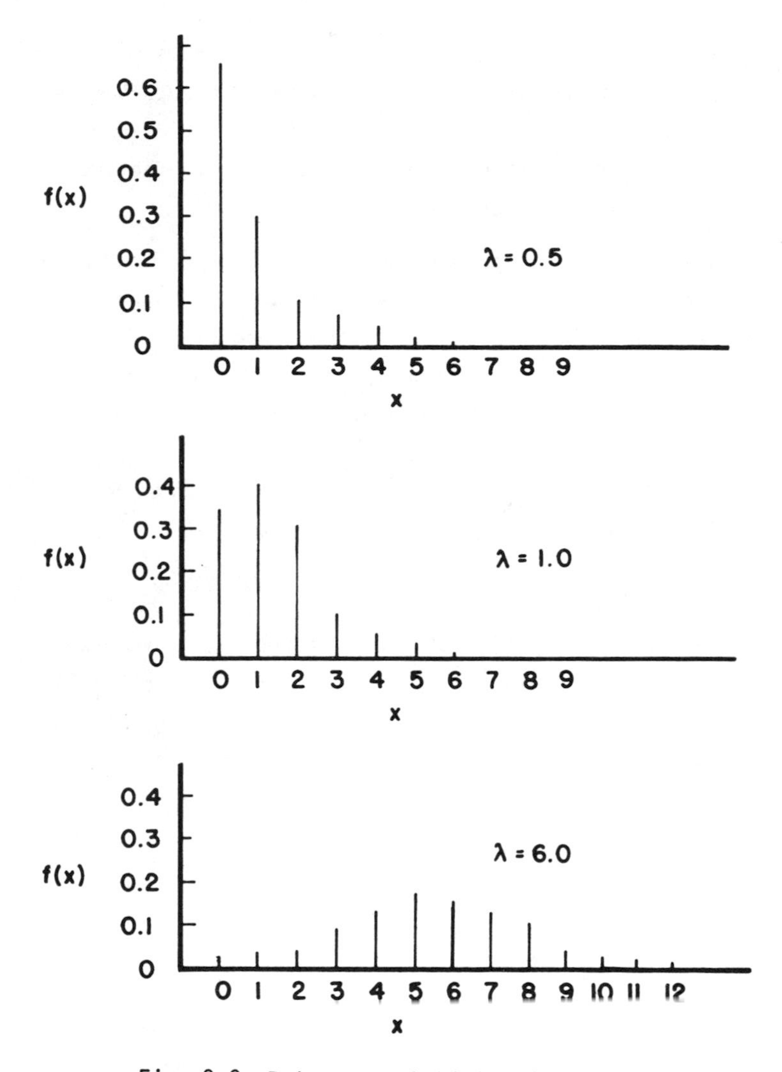

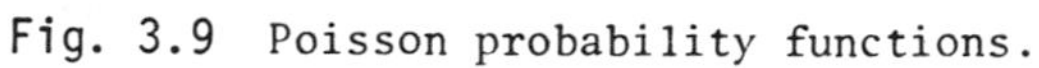
Fig. 3.9 Poisson probability functions.

Example 3.7 In the spray coating of aluminum test panels with paint for use in weathering tests, 8% of the panels are unusable because of improper degreasing prior to paint application. For a sample of 20 such panels, what is the probability of having two defectives? Let X denote the number of defectives.

A binomial distribution is appropriate as a model since the panels are either defective or suitable for use, and the degreasing of each panel can be assumed to be independent of the degreasing of every other panel. Hence, the desired probability, obtained from the binomial with $p = 0.08$ and $n = 20$, is

$$P(X=2) = \frac{20!}{2!18!}(0.08)^2(0.92)^{18}$$

$$= 190(.0064)(0.92)^{18} = 0.2711.$$

[Table I does not cover the value of $p = 0.08$. In cases such as this one has to resort to more extensive tables or evaluate the desired quantities such as $(0.92)^{18}$ by means of logarithms or with a pocket calculator.] However, since p is small and n is large, the Poisson with $\lambda = np = 20(0.08) = 1.6$ can be used to get

$$\begin{aligned} P(X=2) &= P(X \leq 2) - P(X \leq 1) \\ &= 0.783 - 0.525 \\ &= 0.258 \quad . \end{aligned}$$

Thus, we see that the error involved in using the Poisson to approximate the binomial is about 4.7% which may be significant or not depending on the requirements of the experiment.

Example 3.8 It is known that 3 people out of every 2000 have an adverse reaction to a particular drug. What are the probabilities of (a) three adverse reactions and (b) less than two adverse reactions in a group of 1000 test subjects? Let X denote the number of adverse reactions.

(a) The appropriate model is again the binomial with p = 0.0015 and n = 1000. Table I does not cover these values of p and n. The probability of three adverse reactions is given by

$$P(X=3) = \frac{1000!}{997!3!} (0.0015)^3 (0.9985)^{997} = 0.12555.$$

Since calculations of this type are tedious using logarithms, and since the Poisson is applicable, we can obtain a close approximation to the above probability by calculating $P(X=3)$ from the Poisson with $\lambda = np = 1000(0.0015) = 1.5$. From Table II we have

$$\begin{aligned} P(X=3) &= P(X \leq 3) - P(X \leq 2) \\ &= 0.934 - 0.809 \\ &= 0.125 \end{aligned}$$

(b) Since the error in using the Poisson approximation is minute in this case we can estimate the probability of getting less than two adverse reactions using the Poisson as follows:

$$\begin{aligned} P(X < 2) &= P(X \leq 1) \\ &= 0.558 \quad . \end{aligned}$$

Another approach to the Poisson distribution which motivates its use in physical problems follows.

Let the probability that a certain event E occurring in a time interval (t, t + dt) be k dt, where k is a constant. Also assume that the length of the time interval (t, t + dt) is so small that the probability of the event occurring more than once in the time interval is very small and this probability may be neglected. In addition, assume that the occurrence of E in nonoverlapping time intervals are independent events. Now, the problem is to determine the probability $P_x(t)$ that the event E will occur x times within a time t. Note that $P_0(t + dt)$ is the probability that event E will not occur in the time period (0, t + dt). This means that the event E does not occur in the time interval (0,t), nor in the interval (t, t + dt). The probability of the former is $P_0(t)$ and of the latter is (1 - k dt). Since we assumed these probabilities are independent we have

$$P_0(t + dt) = P_0(t)(1 - k\ dt) = P_0(t) - kP_0(t)\ dt$$

$$\frac{P_0(t + dt) - P_0(t)}{dt} = -kP_0(t)$$

The limit of the left-hand side as dt approaches 0 is just the derivative of $P_0(t)$ with respect to t. Thus, a differential equation is obtained, namely,

$$\frac{dP_0(t)}{dt} = -kP_0(t) ,$$

whose solution is $P_0(t) = e^{-kt}$ as can be easily checked, noting that $P_0(0) = 1$.

Now to determine the probability that the event E will occur x times in the interval (0,t + dt), we know that this can happen by the event E occurring x - 1 times during (0,t) and once during (t, t + dt) or E occurring x times during (0,t) and not occurring during (t, t + dt). This is illustrated in Fig. 3.10.

Since these two possibilities are mutually exclusive events we have

$P_x(t + dt) = P$(E occurs exactly x times within time t and 0 times in dt) + P(E occurs x-1 times within time t and 1 time in time dt)

$= P$(E occurs x times within time t) P(E occurs 0 times in dt) + P(E occurs x-1 times within time t) P(E occurs 1 time in dt)

$= P_x(t)(1-k\ dt) + P_{x-1}(t)k\ dt.$

Rearranging yields

$$\frac{P_x(t + dt) - P_x(t)}{dt} = -kP_x(t) + kP_{x-1}(t) \quad .$$

The limit of the left-hand side is just the derivative of $P_x(t)$, and the resulting differential equation is

$$\frac{dP_x(t)}{dt} = -kP_x(t) + kP_{x-1}(t) \quad .$$

A solution, subject to the conditions $P_0(0) = 1$ and $P_x(0) = 0$, for $x > 0$, is

$$P_x(t) = \frac{(kt)^x e^{-kt}}{x!}$$

as can be readily verified.

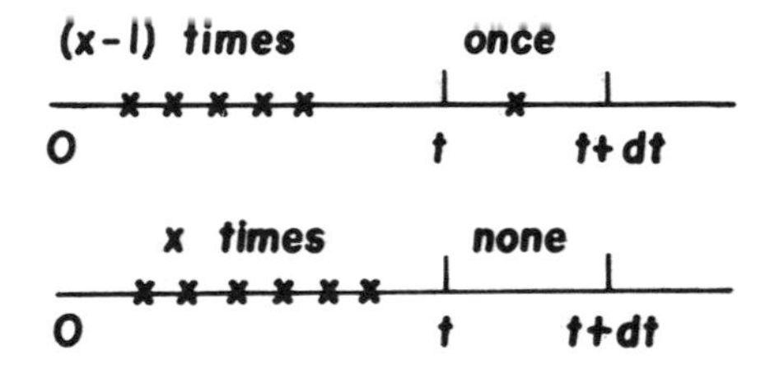

Fig. 3.10 Possible ways event E can occur x times during time interval (0, t + dt).

Therefore the probability function of the number of times the event E occurs in the time interval (0,t) is given by $P_x(t)$ with parameter kt.

The above derivation describes a general physical situation which leads to the Poisson distribution. From the above approach we say that if the probability of each radioactive atom in a mass disintegrating is constant, then the number of atoms disintegrating in a time period t has the Poisson distribution. Also, the Poisson may apply to the number of phone calls per time interval during a given part of a day, to the number of flaws per yard of insulated wire, to the number of misprints per page, to the number of blood cells on individual squares on a haemocytometer, etc. Another specific example is the distribution of the number of failures, x, of electronic equipment within a time t assuming independent failures and a constant failure rate.

From the foregoing derivation of the Poisson probability function, one may wonder about the distribution of the time period between successive occurrences of the event E. We assumed that the probability of the event E occurring in the time interval (t, t + dt) was k dt and found $P_0(t) = e^{-kt}$. Hence, we can write

$$P(\text{time interval between successive occurrences of E} > t)$$
$$= P[\text{E occurs 0 times in } (0,t)] = e^{-kt}.$$

Thus

$$P(\text{time interval between successive occurrences of E} < t)$$
$$= 1 - e^{-kt}.$$

The left-hand side is in the form of a cumulative distribution, therefore differentiating yields the probability density function of the time interval between successive occurrences of the event E. This yields

$$p(t) = ke^{-kt}, \quad t > 0$$

which is the negative exponential probability density function.

In conclusion, if the number of times an event E occurs per interval of time t is distributed as the Poisson with parameter kt, then the distribution of time between successive occurrences of E is distributed as the negative exponential with parameter k.

3.3.3 Negative Binomial Distribution

Assume that the assumptions necessary for a binomial distribution are satisfied, that is, each performance of an experiment results in one of two mutually exclusive outcomes S and F, the outcome of each performance is independent of other performances, and the probabilities of S and F remain fixed at p and q, respectively, for each performance.

In Section 3.3.1 we observed that when the above assumptions were satisfied, the binomial probability function yielded the probability of having S occur exactly x times out of n performances of the experiment. We now consider the situation where n is not fixed. Since the number of performances of the experiment can vary, we will call it x. We wish to obtain an expression for the probability that x performances of the experiment will be required to produce exactly r of the outcomes S. This expression can be obtained from the binomial by reasoning as follows. The only way that exactly r outcomes will be S, out of x performances of an experiment, is for exactly (r - 1) outcomes to be S, out of the first (x - 1) performances of the experiment, and the last performance to be an S. Therefore, letting X be the number of trials needed to produce r of the outcomes S and setting $P(X=x) = P_r(x)$, we have

$$P_r(x) = 0 \text{ for } x < r,$$

$$\begin{aligned} P_r(x) &= P(\text{r-1 out of x-1 are S and the xth outcome is S}) \\ &= P(\text{r-1 out of x-1 are S})P(x^{th} \text{ outcome is S}). \end{aligned}$$

Since the assumptions necessary for a binomial are satisfied, we have

$$P(\text{r-1 out of x-1 are S}) = \binom{x-1}{r-1} p^{r-1} q^{x-r}.$$

Also $P(x^{th}$ outcome is S) = p so that

$$P_r(x) = \binom{x-1}{r-1} p^r q^{x-r}; \; x \geq r. \tag{3.21}$$

The probability function $P_r(x)$ is called the negative binomial.

Example 3.9 Suppose light bulbs are being tested for instant failure. The bulb is turned on, and if it burns out immediately, the bulb is called defective, which we choose to call success. Otherwise, it is classified as nondefective. It is realistic to assume that the performance of each bulb is independent of the performance of the others whenever the production process is under control. Suppose further that it has been established over a long period of time that the manufacturing process produces 1% defective bulbs. What is the probability that 10 bulbs need to be tested to get exactly one defective? In this case $x = 10$, $r = 1$, and $p = 0.01$. Thus letting X denote the number of trials needed we have

$$P(X=10) = P_1(10) = \binom{9}{0} (0.99)^9 (0.01)^1 \simeq 0.08.$$

3.3.4 Hypergeometric Distribution

The hypergeometric distribution is often the model used to obtain probabilities where the sampling is done without replacement. The ensuing example of its use will make it clear when one should consider it as the appropriate model.

Example 3.10 Suppose 5 radiator caps are selected at random from each lot of 50 caps and tested to determine if they hold pressure to within the prescribed limits, as part of a manufacturer's quality-control program. Assume also that if more than 1 out of the 5 caps does not hold the proper pressure and is thus classified as defective, all 50 caps are tested to insure that they are all right. If in a certain lot of 50 caps, 10 are defective, what is the probability that all 50 will end up being tested?

This seems to be closely related to the binomial distribution if we let the testing of a radiator cap be a performance of an experiment, and let a nondefective cap be termed a success, and a defective cap be termed a failure. We are then just asking for the probability of more than 1 failure out of 5, which is very much like the general problem solved by the binomial distribution. However, there is one basic difference which should be noted. The probability of a success differs from test to test. For instance, for the first test the probability of a success is $p_1 = 40/50$, but for the second test it will be $p_2 = 40/49$ if the first was a failure, or $p_2 = 39/49$ if the first was a success, etc.

In order to obtain the desired probability we have to enumerate. The number of ways of selecting x defectives from the 10 that are defective is given as $\binom{10}{x}$ and the number of ways of selecting 5-x nondefectives from the 40 nondefectives is $\binom{40}{5-x}$. Hence the number of ways of selecting x defectives and 5-x nondefectives is $\binom{10}{x}\binom{40}{5-x}$. Since a sample of size 5 can be selected in $\binom{50}{5}$ ways from the lot of 50 caps, we have, by letting X denote the number of defectives,

$$P(X=x) = f(x) = \frac{\binom{10}{x}\binom{40}{5-x}}{\binom{50}{5}}, \qquad x = 0,1,\ldots,5.$$

Since all 50 caps will be tested if more than 1 defective is found, the probability that more than 1 defective is obtained is given by the following probability:

$$P(X \geq 2) = \sum_{x=2}^{5} \frac{\binom{10}{x}\binom{40}{5-x}}{\binom{50}{5}}$$

$$= \frac{\binom{10}{2}\binom{40}{3}}{\binom{50}{5}} + \frac{\binom{10}{3}\binom{40}{2}}{\binom{50}{5}} + \frac{\binom{10}{4}\binom{40}{1}}{\binom{50}{5}} + \frac{\binom{10}{5}\binom{40}{0}}{\binom{50}{5}}$$

$$= 0.251959 \quad .$$

In general, let N be the lot size, n be the sample size, D be the number of defectives in the lot, and x be the number of defectives in the sample, then

$$f(x) = \frac{\binom{D}{x}\binom{N-D}{n-x}}{\binom{N}{n}} , \qquad (3.22)$$

where it is understood that $\binom{a}{b}$ is zero whenever $b > a$. The above probability function is called the hypergeometric distribution.

Example 3.11 In the manufacture of precision tube fittings for high vacuum service, a sample of 5 is selected from every production run of 100 fittings of any given type and size. If any of the 5 randomly selected samples is found to be defective, that is, it will not hold an applied vacuum equivalent to 0.01 psia, then the entire 100 fittings in that particular run is declared suspect. All fittings are then individually checked for vacuum service. The severity of this quality-control criterion is indicative of two things: the manufacturer's desire to uphold his high reputation and his belief that his production and testing techniques are excellent. If, out of the next group of 100 fittings to come off the line, 3 are actually defective, what is the probability that the defect in the processing will be identified and corrected? As we can assume that the defective processing step will be rectified if it can be found, the question is really what is the probability that one or more of the faulty fittings will be in the quality control sample of 5?

For this problem, N = 100, n = 5, and D = 3. The probability is found from

$$P(X \geq 1) = \sum_{x=1}^{5} \frac{\binom{3}{x}\binom{100-3}{5-x}}{\binom{100}{5}}$$

$$= \frac{\binom{3}{1}\binom{97}{4}}{\binom{100}{5}} + \frac{\binom{3}{2}\binom{97}{3}}{\binom{100}{5}} + \frac{\binom{3}{3}\binom{97}{2}}{\binom{100}{5}} + \frac{\binom{3}{4}\binom{97}{1}}{\binom{100}{5}} + \frac{\binom{3}{5}\binom{97}{0}}{\binom{100}{5}}$$

$$= 0.144$$

for which the last two terms are zero because $\binom{3}{4}$ and $\binom{3}{5}$ are both zero.

We will now discuss some continuous distributions, the first of which is somewhat difficult to work with without tables and therefore will be discussed only briefly. This fact should not lead the reader to believe, however, that it is not used frequently as a probability model, only that it will not be dealt with at great length in this text.

3.3.5 Weibull Distribution

A distribution function which is being used more and more as a time-to-failure probability model is the Weibull probability density function given by

$$f(t) = \frac{\alpha}{\lambda}\left(\frac{t}{\lambda}\right)^{\alpha-1} \exp\left[-\left(\frac{t}{\lambda}\right)^{\alpha}\right]; \quad t \geq 0,\ \alpha > 0,\ \lambda > 0. \qquad (3.23)$$

One of its most frequent uses has been in studies on the service life of electron tubes. Some plots of the distribution are given in Fig. 3.11. Note that for $\alpha = 1$ the Weibull reduces to the negative exponential probability density function.

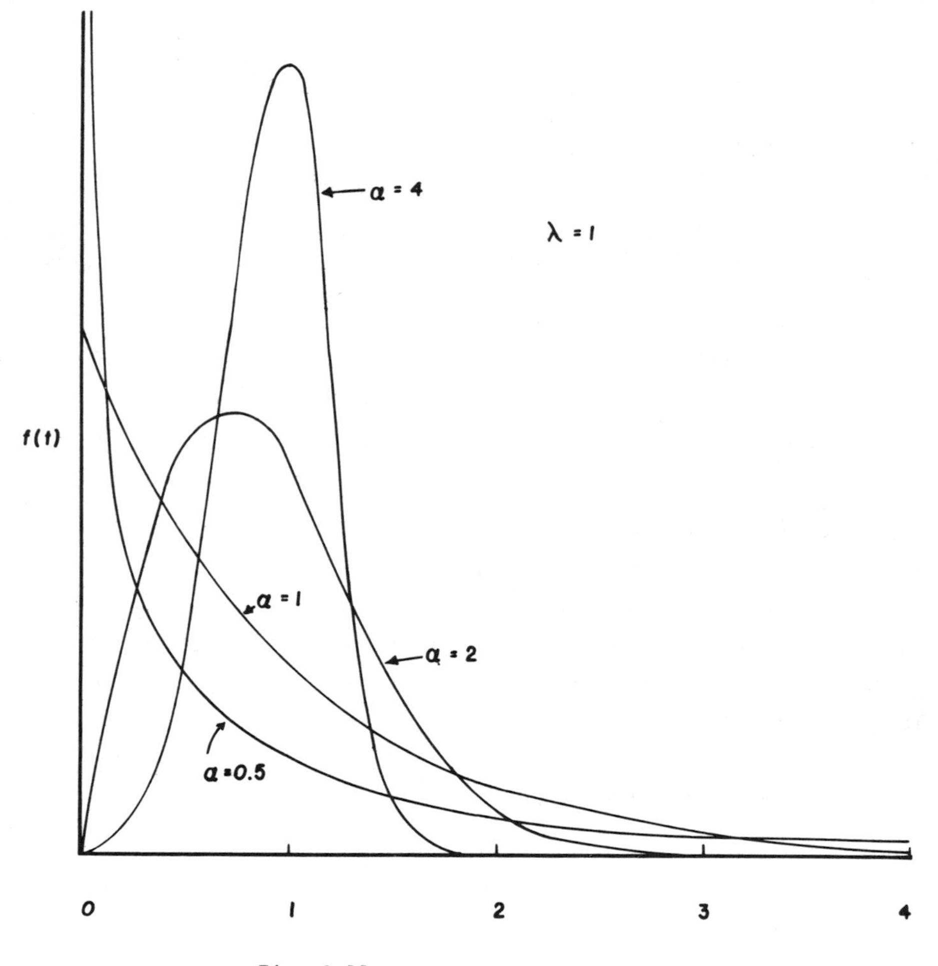

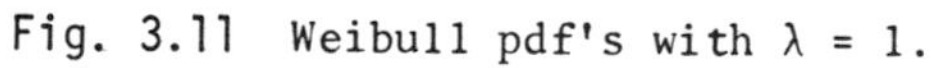
Fig. 3.11 Weibull pdf's with $\lambda = 1$.

3.3.6 Gamma Distribution

The gamma probability density function is

$$f(x) = \frac{x^{\alpha-1} e^{-x/\beta}}{\Gamma(\alpha)\beta^{\alpha}}, \quad x > 0 \qquad (3.24)$$

$$= 0 \qquad\qquad , \quad x \le 0.$$

This distribution has been used very extensively in statistics. The parameters α and β are both positive. If $\alpha = 1$, the gamma density function reduces to the exponential density function,

$$f(x) = (1/\beta)\, e^{-x/\beta}, \quad x > 0.$$

If $\alpha = \nu/2$ and $\beta = 2$, the gamma distribution becomes the chi-square probability density function in Eq. (3.32).

3.3.7 Normal Distribution

The most widely used of all continuous distributions is the normal distribution, or the common bell-shaped error curve. The basic assumption underlying this type of distribution is that any errors in experimental observation are due to variation of a large number of independent causes, each of which produces a small disturbance or error. Many experimental situations are subject to random error and do yield data that can be adequately described by the normal distribution, but this is not always true.

Regardless of the form of distribution followed by the population, the so-called *central limit theorem* allows the normal distribution to be used for descriptive purposes subject to the limitation of a finite variance. This law states that if a population has a finite variance σ^2 and a mean μ, then the distribution of the sample mean, denoted by $\overline{X}$, approaches the normal distribution with variance σ^2/n and mean μ as the size of the sample, n, increases. The proof is left to advanced statistical theory texts.

This theorem is all the more remarkable for its simplicity and profound impact on the treatment and description of data. The only requirement for having the sample mean behave normally is that the population variance be finite. One of the chief problems is to tell when the sample size is large enough to give reasonable compliance with the theorem.

The probability density function f(x) for the normal distribution is given by

$$f(x) = \frac{1}{\sigma\sqrt{2\pi}} e^{-\frac{1}{2}\left(\frac{x-\mu}{\sigma}\right)^2} \quad \text{for } -\infty < x < \infty. \qquad (3.25)$$

Inspection of the equation reveals that two parameters, μ the mean and σ the standard deviation, are required to completely describe the normal distribution. For a constant mean, the spread of the curves varies with the standard deviation and increases with increasing values of σ. Differences in the means at constant σ are reflected merely by changes in the location on the x axis with no change in the shape of the curve. These characteristics are shown in Figs. 3.12 and 3.13.

The normal distribution is a two-parameter distribution. The cumulative distribution function for this probability density is

$$F(x) = P(X \leq x) = \frac{1}{\sigma\sqrt{2\pi}} \int_{-\infty}^{x} e^{-(x-\mu)^2/2\sigma^2} dx. \qquad (3.26)$$

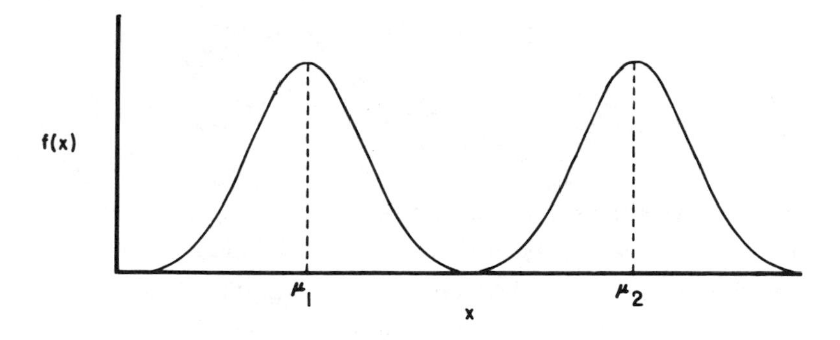

Fig. 3.12 Distributions with equal variance but different means.

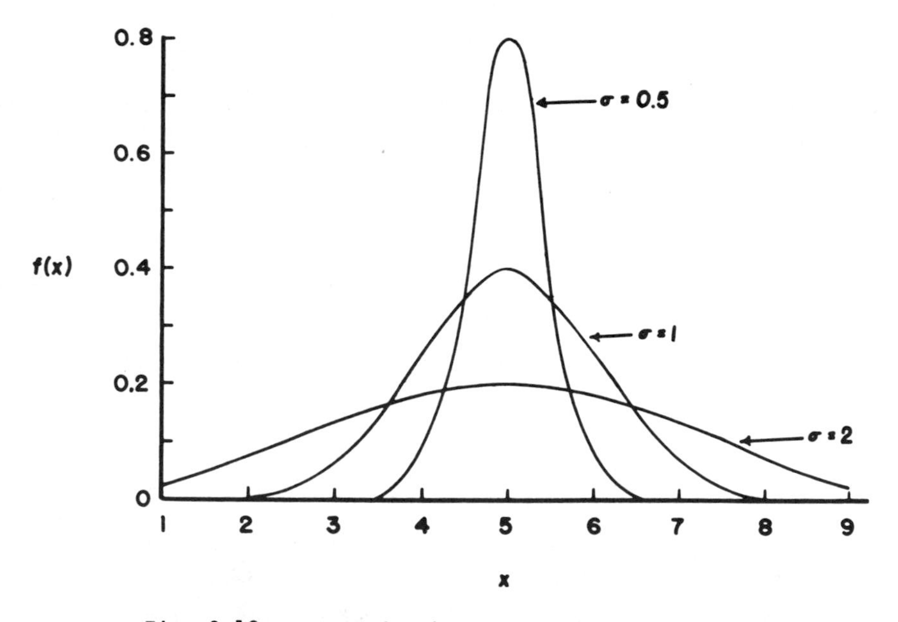

Fig. 3.13 Normal density curves for constant mean (μ=5) with varying σ.

The probability of X falling within the range x_1 to x_2 is

$$P(x_1 \leq X \leq x_2) = F(x_2)-F(x_1) = \frac{1}{\sigma\sqrt{2\pi}} \int_{x_1}^{x_2} e^{-(x-\mu)^2/2\sigma^2} dx. \quad (3.27)$$

Since the two parameters μ and σ are required to establish a particular normal curve, it is convenient to consider a particular member from this family of distributions. The cumulative distribution functions are not easily evaluated and tables are more readily handled if they are made for the particular case of $\mu = 0$ and $\sigma = 1$. Once the integrals have been evaluated for this case they can be easily converted to those corresponding to different values of μ and σ.

The standard normal deviate is defined as the number of standard deviations by which the variable differs from the mean. The standard normal deviate Z, defined by

$$Z = (X-\mu)/\sigma, \quad (3.28)$$

is normally distributed with a mean of 0 and a variance (or standard deviation) of 1. Noting that $dx = \sigma\, dz$, the cumulative distribution function F(z) of the random variable Z becomes

$$F(z) = P\left(\frac{X-\mu}{\sigma} \leq z\right) = \frac{1}{\sqrt{2\pi}} \int_{-\infty}^{z} e^{-t^2/2} dt. \quad (3.29)$$

Thus the distribution functions of all normal variables are identical when expressed in terms of the standard normal deviate.

The standard normal distribution is only the particular case of the normal distribution having a mean of 0 and a variance of 1. Values of the cumulative distribution function F(z) depend only on z. Table III is an abbreviated tabulation of these values. Figure 3.14 shows graphically some of the areas under the standard normal curve.

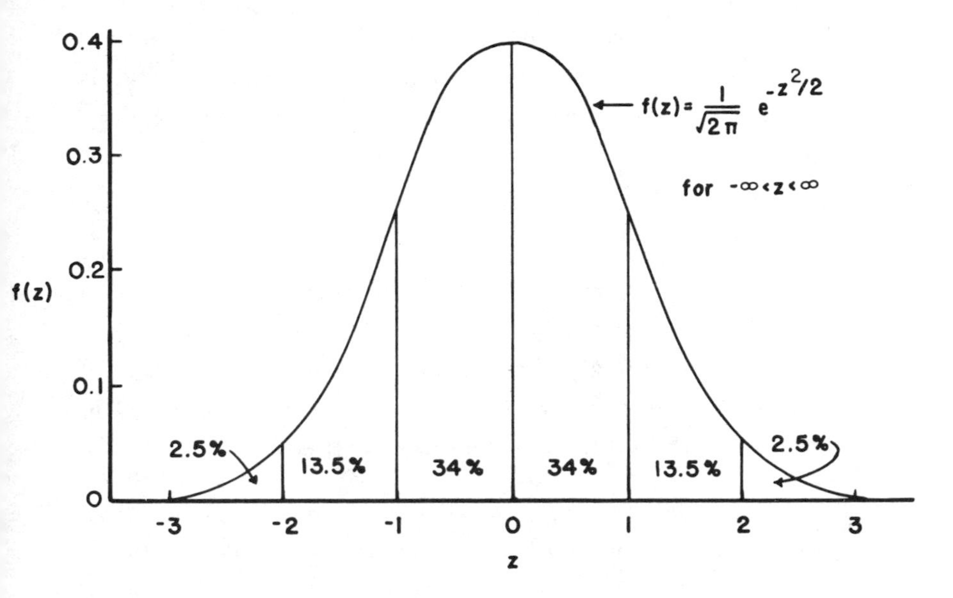

Fig. 3.14 The standard normal: N(0,1) (with approximate areas under the curve).

Let us now consider how Table III can be used. This is best done by way of illustrations. We first show how the table can be used to provide standard or normalized scores on examinations so that they can be compared with scores from other exams having different scales.

Example 3.12 On a particular physics exam, the class average was 72 and the standard deviation was 15. Three students taking a make-up exam (assumed to be the same as the original exam) made actual scores of 60, 93, and 72. What were their normalized scores? Here, $\bar{x} = 72$, $s_x = 15$. Replacing μ and σ by their "estimates" we have z approximately equal to $\frac{(x - \bar{x})}{s_x}$. Hence $z_{60} = \frac{(60 - 72)}{15} = -0.8, z_{93} = 1.4$, and $z_{72} = 0$. We could also use z with a knowledge of $\bar{x}$ and s_x to determine the actual test grades from the normalized scores since $x = \bar{x} + zs_x$. Hence, a z score of -1.0 corresponds to $x = 72 + (-1.0)15 = 57$.

Example 3.13 If one a different physics exam the Z scores of two class members were 0.8 and -0.4 corresponding to grades, respectively, of 88 and 64, what were the mean and standard deviation? Using the relationship $x = \bar{x} + zs_x$ we have two simultaneous equations: $8 = \bar{x} + 0.8s_x$ and $64 = \bar{x} - 0.4s_x$ to solve for $\bar{x}$ and s_x, from which we get $\bar{x} = 72$ and $s_x = 20$.

We now illustrate how probabilities for any member of the family of normal distributions can be obtained from the table for the standard normal. Table III gives the probability of getting a z value between $-\infty$ and some given value of z, where Z ranges from -4.0 to +4.0. The use of the table is illustrated below.

Example 3.14 Using Table III for the standard normal distribution, find the probabilities associated with the ranges of Z given below. This is really done by finding the area under the probability density function for the given z ranges.

(a) $0 \leq z \leq 1.4$. This is obtained by finding $P(Z \leq 1.4) - P(Z < 0)$, which is the area under the curve between $z = 0$ and $z = 1.4$. From the table we get this to be 0.4192.

(b) $-0.78 \leq z < 0$. This probability can be obtained by finding $P(Z < 0) - P(Z < -0.78) = 0.5 - 0.2177 = 0.2823$.

(c) $-0.24 \leq z \leq 1.9$. The result is obtained by taking $P(-0.24 \leq Z \leq 1.9) = P(Z \leq 1.9) - P(Z \leq -0.24) = 0.9713 - 0.4052 = 0.5661$.

(d) $0.75 \leq z \leq 1.96$. To get the probability of Z lying in this interval, we find $P(Z \leq 1.96) - P(Z \leq 0.75) = .9750 - 0.7734 = 0.2016$.

(e) $-\infty < z \leq -0.44$. The probability can be obtained directly from the table as $P(-\infty < Z \leq -0.44) = 0.32997$.

(f) $-\infty < z < 1.2$. For this interval we desire the entire area to the left of 1.2 which is read from the table to be 0.8849.

From the above, it should be clear that the probability of Z being in any interval can be obtained from Table III.

It is sometimes necessary to find values z_1, z_2 for some given value of $P(z_1 \leq Z \leq z_2)$.

Example 3.15

(a) Find z_1 if the probability that Z is between 0 and z_1 is 0.4. This merely requires us to find the value of z_1 for which the area between $z = -\infty$ and $z = z_1$ is 0.9 or 0.1. From Table III, $z_1 = \pm 1.282$ (using linear interpolation between adjacent tabular values).

(b) Find z_1 if $P(-\infty < Z \leq z_1) = 0.92$. From Table III we see that z_1 is between 1.40 and 1.41 and by interpolation we obtain $z_1 = 1.405$.

(c) Find z_1 if $P(-1.6 \leq Z \leq z_1) = 0.03$. Since $P(-1.6 \leq Z \leq z_1)$ $= P(Z \leq z_1) - P(Z \leq -1.6) = P(Z \leq z_1) - 0.05480 = 0.03$, the area to the left of z_1 must be 0.08480. The value of z_1 is thus found to be -1.3733.

As an illustration of the application of a theoretical distribution to experimental data, Fig. 3.15 shows a relative frequency

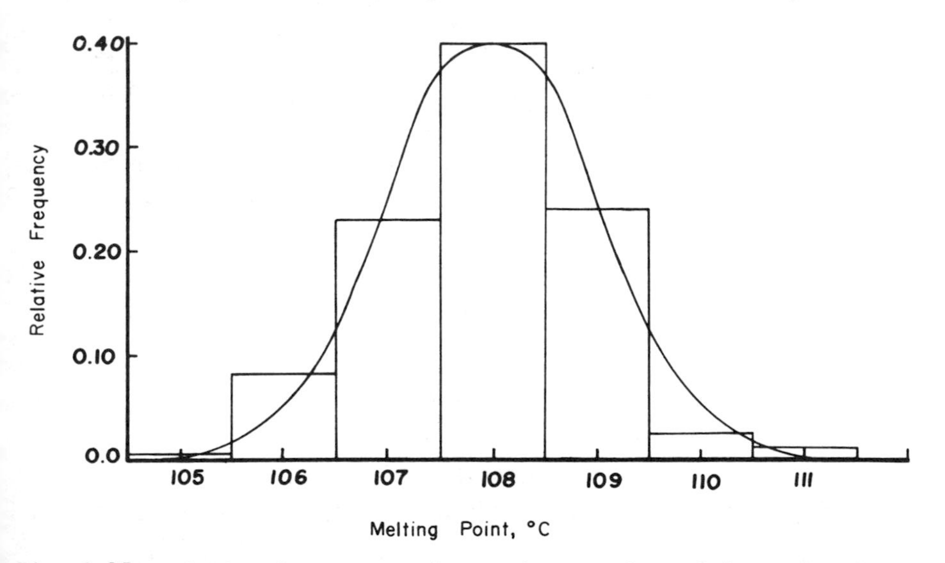

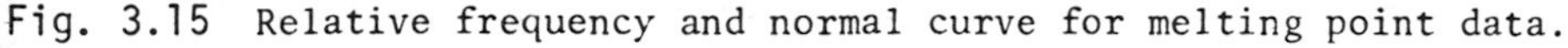
Fig. 3.15 Relative frequency and normal curve for melting point data.

histogram of the data given in Table 3.5 Superimposed on this histogram is a plot of the normal distribution curve using values of $\bar{x}$ and s calculated from the raw data of Table 3.2. The mean of the data, which is a good estimate of the population mean μ, was calculated to be 108.05°C. The standard deviation of the population was estimated from the data to be 0.960°C. In calculating points on the x vs f(x) curve in Fig. 3.15 it is necessary to convert from normal deviate values z to the corresponding values of x. From the definition of z in Eq. (3.28),

$$x = \mu + z\sigma = 108.05 + 0.960z,$$

the x value corresponding to each value of z is calculated. Plotting these values versus f(x) results in the symmetrical normal distribution curve.

There is no evidence that the distribution of the data, as pictured in the relative frequency histogram, deviates significantly from the normal curve. Thus the normal distribution may be used with confidence in describing the melting point population.

Example 3.16 A sample of 36 observations was drawn from a normally distributed population having a mean of 20 and variance of 9. What portion of the population can be expected to have values greater than 26? If the population were sampled repeatedly in groups of 36 observations, what portion of the population of means can be expected to have a value greater than 22?

(a) From Eq. (3.28), we have for the standard normal deviate

$$z_{calc} = \frac{x - \mu}{\sigma_x} = \frac{26 - 20}{\sqrt{9}} = 2 \quad .$$

From Table III we find $P(Z \leq 2) = 0.97725$. Therefore, the probability of obtaining a $Z_{calc} > 2$ is 1 - 0.97725 or 0.02275. Therefore, 2.28% of the population can be expected to have values greater than 26.

(b) The standard error of the mean is defined as

$$\sigma_{\bar{x}} = \sigma_x / \sqrt{n}$$

which for this problem is $3/\sqrt{36} = 0.500$. The z value in this case is

$$z = \frac{\bar{x} - \mu}{\sigma_{\bar{x}}} = \frac{22-20}{0.5} = 4 \quad .$$

From Table III we see that the area under the standard normal curve reaches 0.9999 when z reaches 3.9. Therefore, the probability of another sample of 36 (drawn from this population) having a mean greater than 22 is nil.

3.4 OTHER THEORETICAL DISTRIBUTIONS

Several other theoretical distributions are of importance in statistical methodology and three of the more common will be given without elaboration. The cumulative distribution of each is tabulated at the end of the book, and all three will be mentioned in later chapters as the need arises. Each involves the gamma function defined as

$$\Gamma(\alpha) = \int_0^{\infty} x^{\alpha-1} e^{-x} dx, \quad \text{for } \alpha > 0 \ . \tag{3.30}$$

3.4.1 Chi-Square Distribution

Suppose $Y_1, Y_2, \ldots, Y_\nu$ are independent random variables each distributed normally with mean 0 and variance 1. Let χ^2 (chi-square) be the sum of their squares. The random variable

$$\chi^2 = Y_1^2 + Y_2^2 + Y_3^2 + \ldots + Y_\nu^2 \tag{3.31}$$

has probability density function

$$f(\chi^2) = \frac{1}{2^{\nu/2}\,\Gamma\left(\frac{\nu}{2}\right)} e^{-\chi^2/2}\left(\chi^2\right)^{\nu/2-1} \quad \text{for } 0 \le \chi^2. \tag{3.32}$$

The parameter ν, called degrees of freedom (as discussed later), distinguishes the members of the family of χ^2 distributions.

3.4.2 The "Student's" t-Distribution

The "Student's" t-distribution was originally developed by W. S. Gossett who referred to himself as "Student." It is the basis for the simple t-test so widely used in the analysis of scientific and engineering data.

If Z has a normal distribution with mean 0 and variance 1, and V is distributed as χ^2 with ν degrees of freedom and Z and V are independent, then the random variable

$$t = Z/\sqrt{V/\nu} \tag{3.33}$$

has the probability density function

$$f(t) = \frac{1}{\sqrt{\nu\pi}} \frac{\Gamma\left(\frac{\nu+1}{2}\right)}{\Gamma(\nu/2)} \left(1 + \frac{t^2}{\nu}\right)^{-(\nu+1)/2} \quad \text{for } -\infty < t < \infty. \tag{3.34}$$

It is noted that this distribution depends only on the value of ν which distinguishes between members of the family of t-distributions.

3.4.3 F-Distribution

The F-distribution is the distribution of the random variable F which is defined as

$$F = \frac{U/\nu_1}{V/\nu_2} \tag{3.35}$$

where U and V are independent variables distributed as χ^2 with ν_1 and ν_2 degrees of freedom, respectively. The probability density function of F is given by

$$f(F) = \frac{\Gamma\left(\frac{\nu_1+\nu_2}{2}\right)}{\Gamma\left(\frac{\nu_1}{2}\right)\Gamma\left(\frac{\nu_2}{2}\right)} \left(\frac{\nu_1}{\nu_2}\right)^{\nu_1/2} \frac{F^{(\nu_1-2)/2}}{\left(1 + \frac{\nu_1}{\nu_2} F\right)^{(\nu_1+\nu_2)/2}} . \qquad (3.36)$$

This family of distributions is a two-parameter, ν_1 and ν_2, family, and thus tables and charts based on the F-distribution are inherently more complex than those based on the single parameter χ^2 and t-distributions. Equation (3.36) is restricted to the region where $F \geq 0$. A very important application of the F-distribution occurs in analysis of variance, which is treated in Chapter 8.

PROBLEMS

3.1 The average particulate concentration in micrograms per cubic meter for a station in a petrochemical complex was measured every 6 hours for 30 days. The resulting data are given below. For these data, prepare a table similar to Table 3.3 using appropriate class lengths and boundaries. Plot the class frequency and relative frequency on one graph as in Fig. 3.2 with a double ordinate. Plot the cumulative frequency and cumulative relative frequency polygons as in Fig. 3.3.

5	7	9	12	13	16	17	19	23	24	41
18	24	6	10	16	14	23	19	8	20	26
15	6	11	16	12	22	9	8	15	18	13
7	13	14	8	17	19	11	21	9	55	72
23	24	12	220	25	13	8	9	20	61	48
565	65	10	43	20	45	27	20	72	12	115
130	82	55	26	52	34	66	112	40	34	89
85	95	28	110	16	19	61	67	45	34	32
103	72	67	30	21	122	42	125	50	57	56
25	15	46	30	35	40	16	53	65	78	98
80	65	84	91	71	78	58	26	48	21	

3.2 On a diagnostic examination given recently to a group of 56 high-school science students, the following scores were obtained. Prepare appropriate tables of data and diagrams as in Problem 3.1 for these test scores. 250 points was the maximum possible score.

65	76	81	25	36	103	144	73	184	64	143
84	92	96	67	94	158	203	121	186	40	155
97	106	108	89	118	196	234	162	147	123	161
114	120	150	103	164	246	238	46	111	151	155
134	136	159	148	213	183	145	90	105	55	132
200	218	230	206	236	205	245	147	126	129	148
124	177	137	194	173	159	156	151	157	137	
187	200	214	237	213	149	185	132	160	168	

3.3 The numbers below are the final averages for three courses: ChE 3351, ChE 3311, and ChE 4312-5341. As the last pair are mass transfer courses, they are grouped together. For each set, plot class frequency, relative frequency, cumulative frequency, and cumulative relative frequency.

ChE 3351

89	72	89	81	89	85
87	83	91	64	83	69
90	82	91	91	90	88
91	76	91	78	85	75
88	90	84	70	82	86
75	87	89	77	82	75
89	80	91	82	71	85
74	69	78	94		

ChE 3311

75	59	91	71	70	76
77	80	28	70	73	48
68	71	76	48	65	78
54	76	30	29	62	60
69	85	84	64	66	31
28	46	74	71	74	54
76	72	24	68	61	53
82	80	83	61	66	40
58	59				

ChE 4312-5341

93	97	71	75	90	94
93	96	77	93	95	87
96	79	99	98	98	68
99	76	99	91	96	97
81	88	88	61	87	94
90	85	83	79	95	77
90	90	93	79	96	91
82	84	88	82	90	87
85	84	78	86	82	83
85					

3.4 Air pollution control and monitoring in the city of Lubbock falls under the responsibility of the Lubbock City-County Department of Health. Air samples are drawn from various locations within the city limits and analyzed for both naturally occurring and industrial contaminants. The results of analyses for sulfate (SO_4^{-2}) content in air over a 72-day period are given below. Concentrations are expressed as parts per million in air. Prepare a table of grouped frequencies for the data employing suitable class lengths and boundaries. Plot the class frequency and relative frequency on one graph. Also plot the cumulative frequency and cumulative relative frequency polygons.

3.14	1.87	1.00	1.12	1.03	5.88	0.91	1.86
0.08	1.95	6.38	2.61	0.10	1.96	0.00	0.00
2.00	5.57	2.96	2.68	2.44	2.62	7.61	0.80
2.11	6.99	9.02	3.24	1.25	0.00	8.06	0.00
3.55	7.80	8.90	1.95	1.85	2.07	0.00	6.85
3.51	10.83	14.71	1.24	1.23	4.11	1.75	0.00
3.69	7.17	1.95	0.63	1.24	1.84	1.78	0.00
0.98	1.93	3.30	0.00	1.66	3.60	0.00	0.00
2.01	2.58	2.01	1.24	2.38	0.00	1.83	1.14

3.5 The Department of Chemical Engineering has received a request from the company which manufactured our chain balances for the distribution data on the sensitivity of these balances to compare with the sensitivity of newly manufactured balances. Using the data below from several ChE 3111 laboratory reports, prepare a table of classes, class frequency, cumulative frequency, relative frequency, and cumulative relative frequency. Also, prepare graphs of double ordinates of class frequency and relative frequency versus classes and of cumulative frequency and cumulative relative frequency versus classes.

Sensitivity, mg/division

0.765	3.92
0.0721	0.810
0.791	2.125
0.7725	0.790
0.680	0.7725
0.715	0.818
0.435	0.714
1.012	0.560
0.485	0.600
1.500	0.711
1.000	0.711
0.702	0.900
0.711	0.870

3.6 One of the limiting factors for cooling tower design is the ambient dry-bulb temperature. As part of your design procedure, you have accumulated the average monthly temperature from 1897 to 1964 for July for the proposed plant location. Prepare a table similar to Table 3.3 and submit a plot of the class frequency and relative frequency using appropriate class lengths. Also plot the cumulative frequency and cumulative relative frequency polygons as illustrated in Fig. 3.3.

Temperature (^{o}F)			(Note: 2 years were not recorded)			
77.8	78.5	76.8	76.1	78.0	78.4	78.0
77.1	77.8	76.2	80.7	79.4	79.3	80.5
77.3	77.6	78.4	80.1	79.4	82.7	79.8
81.7	79.2	76.2	78.0	81.1	80.3	78.9
81.1	79.5	79.4	77.9	79.8	79.6	79.2
78.1	79.1	79.2	77.2	76.8	79.8	80.1
79.2	80.1	81.0	79.1	79.9	79.4	77.0
78.2	80.7	79.3	76.8	78.7	80.7	80.9
78.0	80.2	75.8	78.8	79.8	81.8	
77.4	78.9	81.9	80.5	80.4	78.3	

3.7 One of the demands agreed to by a certain oil company after a recent labor strike was to locate a plant in an area that would be suitable for workers with respiratory diseases. The following data give annual precipitation values for Tombstone, Arizona from 1898 to 1961. From these data prepare a table similar to Table 3.3 using appropriate class lengths and boundaries. Plot the class frequency and relative frequency on one graph as in Fig. 3.2 with a double ordinate. Plot the cumulative frequency and cumulative relative frequency polygons as in Fig. 3.3

Annual Precipitation (inches) for Tombstone, Arizona

Year	Precipitation	Year	Precipitation
1898	13.50	1931	23.82
1901	14.84	1932	16.35
1904	11.82	1933	13.54
1905	27.84	1934	11.21
1906	12.14	1935	15.11
1907	19.31	1936	11.07
1909	14.91	1937	12.97
1910	11.77	1938	16.62
1911	19.20	1941	12.29
1912	13.67	1942	9.07

Year	Precipitation	Year	Precipitation
1913	15.97	1943	11.99
1914	19.78	1944	15.05
1915	11.55	1945	9.38
1916	15.99	1946	14.78
1919	21.50	1948	14.07
1920	11.18	1950	10.84
1921	18.55	1954	14.88
1922	10.90	1955	18.77
1923	15.09	1956	8.89
1924	7.36	1957	16.49
1925	14.40	1958	20.44
1927	16.40	1959	14.71
1928	11.90	1960	8.82
1929	14.21	1961	13.32
1930	20.92	1962	12.31

3.8 Contained in the results of the dissertation of H. R. Heichelheim* are values for the compressibility factors for carbon dioxide at 100°C, over the pressure range from 1.3176 to 66.437 atm. Prepare a table and plot the data as in Problem 3.1.

Compressibility Factors

0.9966	0.9969	0.9971
0.9956	0.9957	0.9960
0.9936	0.9938	0.9940
0.9913	0.9912	0.9915
0.9873	0.9874	0.9980
0.9821	0.9823	0.9829
0.9747	0.9750	0.9758

*H. R. Heichelheim, The Compressibility of Gaseous 2,2-Dimethyl Propane by the Burnet Method, pp. 26, Ph.D. Dissertation, Library, The University of Texas, Austin, Texas, 1962. Data reproduced by permission of the author.

0.9648	0.9646	0.9657
0.9507	0.9500	0.9515
0.9447	0.9440	0.9454
0.9380	0.9371	0.9388
0.9306	0.9292	0.9314
0.9223	0.9215	0.9234
0.9130	0.9122	0.9141
0.9029	0.9022	0.9042
0.8916	0.8908	0.8930
0.8793	0.8783	0.8805
0.8656	0.8650	0.8657
0.8506	0.8501	0.8510
0.8343	0.8345	0.8354

3.9 In the experimental evaluation* of a distillation tower for the production of ethanol, the following data were recorded for the steam rate (lb/hr). Prepare a frequency table and plot class frequency and relative frequency on one graph. Also plot cumulative frequency and cumulative relative frequency on another graph.

1170	1350	1640	1800	1800
1260	1440	1730	1710	1710
1350	1440	1800	1530	1530
1440	1170	1260	1350	1350
1530	1440	1350	1170	1170
1620	1170	1440	1800	1800
1260	1260	1170	1710	1710
1350	1440	1530	1530	1530
1440	1495	1620	1350	1170
1260	1540	1170	1170	1440

*Data reprinted from R. Katzen, V. B. Diebold, G. D. Moon, Jr., W. A. Rogers, and K. A. Lemesurier, Chem. Eng. Prog. 64(1), 79-84 (1968), copyright 1968 by the American Institute of Chemical Engineers, by permission of the editor and publisher.

3.10 The diameters of MnO_2 particles used for SO_2 removal in a post-catalytic adsorber for the purification of spacecraft atmospheres were measured in 1/1000ths of an inch. For the particle data below, plot the frequency histogram and cumulative frequency polygon.

49	68	55	29	10	51
60	49	45	31	62	58
69	40	72	60	56	45
33	39	51	50	60	63
35	42	52	59	59	67
33	31	42	52	60	68
35	39	52	37	67	52
44	45	47	36	65	65
39	62	46	40	61	74
37	70	51	34	61	58
32	62	62	61	84	56
44	61	67	51	65	39
35	52	56	54	54	35
31	58	65	61	32	45
37	29	56	42	70	59
43	40	29	59	64	36
40	76	24	56	53	26
30	74	58	56	26	43
75	54	65	61	64	70
73	65	50	58	55	43

3.11 An experiment entitled "Friction Losses in Valves and Fittings" is currently being performed in the unit operations laboratory. The experimental apparatus is equipped with a small orifice which is used to determine the flow rate through the system. For purposes of comparison, several values of the orifice coefficient which have been calculated by different lab groups are presented below. Appropriate frequency tables of data and diagrams should be prepared from these values.

0.626	0.670	0.500	0.648
0.615	0.579	0.517	0.898
0.625	0.627	0.495	0.883
0.640	0.605	0.488	0.882
0.623	0.628	0.526	0.964
0.625	0.588	0.610	0.830
0.608	0.559	0.630	0.654
0.615	0.624	0.633	0.603
0.633	0.615	0.610	0.621
0.622	0.616	0.650	0.621
0.624	0.579	0.640	0.584
0.615	0.670	0.773	0.583
0.616	0.517	0.641	0.632

3.12 A chemical engineer desires to estimate the evaporation rate of water from brine evaporation beds for the month of July. He obtains the following daily evaporation data from local weather bureau records.

Evaporation (inches per day) for July (1964-1968)

0.36	0.34	0.35	0.55	0.33
0.33	0.11	0.30	0.49	0.26
0.29	0.35	0.33	0.37	0.33
0.25	0.14	0.33	0.42	0.31
0.32	0.21	0.44	0.39	0.38
0.21	0.45	0.45	0.45	0.53
0.21	0.02	0.30	0.42	0.52
0.24	0.35	0.32	0.48	0.52
0.23	0.40	0.44	0.44	0.44
0.25	0.33	0.48	0.42	0.56
0.06	0.38	0.56	0.46	0.57
0.34	0.32	0.52	0.48	0.52
0.23	0.42	0.42	0.26	0.37
0.25	0.36	0.43	0.47	0.35

0.18	0.27	0.41	0.14	0.41
0.34	0.29	0.29	0.31	0.59
0.39	0.10	0.20	0.34	0.44
0.30	0.13	0.45	0.45	0.32
0.44	0.04	0.51	0.50	0.34
0.13	0.10	0.37	0.43	0.45
0.22	0.23	0.32	0.39	0.48
0.30	0.27	0.34	0.31	0.51
0.32	0.31	0.16	0.35	0.46
0.23	0.40	0.43	0.34	0.45
0.37	0.43	0.12	0.38	0.47
0.32	0.34	0.20	0.27	0.33
0.29	0.41	0.44	0.33	0.48
0.29	0.44	0.35	0.27	0.45
0.38	0.32	0.41	0.32	0.42
0.28	0.33	0.47	0.29	0.40
0.41	0.39	0.25	0.25	0.33

Plot frequency, relative frequency, cumulative frequency, and cumulative relative frequency diagrams for the data. Data from Texas weather bureau records, 1964-1968, at Dennison Dam Station.

3.13 In a batch process for the manufacture of ethyl cellulose, ethyl chloride is reacted with alkali cellulose to form the cellulose ether. Degree of reaction is dependent on cellulose/ethyl chloride ratio, temperature, time, and agitator speed, each of which has been standardized. Product properties are strongly dependent on percent ethoxyl, a measure of degree of reaction. The following data on ethoxyl content were taken on 201 batches of ethyl cellulose, all manufactured when the process was under normal conditions.

Percent Ethoxyl in Ethyl Cellulose
(Hypothetical Data)

44.0	41.5	49.0	45.25	49.0	46.0
43.5	45.25	46.25	47.25	47.25	42.5
41.75	43.5	43.75	46.5	46.0	45.5
45.25	46.5	44.5	47.75	44.5	48.0
43.75	48.0	50.5	48.0	45.0	42.0
42.75	41.5	40.0	45.75	46.0	42.0
45.75	40.75	43.75	42.0	50.25	44.75
48.5	45.0	46.5	44.0	47.0	47.0
43.0	44.0	47.0	43.25	45.25	47.5
45.0	47.5	44.5	48.0	43.5	48.75
43.0	46.25	47.0	45.25	44.0	45.25
43.0	47.0	45.5	42.0	44.5	46.25
45.0	43.75	47.5	45.5	44.75	47.0
48.0	39.0	44.5	47.0	47.75	43.25
43.25	45.5	44.0	47.0	44.5	47.5
45.0	41.0	44.5	43.0	46.5	46.0
45.0	41.25	49.5	44.0	48.0	44.5
46.5	44.25	45.0	42.25	43.5	43.0
47.0	45.25	45.0	45.0	45.25	45.25
43.0	45.5	45.0	42.0	44.0	42.5
44.75	41.5	42.75	48.5	43.5	43.0
44.5	47.0	41.75	46.25	44.75	45.5
42.75	48.5	41.5	43.0	45.5	45.5
43.0	45.5	46.75	49.5	45.25	43.5
46.0	42.25	42.5	41.5	48.0	43.75
46.0	44.0	47.0	47.75	45.25	47.75
48.0	43.5	43.25	43.0	41.75	49.0
45.0	42.5	42.0	44.75	44.0	46.75
44.0	44.5	44.25	42.5	46.5	46.75

44.5	40.0	45.0	46.0	44.5	46.25
45.75	45.5	44.5	42.0	45.25	44.25
43.0	44.0	46.75	45.25	43.75	47.0
41.0	50.0	50.0	45.5	45.25	44.5
47.5	45.75	50.0			

(a) Tabulate the frequency and relative frequency of each value.

(b) Construct a frequency and relative frequency histogram, selecting appropriate class marks and class lengths.

(c) Construct cumulative frequency and relative frequency diagrams of the data using the same class marks and class lengths as in (b).

3.14 The National Engineers Register contains the results* of a survey made in 1967 as to where Ch.E.'s are and what they are doing. From the data presented below, plot the relative and cumulative relative distribution functions for ages and years of experience. Comment on the striking similarity of the pairs of graphs so obtained.

Years of Professional Experience and Age
Basis 3,112 AIChE Members

Years of experience	No.	%
1 year or less	135	4
2-4	425	14
5-9	473	15
10-14	400	13
15-19	566	18
20-24	379	12
25-29	325	10
30-34	182	6
35-39	84	3
40 or more	83	3
No report	60	2

*Data from Chem. Eng. Prog. 65(4), 18-19(1969).

Age	No.	%
24 or under	150	5
25-29	423	14
30-34	462	15
35-39	452	14
40-44	467	15
45-49	469	15
50-54	350	11
55-59	166	5
60-64	89	3
65-69	53	2
70 and over	29	1
No report	2	-

3.15 A new air pollution regulation requires that the total particulates (includes aerosols, fly ash, dust, and soot) be kept below 70 ± 5 μg/m^3. Using the data of Problem 3.1,

(a) What is the probability that the particulate concentration on any day will fall within this allowed range?
(b) What is the probability of exceeding the upper limit?
(c) What is the probability of operating in the absolutely safe region below 65 μg/m^3?

3.16 The time-weighted average (TWA) values of cotton dust concentration in a textile plant have been measured every semester for 2 years. The data in μg/m^3 are below. Prepare a suitable histogram and cumulative relative frequency polygon for these data by (a) season as assigned by your instructor, or (b) by designated area over all seasons, (c) or for the entire 2-year period for the whole mill. Using the CRF polygon, what portion of the data are at or below the current OSHA standard of 200 μg/m^3 in opening through dyeing?

Area	Spring '81 TWA's				Fall '81 TWA's				Spring '82 TWA's				Fall '82 TWA's			
Shift	A	B	C	D	A	B	C	D	A	B	C	D	A	B	C	D
Wastehouse	731.3	1597.	1222.	1335.	1734.	2143.	664.7	610.8	1979.	439.8	2184.	289.4	615.7	1375.	354.	316.3
Warehouse	114.5	148.5	148.1	56.6	363.3	446.3	406.5	282.3	98.8	166.8	229.1	152.3	483.4	368.3	579.7	348.3
Opening	389.6	634.7	568.8	422.3	967.6	839.9	594.0	615.9	752.4	738.6	871.4	643.3	903.9	962.	1017.6	735.2
Carding	515.2	524.3	596.2	508.7	607.8	470.6	699.3	558.4	621.2	594.6	729.7	547.9	539.	499.2	510.1	357.6
Drawing	307.2	234.4	293.4	274.9	299.3	295.2	313.	251.3	320.7	268.8	323.1	319.6	230.3	246.3	398.1	285.5
Spinning	200.2	178.7	285.5	223.2	278.4	155.9	148.8	219.1	249.0	192.9	212.1	209.0	149.8	139.2	188.3	165.2
Warping	77.5	76.4	151.2	133.0	144.3	148.1	226.1	172.8	193.2	186.6	188.4	167.2	139.1	211.8	153.7	169.5
Dyeing	99.9	181.5	102.0	108.9	91.9	87.1	125.9	346.	168.8	65.7	194.1	253.7	83.9	100.8	74.6	134.2
Sizing	381.2	345.9	327.6	309.8	341.	404.4	368.9	374.4	210.8	119.6	261.0	229.6	164.8	196.2	163.9	131.4
Beaming	114.7	89.5	146.0	131.0	84.2	65.9	83.8	63.5	87.2	130.8	120.6	110.9	90.7	66.1	66.2	74.5
Weaving	75.0	88.6	88.6	68.9	140.2	217.7	89.3	64.7	130.8	96.0	101.9	130.8	67.8	56.8	92.2	58.6

3.17 Repeated measurements of the flow rate of a cotton dust sampler gave the flow rates in liters/min shown below. Prepare an appropriate histogram and cumulative relative frequency polygon for these data.

7.53	7.56	7.53	7.55	7.50	7.51
7.45	7.55	7.58	7.51	7.56	7.56
7.39	7.53	7.47	7.44	7.42	7.45
7.41	7.49	7.43	7.46	7.39	7.52
7.55	7.51	7.54	7.54	7.53	7.53
7.49	7.52	7.51	7.52	7.54	7.53
7.55	7.55	7.54	7.47	7.53	7.51
7.58	7.57	7.57	7.56	7.37	7.36

3.18 Experimental work has indicated that the probability for any group of seniors in the unit operations laboratory to obtain the correct value of α, the specific cake resistance in the filtration experiment, is 0.84. What variability, as measured by the standard deviation, can be expected in their results? If 12 groups of students are involved, what is the expected standard error?

3.19 For the problem above, what are the chances that at least 11 of the groups will obtain the correct value for α?

3.20 If the determination of the correct value of α in Problems 3.18 and 3.19 follows the Poisson distribution, what is the probability of obtaining at least 11 correct α values from the 12 lab groups? As before, the probability for any group to obtain the correct answer is 0.84.

3.21 One of the input gasses for an acetylene unit is purchased and piped from another plant. The gas is usually within 5% (relative) of the 98% purity required for the reactions. If the gas composition ever deviates more than 15% from the required purity, an explosion is very likely to occur. If there have been 3 explosions in the past 10 years, what is the probability, now that the plant is operating again, of its operating safely for the next 5 years? No new safety or warning devices have been installed and the same producer of the gas is being used.

3.22 In an analytical instrumentation class, matched cuvettes must be chosen for a colorimetric determination of flouride concentration. If the chance of one cuvette being the same as a previously chosen standard cuvette is 80%, what is the probability of getting exactly 4 matched cuvettes from a box containing 12?

3.23 The fluidized reactor cooling coils are guaranteed against rupture for 10 years. The plant is in its 9th year of operation. What is the probability of the pipes failing in this year of operation? The probability of failure in any year is 6%.

3.24 For the production of a certain item by compression molding, it is known that the probability of making a defective unit is 0.06. If 5 items are inspected at a time, what is the probability that 1 of the 5 items checked is defective? What is the probability of getting 2 out of 5 that have no defects?

3.25 In the production of humidographs, suppose control limits were set between 2-3. That is, if 2 or fewer defective units in a sample of 10 are found, the quality (20% defective) has not changed. If 3 or more are found defective then it is assumed that a change in quality has occurred and corrective action must be taken. What is the probability of finding 6, 7, or 8 defective units in a sample of 10?

$P = 0.2$

Y	P(Y)
0	0.1074
1	0.2684
2	0.3020
3	0.2013
4	0.0881
5	0.0264
6	0.0055
7	0.0008
8	0.0001
9	0.0000
10	0.0000

3.26 A polyurethane plant in charge of making "flexible foam" for dashboards in automobiles makes one 35,000-lb batch of polymer per 12-hr shift. The plant runs continuously with 4 hr in each shift being required for down time to clean the reactors. The polymer is judged on-grade or off-grade by a rate of reaction versus time test. If the system follows a Poisson distribution with m = 8.5 on a basis of one weeks operation, what is the probability of at least 1.5 million pounds of on-grade output per month?

3.27 In a certain industrial plant, vacancies have occurred in the technical department from 1960 to 1967 in the following manner.

Number of vacancies	Number of months
0	59 (months with 0 vacancies)
1	27 (months with 1 vacancy)
2	9
3	1

Using the equation for the Poisson distribution, calculate the theoretical values for the number of months and compare with the given data.

3.28 In a batch process for producing an oil viscosity additive the probability of successive batches being on-grade is a variable whose distribution follows a Poisson distribution with $m = 6$. What is the probability of the unit going for a week without being cleaned if one batch can be made every day?

3.29 A manufacturer guarantees that a shipment of Raschig rings has no more than 2% broken rings. A stockman takes a random sample of 100 rings. Assuming the Poisson distribution applies, what is the highest number of broken rings the stockman should accept to be 95% confident of the manufacturer's guarantee? 99% confident?

3.30 The table below shows the days in a 50-day period during which varying numbers of automobile accidents occurred in a city. Fit a Poisson distribution to the data.

No. of accidents	No. of days
0	21
1	18
2	7
3	3
4	1

3.31 If the number of graduate students in engineering departments is a random variable depending on student financial status, military status, intellectual drive, and scholastic achievement, and the distribution is closely approximated by the Poisson distribution with $m = 10$, what is the probability of having fewer than 15 graduate students during the spring of any given year?

3.32-3.45 Using the data from Problems 3.1 through 3.14, determine the type distribution followed in each case. Fit the distribution involved to each data set and estimate the distribution parameters.

3.46 Suppose that on a particular 6 engine airplane it has been determined that the probability of an engine failing is 0.05. If the plane can fly on any 3 engines, what is the probability of a crash?

3.47 Given 12 resistors of which 4 are defective, what is the probability that a randomly selected sample of 4 (without replacement) will have at least 2 defectives in it?

3.48 Suppose that 10 television sets are available for testing. What is the probability that if 5 are randomly selected and classified as defective or nondefective that 3 are classified as defective if it is known that there are 2 defective sets in the 10 sets available?

4
DESCRIPTIVE STATISTICS

4.1 INTRODUCTION

In using statistics it is necessary to make a clear distinction between populations and samples taken from the population. A population for example might be represented by a collection of units such as books in a library or by the products resulting from a continuous chemical process. In many cases the population is quite large. In general, we desire estimates of parameters such as the population mean and variance. These estimates are calculated from a sample taken from the population. Depending on the size of the population the sample might represent anywhere from a fraction of 1% to the entire population. To draw conclusions about the population from the sample taken, the method known as statistical inference must be used.

Obviously the sample taken from a population must be representative of the population or else no useful inferences can be made. Therefore, the sample must be a random one. A random sample can be thought of as a collection of values selected from a population in such a way that each value in the population has the same chance of being independently selected as a sample member. This definition of random sample is valid for infinite populations or when sampling from finite populations with replacement. In sampling from a finite population without replacement, a random sample of size n is one selected in such a way that every sample of size n has the same chance of being selected. The idea of a random sample is of utmost importance in statistical inference since most of its theory is based on the underlying assumption that the sample is a random one.

A value computed from a series of observations, or the values in a random sample, is called a statistic. The statistics most commonly encountered in engineering and scientific work are measures of location such as the mean, median, and mode and measures of variability such as the variance, standard deviation, and range of a sample.

We now discuss some statistics that describe a sample and save the topic of statistical inference until Chapter 6.

4.2 MEASURES OF LOCATION

A measure of location is one that indicates where the "center" of the data is located. The first such measure we consider is the sample mean. The sample mean, which is the most important and often used single statistic, is defined as the sum of all the sample values divided by the number of observations in the sample and is denoted by

$$\overline{X} = \sum_{i=1}^{n} X_i/n \qquad (4.1)$$

where $\overline{X}$ is the mean of n values and X_i is any given value in the sample. It is an estimate of the value of the mean of the population from which the sample was drawn.

If a constant, say A, is added to the data (in other words the data have a fixed offset, $W_i = X_i + A$) the mean can be found from

$$\overline{W} = \sum_{i=1}^{n} W_i/n = \sum_{i=1}^{n} (X_i + A)/n = \overline{X} + A \quad . \qquad (4.2)$$

That is, the mean is likewise shifted by an amount A. In like manner, for the case where the data have been multiplied by a constant k, $Z_i = kX_i$, we have

$$\overline{Z} = \sum_{i=1}^{n} Z_i/n = \sum_{i=1}^{n} kX_i/n = k\overline{X} \quad . \tag{4.3}$$

Combining the results in (4.2) and (4.3) the mean of the data transformed according to $U_i = (X_i - A)/B$ is

$$\overline{U} = (\overline{X} - A)/B \quad . \tag{4.4}$$

The result (4.4) is frequently used to "standardize" a given set of data, as will be seen later.

If the items $X_1, X_2, \ldots, X_n$ in a random sample have weights $f_1, f_2, \ldots, f_n$ associated with them then the <u>weighted mean</u> is defined by

$$\overline{X} = \sum_{i=1}^{n} f_i X_i / \sum_{i=1}^{n} f_i \quad .$$

For example, in an empirical frequency distribution such as the one in Table 3.5 the sample mean could be approximated by

$$\overline{X} = \frac{f_1X_1 + f_2X_2 + \ldots + f_cX_c}{f_1 + f_2 + \ldots + f_c} \quad ,$$

where $X_1, X_2, \ldots, X_c$ denote the class marks and $f_1, f_2, \ldots, f_c$ the frequencies of the corresponding classes.

<u>Example 4.1</u> Successive determinations of the opened bottles of HCl were found to be, expressed in normalities, N_i:

N_1	N_2
15.75	15.58
15.64	15.49
15.92	15.72

What was the average concentration of each bottle? From Eq. (4.1), $\overline{X} = \sum_{i=1}^{n} X_i/n$, and for bottle 1,

$$\overline{x}_1 = (15.75 + 15.64 + 15.92)/3 = 15.770,$$

for bottle 2,

$$\overline{x}_2 = (15.58 + 15.49 + 15.72)/3 = 15.597.$$

Another measure of location that is frequently used is the median. The median is defined to be a value such that half of the observations in the sample have values smaller than the median and half have values larger than the median. If the sample has an odd number of observations then the median would be the "middle" value in the sample. If the sample has an even number of observations then a median would be any value between the two middle values in the sample.

A third measure of location is the mode, which is defined as the most frequently occurring value in a sample.

4.3 MEASURES OF VARIABILITY

Having determined the location of our data as expressed by statistics such as the mean, median, and mode, the next thing to be considered is how the data are spread about these central values. Since spread about the median and mode is rarely if ever used in engineering work we will restrict this discussion to the spread of data about the mean. The most popular method of reporting variability is by use of the <u>sample variance</u> defined as

$$S_X^2 = \sum_{i=1}^{n} (X_i - \overline{X})^2/(n - 1) \qquad (4.5)$$

for which useful calculation formulas are

$$S_X^2 = \left(\sum_{i=1}^{n} X_i^2 - n\bar{X}^2 \right) / (n-1) = \left(\sum_{i=1}^{n} X_i^2 - \frac{(\Sigma X)^2}{n} \right) / (n-1). \quad (4.6)$$

The sample variance is the sum of the squares of the deviations of the data points from the mean value of the sample divided by (n - 1). The <u>standard deviation</u> of a sample, denoted by S_X, is defined to be the positive square root of the variance, that is,

$$S_X = \sqrt{S_X^2}. \quad (4.7)$$

Another term frequently encountered is the <u>coefficient of variation</u>, CV, defined by

$$CV = S_X / \bar{X}.$$

Another useful statistic is the sample <u>standard error</u> of the mean, defined by

$$S_{\bar{X}} = \sqrt{S_X^2/n} = S_X/\sqrt{n}.$$

This says that sample means tend to vary less than do the observations themselves.

It should be noted that the sum of the deviations of the sample values about the sample mean is zero, that is,

$$\sum_{i=1}^{n} (X_i - \bar{X}) = 0 \quad .$$

This prompts one to use the squares of the deviations, thus yielding (4.5). One could use the sum of the absolute values of the deviations, that is,

$$\sum_{i=1}^{n} \left| X_i - \bar{X} \right| \quad .$$

However, (4.5) is mathematically more tractable and is the one most widely used.

If the sample $X_1, X_2, \ldots, X_n$ is transformed according to $Z_i = kX_i + A$, then

$$S_Z^2 = \sum_{i=1}^{n} (Z_i - \overline{Z}_i)^2/(n - 1)$$

$$= \sum_{i=1}^{n} (kX_i + A - k\overline{X} - A)^2/(n - 1)$$

$$= k^2 S_X^2 \quad . \qquad (4.8)$$

From (4.8) it is seen that adding a constant value to each data point has no effect on the sample variance.

Equation (4.8) will prove useful in finding the variance of the data standardized according to $U_i = (X_i - A)/B$. According to (4.8) with $k = \frac{1}{B}$ and A replaced with -A/B we have

$$S_U^2 = \frac{1}{B^2} S_X^2 \quad . \qquad (4.9)$$

In particular, if $A = \overline{X}$ and $B = \sqrt{S_X^2} = S_X$, we obtain

$$S_U^2 = \frac{1}{S_X^2} \cdot S_X^2 = 1 \quad .$$

By Eq. (4.4) we also have

$$\overline{U} = \frac{\overline{X} - \overline{X}}{S_X} = 0 \quad .$$

Data from any process which fluctuates over a wide range of values will have a large variance. Conversely, a large variance indicates that the data have a wide spread about the mean. In contrast, if all the sample values are very close together, the sample variance will be quite small. This latter case is usually highly desirable.

For example, in sampling a population one is interested in using the data in the sample to make inferences about the population. The inferences will be more precise if the variance of the sampled population is small. This is seen in Chapter Six in the discussion of the estimation of the mean μ of a population.

The sample variance S_X^2 is merely an estimate of the population variance σ_X^2. The population variance of a population consisting of the values $\{x_1, x_2, \ldots, x_N\}$ is given by

$$\sigma_X^2 = \sum_{i=1}^{N} (x_i - \mu)^2/N \tag{4.10}$$

where μ, the population mean, is given by

$$\mu = \sum_{i=1}^{N} x_i/N.$$

For a sample of size n, $X_1, X_2, \ldots, X_n$, S_X^2 estimates σ_X^2. Similarly, $\overline{X}$ estimates μ. The estimation of parameters will be taken up in Chapter 6 in more detail.

Another measure of variation which is sometimes used is the range. The range of a sample is defined to be the difference of the largest and smallest values in a sample.

Example 4.2 Successive colorimetric determinations of the normality of a $K_2Cr_2O_7$ solution were as follows (expressed as molarities $\times 10^{-4}$): 1.22, 1.23, 1.18, 1.31, 1.25, 1.22, and 1.24. What are the mean, median, mode, standard deviation of a single determination, standard deviation of the mean, range, and coefficient of variation for this sample?

(a) Mean $= \overline{x} = \sum_{i=1}^{n} x_i/n = 1.257 \times 10^{-4}$ M.

(b) The median is found to be 1.23×10^{-4} M from the definition.

(c) The mode is found to be 1.22×10^{-4}M.

(d) $$s_X = \sqrt{s_X^2} = \left(\frac{\sum_{i=1}^{n} x_i^2 - n\bar{x}^2}{n - 1} \right)^{1/2} \text{M}$$

$$= \left(\frac{10.6983 \times 10^{-8} - 10.6887 \times 10^{-8}}{6} \right)^{1/2} \text{M}$$

$$= \left(\frac{96 \times 10^{-12}}{6} \right)^{1/2} \text{M} = 4 \times 10^{-6} \text{ M.}$$

(e) $$s_{\bar{X}} = \sqrt{s_X^2/n} = \sqrt{\frac{96 \times 10^{-12}}{7(6)}} \text{M} = 1.51186 \times 10^{-6} \text{ M.}$$

(f) Range = R = $(1.31 \times 10^{-4} - 1.18 \times 10^{-4})$ M = 13.0×10^{-6} M.

(g) CV = $s_X / \bar{x}$ = 0.03237.

Example 4.3 Obtain the average evaporation in inches per day from the evaporation data of Problem 3.12. What variance can be expected in the daily water evaporation rate? Find the standard error of the mean and the coefficient of variation.

(a) $$\bar{x} = \sum_{i=1}^{n} x_i/n = \frac{54.22 \text{ in./day}}{155} = 0.3498 \text{ in./day}$$

(b) $$s_X^2 = \sum_{i=1}^{n} (X_i - \bar{X})^2/(n - 1) = \left(\sum_{i=1}^{n} x_i^2 - n\bar{x}^2 \right) \Big/ (n - 1)$$

$$= \left[21.0108 - 155(0.3498)^2\right]/154 = 0.01327 \text{ in.}^2/\text{day}^2$$

$$s_X = \sqrt{s_X^2} = 0.1152 \text{ in./day}$$

$$s_{\bar{X}} = \frac{s_X}{\sqrt{n}} = \frac{0.1152}{\sqrt{155}} = 0.009253 \ \frac{\text{in.}}{\text{day}}$$

(c) CV = $s_X / \bar{x}$ = 0.3293.

Example 4.4 Consider the orifice coefficient data of Example 3.1. What are the mean, range, standard deviation, standard error, variance, and coefficient of variation? To solve this using PROC MEANS (*SAS Introductory Guide*), the simplest approach would have been to insert the following two statements:

```
PROC MEANS STD MIN MAX RANGE VAR STDERR CV DATA = CO;
TITLE 'DESCRIPTIVE STATISTICS FOR ORIFICE COEFFICIENT DATA';
```

after the PROC PRINT statement in the original program in Example 3.1. The results (note that the coefficient of variation CV is just the standard deviation expressed as a percentage of the mean $\overline{X}$) are:

```
MEAN = 0.63673077            RANGE = 0.4760      CV = 15.329
STANDARD DEVIATION = 0.09760467                  VARIANCE = 0.00952667
MINIMUM VALUE = 0.4880                           MAXIMUM VALUE = 0.9640
STANDARD ERROR OF MEAN = 0.01353533
```

PROBLEMS

4.1 For convenience, it has been decided that the scores of the aptitude test which were presented in Problem 3.2 will be coded for calculational purposes by the following methods: (a) by dividing all values by 250 and (b) by subtracting 25 from all values. Let us say that this has been done and that the means of the resulting coded data have then been determined. How will you determine the mean of the original data from the means of the coded data? Answer for both coding methods.

4.2-4.13 Calculate the means and variances for the variables given in Problem statements 3.1-3.10 and 3.17 and by season, shift, or area for the data for Problem 3.16.

5
EXPECTED VALUES AND MOMENTS

5.1 INTRODUCTION

The expectation or expected value of a random variable is obtained by finding the average value of the variable over all the possible values of the variable. Expectation of a random variable was introduced in Chapter 3, Section 3.3, in discussing the mean and variance of a population. In this chapter we summarize those results and present a more complete treatment of expectation.

As an example, consider the tossing of two ideal coins. The distribution of X, the number of heads that appear, is given by the binomial probability function

$$P(X=x) = b(x) = \binom{2}{x}\left(\frac{1}{2}\right)^2 , \quad x = 0,1,2 \quad .$$

This indicates that we expect no heads to appear with a relative frequency of 1/4, one head to appear with a relative frequency of 1/2, and two heads to appear with the relative frequency of 1/4 in a large number of trials. Now let us find the average number of heads in 1000 trials. The total number of heads is expected to be

$$250 \times 0 + 500 \times 1 + 250 \times 2 = 1000$$

in the 1000 tosses of 2 coins; thus the average is expected to be one head per trial. This is the expected value, or mean value, of

X, denoted by $\mu = E(X)$. The same result would be obtained if we merely multiplied all possible values of X by their probabilities and added the results; thus,

$$E(X) = 0 \times \frac{1}{4} + 1 \times \frac{1}{2} + 2 \times \frac{1}{4} = 1.0 \quad .$$

This expected value is a theoretical or ideal average. A random variable does not necessarily take on its expected value in a given trial: however, we might reasonably expect the average value of the random variable in a great number of trials to be somewhere near its expected value.

It is often desirable to obtain the expected value of a function of a random variable. In the above example, one may want the expected value of $f(X) = X^2$. This is obtained by multiplying all possible values of X^2 by their probabilities and adding the results; thus,

$$E(X^2) = (0)^2 \times \frac{1}{4} + (1)^2 \times \frac{1}{2} + (2)^2 \times \frac{1}{4} = 1.5 \quad .$$

The variance of a population, denoted by σ_X^2 or σ^2, is defined as the expected value of the square of the difference between X and the population mean μ, that is,

$$\sigma^2 = \sigma_X^2 = E\left[(X-\mu)^2\right] \quad .$$

In the above example the variance of the number of heads is

$$\begin{aligned}
\sigma_X^2 &= E\left[(X - \mu)^2\right] \\
&= E\left[(X - 1)^2\right] \\
&= (0 - 1)^2 \times \frac{1}{4} + (1 - 1)^2 \times \frac{1}{2} + (2 - 1)^2 \times \frac{1}{4} \\
&= \frac{1}{2} \quad .
\end{aligned}$$

In general, if X is a random variable, then a function of X, say g(X), is a random variable also, and its expected value is denoted by $E[g(X)]$.

We now consider the problem of computing certain expected values for discrete and continuous random variables.

5.2 DISCRETE DISTRIBUTIONS

A discrete probability function assigns probabilities to each of all (at most countable) possible outcomes of an experiment. Its distribution function may be looked on as a step function which changes only at those values that the corresponding random variable can take on. For a discrete distribution such that $P(X = x_i) = p_i = f(x_i)$, for $i = 1,2,\ldots,n$, the mean of X is

$$E(X) = \sum_{i=1}^{n} x_i p_i = \sum_{i=1}^{n} x_i f(x_i) \qquad (5.1)$$

where x_i is the value of an outcome and p_i is the corresponding probability. The variance of X is given by

$$\sigma_X^2 = E\left[(X - \mu)^2\right] = \sum_{i=1}^{n} (x_i - \mu)^2 f(x_i) \quad . \qquad (5.2)$$

Three properties of expectations will now be stated. These are that if k is a constant and X is distributed in accord with f(x) and if g(x) and h(x) are any specific functions of X, then

$$E[k] = k, \qquad (5.3)$$

$$E[kg(X)] = kE[g(X)] \quad , \qquad (5.4)$$

and

$$E[g(X) + h(X)] = E[g(X)] + E[h(X)] \qquad (5.5)$$

for discrete distributions. Properties (5.3), (5.4), and (5.5) can be proved very easily. Suppose the probability function of X is given by $f(x_1), f(x_2), \ldots, f(x_n)$, then

$$E[k] = \sum_{i=1}^{n} kf(x_i) = k \sum_{i=1}^{n} f(x_i) = k, \quad \text{as} \quad \sum_{i=1}^{n} f(x_i) = 1$$

which proves (5.3). Now

$$\begin{aligned} E[kg(X)] &= \sum_{i=1}^{n} kg(x_i)P(X = x_i) \\ &= \sum_{i=1}^{n} kg(x_i)f(x_i) \\ &= k \sum_{i=1}^{n} g(x_i)f(x_i) \\ &= kE[g(X)] \quad , \end{aligned}$$

which proves (5.4). Finally

$$\begin{aligned} E[g(X) + h(X)] &= \sum_{i=1}^{n} [g(x_i) + h(x_i)]\, f(x_i) \\ &= \sum_{i=1}^{n} [g(x_i)f(x_i) + h(x_i)f(x_i)] \\ &= \sum_{i=1}^{n} g(x_i)f(x_i) + \sum_{i=1}^{n} h(x_i)f(x_i) \\ &= E[g(X)] + E[h(X)] \end{aligned}$$

which proves (5.5).

Properties (5.3), (5.4), and (5.5) may be used to derive a very useful method of evaluating the variance of a random variable. By (5.3)-(5.5) one has

$$\begin{aligned}
\sigma_X^2 &= E[(X - \mu)^2] \\
&= E(X^2 - 2X\mu + \mu^2) \\
&= E(X^2) - 2\mu E(X) + E(\mu^2) \\
&= E(X^2) - 2\mu\cdot\mu + \mu^2 \\
&= E(X^2) - \mu^2 \\
&= E(X^2) - [E(X)]^2 \quad . \qquad (5.6)
\end{aligned}$$

5.3 CONTINUOUS DISTRIBUTIONS

A continuous distribution is described in terms of a probability density function which assigns probabilities to intervals of values of the corresponding random variable. The probability that the random variable X takes on a value between two specific real values is given by

$$P(a \leq X \leq b) = \int_a^b f(x)\, dx \qquad (5.7)$$

where f(x) is the probability density function which models the population. The expected value of a continuous random variable is obtained by integration rather than the summation technique required for discrete distributions. The expected value of a continuous random variable X is defined by

$$E(X) = \int_{-\infty}^{\infty} xf(x)\, dx \quad . \qquad (5.8)$$

Properties similar to (5.3), (5.4), and (5.5) also hold for continuous distributions, as can be observed from the properties of integrals:

$$E(k) = \int_{-\infty}^{\infty} kf(x)\,dx = k\int_{-\infty}^{\infty} f(x)\,dx = k \quad , \tag{5.9}$$

$$E[kg(X)] = \int_{-\infty}^{\infty} kg(x)f(x)\,dx = k\int_{-\infty}^{\infty} g(x)f(x)\,dx = kE[g(X)] \, , \tag{5.10}$$

$$\begin{aligned} E[g(X) + h(X)] &= \int_{-\infty}^{\infty} [g(x) + h(x)]f(x)\,dx \\ &= \int_{-\infty}^{\infty} g(x)f(x)\,dx + \int_{-\infty}^{\infty} h(x)f(x)\,dx \\ &= E[g(X)] + E[h(X)] \quad . \end{aligned} \tag{5.11}$$

5.4 JOINT DISTRIBUTIONS AND INDEPENDENCE OF RANDOM VARIABLES

Consider the experiment of tossing two fair dice. Let X and Y denote the outcomes of the first and second die, respectively. The probability functions for X and Y are

$$f(x_i) = P(X = x_i) = \frac{1}{6}, \; x_i = 1,2,\ldots,6$$

and

$$g(y_i) = P(Y = y_i) = \frac{1}{6} \, , \; y_i = 1,2,\ldots,6 \quad .$$

We assume here that the outcomes of the two dice occur <u>independently</u> of each other. Thus we can make statements such as

$$\begin{aligned} P(X = 1 \text{ and } Y = 3) &= P(X = 1)P(Y = 3) \\ &= f(1)g(3) \\ &= \frac{1}{6} \cdot \frac{1}{6} \\ &= \frac{1}{36} \quad , \end{aligned} \tag{5.12}$$

since the events X = 1 and Y = 3 are independent (see Chapter 2 for independence of two events A and B). To make joint probability statements involving X and Y we use the model

$$h(x_i, y_j) = P(X = x_i \text{ and } Y = y_j) = P(X = x_i)P(Y = y_j)$$
$$= f(x_i)g(y_j) \qquad (5.13)$$

for all values of x_i and y_j. The function h(x,y) is called the joint probability function of X and Y. Equation (5.13) prompts the following definition.

Definition 5.1 The random variables X and Y are said to be statistically independent if their joint probability function h(x,y) can be written as h(x,y) = f(x)g(y) for all values of x and y.

Definition 5.1 holds also if X and Y are continuous random variables. That is, two continuous random variables are statistically independent if their joint probability density function h(x,y) factors into f(x)g(y) for all real values x and y, where f(x) and g(y) denote the density functions of X and Y, respectively.

From Definition 5.1 it follows that if X and Y are independent then for any functions q(x) and r(y)

$$E[q(X)r(Y)] = E[q(X)]\,E[r(Y)] \quad . \qquad (5.14)$$

Equation (5.14) follows from

$$E[q(X)r(Y)] = \sum_{i=1}^{n}\sum_{j=1}^{m} q(x_i)r(y_j)h(x_i, y_j)$$
$$= \sum_{i=1}^{n}\sum_{j=1}^{m} q(x_i)r(y_j)f(x_i)g(y_j)$$
$$= \left(\sum_{i=1}^{n} q(x_i)f(x_i)\right)\left(\sum_{j=1}^{m} r(y_j)g(y_j)\right)$$
$$= E[q(X)] \cdot E[r(Y)]$$

for discrete random variables. For continuous random variables the similar result is obtained by replacing summations by integrals.

A special case of (5.14) is obtained when $g(X) = X$ and $r(Y) = Y$, in which case we have

$$E(XY) = E(X)E(Y)$$

if X and Y are independent random variables.

The idea of random sample has been discussed in Chapter 3. The outcomes in drawing a random sample from a population occur at random. Thus the random sample of size n is denoted by the random variables $X_1, X_2, \ldots, X_n$, which are taken to be statistically independent. That is, the joint distribution of $X_1, X_2, \ldots, X_n$ is given by

$$h(x_1, x_2, \ldots, x_n) = f_1(x_1)f_2(x_2)\ldots f_n(x_n) \tag{5.15}$$

for all real values $x_1, x_2, \ldots, x_n$, where $f_i(x_i)$ denotes the probability function or density function of X_i. The result in (5.15) is an extension of the independence of two random variables given in Definition 5.1.

Since each of the random variables $X_1, X_2, \ldots, X_n$ in a random sample comes from the same underlying population, each of the functions $f_i(x)$ in (5.15) may be replaced by the same function, say $f(x)$. Thus for a random sample $X_1, X_2, \ldots, X_n$ the corresponding joint distribution is described by

$$h(x_1, x_2, \ldots, x_n) = f(x_1)f(x_2)\ldots f(x_n) = \prod_{i=1}^{n} f(x_i). \tag{5.16}$$

The results given by Eqs. (5.5) and (5.11) can be easily extended to the case of a finite number of functions, that is,

$$E[g_1(X) + g_2(X) + \ldots + g_n(X)] = E[g_1(X)] + \ldots + E[g_n(X)]. \tag{5.17}$$

Now, suppose $X_1, X_2, \ldots, X_n$ is a random sample from some population with mean μ and variance $\sigma_X^2 = \sigma^2$. The sample mean has been defined to be

$$\overline{X} = \sum_{i=1}^{n} X_i/n \quad .$$

By (5.17) we have

$$E(\overline{X}) = E\left(\frac{\sum_{i=1}^{n} X_i}{n}\right) = \frac{\sum_{i=1}^{n} E(X_i)}{n} = \frac{\sum_{i=1}^{n} \mu}{n} = \frac{n\mu}{n} = \mu \qquad (5.18)$$

which shows that the expected value or mean value of the sample means is equal to the population mean. The sample variance has been defined to be

$$S^2 = S_X^2 = \sum_{i=1}^{n} (X_i - \overline{X})^2/(n - 1) \quad .$$

The expected value of S^2 is

$$E(S^2) = E\left[\frac{\sum_{i=1}^{n} (X_i - \overline{X})^2}{n - 1}\right]$$

$$= \frac{1}{(n - 1)} E\left[\sum_{i=1}^{n} X_i^2 - \left(\sum_{i=1}^{n} X_i\right)^2 \Big/ n\right] .$$

$$= \frac{1}{n - 1}\left\{\sum_{i=1}^{n} E(X_i^2) - \frac{1}{n} E\left[\left(\sum_{i=1}^{n} X_i\right)^2\right]\right\}.$$

However, $E(X_i^2) = \sigma^2 + \mu^2$ by Eq. (5.6). Furthermore

$$E\left[\left(\sum_{i=1}^{n} X_i\right)^2\right] = E\left[\left(\sum_{i=1}^{n} X_i\right)\left(\sum_{j=1}^{n} X_j\right)\right]$$

$$= E\left(\sum_{i=1}^{n}\sum_{j=1}^{n} X_i X_j\right)$$

$$= E\left(\sum_{i=1}^{n} X_i^2 + \sum_{i=1}^{n}\sum_{\substack{j=1 \\ i \neq j}}^{n} X_i X_j\right)$$

$$= \sum_{i=1}^{n} E(X_i^2) + \sum_{i=1}^{n}\sum_{\substack{j=1 \\ i \neq j}}^{n} E(X_i)E(X_j)$$

$$= \sum_{j=1}^{n} (\sigma^2 + \mu^2) + \sum_{i=1}^{n}\sum_{\substack{j=1 \\ i \neq j}}^{n} \mu^2$$

$$= n(\sigma^2 + \mu^2) + n(n - 1)\mu^2$$

$$= n\sigma^2 + n^2\mu^2 \quad .$$

Thus

$$E(S_X^2) = \frac{1}{n - 1}\left[n(\sigma^2 + \mu^2) - \frac{1}{n}(n\sigma^2 + n^2\mu^2)\right]$$

$$= \frac{1}{n - 1}\left[(n - 1)\sigma^2\right]$$

$$= \sigma^2 \quad . \tag{5.19}$$

The results given by (5.18) and (5.19) say that the average values of $\overline{X}$ and S^2 are given by μ and σ^2, respectively. In estimation theory we say that $\overline{X}$ and S^2 are "unbiased" for μ and σ^2. (This topic is considered further in Chapter 6). We note in passing that the reason for dividing by (n - 1) in the definition of S^2 was so that $E(S^2) = \sigma^2$.

5.5 MOMENTS

In general the expected value $E[(X - a)^k]$, for k a positive integer and any number a, is referred to as the kth moment about a of the random variable X. In particular, if a = 0, then we have $E(X^k)$, which is called the kth moment of X about the origin or simply the kth moment and is denoted by μ'_k. Letting k = 1 yields the first moment of X about the origin, that is, $\mu'_1 = \mu$. The general form for moments about the origin for discrete distributions is

$$\mu'_k = E(X^k) = \sum_{i=1}^{N} x_i^k f(x_i), \qquad (5.20)$$

and for continuous distributions it is

$$\mu'_k = E(X^k) = \int_{-\infty}^{\infty} x^k f(x)\, dx \quad . \qquad (5.21)$$

For many statistical analyses, values of the kth moment of X about the mean are most useful. These moments are called central moments and are denoted by μ_k. For discrete distributions, the general form for the kth central moment is

$$\mu_k = E(X - \mu)^k = \sum_{i=1}^{N} (x_i - \mu)^k f(x_i) \quad . \qquad (5.22)$$

For continuous distributions, the kth central moment is

$$\mu_k = E(X - \mu)^k = \int_{-\infty}^{\infty} (x - \mu)^k f(x)\, dx \quad . \qquad (5.23)$$

For any distribution, the 1st moment about the mean is μ_1, where

$$\mu_1 = E(X - \mu) = E(X) - \mu = \mu - \mu = 0.$$

The second moment about the mean, μ_2, is the variance of X, denoted by σ_X^2 (or more often σ^2). A convenient computational form, according to Eq. (5.6), is

$$\sigma^2 = \mu_2' - (\mu_1')^2 = \mu_2' - \mu^2 \quad . \tag{5.24}$$

5.6 EXAMPLES

We now consider some examples to illustrate the main results in this chapter.

Example 5.1 The grades from the first hour exam in this course were

i	1	2	3	4	5	6	7	8	9	10	11
x_i	100	84	83	75	70	67	63	52	50	35	28
$p(x_i)$	$\frac{3}{14}$	$\frac{1}{14}$	$\frac{1}{14}$	$\frac{2}{14}$	$\frac{1}{14}$	$\frac{1}{14}$	$\frac{1}{14}$	$\frac{1}{14}$	$\frac{1}{14}$	$\frac{1}{14}$	$\frac{1}{14}$

Assuming that this data is the entire population, calculate the mean and variance of the test grades from the appropriate moments.

We obtain

$$\mu = \mu_1' = \sum_{i=1}^{11} x_i p(x_i) = 70.14;$$

$$\sigma^2 = \mu_2 = \sum_{i=1}^{11} x_i^2 p(x_i) - (\mu_1')^2 = 492.24$$

The standard deviation is 22.19.

Example 5.2 Given the function $f(x) = ce^{-x/4}$, what must be the value of c if f(x) is to be a probability density function? The domain of the function is $0 < x < \infty$. Thus

$$\int_0^\infty f(x)\, dx = 1 \Rightarrow c\int_0^\infty e^{-x/4}\, dx = 1 \Rightarrow c = 1/4 \quad .$$

Example 5.3 Find the mean and variance of X if $f(x) = 3/x^4$ for the range $1 < x < \infty$.

$$\mu_1' = \int_1^\infty x\cdot\frac{3}{x^4}\,dx = \int_1^\infty 3\,dx/x^3 = 3/2 = \mu$$

$$\mu_2' = \int_1^\infty x^2\cdot\frac{3}{x^4}\,dx = \int_1^\infty 3\,dx/x^2 = 3$$

$$\sigma^2 = \mu_2 = \mu_2' - \mu^2 = 3 - 9/4 = 3/4 \quad .$$

Example 5.4 Find the mean μ and variance σ^2 of the binomial distribution with parameters n and p.

The binomial probability function is given by Eq. (3.14) as

$$b(x) = \binom{n}{x} p^x (1 - p)^{n - x}, \quad x = 0,1,2,\ldots,n \quad .$$

The mean and variance could be found directly by using the function b(x) and the definition of mean and variance. However, in order to illustrate some of the ideas in this chapter we use a somewhat different (but easier) approach. The binomial random variable is the number of successes in a series of n independent trials of an experiment which can yield one of two outcomes on each trial. Let $Y_1,Y_2,\ldots,Y_n$ denote the random outcomes of the n trials. Thus $P(Y_i = 1) = p$ and $P(Y_i = 0) = q = 1 - p$, for each i. The mean and variance of each Y_i are

$$E(Y_i) = 1\cdot p + 0\cdot(1 - p) = p$$

and

$$\sigma_Y^2 = E(Y_i^2) - [E(Y_i)]^2$$

$$= (1)^2\cdot p + (0)^2(1 - p) - p^2$$

$$\sigma_Y^2 = p(1 - p)$$

$$= pq \quad .$$

Now

$$X = \sum_{i=1}^{n} Y_i$$

and the mean and variance of X are

$$\mu = E(X) = \sum_{i=1}^{n} E(Y_i) = \sum_{i=1}^{n} p = np$$

and

$$\begin{aligned}
\sigma^2 &= E\left[(X - \mu)^2\right] \\
&= E(X^2) - \mu^2 \qquad \left[\text{by } (5.6)\right] \\
&= E\left(\sum_{i=1}^{n} Y_i\right)^2 - \mu^2 \quad . \\
&= E\left(\sum_{i=1}^{n} Y_i^2 + \sum_{i=1}^{n} \sum_{\substack{j=1 \\ i \neq j}}^{n} Y_i Y_j\right) - \mu^2 \quad . \\
&= \sum_{i=1}^{n} p + \sum_{i=1}^{n} \sum_{\substack{j=1 \\ i \neq j}}^{n} p^2 - \mu^2 \\
&= np + n(n - 1)p^2 - n^2p^2 \\
&= np - np^2 \\
&= np(1 - p) \\
&= npq \quad .
\end{aligned}$$

In this example it should be noted that $Y_1, Y_2, \ldots, Y_n$ is actually a random sample of size n from the population modeled by the point binomial probability function

$$f(x) = p^x (1 - p)^{1-x}, \; x = 0,1 \quad .$$

Example 5.5 Past history indicates that the probability of a given batch of chemical being on-grade is 0.95. Let X denote the number of off-grade batches obtained in 200 runs. What are the mean and variance of X?

In this example p = 1 -0.95 = 0.05 and n = 200. Thus the mean and the variance of X are

$$\mu = E(X) = np = 200(0.05) = 10$$

and

$$\sigma^2 = npq = 200(0.05)(0.95) = 9.5 \quad .$$

PROBLEMS

5.1 Given the function $F(x) = x^2/\theta$ for which $\theta > 0$ and $0 < x < k$. For what value of k is F(x) a cumulative distribution function? What is the associated probability density function f(x)?

5.2 If f(x) is the probability density function given by $f(x) = ax^2$ where $0 \leq x \leq 1$, find a.

5.3 A random variable X has values in the range $2 \leq x \leq 4$, with probability density function $f(x) = A(3 + 2x)$. What is the value of A? What are the mean and variance of X? What is the probability that X is between 3 and 4? Less than 3?

5.4 Given that $f(x) = 1/\sqrt{2\pi}\, \exp(-x^2/2)$ for $-\infty < x < \infty$, what is the corresponding cumulative distribution function?

5.5 What are the required conditions for f(x) to be a probability density function? If $f(x) = c/x^2$ for $x > 600$, find the value of x_0 for which $P(X \geq x_0) = 1/2$.

5.6 Find the mean and variance of X whose probability density function is $f(x) = 2e^{-2x}$ over the range $0 < x < \infty$.

5.7 Find the mean and variance of the Poisson distribution.

5.8 The life for a particular transistor in an oscilloscope can be expressed by the probability density function $f(x) = 100/x^2$ where $100 < x < \infty$. What is the probability that not one of the 4 transistors in our scope will have to be replaced within the first 250 hours of operation?

5.9 A probability density function is $f(x) = 2x$, $0 \le x \le 1$. Find:

(a) the cumulative distribution function $F(x)$,

(b) $P(1/4 \le X \le 1/2)$,

(c) the mean and variance of X.

5.10 Given that X_1, X_2, and X_3 are statistically independent and each has a mean of 1 and a variance of 2, find E(Y) and E(Z) for

$$Y = 2X_1 - X_2 - X_3$$

$$Z = X_2 + X_3 \quad .$$

5.11 A probability density function may be expressed as $f(x) = C(\theta)(1-\sin x)^{\theta}\cos x$ for $0 < x < \pi/2$. Find $C(\theta)$.

5.12 The following values were obtained for a probability density function: $\mu_1' = 3/2$, $\mu_2' = 3$, $\sigma^2 = 3/4$. If the range of X is $1 < x < \infty$, what is the probability density function? You have reason to believe that f(x) may be expressed as an inverse integral power of x.

5.13 Given that the variance of each $\overline{Y}_i$ is 12. What is the variance of

$$X = 3\overline{Y}_1 - 2\overline{Y}_2 + 3\overline{Y}_3 - \overline{Y}_4 \quad ?$$

5.14 If $X_1, X_2, \ldots, X_n$ denotes a random sample from a continuous distribution, show that the variance of $\overline{X}$ is σ^2/n.

5.15 If $X_1, X_2, \ldots, X_n$ are statistically independent with density functions $f_1, \ldots, f_n$, show that the variance of $\sum_{i=1}^{n} X_i$ is $\sum_{i=1}^{n} \sigma_{X_i}^2$, where $\sigma_{X_i}^2$ denotes the variance of X_i.

6
STATISTICAL INFERENCE: ESTIMATION

6.1 INTRODUCTION

The area of statistical inference may be divided into two broad categories: estimation and hypothesis testing. In particular, these two categories pertain to the estimation of parameters or testing hypotheses involving the parameters of a given distribution which models a corresponding underlying population. The parameters that are involved in a distribution are, in general, unknown. Consequently it is of interest and importance to be able to obtain information regarding them. This information is obtained in part by estimating them and involves the calculation of numerical values from sample data which estimate the true values of the corresponding population parameters. These numerical values are called estimates of the actual population parameters. In this chapter we will develop statistical procedures for estimating the parameters of various populations. The corresponding ideas concerned with hypothesis testing will be taken up in Chapter 7.

6.2 STATISTICAL ESTIMATION

Engineers are frequently faced with the problem of using data to calculate quantities that they hope will describe the behavior of a process under consideration. This calculated estimate is subject to error because of random fluctuations of the process as well as random measurement errors. The ability to estimate the magnitude of the errors involved is important. This is where the method of statistical estimation can be most useful.

Statistical estimation procedures use sample data to obtain the "best" possible estimates of the corresponding population parameters. In addition, the estimates furnish a quantitative measure of the probable error involved in the estimation. There are two types of estimates: point estimates and interval estimates. Point estimates use sample data to calculate a single "best" value which estimates a population parameter. Point estimates alone give no idea of the magnitude of the error involved in the estimation process. On the other hand, an interval estimate gives a range of values that can be expected to include the correct value a certain specified percentage of the time. This gives the engineer a measure of the probable error involved in the estimation.

6.3 POINT ESTIMATES

The usefulness of a point estimate depends on the criteria by which it is judged. The properties of estimators which are used to help us pick that particular estimator that is most useful for our purposes are unbiasedness, efficiency, and consistency. Ideally, an estimator should be unbiased, highly efficient, and consistent.

We will take an estimator to be a statistic (actually a random variable) and an estimate to be a particular value of the estimator. An estimator (or estimate) of θ is conventionally denoted by $\hat{\theta}$ (^ is a circumflex) in statistical literature. For example, estimates of μ, σ^2, and $\sigma_{\bar{X}}^2$ would be denoted by $\hat{\mu}$, $\hat{\sigma}^2$, and $\hat{\sigma}_{\bar{X}}^2$, respectively. We will use this notation together with the usual notation for random variables, that is, upper case Latin letters, to denote estimators.

An estimator is unbiased if on the average it predicts the correct value. More precisely, we have the following definition.

Definition 6.1 An estimator $T = \hat{\theta}$ is an unbiased estimator of the parameter θ if its expected value is equal to that of the parameter itself, that is, $E(T) = \theta$.

For example, the sample mean $\overline{X} = \hat{\mu}$ is an unbiased estimator of the population mean μ since

$$E(\overline{X}) = \mu, \tag{6.1}$$

which was shown in Section 5.4 of Chapter 5. In the same section it was shown that the expected value of the sample variance

$$\hat{\sigma}^2 = S^2 = \frac{\sum_{i=1}^{n} (X_i - \overline{X})^2}{n - 1} \tag{6.2}$$

is σ^2, that is, $E(S^2) = \sigma^2$. Thus S^2 is an unbiased estimator of σ^2.

An estimator T will be said to be efficient for θ if it yields estimates which are "close" to the value of θ. Now one way of measuring the degree of closeness of the values of T to θ is by considering the variation of T about its expected value θ, that is, by considering the variance σ_T^2 of T. The variance of an unbiased estimator T measures the spread of values of the estimator about the parameter θ. The estimator that has the smallest variance is said to be the most efficient estimator. That is, the estimator T is the most efficient estimator for θ if $E(T) = \theta$ and if $\sigma_{T_1}^2 \geq \sigma_T^2$ where T_1 denotes any other unbiased estimator for θ. An efficient estimator is sometimes called the best estimator and is an estimator that is unbiased and has the smallest possible variance.

If $X_1, X_2, \ldots, X_n$ denotes a sample of size n from a normal population with mean μ and variance σ^2, then the sample mean $\overline{X}$ and the sample variance S^2 are the best estimators for μ and σ^2. The efficiency of any other unbiased estimator T of μ relative to $\overline{X}$ is given by

$$\text{Eff}(T,\overline{X}) = \frac{\sigma^2_{\overline{X}}}{\sigma^2_T} \times 100\% \quad .$$

The variance of $\overline{X}$ is given by $\sigma^2_{\overline{X}} = \frac{\sigma^2}{n}$. This can be easily shown by the methods of Section 5.4 (see also Problem 5.14). Suppose that

$$T = \frac{X_1 + X_2 + \ldots + X_{n-1}}{n - 1} ,$$

that is, T is the sample mean based on the first n - 1 observations. Then $E(T) = \mu$ and $\sigma^2_T = \sigma^2/(n-1)$. Thus

$$\begin{aligned}\text{Eff}(T,\overline{X}) &= \frac{\sigma^2/n}{\sigma^2/(n-1)} \times 100\% \\ &= \frac{n-1}{n} \times 100\%.\end{aligned}$$

If n = 10, then $\text{Eff}(T,\overline{X}) = 90\%$.

The following definition summarizes the idea of best estimator.

Definition 6.2 An estimator T of θ is said to be best for estimating θ if

$$E(T) = \theta$$

and

$$\sigma^2_T \leq \sigma^2_{T_1}$$

where T_1 is any other estimator such that $E(T_1) = \theta$.

The third property of estimators to be considered is that of consistency. An estimator is consistent for θ if it yields values which get closer and closer to the true value of the parameter θ as the size of the sample is increased. As an example, as the size of the sample increases, the sample mean $\overline{X} = \hat{\mu}$ approaches the population mean μ. Therefore, we can say that $\overline{X}$ is a consistent estimator for μ. To say that $\overline{X}$ approaches μ as the sample size increases means that $P(|\overline{X} - \mu| > \varepsilon)$ tends to zero as the sample size n increases without bound where ε denotes any small positive number. This says that the values of $\overline{X}$ are concentrated very close to μ, with a very high probability (close to 1), if the sample size is large. This follows also from the central limit theorem since $\overline{X}$ has an approximate normal distribution with mean μ and variance σ^2/n. However, σ^2/n tends to zero as n tends to ∞ which also says that the values of $\overline{X}$ are close to μ with a high probability. In the case of a normal population with mean μ and variance σ^2, the sample mean $\overline{X}$ has an exact normal distribution with mean μ and variance σ^2/n. Thus

$$\begin{aligned}
P(|\overline{X} - \mu| > \varepsilon) &= 1 - P(|\overline{X} - \mu| \leq \varepsilon) \\
&= 1 - P(-\varepsilon \leq \overline{X} - \mu \leq \varepsilon) \\
&= 1 - P\left(\frac{-\varepsilon}{\sigma/\sqrt{n}} \leq \frac{\overline{X}-\mu}{\sigma/\sqrt{n}} \leq \frac{\varepsilon}{\sigma/\sqrt{n}}\right) \\
&= 1 - \int_{\frac{-\varepsilon}{\sigma/\sqrt{n}}}^{\frac{\varepsilon}{\sigma/\sqrt{n}}} \frac{1}{\sqrt{2\pi}} e^{-x^2/2}\, dx . \qquad (6.3)
\end{aligned}$$

The integral in Eq. (6.3) tends to 1 as n increases without bound since the limits of integration approach $-\infty$ to $+\infty$. Thus $P(|\overline{X} - \mu| > \varepsilon)$ tends to 0 as n tends to ∞ and $\overline{X}$ is consistent for μ.

The requirement of consistency is not a demanding one since most reasonable estimators satisfy it. We will not pursue it any further but rather go to the more interesting and useful idea of interval estimation.

6.4 INTERVAL ESTIMATES

From the preceding section it has been seen that a point estimate is a single value that gives information about the parameter for which it is an estimate. T is a good estimator for θ if the values of T fall close to θ, where the degree of closeness is measured by σ_T^2. After an estimate has been made, the engineer may be asked about how good it is. It seems that if $E(T) = \theta$ and σ_T^2 is small then one should be able to determine an interval of values about a given estimate t of θ such that the interval is "reasonably short" and one is "reasonably sure" that the true value of θ is in that interval. Such an interval, denoted by (t_L, t_U), is an interval estimate of θ, that is, $t_L < \theta < t_U$. The end points of the interval depend on the point estimate t and on the distribution of the estimator T. Since σ_T^2, the variance of T, measures how closely the values of T are concentrated about θ it follows that the length of an interval estimate will vary according to the magnitude of σ_T^2.

The most important issue involving interval estimates is a quantification of what we mean when we say that we should be "reasonably sure" that the interval estimate contains the true value of the unknown parameter θ. Every sample from the population for which θ is a parameter yields a value of T, the estimator of θ, which, in turn, yields an interval estimate (t_L, t_U) of θ. Now every sample will produce a different estimate of θ and thus a different interval estimate of θ. The distribution of T will allow us to find a t_L and a t_U for all possible samples such that a specified proportion of the intervals will contain θ. For example, t_L and t_U may be computed so that 95% of the intervals contain θ. Thus choosing a sample and computing (t_L, t_U) is comparable to choosing one interval from the entire collection of intervals of which 95% contain θ. The particular interval (t_L, t_U) is called a <u>95% confidence interval</u>, and we say we are 95% confident that the interval (t_L, t_U) contains θ. The quantities t_L and t_U are called the confidence limits.

In general the end points t_L and t_U are particular values of the random variables T_L and T_U. The interval (T_L, T_U) is a <u>random interval</u> of which (t_L, t_U) is a particular outcome. When we say that (t_L, t_U) is a 95% confidence interval for θ we actually mean, prior to taking the sample, that $P(T_L \leq \theta \leq T_U) = 0.95$. However, once the sample has been observed and (t_L, t_U) has been computed then we say that (t_L, t_U) is a 95% confidence interval, that is, 95% of all possible outcomes of (T_L, T_U) contain θ. In general we may desire (t_L, t_U) to be such that the proportion $1 - \alpha$, $0 < \alpha < 1$, of all intervals contain θ. The proportion $1 - \alpha$ is called the confidence coefficient and (t_L, t_U) is called a $(1 - \alpha) \times 100\%$ confidence interval for θ.

Ideally one wants α to be small (between 0.01 and 0.10) and $t_U - t_L$ to be small. In the ensuing sections we present the most efficient estimators for certain parameters corresponding to certain populations and the corresponding confidence intervals.

The next three sections discuss important distributions which arise in constructing confidence intervals on the parameters of normal populations and binomial populations. These distributions were introduced in Section 3.4. However, a more detailed discussion is presented here for ready reference.

The parameters of the normal distribution are the mean μ and the variance σ^2. Sections 6.8-6.10 contain statistical inference procedures for the mean(s) of a normal population(s). These procedures are given in the form of confidence intervals, derived under the assumption of normality. These methods work rather well for other types of distributions: the assumption of normality is not critical in the interval estimation of μ.

Sections 6.11-6.13 discuss techniques for the estimation of the variance(s) under the assumption of normality. The lack of normality in this case is quite serious in that it can affect the results quite adversely. If non-normality is expected one should use other techniques, such as those found in Conover [1].

6.5 CHI-SQUARE DISTRIBUTIONS

The chi-square (χ^2) distribution was introduced in Section 3.4.1. It will be used to make inferences concerning population variances and, for other inferential purposes in Chapter 7, not directly involving parameters such as the problem of testing a hypothesis of independence of two attributes.

The random variable χ^2 has a chi-square distribution with ν degrees of freedom (d.f.) if it has the density function

$$f(\chi^2) = \frac{(\chi^2)^{\nu/2 - 1} e^{-\chi^2/2}}{\left(\frac{\nu}{2} - 1\right)! 2^{\nu/2}} = \frac{(\chi^2)^{\nu/2 - 1} e^{-\chi^2/2}}{2^{\nu/2}\Gamma(\nu/2)},$$

$$\chi^2 > 0, \ \nu > 0 \tag{6.4}$$

which is a continuous function. The parameter ν can be any positive real value; however, for our purposes ν can be considered to take on integer values, that is, $\nu = 1,2,\ldots$. These values of ν arise quite naturally in many situations. For example, if $X_1,X_2,\ldots,X_n$ denotes a random sample of size n from a standard normal population, then the random variable

$$V = \sum_{i=1}^{n} X_i^2$$

has a χ^2-distribution with $\nu = n$ d.f. If the sample is from a normal population with mean μ and variance σ^2 then the random variable

$$V = \sum_{i=1}^{n} (X_i - \overline{X})^2/\sigma^2$$

has a χ^2-distribution with $\nu = n - 1$ d.f. All these results can be proved using the notions from advanced statistical theory.

The mean and variance of the χ^2-distribution (6.4) are

$$E(\chi^2) = \nu$$

and

$$\sigma^2_{\chi^2} = \mathrm{Var}(\chi^2) = 2\nu \quad .$$

Thus as the mean ν increases so does the variance, and the region of "heavy concentration" of the distribution is shifted to the right. Figure 6.1 gives the graph of a typical χ^2-density function ($\nu = 8$ d.f.). The quantity $\chi^2_{\nu,\alpha}$ is that number that has the fraction α of the area under the curve to the left. That is

$$P(\chi^2 \leq \chi^2_{\nu,\alpha}) = \alpha. \tag{6.5}$$

For example, $\chi^2_{\nu,\alpha/2}$ and $\chi^2_{\nu,1-\alpha/2}$ in Fig. 6.1 are such that $P(\chi^2_{\nu,\alpha/2} \leq \chi^2 \leq \chi^2_{\nu,1-\alpha/2}) = 1 - \alpha$. Equation (6.5) is actually the distribution function of a χ^2 random variable and is tabulated in Table V for various values of α and ν.

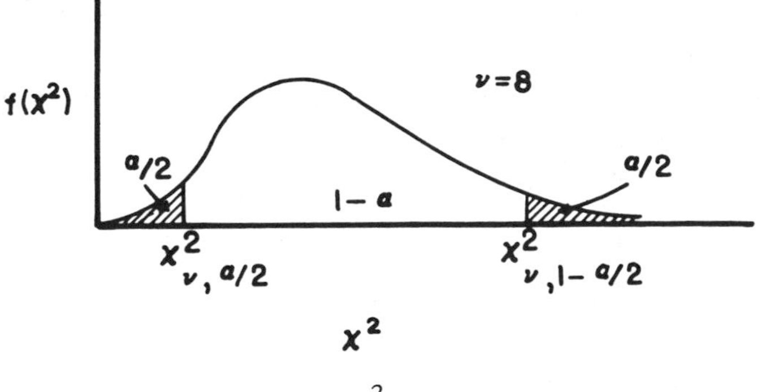

Fig. 6.1 A χ^2-distribution.

For values of ν greater than 30, the χ^2-distribution may be approximated from the standard normal distribution as follows:

$$\chi^2_{\nu,\alpha} = \frac{1}{2}(z_\alpha + \sqrt{2\nu - 1})^2 \quad , \tag{6.6}$$

where z_α is the equivalent percentile of the standard normal variable. For example, $z_{0.975} = 1.96$ so that for $\nu = 30$ we find from Eq. (6.6) that $\chi^2_{30,0.975} = 46.7$ compared to the value $\chi^2_{30,0.975} = 47.0$ found in Table V.

6.6 THE t-DISTRIBUTION

The t-distribution, introduced in Section 3.4.2, will be used in this chapter in regard to the interval estimation of the mean μ when the variance is unknown.

A random variable T has the t-distribution with ν degrees of freedom if its density function is

$$f(t) = \frac{\Gamma[(\nu + 1)/2]}{\sqrt{\pi\nu}\,\Gamma(\nu/2)\,[1 + t^2/\nu]^{(\nu + 1)/2}} \quad , \quad -\infty < t < \infty \quad .$$

The density function contains the parameter ν, which for our purposes can be assumed to take on the values $\nu = 1,2,\ldots$.

The density function f(t) resembles the standard normal in that its graph is also bell-shaped with a mean of zero (if $\nu > 1$). However, its variance exceeds that of the normal, for all ν, that is it has "heavier" tails. It can be shown that

$$\lim_{\nu \to \infty} f(t) = \frac{1}{\sqrt{2\pi}} e^{-t^2/2} \quad ,$$

that is, for large values of ν the t-distribution can be approximated by the standard normal. The approximation is actually quite good for $\nu > 30$.

Let $X_1, X_2, \ldots, X_n$ denote a random sample from a normal population with mean μ and variance σ^2. A well-known result in statistical theory is that the random variable

$$\frac{(\overline{X} - \mu)\sqrt{n}}{\sqrt{\sum_{i=1}^{n} (X_i - \overline{X})^2/(n - 1)}} \qquad (6.7)$$

has a t-distribution with $\nu = n - 1$ d.f.

The relation of the t-distribution to the normal and χ^2-distributions is as follows. If Z has a standard normal distribution and U has a χ^2-distribution with ν d.f., and if Z and U are independent, then the random variable

$$\frac{Z}{\sqrt{U/\nu}} \qquad (6.8)$$

has a t-distribution with ν d.f. The result (6.7) follows from (6.8) by letting

$$Z = \frac{\overline{X} - \mu}{\sigma/\sqrt{n}},$$

$$U = \sum_{i=1}^{n} \frac{(X_i - \overline{X})^2}{\sigma^2},$$

and

$$\nu = n - 1.$$

Figure 6.2 illustrates a typical t curve in comparison to the standard normal curve.

The quantity $t_{\nu,\alpha}$ denotes that number such that a fraction α of the area is to the left of $t_{\nu,\alpha}$. That is, $P(T \leq t_{\nu,\alpha}) = \alpha$. For example,

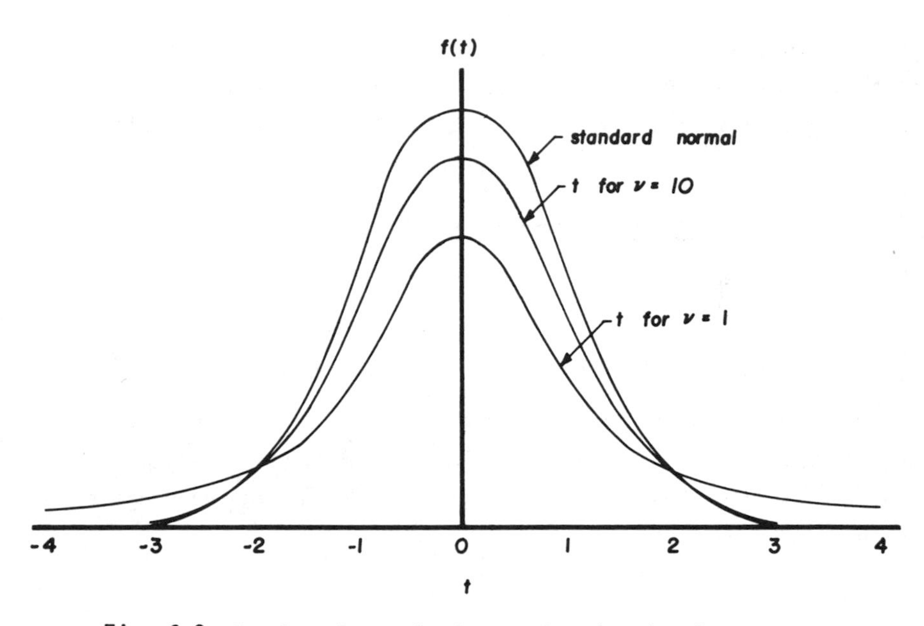

Fig. 6.2 Graphs of standard normal and t-density curves.

$t_{\nu,\alpha/2}$ and $t_{\nu,1-\alpha/2}$ in Fig. 6.2 are such that $P(t_{\nu,\alpha/2} \leq T \leq t_{\nu,1-\alpha/2}) = 1 - \alpha$. This notation will be used later.

The cumulative t-distribution function $P(T \leq t_{\nu,\alpha})$ is contained in Table IV for several values of ν and α.

6.7 THE F-DISTRIBUTION

The F-distribution, introduced in Section 3.4.3, will be used in the comparison of two variances and in analysis of variance and related ideas. A random variable F has an F-distribution with ν_1 and ν_2 degrees of freedom if its density function is

$$g(f) = \left(\frac{\Gamma\left(\frac{\nu_1+\nu_2}{2}\right)\left(\frac{\nu_1}{\nu_2}\right)^{\nu_1/2}}{\Gamma\left(\frac{\nu_1}{2}\right)\Gamma\left(\frac{\nu_2}{2}\right)} \right) \left(\frac{f^{(\nu_1-2)/2}}{(1+\nu_1 f/\nu_2)^{(\nu_1+\nu_2)/2}} \right), \quad f > 0 \quad .$$

The two parameters ν_1 and ν_2 can take on noninteger values: however, we will use $\nu_1 = 1,2,\ldots$ and $\nu_2 = 1,2,\ldots$

The F-distribution can arise in a variety of ways. For example, if V has a χ^2-distribution with ν_1 d.f. and U has a χ^2-distribution with ν_2 d.f., and if U and V are independent, then

$$F = \frac{V/\nu_1}{U/\nu_2} = \frac{\chi_1^2/\nu_1}{\chi_2^2/\nu_2}$$

has an F-distribution with ν_1 and ν_2 d.f. Similarly,

$$F = \frac{U/\nu_2}{V/\nu_1}$$

has an F-distribution with ν_2 and ν_1 d.f. Note that the "first" d.f. are associated with the numerator and the "second" d.f. with the denominator.

If $X_1, X_2, \ldots, X_{n_1}$ and $Y_1, Y_2, \ldots, Y_{n_2}$ are independent random samples from the two normal populations with parameters μ_1, σ_1^2 and μ_2, σ_2^2, respectively, then

$$F = \frac{S_1^2/\sigma_1^2}{S_2^2/\sigma_2^2} = \frac{\sum_{i=1}^{n_1} \dfrac{(X_i - \overline{X})^2}{(n_1 - 1)} \Big/ \sigma_1^2}{\sum_{i=1}^{n_2} \dfrac{(Y_i - \overline{Y})^2}{(n_2 - 1)} \Big/ \sigma_2^2}$$

has an F-distribution with $\nu_1 = n_1 - 1$ and $\nu_2 = n_2 - 1$ d.f. If $\sigma_1^2 = \sigma_2^2$ then F becomes S_1^2/S_2^2, the ratio of two sample variances.

Figure 6.3 gives the graph of a typical F-density function. The value $F_{\nu_1,\nu_2,\alpha}$ is such that

$$P(F \le F_{\nu_1,\nu_2,\alpha}) = \alpha. \tag{6.9}$$

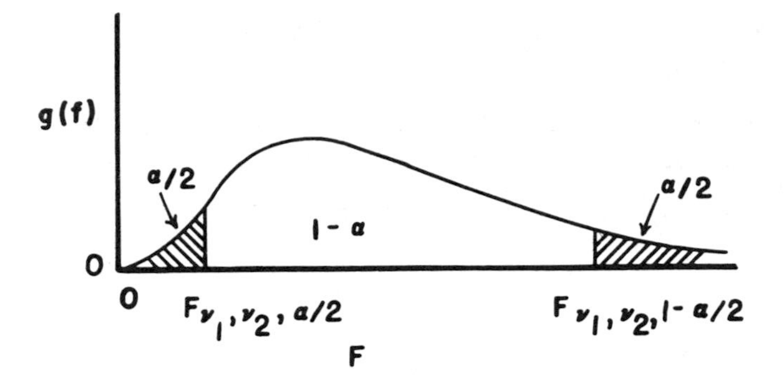

Fig. 6.3 An F-distribution.

Figure 6.3 illustrates the values $F_{\nu_1,\nu_2,\alpha/2}$ and $F_{\nu_1,\nu_2,1-\alpha/2}$ such that $P(F_{\nu_1,\nu_2,\alpha/2} \leq F \leq F_{\nu_1,\nu_2,1-\alpha/2}) = 1 - \alpha$. The distribution corresponding to (6.9) is tabulated in Table VI for several values of α and ν_1, ν_2. The parameter ν_1 corresponds to the numerator d.f.

6.8 ESTIMATION OF THE MEAN

Suppose $X_1, X_2, \ldots, X_n$ is a sample from a normal population with mean μ and known variance σ^2. The most efficient estimator of the population mean μ is given by the sample mean $\hat{\mu} = \bar{X} = \sum_{i=1}^{n} X_i/n$.

Thus the best point estimate of μ is $\bar{x}$. Recall that the expected value and variance of $\bar{X}$ are $E(\bar{X}) = \mu$ and $\sigma_{\bar{X}}^2 = \sigma^2/n$, respectively.

The statistic $\bar{X}$ has a normal distribution with mean μ and variance σ^2/n. Thus the random variable

$$Z = \frac{\bar{X} - \mu}{\sigma/\sqrt{n}}$$

has the standard normal distribution, that is $\mu_Z = E(Z) = 0$ and $\sigma_Z^2 = 1$. Thus,

$$P(-z_{1-\alpha/2} < Z < z_{1-\alpha/2}) = 1 - \alpha \tag{6.10}$$

where $z_{1-\alpha/2}$ is that value in the standard normal distribution that has $1 - \alpha/2$ area to the left, that is, $P(Z < -z_{1-\alpha/2}) = \alpha/2$, and $z_{\alpha/2} = -z_{1-\alpha/2}$.

Substituting for Z in (6.10) gives

$$P\left(-z_{1-\alpha/2} < \frac{\overline{X} - \mu}{\sigma/\sqrt{n}} < z_{1-\alpha/2}\right) = 1 - \alpha. \tag{6.11}$$

Multiplying each member of the inequality by $\sigma/\sqrt{n}$ and subtracting $\overline{X}$ from each member yields

$$P\left(-\overline{X} - z_{1-\alpha/2}\frac{\sigma}{\sqrt{n}} < -\mu < -\overline{X} + z_{1-\alpha/2}\frac{\sigma}{\sqrt{n}}\right) = 1 - \alpha.$$

Then multiplying each member by -1 we have

$$P\left(\overline{X} - z_{1-\alpha/2}\frac{\sigma}{\sqrt{n}} < \mu < \overline{X} + z_{1-\alpha/2}\frac{\sigma}{\sqrt{n}}\right) = 1 - \alpha \tag{6.12}$$

The interval

$$(T_L, T_U) = (\overline{X} - z_{1-\alpha/2}\frac{\sigma}{\sqrt{n}}, \overline{X} + z_{1-\alpha/2}\frac{\sigma}{\sqrt{n}})$$

is a random interval. For a particular sample we have $\overline{x}$ as a point estimate of μ and

$$(t_L, t_U) = \left(\overline{x} - z_{1-\alpha/2}\frac{\sigma}{\sqrt{n}}, \overline{x} + z_{1-\alpha/2}\frac{\sigma}{\sqrt{n}}\right) \tag{6.13}$$

is a $(1-\alpha)100\%$ confidence interval for μ and the end points of the interval are called $(1-\alpha)100\%$ confidence limits for μ.

Note that (6.11) is a "bona fide" probability statement. However, once the sample is taken and the random interval takes on the form in (6.13) we use the term confidence interval to mean that $(1-\alpha)100\%$ of all such intervals cover the mean μ.

The confidence interval in (6.13) was derived under the assumption of normality. However, the central limit theorem states that, regardless of what the underlying distribution is (so long as σ^2 is finite), the distribution of $\overline{X}$ is approximately normal (Section 3.3.6). Thus (6.13) gives confidence limits for μ which are generally used without regard to the underlying distribution so long as n is sufficiently large. Furthermore $\overline{X}$ is a rather efficient estimator of μ regardless of the underlying distribution.

Example 6.1 Referring to the data of Example 4.2, suppose that the true variance of an individual observation had been 49×10^{-10} M. Since 7 samples were analyzed,

$$\sigma_{\overline{X}} = \sqrt{\sigma^2/n} = \sqrt{\frac{49 \times 10^{-10}}{7}} = \frac{7 \times 10^{-5}}{\sqrt{7}} = 2.646 \times 10^{-5} \text{ M}$$

At the 95% confidence level, $z_{1-\alpha/2} = z_{0.975} = 1.96$.

Recalling that $\overline{x} = 12.36 \times 10^{-5}$, the 95% confidence interval for the mean μ of the $K_2Cr_2O_7$ determinations can be found from Eq. (6.13) as

$$12.36 \times 10^{-5} \text{ M} - 2.646 \times 10^{-5} \text{ M}(1.96) <$$

$$\mu < 12.36 \times 10^{-5} \text{ M} + 2.646 \times 10^{-5} \text{ M}(1.96)$$

or

$$7.17 \times 10^{-5} \text{ M} < \mu < 17.55 \times 10^{-5} \text{ M}$$

or approximately

$$0.72 \times 10^{-4} \text{ M} < \mu < 1.76 \times 10^{-4} \text{ M}.$$

The confidence limits for μ given in (6.13) were obtained under the assumption that σ^2, the population variance, is known. If σ^2 is not known, then the usual procedure is to replace it by its estimate, $s^2 = \sum_{i=1}^{n} (x_i - \overline{x})^2/(n - 1)$. However, the distribution of $Z = \frac{\overline{X} - \mu}{(S/\sqrt{n})}$ is no longer normal so that the development which led to (6.13) is no longer valid. We are saved by observing that the distribution of $T = \frac{(\overline{X} - \mu)}{S_{\overline{x}}}$ is the t-distribution with n - 1 degrees of freedom, which is discussed in Section 6.6. The t-distribution has a curve which is symmetrical, just like the normal curve. Furthermore, if $n > 30$ one could just as well use the normal distribution since the t and normal curves agree very closely for all such sample sizes.

Proceeding as in (6.10) we have

$$P(-t_{n-1,1-\alpha/2} < T < t_{n-1,1-\alpha/2}) = 1 - \alpha$$

where $t_{n-1,1-\alpha/2}$ is that value in the t-distribution with n - 1 degrees of freedom that has $1-\alpha/2$ to the left, that is,

$$P(T < t_{n-1,1-\alpha/2}) = 1-\alpha/2 \text{ or } P(T < -t_{n-1,1-\alpha/2}) = \alpha/2.$$

Replacing T by $(\overline{X} - \mu)/S_{\overline{x}}$ we have

$$P\left(-t_{n-1,1-\alpha/2} < \frac{\overline{X} - \mu}{S_{\overline{x}}} < t_{n-1,1-\alpha/2}\right) = 1 - \alpha$$

which yields

$$P\left(\overline{X} - t_{n-1,1-\alpha/2}\frac{S_x}{\sqrt{n}} < \mu < \overline{X} + t_{n-1,1-\alpha/2}\frac{S_x}{\sqrt{n}}\right) = 1 - \alpha. \quad (6.14)$$

The result in (6.14) can be obtained from (6.12) by replacing σ and z in (6.12) by S and t, respectively. From (6.14) we obtain

$$\left(\overline{X} - t_{n-1,1-\alpha/2}\frac{S_x}{\sqrt{n}},\ \overline{X} + t_{n-1,1-\alpha/2}\frac{S_x}{\sqrt{n}}\right) \quad (6.15)$$

as the $(1-\alpha)100\%$ confidence limits for μ when σ^2 is unknown. The result (6.15) is very popular and useful even when the sampled population is only approximately normal.

Example 6.2 The following data were obtained for the calibration of the Ruska Dead Weight Gage used with our Burnett P-V-T apparatus. The weights corresponding to 100 psi had the following apparent masses*:

26.03570	26.03575	26.03599
26.03581	26.03551	26.03533
26.03529	26.03588	26.03570
26.03573	26.03586	

What is the 95% confidence limit for the apparent mass of the "100 psi" weight?

The data is first coded according to $W_i = (X_i - 26.03500)10^5$. For n = 11, the following values were obtained: $\sum_i w_i = 755$; $\Sigma w_i^2 = 56{,}787$; $\bar{w} = \Sigma w_i/n = 68.6$; $\bar{x} = 26.03500 + \bar{w}(10^{-5}) = 26.03568$ psi; and $s_W = |(\Sigma(w_i)^2 - \bar{w}\Sigma w_i)/(n - 1)|^{1/2} = 22.36$ psi. s_X is therefore found from s_W by $s_X = s_W(10^{-5}) = 22.36(10^{-5})$ psi. The standard error of the mean, $s_{\bar{X}}$, is then $6.741(10^{-5})$ psi. Since the degrees of freedom are n - 1 = 10, we find $t_{10,\ 0.025} = -2.228$ from Table IV. The desired 95% confidence limits for the mean μ in psi (remembering that $\bar{x}$ and $s_{\bar{X}}$ are used for particular or calculated values are, by (6.15),

$$\bar{x} - s_{\bar{x}} t_{10,1-\alpha/2} < \mu < \bar{x} + s_{\bar{x}} t_{10,1-\alpha/2}$$

$$26.03568 - 2.228(6.741 \times 10^{-5}) < \mu < 26.03568 + 2.228(6.741 \times 10^{-5})$$

$$26.03553 < \mu < 26.03583$$

*Miks, C.E., "Test Report, Ruska Dead Weight Gage", Ruska Instruments Corp., Houston, Tex. (1964). Gage and report owned by Chemical Engineering Dept., Texas Tech University, Lubbock, Texas. Data reprinted by permission of the Department.

Example 6.3 Chief Design Engineer Gant for the Frigid-Flow Corporation has received final test results on the company's new heat exchanger. The values given below are overall heat-transfer coefficients (U_c's) for the exchanger as determined by Gant's testing section. Frigid-Flow employs a 5% "minimizing factor" in their calculations before offering specifications to buyers. At the 99% confidence level, what minimum value for the exchanger's overall heat-transfer coefficient can Gant suggest for Frigid-Flow?

Values of U_c, Btu/hr ft^2 ^oF									
60	63	60	68	70	72	65	61	69	67

The sample mean and variance are

$$\bar{x} = \sum_{i=1}^{n} x_i/n = 65.5 \text{ Btu/hr ft}^2\ ^oF$$

and

$$s_X = \sqrt{s_X^2} = \left(\frac{43073 - 42903}{9}\right)^{1/2} = 4.3525 \text{ Btu/hr ft}^2\ ^oF.$$

Furthermore

$$s_{\bar{X}} = \frac{4.3525}{\sqrt{10}} = 1.3764 \text{ Btu/hr ft}^2\ ^oF$$

At the 99% confidence level, $\alpha/2 = 0.005$. Since the degrees of freedom are 9, the t-table yields $t_{9,0.995} = 3.250$.

Then the 99% confidence limits for the mean U_c are

$$\bar{x} - t_{9,1-\alpha/2}\, s_{\bar{x}} < U_c < \bar{x} + t_{9,1-\alpha/2}\, s_{\bar{x}}$$

$$65.5 - 3.25\ (1.3764) < U_c < 65.5 + 3.25\ (1.3764)$$

$$61.03 \text{ Btu/hr ft}^2\ ^oF < U_c < 69.97 \text{ Btu/hr ft}^2\ ^oF.$$

Using the "minimizing factor" we have

$$(0.05)\ 61.03 = 3.05 \text{ Btu/hr ft}^2\ {}^\circ\text{F}.$$

Thus the "minimum" value is

$$61.03 - 3.05 = 57.98 \text{ Btu/hr ft}^2\ {}^\circ\text{F}.$$

Gant will suggest a value of 58 Btu/hr ft^2 $^\circ$F for Frigid-Flow.

6.9 COMPARISON OF TWO MEANS

The next problem we consider is that of comparing the means of two populations. Consider two normal populations, the first with mean μ_1 and variance σ_1^2 and the second with mean μ_2 and variance σ_2^2. The means μ_1 and μ_2 can be compared by considering their difference $\mu_1 - \mu_2$. Thus the aim here will be to construct a confidence interval for the difference $\mu_1 - \mu_2$. Our development will be broken down into three cases: (1) σ_1^2 and σ_2^2 <u>known</u>, (2) σ_1^2 and σ_2^2 <u>unknown</u> but <u>equal</u>, and (3) σ_1^2 and σ_2^2 <u>unknown</u> and <u>unequal</u>.

<u>Case 1</u> (σ_1^2 and σ_2^2 known) Suppose $X_{11}, X_{12}, \ldots, X_{1n_1}$ is a random sample of size n_1 from a normal population with mean μ_1 and variance σ_1^2, and $X_{21}, X_{22}, \ldots, X_{2n_2}$ is a sample from a normal population with mean μ_2 and variance σ_2^2. Note that the first subscript in X_{ij} identifies the population (1 or 2) and the second subscript identifies the observation within the sample. The best estimators for μ_1 and μ_2 are given by

$$\overline{X}_1 = \sum_{i=1}^{n_1} X_{1i}/n_1$$

and

$$\overline{X}_2 = \sum_{i=1}^{n_2} X_{2i}/n_2 ,$$

respectively. Since the variances are assumed known there is no need to estimate them. Now

$$E(\overline{X}_1 - \overline{X}_2) = \mu_1 - \mu_2 \quad ,$$

and the variance of $\overline{X}_1 - \overline{X}_2$ is given by

$$\sigma^2_{\overline{X}_1-\overline{X}_2} = \sigma^2_{\overline{X}_1} + \sigma^2_{\overline{X}_2} = \frac{\sigma_1^2}{n_1} + \frac{\sigma_2^2}{n_2} \tag{6.16}$$

since $\overline{X}_1$ and $\overline{X}_2$ are independent. Thus

$$Z = \frac{\overline{X}_1 - \overline{X}_2 - (\mu_1 - \mu_2)}{\sigma_{\overline{X}_1-\overline{X}_2}} \tag{6.17}$$

has mean 0 and variance 1, furthermore, Z has a standard normal distribution. Therefore, we may write

$$P(-z_{1-\alpha/2} < Z < z_{1-\alpha/2}) = 1 - \alpha \quad ,$$

and replacing Z according to Eq. (6.17) we have

$$P\left(-z_{1-\alpha/2} < \frac{\overline{X}_1 - \overline{X}_2 - (\mu_1 - \mu_2)}{\sigma_{\overline{X}_1-\overline{X}_2}} < z_{1-\alpha/2}\right) = 1 - \alpha \tag{6.18}$$

By solving the inequality in Eq. (6.18) for $\mu_1 - \mu_2$, in exactly the same way as Eq. (6.12) was obtained, we have

$$P(\overline{X}_1-\overline{X}_2 - z_{1-\alpha/2}\,\sigma_{\overline{X}_1-\overline{X}_2} < (\mu_1-\mu_2) < \overline{X}_1-\overline{X}_2 + z_{1-\alpha/2}\,\sigma_{\overline{X}_1-\overline{X}_2}) = 1 - \alpha, \tag{6.19}$$

where

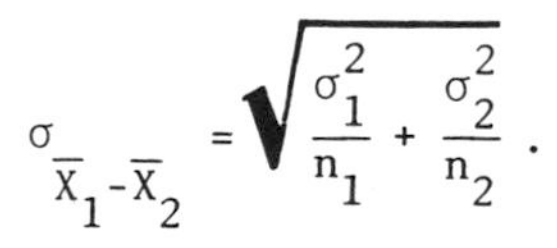

$$\sigma_{\overline{X}_1-\overline{X}_2} = \sqrt{\frac{\sigma_1^2}{n_1} + \frac{\sigma_2^2}{n_2}} \; .$$

Thus a$(1-\alpha)100\%$ confidence interval for $\mu_1 - \mu_2$, where σ_1^2 and σ_2^2 are known, is given by

$$\left(\bar{x}_1-\bar{x}_2 - z_{1-\alpha/2}\sqrt{\frac{\sigma_1^2}{n_1}+\frac{\sigma_2^2}{n_2}},\ \bar{x}_1-\bar{x}_2 + z_{1-\alpha/2}\sqrt{\frac{\sigma_1^2}{n_1}+\frac{\sigma_2^2}{n_2}}\right). \quad (6.20)$$

Example 6.4 Consider the data presented in Example 4.1. Information received from the manufacturer indicates that the average variance in concentration for single analyses for last year's production runs was 0.016. Construct a 95% confidence interval for $\mu_1 - \mu_2$ where μ_1 and μ_2 are the true means of the analyses for the two bottles.

We take the variance of the analysis for each bottle (population) to be 0.016, i.e., $\sigma_1^2 = \sigma_2^2 = 0.016$. From Example 4.1 we have $\bar{x}_1 = 15.770$ and $\bar{x}_2 = 15.597$. Furthermore, $z_{1-\alpha/2} = z_{0.975} = 1.96$

and

$$\sigma_{\bar{X}_1-\bar{X}_2} = \sqrt{\frac{\sigma_1^2}{n_1}+\frac{\sigma_2^2}{n_2}}$$

$$= \sqrt{\frac{0.016}{3}+\frac{0.016}{3}}$$

$$= \sqrt{\frac{0.032}{3}}$$

$$= 0.10327 \quad .$$

Thus, a 95% confidence interval for $\mu_1 - \mu_2$, according to (6.20), is

$$\bar{x}_1 - \bar{x}_2 - 1.96(0.10327) < \mu_1 - \mu_2 < \bar{x}_1 - \bar{x}_2 + 1.96(0.10327)$$

or

$$-0.029 < \mu_1 - \mu_2 < 0.375 \quad .$$

This means we are 95% confident that the interval (-0.029, 0.375) contains the unknown difference $\mu_1 - \mu_2$ in it. Since the interval contains 0, we have no real reason to infer that $\mu_1 \neq \mu_2$ (or $\mu_1 - \mu_2 \neq 0$), that is, that the two reagent bottles were not filled from the same bottle. In light of this reasoning we see that the confidence interval may be used to "test" the hypothesis that $\mu_1 = \mu_2$. These remarks will be examined in more detail in Chapter 7, where testing hypotheses will be related to confidence intervals.

We now consider confidence intervals on $\mu_1 - \mu_2$ when σ_1^2 and σ_2^2 are unknown, but when $\sigma_1^2 = \sigma_2^2 = \sigma^2$ is a reasonable assumption.

Case 2 If the variances are unknown but can be presumed equal then each sample can be used to obtain an estimator for the common variance σ^2. These estimators are given by

$$S_1^2 = \sum_{i=1}^{n_1} (X_{1i} - \bar{X}_1)^2/(n_1 - 1)$$

and

$$S_2^2 = \sum_{i=1}^{n_2} (X_{2i} - \bar{X}_2)^2/(n_2 - 1) .$$

Since S_1^2 and S_2^2 are both estimators for σ^2, we may "pool" them to obtain a better estimator for σ^2, given by

$$S_p^2 = \frac{(n_1-1)\, S_1^2 + (n_2-1)\, S_2^2}{n_1 + n_2 - 2} = \frac{\sum_{i=1}^{n_1} (X_{1i} - \bar{X}_1)^2 + \sum_{i=1}^{n_2} (X_{2i} - \bar{X}_2)^2}{n_1 + n_2 - 2} . \tag{6.21}$$

The denominator $n_1 + n_2 - 2$ makes S_p^2 unbiased for σ^2, that is, $E(S_p^2) = \sigma^2$. Furthermore, the variance of $\bar{X}_1 - \bar{X}_2$ is given by

$$\sigma^2_{\bar{X}_1-\bar{X}_2} = \frac{\sigma^2}{n_1} + \frac{\sigma^2}{n_2} \quad (\text{since } \sigma_1^2 = \sigma_2^2 = \sigma^2)$$

and the best estimator for $\sigma^2_{\bar{X}_1-\bar{X}_2}$ is given by

$$S^2_{\bar{X}_1-\bar{X}_2} = S_p^2\left(\frac{1}{n_1} + \frac{1}{n_2}\right). \tag{6.22}$$

The random variable

$$T = \frac{\bar{X}_1 - \bar{X}_2 - (\mu_1 - \mu_2)}{S_{\bar{X}_1-\bar{X}_2}},$$

which is actually (6.17) with $\sigma_{\bar{X}_1-\bar{X}_2}$ replaced by $S_{\bar{X}_1-\bar{X}_2}$, has a t-distribution with $\nu = n_1 + n_2 - 2$ degrees of freedom. Thus we can say that

$$P(-t_{\nu,1-\alpha/2} < T < t_{\nu,1-\alpha/2}) = 1 - \alpha \tag{6.23}$$

or

$$P\left(-t_{\nu,1-\alpha/2} < \frac{\bar{X}_1-\bar{X}_2 - (\mu_1-\mu_2)}{S_{\bar{X}_1-\bar{X}_2}} < t_{\nu,1-\alpha/2}\right) = 1 - \alpha, \tag{6.24}$$

and consequently

$$P(\bar{X}_1-\bar{X}_2 - t_{\nu,1-\alpha/2}\, S_{\bar{X}_1-\bar{X}_2} \leq \mu_1-\mu_2 \leq \bar{X}_1-\bar{X}_2 + t_{\nu,1-\alpha/2} S_{\bar{X}_1-\bar{X}_2}) = 1 - \alpha. \tag{6.25}$$

A $(1-\alpha)100\%$ confidence interval in Case 2 is thus

$$\bar{x}_1-\bar{x}_2 - t_{\nu,1-\alpha/2}\, s_p\sqrt{\frac{1}{n_1}+\frac{1}{n_2}} < \mu_1-\mu_2 < \bar{x}_1-\bar{x}_2 + t_{\nu,1-\alpha/2}\, s_p\sqrt{\frac{1}{n_1}+\frac{1}{n_2}} \tag{6.26}$$

where s_p is defined by Eq. (6.21) and $\nu = n_1 + n_2 - 2$.

The results for Cases 1 and 2 are quite robust, that is, they are valid even when the underlying population is only approximately normal.

Example 6.5 In order to compare the effects of two solid catalyst component concentrations on NO_2 reductions, six groups of observations were made. Each group consisted of three replicates of five observations each, that is, a total of 15 determinations were made for each concentration. The concentrations (in mass percent) were 0.5 and 1.0%. The reduction data are summarized below.

Group	Replicate means (5 observations each)			Group mean (based on 15 observations)
	A	B	C	
1	5.18	5.52	5.42	$5.37333 = \bar{x}_1$
2	5.58	5.62	5.82	$5.67333 = \bar{x}_2$

Construct a 95% confidence interval on $\mu_1 - \mu_2$, the difference of the means corresponding to concentrations 0.5% and 1.0%.

The replicate means are treated as individual observations. Thus $\bar{x}_1 = (5.18 + 5.52 + 5.42)/3 = 5.37$ and $\bar{x}_2 = 5.67$. Furthermore, the sample variance of 5.18, 5.52, and 5.42 is $s_1^2 = 0.0305334$ and similarly $s_2^2 = 0.0165334$. It is assumed that $\sigma_1^2 = \sigma_2^2$ so that

$$s_p^2 = \frac{2s_1^2 + 2s_2^2}{3 + 3 - 2} = \frac{0.0610668 + 0.0330667}{4} = 0.0235334 .$$

Since the confidence level is 95% and $n_1 = 3$, $n_2 = 3$, we have from the t-distribution table with 4 d.f., $t_{4,0.975} = 2.776$. Thus by using the result given in (6.26) we have

$$\bar{x}_1 - \bar{x}_2 - 2.776\sqrt{0.1534056}\sqrt{\frac{1}{3} + \frac{1}{3}} < \mu_1 - \mu_2 < \bar{x}_1 - \bar{x}_2 + 2.776\sqrt{0.1534056}\sqrt{\frac{2}{3}}$$

or

$$-0.1258 < \mu_1 - \mu_2 < 0.7258 \ .$$

The 95% confidence interval contains 0 and on this basis the results of the experiment do not warrant inferring that $\mu_1 \neq \mu_2$. Thus we conclude that one catalyst component is not more (or less) effective than the other. From this example we again see that we can use a confidence interval to "test" the hypothesis that $\mu_1 = \mu_2$.

We now consider the third case in constructing confidence intervals for $\mu_1 - \mu_2$.

Case 3 Suppose that $X_{11}, X_{12}, \ldots, X_{1n_1}$ and $X_{21}, X_{22}, \ldots, X_{2n_2}$ are random samples from normal populations with mean and variance (μ_1, σ_1^2) and (μ_2, σ_2^2). σ_1^2 and σ_2^2 are not known and are not assumed to be equal.

If the two unknown population variances cannot be presumed equal, the individual sample variances, s_1^2 and s_2^2, cannot be pooled and construction of the $(1-\alpha)100\%$ confidence interval is based on the random variable

$$T = \frac{\bar{X}_1 - \bar{X}_2 - (\mu_1 - \mu_2)}{(S_1^2/n_1 + S_2^2/n_2)^{1/2}} \tag{6.27}$$

where T has an approximate t-distribution with f degrees of freedom where f is defined by

$$f = \frac{(s_1^2/n_1 + s_2^2/n_2)^2}{\frac{(s_1^2/n_1)^2}{n_1-1} + \frac{(s_2^2/n_2)^2}{n_2-1}} \qquad (6.28)$$

We have

$$P(-t_{f,1-\alpha/2} < T < t_{f,1-\alpha/2}) = 1 - \alpha \qquad (6.29)$$

where $t_{f,1-\alpha/2}$ is obtained from the t-table with degrees of freedom f given by (6.28). (If $\sigma_1^2 = \sigma_2^2 = \sigma^2$ recall that the d.f. were $n_1 + n_2 - 2$.) By substituting T defined in (6.27) into (6.29) we obtain the $(1-\alpha)100\%$ confidence limits

$$\bar{x}_1 - \bar{x}_2 \pm t_{f,1-\alpha/2}\sqrt{\frac{s_1^2}{n_1} + \frac{s_2^2}{n_2}} \; . \qquad (6.30)$$

The fact that T defined by Eq. (6.27) has an approximate t-distribution causes (6.30) to approximate a $(1-\alpha)100\%$ confidence interval.

We postpone an example for Case 3 until Chapter 7 where we look at the analogous problem concerning the hypothesis that $\mu_1 = \mu_2$.

6.10 ESTIMATION INVOLVING PAIRED OBSERVATIONS

In some studies involving data which consist of two samples one cannot make the assumption that the two samples are independent of each other, as was done in Section 6.5 This happens when, by the very nature of the study, the observations occur in pairs. For example, in studying the weather resistance of two types of paints, each of 5 shingles is painted with both paints, one-half with one paint and the other half with the other paint. An observation from one shingle is an ordered pair. Another example involves the question of whether two methods of measuring the strength of materials are numerically equivalent (except for statistical imprecision). Each specimen is scored by each method, giving rise to a pair of values.

Suppose the samples $X_1, X_2, \ldots, X_n$ and $Y_1, Y_2, \ldots, Y_n$ occur in a way that results in the natural pairings (X_1,Y_1), $(X_2,Y_2), \ldots, (X_n,Y_n)$ where X_i and Y_i are not independent, but (X_i,Y_i) is independent of (X_j,Y_j) for each $i \neq j$. Assume that X and Y have normal distributions with means μ_X and μ_Y, respectively. The fundamental problem here is that of obtaining an inference concerning $\mu_X - \mu_Y$, e.g., can we say that μ_X and μ_Y are different? The appropriate procedure is to calculate the difference $D_i = X_i - Y_i$, $i = 1,2,\ldots,n$ and observe that $\mu_D = \mu_X - \mu_Y$. We have thus reduced a two-sample problem to a one-sample problem which is exactly like the one in Section 6.4. To construct a confidence interval on $\mu_X - \mu_Y$ we need only to construct one on μ_D. Let σ_D^2 denote the population variance of $D = X - Y$. According to the confidence interval (6.15) in Section 6.4, a $(1-\alpha)100\%$ confidence interval on $\mu_D = \mu_X - \mu_Y$ is given by

$$\bar{d} - t_{n-1,1-\alpha/2} \frac{s_D}{\sqrt{n}} < \mu_D < \bar{d} + t_{n-1,1-\alpha/2} \frac{s_D}{\sqrt{n}} \tag{6.31}$$

where

$$\bar{d} = \sum_{i=1}^{n} d_i/n = \sum_{i=1}^{n} (x_i - y_i)/n$$

$$= \bar{x} - \bar{y} \tag{6.32}$$

$$s_D^2 = \sum_{i=1}^{n} (d_i - \bar{d})^2/(n-1) \quad , \tag{6.33}$$

and $t_{\nu,1-\alpha/2}$ is the usual tabular t-value using n-1 degrees of freedom.

Example 6.6 In studying weather resistance of two types of paint, each of 5 shingles was painted with both paints, one-half of the shingle with one of the paints and the other half of the shingle with the other paint. After a month's exposure, wear, measured in suitable units, was found to be (1.57, 1.45), (1.46, 1.59), (1.53, 1.27), (1.30, 1.48), (1.37, 1.40). Is it plausible that μ_D is greater than zero? less than zero? Between what two limits may μ_D reasonably be expected to lie?

We have

$$d_i = x_i - y_i \ ,$$

$$d_1 = +\ 0.12 \ ,$$

$$d_2 = -\ 0.13 \ ,$$

$$d_3 = +\ 0.26 \ ,$$

$$d_4 = -\ 0.18 \ ,$$

$$d_5 = -\ 0.03 \ ,$$

and

$$\bar{d} = 0.008 \ .$$

By (6.33) we have

$$s_D = 0.1815764$$

and

$$s_{\bar{D}} = s_D/\sqrt{n} = 0.0812034 \ .$$

Using n-1 = 4 d.f., we have $t_{4,0.975} = 2.776$ and the 95% confidence interval is

$$\bar{d} - t_{4,0.975}\, s_{\bar{D}} \le \mu_D \le \bar{d} + t_{4,0.975}\, s_{\bar{D}}$$

$$-\ 0.2174 \le \mu_D \le 0.2334 \ .$$

From the 95% confidence interval on μ_D, we come to the conclusion that μ_D is "very nearly" 0. If we had conjectured $\mu_D = 0$ prior to collecting the data then the resulting confidence interval does not seem to contradict that conjecture. It should be noted that by using both types of paint on each shingle we have eliminated any possible errors due to differences in the individual shingles.

Example 6.7 The question is whether two methods of measuring the strength of materials are numerically equivalent, except for statistical imprecision (i.e., we recognize that the two techniques cannot be expected to yield identical results even for identical specimens, but we are wondering whether "on the average" nearly identical strength values tend to be scored for nearly identical specimens). Fifteen matched test specimens have been scored by the two methods, with the following results: (338, 327), (156, 232), (243, 248), (267, 246), (195, 192), (203, 222), (262, 261), (225, 223), (214, 216), (292, 285), (218, 230), (151, 142), (168, 181), (223, 234), (260, 236). One of the above thirty measurements is in error. Attempt to locate this measurement on sight. Delete the maverick pair and, on the basis of the remaining 14 d's, compute $\bar{d}$ and $s_{\bar{D}}$. What about the question of the equivalence of the two methods?

Pair No. 2 is obviously the maverick pair. If we throw out the maverick pair, we find $t_{13,0.975} = 2.160$ from Table IV. Also we find $\bar{d} = 1.14285$,

$$s_D = 12.672545, \quad s_{\bar{D}} = 3.38688$$

so

$$\bar{d} - t_{n-1,1-\alpha/2}\, s_{\bar{D}} \le \mu_D \le \bar{d} + t_{n-1,1-\alpha/2}\, s_{\bar{D}}$$

$$-6.173 \le \mu_D \le 8.458 .$$

Note that the degrees of freedom used for s_D and t are 13 now that the maverick pair has been discarded. The resulting confidence interval on μ_D appears to indicate there is no reason to doubt that the two methods are numerically equivalent.

6.11 THE VARIANCE

The next thing to be considered is the variability of the data that are obtained and used. In other words, how far do the data scatter about a central point? If the scatter is too large, a given observation is less reliable than for the case where the data points are very close together. A measure of the scatter or variability of data is the variance. Statistical techniques which are derived concerning interval estimation of the variance are based on the χ^2-distribution and the F-distribution. More specifically, interval estimation on a single variance σ^2 is based on the χ^2-distribution and estimation concerning two variances is based on the F-distribution. The next two sections consider these two ideas.

6.12 ESTIMATION OF A VARIANCE

Let $X_1, X_2, \ldots, X_n$ denote a random sample from a normal population with mean μ and variance σ^2. According to Section 6.5 we know that the random variable

$$V = (n-1)\frac{S^2}{\sigma^2} = \sum_{i=1}^{n} (X_i - \overline{X})^2/\sigma^2 , \qquad (6.34)$$

where S^2 is the sample variance of $X_1, X_2, \ldots, X_n$ has a χ^2 distribution with (n-1) degrees of freedom. Hence

$$P(V < \chi^2_{n-1,\alpha/2}) = \alpha/2$$

and

$$P(V > \chi^2_{n-1,1-\alpha/2}) = \alpha/2 .$$

Thus

$$P(\chi^2_{n-1,\alpha/2} < V < \chi^2_{n-1,1-\alpha/2}) = 1 - \alpha . \qquad (6.35)$$

Replace V in (6.35) according to (6.34) to obtain

$$P(\chi^2_{n-1,\alpha/2} < (n-1)S^2/\sigma^2 < \chi^2_{n-1,1-\alpha/2}) = 1 - \alpha . \qquad (6.36)$$

Now

$$\chi^2_{n-1,\alpha/2} < (n-1)S^2/\sigma^2 < \chi^2_{n-1,1-\alpha/2}$$

is equivalent to $\chi^2_{n-1,\alpha/2} < (n-1)S^2/\sigma^2$ and $(n-1)S^2/\sigma^2 < \chi^2_{n-1,1-\alpha/2}$ or to $\sigma^2 < (n-1)(S^2/\chi^2_{n-1,\alpha/2})$ and $(n-1)(S^2/\chi^2_{n-1,1-\alpha/2}) < \sigma^2$. Therefore, (6.36) may be written as

$$P\left(\frac{(n-1)S^2}{\chi^2_{n-1,1-\alpha/2}} < \sigma^2 < \frac{(n-1)S^2}{\chi^2_{n-1,\alpha/2}}\right) = 1 - \alpha. \qquad (6.37)$$

Equation (6.37) yields a (1-α)100% confidence interval for σ^2 which is

$$\frac{(n-1)s^2}{\chi^2_{n-1,1-\alpha/2}} \le \sigma^2 \le \frac{(n-1)s^2}{\chi^2_{n-1,\alpha/2}} . \qquad (6.38)$$

Example 6.8 Five similar determinations of the cold water flow rate to a heat exchanger were, in GPM, 5.84, 5.76, 6.03, 5.90, and 5.87. Compute a 95% confidence interval for the imprecision (on a per measurement basis) affecting this measuring operation.

We have

$$\bar{x} = \sum_{i=1}^{n} x_i/n = 5.88$$

and

$$s_X^2 = (\Sigma x_i^2 - n\bar{x}^2)/(n-1) = 0.00975 \quad .$$

Since $n = 5$, $\nu = n-1 = 4$. The confidence interval is obtained from Eq. (6.38). The values of $\chi^2_{4,0.025}$ and $\chi^2_{4,0.975}$ are 0.484 and 11.1, respectively, according to Table V. Therefore, the calculated 95% confidence interval on σ^2 is

$$\frac{4(0.00975)}{11.1} \leq \sigma^2 \leq \frac{4(0.00975)}{0.484}$$

or

$$0.00351 \leq \sigma^2 \leq 0.08057 \quad .$$

If the mean μ is known then $\bar{x}$ is replaced by μ in (6.34) to yield

$$V = \sum_{i=1}^{n} (x_i - \mu)^2/\sigma^2$$

where V has a χ^2-distribution with n d.f. Thus, in this case the confidence interval in (6.36), on replacing $(n-1)s^2$ with $\sum_{i=1}^{n} (x_i - \mu)^2$, becomes

$$\sum_{i=1}^{n} \frac{(x_i - \mu)^2}{\chi^2_{n,1-\alpha/2}} \leq \sigma^2 \leq \sum_{i=1}^{n} \frac{(x_i - \mu)^2}{\chi^2_{n,\alpha/2}} \tag{6.39}$$

where $\chi^2_{n,1-\alpha/2}$ and $\chi^2_{n,\alpha/2}$ are tabular values corresponding to n d.f. The result (6.39) is not very useful, however, since μ is generally unknown.

6.13 COMPARISON OF TWO VARIANCES

The F-distribution is used to compare the variances of two populations. Suppose we let S_1^2 and S_2^2 denote the sample variances corresponding to the independent samples $X_{11}, X_{12}, \ldots, X_{1n_1}$ and

$X_{21}, X_{22}, \ldots, X_{2n_2}$, respectively, from two normal populations with means μ_1 and μ_2 and variances σ_1^2 and σ_2^2. By Section 6.7 the random variable

$$F = \frac{S_1^2/\sigma_1^2}{S_2^2/\sigma_2^2} = \frac{\chi_1^2/(n_1-1)}{\chi_2^2/(n_2-1)} \tag{6.40}$$

where

$$(n_1 - 1)S_1^2 = \sum_{i=1}^{n_1} (X_{1i} - \overline{X}_1)^2 = \chi_1^2 \sigma_1^2$$

and

$$(n_2 - 1)S_2^2 = \sum_{i=1}^{n_2} (X_{2i} - \overline{X}_2)^2 = \chi_2^2 \sigma_2^2.$$

has an F-distribution with $\nu_1 = (n_1 - 1)$ and $\nu_2 = (n_2 - 1)$ d.f., respectively. Recall that $(n_1 - 1)$ is the d.f. corresponding to the numerator in (6.40). Let $F_{\nu_1,\nu_2,\alpha/2}$ and $F_{\nu_1,\nu_2,1-\alpha/2}$ denote the tabular F-values with $\alpha/2$ and $1-\alpha/2$ area to the left, respectively. That is,

$$P(F_{\nu_1,\nu_2,\alpha/2} < F < F_{\nu_1,\nu_2,1-\alpha/2}) = 1 - \alpha. \tag{6.41}$$

Replacing F in (6.41) according to (6.40) gives

$$P\left(F_{\nu_1,\nu_2,\alpha/2} < \frac{\sigma_2^2}{\sigma_1^2} \cdot \frac{S_1^2}{S_2^2} < F_{\nu_1,\nu_2,1-\alpha/2}\right) = 1 - \alpha$$

or

$$P\left(\frac{S_2^2}{S_1^2} F_{\nu_1,\nu_2,\alpha/2} < \frac{\sigma_2^2}{\sigma_1^2} < \frac{S_2^2}{S_1^2} F_{\nu_1,\nu_2,1-\alpha/2}\right) = 1 - \alpha.$$

Therefore a $(1-\alpha)100\%$ confidence interval for σ_2^2/σ_1^2 is given by

$$\frac{s_2^2}{s_1^2} F_{n_1-1,n_2-1,\alpha/2} < \frac{\sigma_2^2}{\sigma_1^2} < \frac{s_2^2}{s_1^2} F_{n_1-1,n_2-1,1-\alpha/2}. \qquad (6.42)$$

Similarly, a $(1-\alpha)100\%$ confidence interval for σ_1^2/σ_2^2 is given by

$$\frac{s_1^2}{s_2^2} F_{n_2-1,n_1-1,\alpha/2} < \frac{\sigma_1^2}{\sigma_2^2} < \frac{s_1^2}{s_2^2} F_{n_2-1,n_1-1,1-\alpha/2}. \qquad (6.43)$$

Confidence intervals corresponding to (6.42) and (6.43) could be derived in the case where μ_1 or μ_2 (or both) are known; however, the result would not be very useful so we shall not pursue it.

Example 6.9 Reaction temperatures in degrees centigrade (measured on two different days) for two catalyst concentrations were

x_1	x_2
310.95	308.94
308.86	308.23
312.80	309.98
309.74	311.59
311.03	309.46
311.89	311.15
310.93	311.29
310.39	309.16
310.24	310.68
311.89	311.86
309.65	310.98
311.85	312.29
310.73	311.21

Find a 98% confidence interval for σ_1/σ_2.

We have

$$\bar{x}_1 = 310.8423^\circ C \qquad \bar{x}_2 = 310.5246^\circ C$$

$$s_1^2 = 1.1867 \qquad s_2^2 = 1.5757 \quad .$$

Now, since the d.f. are $n_1 - 1 = 12$ and $n_2 - 1 = 12$ we have, from Table VI, $F_{12,12,0.01} = 0.241$ and $F_{12,12,0.99} = 4.16$. Thus a 98% confidence interval on σ_1^2/σ_2^2 is (from Eq. (6.43)):

$$\frac{1.1867}{1.5757}\,(0.241) \le \frac{\sigma_1^2}{\sigma_2^2} \le \frac{1.1867}{1.5757}\,(4.16)$$

or

$$0.18104 \le \frac{\sigma_1^2}{\sigma_2^2} \le 3.13300 \quad .$$

The corresponding 98% confidence interval on σ_1/σ_2 is

$$0.42548 \le \sigma_1/\sigma_2 \le 1.7700 \quad .$$

Note that since the confidence interval contains the value 1 we cannot "justifiably" say that $\sigma_1/\sigma_2 \ne 1$ or that $\sigma_1 \ne \sigma_2$.

6.14 ESTIMATION OF A PROPORTION P

Consider a population that contains two types of objects. This type of population was discussed in Section 3.3.1. Each of the objects can be represented by one of the values, 0 or 1. Such a population is labeled a binomial population and we associate success and failure with the values 1 and 0. The probability function corresponding to the population is

$$P(X=x) = f(x) = P^x(1-P)^{1-x}, \quad x = 0,1;\ 0 < P < 1.$$

The parameter P is the proportion of 1's in the population. The only values X can take on are 0 and 1.

If a random sample $X_1, X_2, \ldots, X_n$ is chosen from the population then $S = X_1 + X_2 + \ldots + X_n$ has the binomial probability function

$$P(S=x) = b(x) = \binom{n}{x} P^x (1-P)^{n-x}, \quad x = 0,1,2,\ldots,n. \tag{6.44}$$

The random variable S is actually the number of successes (1's) in the sample. In this section we consider the problem of estimating the parameter P, that is, the probability of obtaining a success in one draw from the population.

It was shown in Example 5.4, Section 5.6, that

$$E(S) = E\left(\sum_{i=1}^{n} X_i\right) = nP$$

and

$$\sigma_S^2 = nP(1-P) = nPQ \quad .$$

The best estimator for P is given by

$$\hat{P} = S/n = \sum_{i=1}^{n} X_i/n = \overline{X},$$

that is, the sample mean. We have

$$E(\hat{P}) = E(S/n) = \frac{1}{n} E(S) = P$$

and

$$\sigma_{\hat{P}}^2 = \sigma_{\overline{X}}^2 = \frac{\sigma_X^2}{n} = \frac{PQ}{n} \quad . \tag{6.45}$$

Thus no other estimator of P has smaller variance than PQ/n.

We turn now to the problem of constructing an interval estimate for P. To find confidence intervals for P one must solve each of the equations

$$\sum_{x=s}^{n} \binom{n}{x} P^x (1-P)^{n-x} = \frac{\alpha}{2} \qquad (6.46)$$

and

$$\sum_{x=0}^{s} \binom{n}{x} P^x (1-P)^{n-x} = \frac{\alpha}{2} \qquad (6.47)$$

for P, where s is the number of successes in the sample. The solution, say P_1, of (6.46) is the lower confidence limit and the solution, say P_2, of (6.47) is the upper confidence limit. To solve (6.46) and (6.47) one must resort to extensive binomial tables which cover a wide range of values for P.

Example 6.10 Let P denote the unknown proportion of batches of chemical which are on-grade. Suppose 100 batches of chemical are obtained and it is determined that 90 are on-grade. The best estimate of P is $\hat{P} = 90/100 = 0.9$. To find a 95% confidence interval for P it is necessary to solve (6.46) and (6.47) for P with s = 90, n = 100, and $\alpha = 0.05$. The solution of (6.46) is the lower confidence limit and the solution of (6.47) is the upper confidence limit. This example illustrates that determining exact confidence limits is a very tedious chore. The exact confidence limits are not given for this example but will be determined approximately later by a very popular technique which will now be discussed.

According to the central limit theorem the random variable $\bar{X}$ has an approximate normal distribution with mean μ and variance σ^2/n even if the distribution of X is discrete. Thus the random variable S/n has an approximate normal distribution with mean P and variance PQ/n. Therefore,

$$Z = \frac{S/n - P}{\sqrt{\frac{PQ}{n}}} = \frac{\hat{P} - P}{\sigma_{\hat{P}}} \tag{6.48}$$

has an approximate standard normal distribution. We modify the random variable (6.48) further by replacing PQ in the denominator with $\hat{P}(1-\hat{P})$. Thus we have

$$Z = \frac{\hat{P} - P}{\sqrt{\hat{P}\hat{Q}/n}} = \frac{\hat{P} - P}{S_{\hat{P}}}$$

Proceeding in the usual fashion we set

$$P(z_{\alpha/2} < Z < z_{1-\alpha/2}) = 1 - \alpha$$

which is the same as

$$P(-z_{1-\alpha/2} < Z < z_{1-\alpha/2}) = 1 - \alpha$$

and can be directly converted to

$$P\left(-z_{1-\alpha/2} < \frac{\hat{P} - P}{S_{\hat{P}}} < z_{1-\alpha/2}\right) = 1 - \alpha \quad . \tag{6.49}$$

This last equation is of exactly the same form as some previously obtained; e.g., see Eqs. (6.11) and (6.18). Equation (6.49) can be written as

$$P\left(\hat{P} - z_{1-\alpha/2}\sqrt{\frac{\hat{P}(1-\hat{P})}{n}} < P < \hat{P} + z_{1-\alpha/2}\sqrt{\frac{\hat{P}(1-\hat{P})}{n}}\right) = 1 - \alpha,$$

from which we obtain an approximate $(1-\alpha)100\%$ confidence interval for P:

$$\hat{P} - z_{1-\alpha/2}\sqrt{\frac{\hat{P}(1-\hat{P})}{n}} < P < \hat{P} + z_{1-\alpha/2}\sqrt{\frac{\hat{P}(1-\hat{P})}{n}} \quad . \tag{6.50}$$

Example 6.11 In Example 6.10 we had a sample of 100 batches of chemical of which 90 were on-grade. Find a 95% confidence interval for P, the true unknown probability that a batch is on-grade.

The best estimate of $\hat{P}$ is 0.90. Furthermore, n = 100 and $z_{0.975} = -z_{0.025} = 1.96$. Thus a 95% confidence interval for P is

$$0.90 - 1.96\sqrt{\frac{0.9(0.1)}{100}} < P < 0.90 + 1.96\sqrt{\frac{0.9(0.1)}{100}} \quad ,$$

$$0.90 - 1.96(0.03) < P < 0.90 + 1.96(0.03) \quad ,$$

$$0.90 - 0.0588 < P < 0.90 + 0.588 \quad ,$$

$$0.841 < P < 0.959 \quad .$$

Since the interval contains the value 0.90 we cannot safely say that $P \neq 0.90$. (The exact confidence limits in this example are 0.82 and 0.95.)

Example 6.12 Each compression ring in a population (lot) of compression rings either fits properly or it does not. Let P denote the true proportion of those rings that fit properly. In a sample of 400 rings 10% were observed to be faulty, that is, 10% did not seat properly. Find the 99% confidence interval for the proportion faulty in the lot from which the sample was taken.

The best estimate of P is $\hat{P} = 0.10$. Since n = 400 and $z_{0.005} = -2.58$ the 99% confidence limits for P are, by (6.50),

$$0.10 \pm 2.58\sqrt{\frac{0.1(0.9)}{400}} \quad ,$$

which simplifies to

$$0.10 \pm 2.58\left(\frac{0.3}{20}\right)$$

and reduces to

$$(0.061,\ 0.139) \quad .$$

6.15 COMPARISON OF TWO PROPORTIONS

Consider the problem of comparing the parameters P_1 and P_2 of two binomial populations. Let $X_1, X_2, \ldots, X_{n_1}$ and $Y_1, Y_2, \ldots, Y_{n_2}$ denote random samples from the two binomial populations with parameters P_1 and P_2, respectively. The best estimates of P_1 and P_2 are

$$\hat{P}_1 = \sum_{i=1}^{n_1} X_i/n_1 \text{ and } \hat{P}_2 = \sum_{i=1}^{n_2} Y_i/n_2 \quad .$$

Furthermore,

$$S^2_{\hat{P}_1} = \frac{\hat{P}_1(1-\hat{P}_1)}{n_1} \text{ and } S^2_{\hat{P}_2} = \frac{\hat{P}_2(1-\hat{P}_2)}{n_2} \quad .$$

To compare P_1 and P_2 we consider the difference $P_1 - P_2$. The best estimate of $P_1 - P_2$ is $\hat{P}_1 - \hat{P}_2$. Since $\hat{P}_1$ and $\hat{P}_2$ are independent, an estimate of the variance of $\hat{P}_1 - \hat{P}_2$ is

$$S^2_{\hat{P}_1-\hat{P}_2} = S^2_{\hat{P}_1} + S^2_{\hat{P}_2} = \frac{\hat{P}_1(1-\hat{P}_1)}{n_1} + \frac{\hat{P}_2(1-\hat{P}_2)}{n_2} \quad .$$

The random variables $\hat{P}_1$ and $\hat{P}_2$ are each approximately normally distributed, and consequently $\hat{P}_1 - \hat{P}_2$ has an approximate normal distribution with mean $P_1 - P_2$ and variance

$$\sigma^2_{\hat{P}_1-\hat{P}_2} = \frac{P_1(1-P_1)}{n_1} + \frac{P_2(1-P_2)}{n_2} \quad .$$

Thus

$$\frac{\hat{P}_1 - \hat{P}_2 - (P_1-P_2)}{\sqrt{\dfrac{P_1(1-P_1)}{n_1} + \dfrac{P_2(1-P_2)}{n_2}}} \tag{6.51}$$

has approximately the standard normal distribution. Replacing P_1 and P_2 by $\hat{P}_1$ and $\hat{P}_2$ in the denominator of (6.51) it follows that

$$Z = \frac{\hat{P}_1 - \hat{P}_2 - (P_1 - P_2)}{\sqrt{\dfrac{\hat{P}_1(1-\hat{P}_1)}{n_1} + \dfrac{\hat{P}_2(1-\hat{P}_2)}{n_2}}}$$

has approximately the standard normal distribution. Thus

$$P(-z_{1-\alpha/2} < Z < z_{1-\alpha/2}) = 1 - \alpha$$

from which we obtain the $(1-\alpha)100\%$ confidence limits for $P_1 - P_2$:

$$\hat{P}_1 - \hat{P}_2 \pm z_{1-\alpha/2}\sqrt{\frac{\hat{P}_1(1-\hat{P}_1)}{n_1} + \frac{\hat{P}_2(1-\hat{P}_2)}{n_2}} \tag{6.52}$$

Example 6.13 Two different methods (say 1 and 2) of manufacturing compression rings are being used. One hundred rings were sampled for each method. It was found that 10 rings using method 1 and 3 rings from method 2 did not seat properly. Construct a 99% confidence interval for $P_1 - P_2$.

The best estimate for $P_1 - P_2$ is $\hat{P}_1 - \hat{P}_2 = 0.10 - 0.03 = 0.07$. We have $n_1 = 100 = n_2$ and $z_{0.995} = 2.58$. Thus

$$s^2_{\hat{P}_1-\hat{P}_2} = s^2_{\hat{P}_1} + s^2_{\hat{P}_2} = \frac{0.10(0.90)}{100} + \frac{0.03(0.97)}{100}$$

$$= 0.001191$$

and

$$s_{\hat{P}_1-\hat{P}_2} = \sqrt{0.001191} = 0.0345.$$

The 99% confidence limits are, by (6.44),

$$\hat{P}_1 - \hat{P}_2 \pm z_{1-\alpha/2}\, s_{\hat{P}_1-\hat{P}_2} ,$$

$$0.07 \pm 2.58(0.0345) ,$$

$$0.07 \pm 0.080 ,$$

$$-0.01,\ 0.15 .$$

Since the interval contains 0 we cannot safely infer that $P_1 - P_2 \neq 0$ or that $P_1 \neq P_2$.

PROBLEMS

6.1 Data taken from the plate and frame filter press located in the unit operations laboratory are used to determine α, the specific cake resistance, of a calcium carbonate slurry. Several values of α, expressed in ft/lb, have been calculated from data taken during the fall semester. Based on these values, predict the interval within which 90% of all such values calculated in the future must fall.

Specific cake resistance - ft/lb	
2.49×10^{11}	2.67×10^{11}
2.40×10^{11}	2.60×10^{11}
2.43×10^{11}	2.50×10^{11}
2.30×10^{11}	2.54×10^{11}
2.53×10^{11}	2.55×10^{11}

6.2 The endurance limit of a material is a function of its surface roughness. If this department asked you to design a new elevator, specifying an endurance limit for a bearing material of 105,000 psi, what specification for surface roughness would you submit to the contractor? The firm limits surface finish of all bearings to nine microinches roughness as an economic factor. The following data are available for 12 different bearings .

Surface roughness, microinches	Endurance limit, psi
18	104,800
16	107,000
6	115,000
17	106,000
15	107,300
12	109,000
7	114,000
14	108,000
13	108,900
8	112,500
11	110,000
20	104,000

6.3 Due to the burning of cotton plant wastes (hulls, leaves, etc.), sulfate content in the air over Lubbock is highest during the month of November. If the data given below are the mean values of sulfate content during the month of November (analyses of air performed daily) over the past 10 years, what value for the sulfate content in Lubbock air can the Lubbock City-County Department of Health predict at a 95% confidence level during next November?

Sulfate content, μg/m^3 of air									
10.83	8.90	14.71	12.35	11.86	13.80	11.75	9.68	9.33	10.9

6.4 Twenty companies have submitted bids for a batch reactor which the Department of Chemical Engineering is planning to install in the unit operations lab. As part of the financial review, you are asked to submit a price range based on these bids within which the price of a suitable reactor will fall. The reactor must have a minimum capacity of 8 gal and a maximum cost of $550. Your answer should be based on the 95% confidence interval of reactor cost.

Company	Capacity, gal	Cost
A	7.5	$425
B	8.0	500
C	9.0	525
D	10.0	550
E	6.0	450
F	16.5	675
G	12.0	575
H	7.0	475
I	8.5	510
J	8.75	525
K	8.25	510
L	9.25	540
M	10.25	625
N	5.75	350
O	6.00	375
P	9.00	500
Q	8.50	525
R	8.00	450
S	9.25	550
T	10.00	600

6.5 The following values have been obtained for the critical moisture content of sawdust. Based on this data, your lab partner submits a confidence interval of $1.0779 < \mu < 1.1194$. What must be the value of z to obtain such an interval? What percent confidence interval is represented?

Critical moisture content of sawdust, lbs moisture/lb bone-dry solid
1.1000, 1.0500, 1.0800, 1.1200, 1.0900, 1.1300, 1.1100, 1.0700

6.6 The Gulp-a-Cup Coffee Company utilizes spray drying in their coffee production process. In the past, Gulp-a-Cup has utilized banks of single atomizer nozzles with external mixing of the gas and liquid phases in all their drying chambers. Two-fluid nozzles employing internal mixing were recently installed in chamber #12 of the plant and trial runs were made to determine the optimum drying conditions. Data given below are for entrance gas pressure. Above 52 psig the coffee particle size was too fine. Below 48 psig the coffee was not dried sufficiently. What new value for gas pressure will Gulp-a-Cup furnish their operators for chamber #12? Use the 99% confidence limit.

Entrance gas pressure, psig									
52.00	51.00	51.80	51.75	51.30	50.85	50.25	49.00	48.65	48.00

6.7 If the interval estimate for the optimum reference gas flow rate in ml/min during a series of chromatographic determinations of ketones in a sample mixture was

$$11.36 < \mu < 14.14 ,$$

what was the confidence level of the estimate? Data are given below:

Reference gas flow, ml/min					
10.0	10.5	12.0	15.0	13.0	16.0

6.8 The lab instructor of ChE 3111 made a binary liquid unknown for his students and told them to determine the weight percent of the more volatile component. At the end of the experiment, all the students became upset when the instructor informed them that all their results were outside the limits of analysis:

$$\text{Conc.} = 91\% \pm 0.1\%$$

Any concentration outside the above range made an "F". Values inside the range received grades of 100. If the instructor's conclusions were based on the following data, what value of α corresponds to a grade of 100? Was the instructor fair?

% more volatile component	
91.0	91.0
91.2	90.8
91.1	90.9
91.5	90.5

6.9 In a process for manufacture of ester gum (chemically glyceryl abietate), rosin is reacted with glycerine, and water is taken off as the reaction proceeds. The mean color of the product using the standard source of rosin is 65. A new source of rosin is being considered on the basis of its giving an improved (lower) color product. Test runs were made using rosin from this new source with the following color results: 55, 62, 54, 57, 65, 64, 60, 63, 58, 67, 63, 61. Each of these values may be considered independent. Find a 95% confidence interval for color of the product made from the new rosin source.

6.10 A random sample of 100 students who smoke cigarettes on this campus showed that the average student smokes 17.8 cigarettes per day. The variance in cigarettes smoked per day for any individual was 480.

(a) What is the 95% confidence limit for the actual average number of cigarettes smoked per day by these students?
(b) Discuss the application of these statistics to the population of university smokers.

6.11 A company engaged in the manufacture of cast iron has employed a system of raw material and processing procedures which has produced a product whose overall population average silicon content was 0.85%. A new contract was put into effect in which a new supplier of raw material supplanted the old one. During the first month of operation using the new material, random samples of the product silicon content were found to be

1.13	0.87
0.80	0.92
0.85	0.81
0.60	0.97
0.97	0.48
0.92	1.00
0.94	0.92
0.72	0.61
1.17	0.81
0.87	0.71
0.36	0.97
0.68	0.89
0.73	1.16
0.82	0.68
0.79	1.00

$\Sigma x = 25.15$ $\Sigma x^2 = 22.0680$

What are your 99% confidence limits on the silicon content of the iron using the new raw material?

6.12 Five analyses of the methane content of a natural gas showed: 92.4, 92.8, 92.3, 93.0, 92.5%. What is the 95% CI for the true methane content? Can it be safely said that the true methane content averages at least 91.8%? (Evaluate this part by the appropriate t-test.) For these data, $S_X^2 = 0.08526$.

6.13 Use the data in problem 8.32 to calculate the 95% confidence limit for the flow rate of each of the six elutriators.

6.14 Using the data of Problem 3.16 and the cumulative relative frequency polygon you obtained, what is the 95% confidence interval for the cotton dust concentration? Obtain this value by calculation using any computerized method and compare the results. Are they different? If so, why?

6.15 The following cotton dust concentrations in μg/m^3 were measured in accordance with the provisions of the OSHA standard for two shifts in the spinning room of a cotton mill. Is there any significant difference (α = 0.05) in the average concentration for the two shifts?

C	D
96.7	58.7
149.7	75.9
107.6	101.3
73.5	117.8
213.1	141.9
195.5	143.3

6.16 The hydrogen gas content available from two sources of supply have been analyzed with the results below.

$$\bar{X}_1 = 69.1 \qquad \bar{X}_2 = 66.2$$
$$n_1 = 7 \qquad n_2 = 5$$
$$S^2_{X_1} = 26.1 \qquad S^2_{X_2} = 11$$

What is the 95% CI for the difference in hydrogen content for these sources?

6.17 An experiment conducted to compare the tensile strengths of the types of synthetic fibers gave the breaking loads shown below in thousands of pounds force per square inch.

Fiber A:	14	4	10	6	3	11	12		
Fiber B:	16	17	13	12	7	16	11	8	7

Calculate the 99% confidence interval for the difference between the means.

6.18 Two gaskets were cut from each of eight different sheet stocks. One gasket of each pair was randomly selected for use in dilute HCl service. The other gasket of each pair was for concentrated HCl service. All gaskets were subjected to accelerated life tests for their respective intended uses. From the data so obtained, the estimated average service life in weeks are given below:

Material	Dilute HCl	Concentrated HCl
1	35	30
2	40	32
3	27	28
4	25	27
5	36	33
6	48	38
7	53	41
8	48	39

(a) Calculate the mean difference in expected service life in the two environments and a 95% confidence interval for this mean difference.

(b) Explain the advantages obtained from pairing the test samples in the manner described.

6.19 Seven samples of a catalyst have been analyzed for carbon content by two technicians, i.e. each technician made one analysis of each sample. The results were:

Sample	$X_1 - X_2 = D$
1	-0.5
2	-0.5
3	-0.2
4	0
5	-0.9
6	-0.1
7	-0.6

Is there any difference between the two technicians? Make whatever comments are appropriate regarding whether one technician is better than the other.

6.20 Using the data in problem 6.17 calculate a 95% confidence interval for the ratio of variances σ_A^2/σ_B^2.

6.21 The following data give the yields of a product that resulted from trying catalysts from two different suppliers in a process.

Catalyst I: 36, 33, 35, 34, 32, 34
Catalyst II: 35, 39, 37, 38, 39, 38, 40

Calculate (a) the 90% confidence interval for σ_I^2/σ_{II}^2 and (b) the 95% confidence interval. What do you infer from the differences in the estimated ranges of the ratio?

6.22 A sulfuric acid plant produces acid with a long term mean concentration of 60%. What is the maximum value of the standard error allowable to have a 99% confidence interval that the acid concentration is between 56 and 65%?

6.23 The weight of a particular medication capsule has a mean of 40 mg and a standard deviation of 2 mg. If the capsule weight distribution is NID, what is the probability that a capsule chosen at random will weigh between 39 and 40 mg?

6.24 A computer manufacturer is considering using one of the two types of components in their home computers. Ninety components of Type 1 are tested and five fail. One hundred of Type 2 are tested and seven fail. Based on this data, compute a 95% confidence interval for the difference $P_1 - P_2$, where P_1 and P_2 are the true proportions of failures for the Type 1 and Type 2 components, respectively.

6.25 In a sample of 140 batches of chemical A, 120 were on-grade. In a sample of 110 batches of chemical B, only 90 were on-grade. Find a 99% confidence interval for $P_A - P_B$ where P_A and P_B are the true proportions of A and B that are on-grade.

REFERENCE

1. Conover, W. J., Practical Nonparametric Statistics, Ch. 5, 2nd ed., John Wiley & Sons, Inc., New York, 1980.

7
STATISTICAL INFERENCE: HYPOTHESIS TESTING

7.1 INTRODUCTION

The second area of statistical inference is that of hypothesis testing. A statistical hypothesis is simply a statement concerning the probability distribution of a random variable. Before the hypothesis is formulated, it is necessary to choose a probability model for the population. There are two types of general hypotheses: simple and composite. A simple hypothesis is one which states that the data in question are represented by a distribution which is a specific member of a particular family of distributions. For example, the hypothesis that $\mu = 6$, $\sigma^2 = 1.7$ for a normal population uniquely specifies a particular normal distribution. If the value of the variance is unknown and we stated the hypothesis as $\mu = 6$, we would have a composite hypothesis. This is because the hypothesis states that the distribution involved can be any member of the family determined by the parameters $\mu = 6$ and $\sigma^2 > 0$.

Once the hypothesis has been stated, appropriate statistical procedures are used to determine whether it is an acceptable conjecture or an unacceptable one. In general we cannot prove that an hypothesis is absolutely true or false. If the information furnished by the data supports the hypothesis, then we do not reject it. If the data does not support the hypothesis, we reject it. The hypothesis being tested is referred to as the <u>null hypothesis</u> and will be denoted by H_0. Another hypothesis is generally stipulated which complements the null hypothesis. It will be referred to as the alternate hypothesis and is denoted by H_A. If we accept H_0, we automatically reject H_A and vice versa. Regardless of which we do, we are susceptible to the chance of having committed an error if the wrong choice is made. The possible errors will be discussed in Section 7.2.

7.2 TYPES OF ERRORS

Many statistical hypotheses are of the type that specify a value, or range of values, of one or more parameters of the distribution in question. Such hypotheses can be tested by using the sample characteristics obtained from the data. The values of these characteristics vary from sample to sample and thus they may by chance be quite different from those of the population. Since this is true, it is possible to make an error in accepting or rejecting any given hypothesis. A type 1 error occurs when a true hypothesis H_0 is rejected. The Greek letter α will denote the probability of rejecting H_0 when it is true. The term α is also referred to as the level of significance of the test. A type 2 error consists of accepting H_0 when it is false. The Greek letter β is used to represent the probability of making a type 2 error. Ideally we would prefer to use a test that minimizes both types of errors. Unfortunately, as α decreases, β tends to increase. The reverse is also true. The power of a test is defined as the probability of rejecting H_0 when it is false, that is, power = $1 - \beta$.

The choice of values for α and β is not always easy to make. No hard and fast rules can be made for the choice of α and β. This decision must be made for each problem as it arises based on economics and quality-control considerations.

Consider the routine inspection of thermostats for baby incubators. If from a suitably sized sample we conclude that the lot is satisfactory based on some predetermined significance level, say

$\alpha = 0.05$, we still run the risk of being wrong some of the time in accepting the thermostats. If, however, we reject the hypothesis that the lot is satisfactory at the 5% level, then the probability of producing an incubator with unacceptable temperature control has been virtually eliminated whether the thermostats are defective or not. If the thermostats are satisfactory and the lot has been rejected, then a type 1 error has occurred. Consider the alternative: The lot in question contains a significant number of defective thermostats. If we accept the lot under these circumstances, we will certainly produce faulty incubators unless this defect is caught later in the quality-control process. As the production of improperly regulatable incubators is unacceptable, the power of the test used should be quite high, that is, the hypothesis that the thermostats are acceptable should be rejected with a high probability if they are in fact faulty. We must, in other words, minimize the probability of committing a type 2 error.

Much more extreme examples are available from spacecraft design work where the reliability must be at least 0.9999. To achieve this, β must be almost vanishingly small so that no defective parts will be used.

7.3 TESTING OF HYPOTHESES

To test a hypothesis, sample data are collected and used to calculate a test statistic. Depending on the value of this statistic the null hypothesis H_0 is accepted or rejected. The critical region for H_0 is defined as the range of values of the test statistic that corresponds to a rejection of the hypothesis at some fixed probability of committing a type 1 error. The test statistic itself is determined by the specific probability distribution sampled and by the parameters selected for testing.

The general procedure used for testing a hypothesis is as follows:

1. Assume an appropriate probability model for the outcomes of the random experiment under investigation. This choice should be based on previous experience or intuition.
2. Formulate a null hypothesis and an alternate hypothesis. This must be carefully done in order to permit meaningful conclusions to be drawn from the test.
3. Specify the test statistic.
4. Choose a level of significance α for the test.
5. Determine the distribution of the test statistic and the critical region for the test statistic.
6. Calculate the value of the test statistic from a random sample of data.
7. Accept or reject H_0 by comparing the calculated value of the test statistic with the values in the critical region.

7.4 ONE-TAILED AND TWO-TAILED TESTS

The statistical tests we are going to consider will be either one-tailed or two-tailed (sometimes referred to as one-sided or two-sided), depending on the exact nature of the null hypothesis H_0 and the alternate hypothesis H_A.

Consider a test of the null hypothesis H_0: $\mu = \mu_0$ when sampling from a normal distribution with known variance σ^2, against H_A: $\mu \neq \mu_0$. We would expect our test statistic to be based on the sample mean $\overline{X}$, since it is the best estimator for μ. Also, intuitively, if $\overline{X}$ has a value quite different from the value μ_0 specified by the null hypothesis we would tend not to believe H_0. Hence, we use the test statistic

$$Z = \frac{(\overline{X} - \mu_0)}{\sigma/\sqrt{n}} ,$$

where n is the sample size. Since $\overline{X}$ has a normal distribution with mean μ_0 (if H_0 is true) and variance σ^2/n, Z has the standard normal distribution. Thus, whenever the value of Z falls far enough out under either tail of the normal distribution the hypothesis H_0: $\mu = \mu_0$ is rejected.

The distribution of Z when H_0 is true is the standard normal. This information can be used to find α, the probability of committing a type 1 error. Since we want α to be fixed, we must find the values $z_{\alpha/2}$ and $z_{1-\alpha/2}$ in the standard normal table such that

$$P(Z > z_{1-\alpha/2} \text{ or } Z < z_{\alpha/2}) = \alpha \quad .$$

Since the standard normal distribution is symmetric about zero, the "critical" values for rejecting H_0 can be chosen symmetrically about zero. Recall that

$$P(Z > z_{1-\alpha/2}) = \alpha/2 = P(Z < z_{\alpha/2}) \quad .$$

The two-sided rejection region (hatched) for testing H_0: $\mu = \mu_0$ against H_A: $\mu \neq \mu_0$ is illustrated in Fig. 7.1. Note that the inequality symbols in the alternate hypotheses point towards the rejection region(s) for this or any other null hypothesis.

Suppose, however, that we wanted to test H_0: $\mu \geqq \mu_0$ against the alternative H_A: $\mu < \mu_0$, again sampling from a normal distribution with known variance σ^2.

In this case, if $\overline{X}$ is larger than μ_0, we would tend to believe H_0 rather than H_A. Thus, we want to reject H_0 only when $\overline{X}$ is sufficiently smaller than μ_0. After normalizing, this corresponds to having $Z < z_\alpha$ where z_α is such that $P(Z < z_\alpha) = \alpha$ with α fixed. Thus, our rejection region is under the left tail of the standard normal distribution, and we have a one-tailed test. This critical region (hatched) is illustrated in Fig. 7.2.

Similarly, for H_0: $\mu \leqq \mu_0$ versus H_A: $\mu > \mu_0$ we would have a rejection region of size α under the right tail of the standard normal distribution.

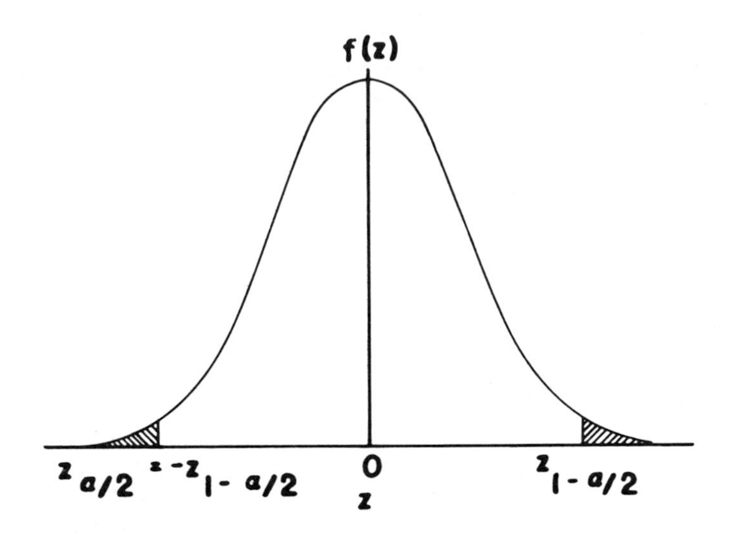

Fig. 7.1 The standard normal distribution showing rejection region for a two-tailed test.

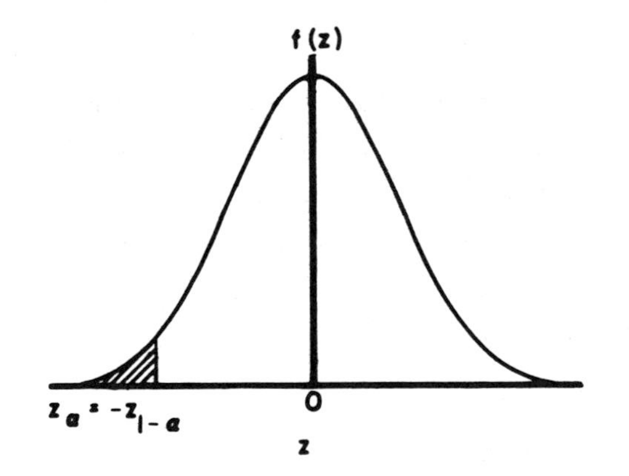

Fig. 7.2 The standard normal distribution showing rejection region for a one-tailed test

If we desire to test H_0: $\mu = \mu_0$ against either one of the three alternatives H_A: $\mu \neq \mu_0$, H_A: $\mu > \mu_0$, or H_A: $\mu < \mu_0$, when sampling from a normal distribution with unknown variance σ^2 we can no longer use the statistic $Z = (\bar{X} - \mu_0)/(\sigma/\sqrt{n})$. The reason for this is that we could not carry out the test since we would not be able to evaluate $Z = (\bar{X} - \mu_0)/(\sigma/\sqrt{n})$ given the sample values, since σ would be unknown. If the variance is unknown, as will be seen later, Z is replaced with $T = (\bar{X} - \mu_0)/(S/\sqrt{n})$ where S is the sample standard deviation. The distribution of T is the t-distribution with (n - 1) degrees of freedom.

Example 7.1 Suppose X has a normal distribution with mean μ and variance $\sigma^2 = 4$. Consider the simple null hypothesis H_0: $\mu = 1$ against the simple alternative H_A: $\mu = 3$. If a sample of size 1 is chosen from the population, find the critical region for testing the hypothesis at the $\alpha = 0.05$ level of significance and find the power of the test.

The choice of sample size 1 is not realistic, however, the purpose here is to illustrate the basic ideas. Intuitively, if the value in the random sample is sufficiently larger than 1 then we would reject H_0. Thus the procedure would be to reject H_0: $\mu = 1$ if $X_1 - 1 > c$ where c is to be determined. Since $\alpha = 0.05$, c must be determined so that

$$P(X_1 - 1 > c) = 0.05 \qquad (7.1)$$

where the probability is computed under the assumption that H_0: $\mu = 1$ is true. $P(X_1 - 1 > c)$ is the probability of rejecting H_0 when H_0 is in fact true. If H_0: $\mu = 1$ is true, then since $\sigma^2 = 4$ and $n = 1$,

$$Z = \frac{X_1 - \mu_0}{\sigma/\sqrt{n}} = \frac{X_1 - 1}{2}$$

has a standard normal distribution. Thus, by (7.1),

$$P(X_1 - 1 > c) = P\left(\frac{X_1 - 1}{2} > \frac{c}{2}\right)$$

$$= P\left(Z > \frac{c}{2}\right) = 0.05 \quad .$$

The last equality implies that $c/2 = 1.645$, from the standard normal table. Hence $c = 3.290$ and the procedure would be to reject H_0: $\mu = 1$ if $X_1 - 1 > 3.290$ or $X_1 > 4.290$. The critical region is the set of all real values x_1 such that $x_1 > 4.290$. To carry out the test a sample of size 1 is taken and H_0: $\mu = 1$ is rejected or not rejected according as $X_1 > 4.290$ or $X_1 \leq 4.29$ [note $P(X_1 = 4.29) = 0$].

To find the power of the test we must find the probability of rejecting H_0: $\mu = 1$ when it is false or when H_A: $\mu = 3$ is true. If $\mu = 3$, then we must find $P(X_1 > 4.29)$ where X_1 has a normal distribution with mean 3 and variance 4. This probability is given by

$$P(X_1 > 4.29) = P\left(\frac{X_1 - 3}{2} > \frac{4.29 - 3}{2}\right)$$

$$= P(Z > 0.645)$$

$$= 0.740 \quad .$$

The value 0.740 was obtained from Table III by linear interpolation. Thus, the probability of rejecting H_0: $\mu = 1$ when H_A: $\mu = 3$ is true is 0.740.

We now consider some important and popular hypothesis testing problems. These will include all the problems that were discussed in Chapter 6 in terms of interval estimation.

7.5 TESTS CONCERNING THE MEAN

The tests pertaining to the mean of a population and those comparing two means will be developed under the assumption of normality.

The tests are also valid in cases where only approximate normality exists.

To test the hypothesis that a random sample of size n comes from a population with mean $\mu = \mu_0$ we use the statistic

$$Z = \frac{\overline{X} - \mu_0}{\sigma/\sqrt{n}} ,$$

where we have assumed that the population variance σ^2 is known. The problem is that of testing the hypothesis H_0: $\mu = \mu_0$ against the two-sided alternative H_A: $\mu \neq \mu_0$. The hypothesis H_0 is rejected if the observed value of Z is "too large" or "too small." Since Z has a standard normal distribution, H_0: $\mu = \mu_0$ is rejected if $Z = (\overline{X} - \mu)/(\sigma/\sqrt{n}) > z_{1-\alpha/2}$ or $Z < z_{\alpha/2} = -z_{1-\alpha/2}$. The level of significance α is such that

$$P\left(\frac{\overline{X} - \mu_0}{\sigma/\sqrt{n}} > z_{1-\alpha/2} \text{ or } \frac{\overline{X} - \mu_0}{\sigma/\sqrt{n}} < -z_{1-\alpha/2}\right) = \alpha. \tag{7.2}$$

The probability implied in Eq. (7.2) is that of rejecting H_0: $\mu = \mu_0$ when it is true. The critical region consists of all those values of $\overline{X}$ such that $Z > z_{1-\alpha/2}$ or $Z < -z_{1-\alpha/2}$. This means that

$$\frac{\overline{X} - \mu_0}{\sigma/\sqrt{n}} > z_{1-\alpha/2} \quad \text{or} \quad \frac{\overline{X} - \mu_0}{\sigma/\sqrt{n}} < -z_{1-\alpha/2} ,$$

which reduces to

$$\overline{X} - z_{1-\alpha/2}\,\frac{\sigma}{\sqrt{n}} > \mu_0 \quad \text{or } \overline{X} + z_{1-\alpha/2}\,\frac{\sigma}{\sqrt{n}} < \mu_0 \quad . \tag{7.3}$$

We will reject H_0: $\mu = \mu_0$ if either inequality in (7.3) is satisfied. Consequently, we will not reject (i.e., we will accept) H_0 if μ_0 satisfies neither inequality in (7.3), that is, if

$$\overline{X} - z_{1-\alpha/2}\frac{\sigma}{\sqrt{n}} < \mu_0 < \overline{X} + z_{1-\alpha/2}\frac{\sigma}{\sqrt{n}} \quad ,$$

which is just another way of saying that H_0: $\mu = \mu_0$ is not rejected if μ_0 is contained in the $(1-\alpha)100\%$ confidence interval for μ. Thus, every confidence interval in Chapter 6 can be used to test a two-sided hypothesis.

To test H_0: $\mu \leq \mu_0$ against H_A: $\mu > \mu_0$ we reject H_0 if $Z = (\overline{X} - \mu_0)/(\sigma/\sqrt{n}) > z_{1-\alpha}$, i.e.,

$$P\left(\frac{\overline{X} - \mu_0}{\sigma/\sqrt{n}} > z_{1-\alpha}\right) = \alpha$$

for the one-sided hypothesis. But $(\overline{X} - \mu_0)/(\sigma/\sqrt{n}) > z_{1-\alpha}$ reduces to $\overline{X} - z_{1-\alpha}\frac{\sigma}{\sqrt{n}} > \mu_0$. Thus corresponding to this situation is the one-sided $(1-\alpha)100\%$ confidence interval

$$\left(\overline{X} - z_{1-\alpha}\frac{\sigma}{\sqrt{n}}, \infty\right) \quad .$$

If the hypothesized μ_0 falls in $(\overline{X} - z_{1-\alpha}\frac{\sigma}{\sqrt{n}}, \infty)$, then $(\overline{X} - \mu_0)(\sigma/\sqrt{n}) < z_{1-\alpha}$ and H_0: $\mu \leq \mu_0$ is accepted.

Since all the parameters for which confidence intervals were constructed in Chapter 6 have hypothesis testing counterparts, we shall simply summarize the results and give some examples. They can all be developed in exactly the same way as was done in the first part of this section.

The procedures for testing the mean μ when the variance is known are summarized in Table 7.1.

TABLE 7.1

Testing μ When σ^2 Is Known

Statistic: $Z = \dfrac{\overline{X} - \mu_0}{\sigma/\sqrt{n}}$

Null hypothesis	Alternative	Rejection region
H_0: $\mu = \mu_0$	H_A: $\mu \neq \mu_0$	$Z > z_{1-\alpha/2}$ or $Z < -z_{1-\alpha/2}$
H_0: $\mu \leq \mu_0$	H_A: $\mu > \mu_0$	$Z > z_{1-\alpha}$
H_0: $\mu \geq \mu_0$	H_A: $\mu < \mu_0$	$Z < -z_{1-\alpha}$

Example 7.2 Referring to the data of Example 4.2, suppose that the true variance of an individual observation had been 49×10^{-10} M. Seven samples were analyzed for which $\overline{x} = 12.36 \times 10^{-4}$ M. Furthermore

$$\frac{\sigma}{\sqrt{n}} = \sqrt{\frac{49 \times 10^{-10}}{7}} = 2.646 \times 10^{-5} \text{ M.}$$

Can we assume that the mean μ of the $K_2Cr_2O_7$ is 10×10^{-4} M?

We will follow the procedure set forth in Section 7.3 for testing a hypothesis:

1. We assume the data came from a normal population (or approximately so) with known variance.
2. The hypothesis and alternative are H_0: $\mu = 10 \times 10^{-4}$ M and H_A: $\mu \neq 10 \times 10^{-4}$ M.
3. The test statistic is

$$Z = \frac{\overline{X} - 10^{-3}}{\sigma/\sqrt{n}}.$$

4. Let $\alpha = 0.05$.
5. Z has a standard normal distribution and the critical region is given in Table 7.1 as $|Z| > z_{1-\alpha/2} = z_{0.975} = 1.96$.
6. The value of the test statistic is

$$\frac{12.36 \times 10^{-4} - 10 \times 10^{-4}}{2.646 \times 10^{-5}} = 0.47$$

7. Since the value 0.47 does not exceed $z_{0.975} = 1.96$, we accept $H_0 : \mu = 10 \times 10^{-4}$ M (we do not have sufficient evidence to reject H_0). This example was considered in Chapter 6, Example 6.1, in terms of confidence intervals.

To test hypotheses on μ when σ^2 is unknown we use the t-statistic $T = (\overline{X} - \mu_0)/(S/\sqrt{n})$ and replace z with t and σ^2 with the sample variance S^2 in Table 7.1 to obtain Table 7.2.

TABLE 7.2

Testing μ When σ^2 Is Unknown

Statistic: $T = \dfrac{\overline{X} - \mu_0}{S/\sqrt{n}}$, (n-1 d.f.)

Null hypothesis	Alternative	Reject H_0 if
$H_0: \mu = \mu_0$	$H_A: \mu \neq \mu_0$	$T > t_{n-1,1-\alpha/2}$ or $T < -t_{n-1,1-\alpha/2}$
$H_0: \mu \leq \mu_0$	$H_A: \mu > \mu_0$	$T > t_{n-1,1-\alpha}$
$H_0: \mu \geq \mu_0$	$H_A: \mu < \mu_0$	$T < -t_{n-1,1-\alpha}$

Example 7.3 The following data were obtained for the calibration of the Ruska dead weight gauge used with our Burnett P-V-T apparatus (see Example 6.2, Chapter 6). The weights corresponding to the 1000 psi loading had the following apparent masses:

26.03570	26.03575	26.03599
26.03581	26.03551	26.03533
26.03529	26.03588	26.03570
26.03573	26.03586	

Can we say that the average apparent mass μ does not exceed 26.5?

1. Assume approximate normality.
2. H_0: $\mu \geq 26.5$ against H_A: $\mu < 26.5$ (actually the hypothesis was set up before any data were observed).
3. Since σ^2 is unknown use $T = (\overline{X} - 26.5)/S_{\overline{X}}$.
4. Let $\alpha = 0.05$.
5. T has the t-distribution with $\nu = 10$ degrees of freedom.
6. $\overline{x} = 26.03568$ and $s/\sqrt{11} = 6.741 \times 10^{-5}$ by Example 6.2, Chapter 6. Thus the calculated t-value is

$$\frac{26.03568 - 26.5}{6.741 \times 10^{-5}} = -\frac{0.46432}{6.741} \times 10^5 .$$

7. The critical region is, by Table 7.2,

$$\frac{\overline{X} - 26.5}{S/\sqrt{n}} < -t_{10,0.95} = -1.812.$$

As the calculated t-value is $-\frac{0.46432}{6.741} \times 10^5 < -1.812$, H_0 is rejected. Thus we can say that μ probably does not exceed 26.5. The chances that we are making an error in so saying are 5 out of 100, which corresponds to $\alpha = 0.05$.

Example 7.4 Fifty determinations of a certain concentration yielded the following values:

54.20	53.82	55.13	55.77	56.52
51.73	54.15	51.12	52.22	56.91
52.56	53.10	53.73	54.55	52.35
53.55	51.56	55.01	56.78	52.02
56.15	53.43	55.57	56.00	58.16
57.50	53.77	53.95	57.27	57.73
52.94	55.88	53.39	54.89	55.33
54.25	54.96	54.30	57.05	54.13
54.46	58.51	52.89	56.25	56.60
53.08	54.65	57.35	56.35	55.21

From the data we obtain

$$\bar{x} = \sum_{i=1}^{n} x_i/n = 54.76,$$

$$s^2 = \frac{\sum_{i=1}^{n} (x_i - \bar{x})^2}{n - 1} = 4.216 \quad ,$$

and

$$s = \sqrt{s^2} = 2.053 \quad .$$

To test H_0: $\mu \geq 55$, the test statistic is, according to Table 7.2, $T = (\bar{X} - \mu_0)/(S/\sqrt{n})$. Thus

$$t = \frac{54.76 - 55.0}{2.053/\sqrt{50}} = \frac{-0.024}{0.29} = -0.93$$

Since $t_{49,.0.01} = -2.405$, we do not reject H_0 at the 1% significance level.

7.6 TESTS ON THE DIFFERENCE OF TWO MEANS

There are three cases to be considered in comparing the means of two populations, namely, the same ones that were considered in Section 6.9.

Case 1 (σ_1^2 and σ_2^2 known) Suppose we have two random samples from two different populations with parameters μ_1, σ_1^2 and μ_2, σ_2^2 where σ_1^2 and σ_2^2 are known. The test statistic used to test H_0: $\mu_1 = \mu_2$ ($\mu_1 - \mu_2 = 0$) against H_A: $\mu_1 \neq \mu_2$ is the standard normal random variable

$$Z = \frac{\bar{X}_1 - \bar{X}_2 - (\mu_1 - \mu_2)}{\sqrt{\dfrac{\sigma_1^2}{n_1} + \dfrac{\sigma_2^2}{n_2}}} \tag{7.4}$$

where n_1 and n_2 are the sample sizes. The statistic in (7.4) can actually be used to test the more general hypothesis H_0: $\mu_1 - \mu_2 = d$, where d is a specified number. However, the case $\mu_1 - \mu_2 = 0$ is of special interest. The test rejects H_0: $\mu_1 = \mu_2$ if $Z > z_{1-\alpha/2}$ or $Z < z_{\alpha/2} = -z_{1-\alpha/2}$. This result and other possibilities are included in Table 7.3

A confidence interval corresponding to each H_0 in Table 7.3 can be derived. We will leave those derivations to the interested reader and turn our attention to Case 2.

TABLE 7.3

Testing $\mu_1 - \mu_2$ When σ_1^2 and σ_2^2 Are Known

Statistic: $$Z = \frac{\overline{X}_1 - \overline{X}_2}{\sqrt{\frac{\sigma_1^2}{n_1} + \frac{\sigma_2^2}{n_2}}}$$

Null hypothesis	Alternative	Rejection Region
$H_0: \mu_1 - \mu_2 = 0$	$H_A: \mu_1 - \mu_2 \neq 0$	$Z > z_{1-\alpha/2}$ or $Z < -z_{1-\alpha/2}$
$H_0: \mu_1 - \mu_2 \leq 0$	$H_A: \mu_1 - \mu_2 > 0$	$Z > z_{1-\alpha}$
$H_0: \mu_1 - \mu_2 \geq 0$	$H_A: \mu_1 - \mu_2 < 0$	$Z < -z_{1-\alpha}$

Case 2 (σ_1^2 and σ_2^2 unknown, but assumed equal) In this case, the population variances are unknown but assumed equal. The common variance σ^2 is estimated by the pooled variance

$$s_p^2 = \frac{\sum_{i=1}^{n_1} (X_{1i} - \overline{X}_1)^2 + \sum_{i=1}^{n_2} (X_{2i} - \overline{X}_2)^2}{n_1 + n_2 - 2}$$

as given in Eq. (6.21). The test statistic is

$$T = \frac{\overline{X}_1 - \overline{X}_2 - (\mu_1 - \mu_2)}{s_p\sqrt{\frac{1}{n_1} + \frac{1}{n_2}}} \qquad (7.5)$$

which follows the t-distribution with $n_1 + n_2 - 2$ degrees of freedom. A summary of test criteria for Case 2 is presented in Table 7.4.

TABLE 7.4

Testing $\mu_1 - \mu_2$ When σ_1^2 and σ_2^2 Are Unknown but Equal

Statistic: $T = \dfrac{\bar{X}_1 - \bar{X}_2}{S_p \sqrt{\dfrac{1}{n_1} + \dfrac{1}{n_2}}}$ $\nu = n_1 + n_2 - 2$ d.f.

Null hypothesis	Alternative	Rejection region
H_0: $\mu_1 - \mu_2 = 0$	H_A: $\mu_1 - \mu_2 \neq 0$	$T > t_{\nu,1-\alpha/2}$ or $T < -t_{\nu,1-\alpha/2}$
H_0: $\mu_1 - \mu_2 \leq 0$	H_A: $\mu_1 - \mu_2 > 0$	$T > t_{\nu,1-\alpha}$
H_0: $\mu_1 - \mu_2 \geq 0$	H_A: $\mu_1 - \mu_2 < 0$	$T < -t_{\nu,1-\alpha}$

Example 7.5 Consider the data presented in Example 4.1 From those data, is it reasonable to believe that the two bottles of reagent were filled from the same production run? Information received from the manufacturer indicates that the average variance in concentration for single analyses for the last year's production runs was 0.016.

We set up the hypothesis that the two bottles were indeed filled out of the same batch. This signifies that the means of the analyses for the two bottles should be equal and we write H_0: $\mu_1 = \mu_2$, against H_A: $\mu_1 \neq \mu_2$.

We take the variance for both samples, σ_1^2 and σ_2^2, to be 0.016 and we have a Case 1 situation. From Table 7.3, we choose the appropriate statistic which when evaluated yields

$$z = \frac{15.770 - 15.597 - (\mu_1 - \mu_2)}{\sqrt{0.016}\ (\sqrt{1/3 + 1/3})}$$

since $\bar{x}_1 = 15.770$, $\bar{x}_2 = 15.597$, $n_1 = n_2 = 3$, and $\sigma_1^2 = \sigma_2^2 = 0.016$. If H_0: $\mu_1 = \mu_2$ is true, z reduces to

$$z = \frac{15.770 - 15.597}{0.12649(0.8165)} = \frac{0.173}{0.10327} = 1.675 \quad .$$

If we choose $\alpha = 0.05$ then $z_{1-\alpha/2} = z_{0.975} = 1.96$ and since $-1.96 < 1.675 = z < 1.96$, we cannot reject H_0: $\mu_1 = \mu_2$ and so conclude that $\mu_1 = \mu_2$. From Example 6.4 in Chapter 6, a 95% confidence interval for $\mu_1 - \mu_2$ is $-0.029 < \mu_1 - \mu_2 < 0.375$. Since the hypothesized difference, $\mu_1 - \mu_2 = 0$, falls in the 95% confidence interval, we reach the same conclusion regarding H_0 by that route.

Example 7.6 In order to compare the effects of two solid catalyst component concentrations on NO_2 reduction, six groups of observations were made. Each group consisted of three replicates of five observations each, that is, a total of 15 determinations were made for each concentration. The concentrations (in weight %) were 0.5% and 1.0%. The reduction data are summarized below:

Group	Replicate means, $\bar{X}_i$ (5 observations each)			Group means, $\bar{\bar{X}}_i$ (based on 15 observations)
	A	B	C	
1	5.18	5.52	5.42	5.37
2	5.58	5.62	5.82	5.67

Can we conclude that there is no significant difference between the means μ_1 and μ_2? The null hypothesis is thus H_0: $\mu_1 - \mu_2 = 0$.

In this case we treat the group means A, B, C as individual observations (see Example 6.5). Thus $\bar{x}_1 = 5.37$, $\bar{x}_2 = 5.67$, and $n_1 = n_2 = 3$. We assume $\sigma_1^2 = \sigma_2^2$ so we have a Case 2 situation. As σ^2 is unknown, we cannot use Z for our test statistic but must use T, which we choose from Table 7.4

From Example 6.5 we already have $s_p^2 = 0.07875$ so

$$T = \frac{\bar{x}_1 - \bar{x}_2 - (\mu_1 - \mu_2)}{s_p \sqrt{\frac{1}{3} + \frac{1}{3}}}$$

$$= \frac{5.37 - 5.67 - 0}{\sqrt{0.07875}\ \sqrt{2/3}} = -1.309 \quad .$$

If we choose $\alpha = 0.05$ then $t_{4,0.025} = -2.776$, and since $t = -1.309$ does not fall below -2.776 we accept H_0: $\mu_1 - \mu_2 = 0$. On calculating the 95% confidence interval for $\mu_1 - \mu_2$, we have $-0.9361 < \mu_1 - \mu_2 < 0.2361$ which leads us to the same conclusion, i.e., there is no significant difference between μ_1 and μ_2 at the 5% level of significance.

Case 3 In this case σ_1^2 and σ_2^2 are not known and cannot be presumed equal. To test hypotheses involving $\mu_1 - \mu_2$ we use the statistic

$$T_f = \frac{\bar{X}_1 - \bar{X}_2 - (\mu_1 - \mu_2)}{\sqrt{\frac{S_1^2}{n_1} + \frac{S_2^2}{n_2}}} \tag{7.6}$$

where n_1 and n_2 are the sample sizes and S_1^2 and S_2^2 are the sample variances. The degrees of freedom associated with the random variable T is f where

$$f = \frac{(s_1^2/n_1 + s_2^2/n_2)^2}{\dfrac{(s_1^2/n_1)^2}{n_1 - 1} + \dfrac{(s_2^2/n_2)^2}{n_2 - 1}} \tag{7.7}$$

Thus T_f in (7.6) has a t-distribution with f d.f. Since this result is not used as frequently as that in Cases 1 and 2 we do not give the summary results in tabular form but rather consider one example.

Example 7.7 In order to test the effect of a solid catalyst component on NO_2 reduction, six groups of observations were made. Each group consisted of three replicates of five observations each. The component concentrations by groups were:

Group	Concentration, wt.%
1	0.0
2	0.05
3	0.1
4	0.5
5	1.0
6	2.5

The following reduction data were obtained on standardized activity tests:

Group	Replicate means, $\bar{X}_i$			Group mean,
	A	B	C	$\bar{\bar{X}}_i$
1	4.96	4.98	4.82	4.92
2	4.86	4.84	4.88	4.86
3	4.94	5.14	5.10	5.03
4	5.18	5.52	5.42	5.37
5	5.58	5.62	5.82	5.67
6	6.00	5.82	5.72	5.87

From these data, the variances, sample standard deviations, and standard errors of the group means are calculated. For these data, $s_{\bar{X}}$ has the connotation associated with s_X, the sample standard deviation, because the calculations were made using the replicate means as individual data points. By the same token, $s_{\bar{\bar{X}}}$ is the standard error of the group means.

Group	$\sum_{i=1}^{3} \bar{x}_i^2$	$s_{\bar{X}}^2$	$s_{\bar{X}}$	$s_{\bar{\bar{X}}}$
1	72.6344	0.0076	0.0872	0.05036
2	70.8596	0.0004	0.0200	0.01155
3	75.8232	0.03975	0.1994	0.11513
4	86.6792	0.08425	0.2903	0.16761
5	96.5932	0.07325	0.2707	0.15676
6	103.5572	0.09325	0.3054	0.17630

The null hypothesis H_0: $\bar{\mu} = \bar{\mu}_1$ (Group 1 was the control group) is used to determine whether or not the addition of the catalyst component had a significant effect on NO_2 reduction. By visual inspection, we conclude that all the group variances are not equal. Certainly the variances of Groups 1, 2, and 3 are different from each other and from any of the variances of the other three groups. Many users of statistics might stretch a point and say that the variances of Groups 4, 5, and 6 are close enough to be considered equal. Yet as all the variances are not equal or even nearly so, we will use Eq. (7.7) to calculate the test statistic T_f needed to test the hypothesis.

We must also determine the number of degrees of freedom f with which T_f is distributed and the corresponding tabular value of $t_{f,0.95}$ which will be used in testing the hypothesis. The 95% significance level in a two-tailed test was chosen because one group mean $\bar{x}_2$ falls below $\bar{x}_1$ while all the other $\bar{x}_i$ fall above $\bar{x}_1$. Using

Group 1 as the control group and the data of Group 5 for illustration,

$$f = \frac{(s_5^2/n_5 + s_1^2/n_1)^2}{\dfrac{(s_5^2/n_5)^2}{n_5 - 1} + \dfrac{(s_1^2/n_1)^2}{n_1 - 1}}$$

$$f = \frac{\left(\dfrac{0.07325}{3} + \dfrac{0.0076}{3}\right)^2}{\dfrac{\left(\dfrac{0.07325}{3}\right)^2}{2} + \dfrac{\left(\dfrac{0.0076}{3}\right)^2}{2}}$$

$$f = (0.02695)^2/0.0030129 = 2.4106 \simeq 2.41$$

$$T = \frac{x_5 - x_1}{\sqrt{\dfrac{s_5^2}{n_5} + \dfrac{s_1^2}{n_1}}} = \frac{5.67 - 4.92}{\sqrt{0.02695}} = 4.568 \quad .$$

Linearly interpolating between $t_{2,0.975}$ and $t_{3,0.975}$ in Table IV, we have $t_{2.41,0.975} = \pm 3.642$. As the rejection region for H_0 is

$$T \leq -t_{2.41,0.975} \text{ and } T \geq t_{2.41,0.975}$$

the hypothesis is rejected. Results for all groups are shown below.

Group	f	$t_{f,0.975}$	H_0
2	2.210	±4.067	Accept
3	2.738	±3.476	Accept
4	2.358	±3.902	Accept
5	2.411	±3.642	Reject
6	2.324	±3.940	Reject

From examination of the original data and by comparing corresponding values of $t_{f,0.975}$ we suspect that the inclusion of the catalyst component could have a beneficial effect on NO_2 reduction if present in concentrations at or above 1.0 wt.%. Attempts to draw any further conclusions from these results should not be made.

The use of PROC TTEST (SAS for Linear Models. SAS Institute Inc. Cary, NC: SAS Institute Inc. 1981. 231 pp.) to evaluate the difference of two means is quite simple. This procedure does not depend on your knowing in advance whether the variances can be presumed equal or not as it calculates the value of T both ways as illustrated in the following example.

Example 7.8 Two of the orifice holes in a Saybolt viscometer are used to make simultaneous readings of time for flow of 60-ml samples. The times (in seconds) for each orifice for a sample of CCl_4 are as shown in the SAS program below. The students running the experiment ask the instructor if the orifices should be cleaned and the trials rerun. If the instructor operates on a 95% confidence level what should his answer be?

```
(put initial JCL cards/statements here)
DATA;
INPUT HOLE 2 TIME 5-8;
CARDS;
1   27.8
1   27.9
1   28.0
1   27.8
2   29.5
2   29.7
2   29.4
2   29.8
PROC TTEST;
CLASS HOLE;
TITLE 'EVALUATION OF VISCOMETER';
(put final JCL cards/statements here)
```

The results are as shown below.

EVALUATION OF VISCOMETER

TTEST PROCEDURE

VARIABLE: TIME

HOLE	N	MEAN	STD DEV	STD ERROR	MINIMUM	MAXIMUM
1	4	27.87500000	0.09574271	0.04787136	27.80000000	28.00000000
2	4	29.60000000	0.18257419	0.09128709	29.40000000	29.80000000

VARIANCES	T	DF	PROB > |T|
UNEQUAL	-16.7350	4.5	0.0001
EQUAL	-16.7350	6.0	0.0001

The variances for case 2 and case 3 t-tests have been used to calculate the degrees of freedom and the corresponding T values for comparing the means. For the hypothesis that the variances are equal, the program calculates f = 3.64 with a probability of 0.3172 indicating no evidence that the variances are unequal. Thus the calculated T for case 2 (equal but unknown population variances) would be used in evaluating the null hypothesis that there is no difference in the mean times of the two viscometer holes. The last column shows the probability that the absolute value of T will exceed the corresponding computed value, e.g. $|-16.735|$ in this example.

7.7 PAIRED t-TEST

In the case of paired observations $(X_1,Y_1),\ldots,(X_n,Y_n)$ we wish to test H_0: $\mu_1 = \mu_2$ or H_0: $\mu_D = 0$ where $\mu_D = \mu_1 - \mu_2$. The procedure is to consider the differences $D_i = X_i - Y_i$, $i = 1,2,\ldots,n$ and observe that $\mu_D = \mu_X - \mu_Y$. The H_0: $\mu_D = 0$ can be tested by using the statistic

$$T = \frac{\overline{D} - \mu_D}{S_D/\sqrt{n}}$$

which has a t-distribution with n - 1 d.f. Recall from Section 6.10 that

$$\overline{D} = \sum_{i=1}^{n} (X_i - Y_i)/n = \overline{X} - \overline{Y}$$

and

$$S_D^2 = \sum_{i=1}^{n} (D_i - \overline{D})^2/(n - 1) \quad .$$

The test procedures are summarized in Table 7.5 below.

TABLE 7.5

Testing $\mu_D = 0$ When Observations Are Paired

$$\text{Statistic:} \quad T = \frac{\overline{X} - \overline{Y}}{S_D/\sqrt{n}}, \quad n - 1 \text{ d.f.}$$

Null hypothesis	Alternative	Rejection region
$H_0: \mu_D = 0$	$H_A: \mu_D \neq 0$	$T > t_{n-1,1-\alpha/2}$ or $T < -t_{n-1,1-\alpha/2}$
$H_0: \mu_D \leq 0$	$H_A: \mu_D > 0$	$T > t_{n-1,1-\alpha}$
$H_0: \mu_D \geq 0$	$H_A: \mu_D < 0$	$T < -t_{n-1,1-\alpha}$

Example 7.9 Let us reexamine the data in Example 6.6. In studying weather resistance of two types of paint, each of 5 shingles was painted with both paints, one-half with one paint and the other half with the other paint. After a month's exposure, the wear, measured in suitable units, was found to be (1.57, 1.45), (1.46, 1.59), (1.53, 1.27), (1.30, 1.48), (1.37, 1.40). Is it plausible that $\mu_D = \mu_1 - \mu_2$ is greater than zero?

From Example 6.6, if $D_i = X_i - Y_i$, we have $d_1 = 0.12$, $d_2 = -0.13$, $d_3 = 0.26$, $d_4 = -0.18$, $d_5 = -0.03$. Also, $\overline{d} = 0.008$ and $S_D = 0.181576$. We want to test $H_0: \mu_D \leq 0$ against $H_A: \mu_D > 0$. The statistic to be used is

$$T = \frac{\overline{D} - \mu_D}{S_D/\sqrt{n}} = \frac{\overline{D}}{S_D/\sqrt{n}} .$$

From Table 7.5, we must reject $H_0: \mu_D \leq 0$ if $T = \dfrac{\overline{D}}{(S_D/\sqrt{n})} > t_{4,1-\alpha}$.

Let $\alpha = 0.05$ so that $t_{4,0.95} = 2.132$. Since

$t = \dfrac{0.008}{0.081203} = 0.0985 < t_{4,0.95} = 2.132$, we do not have sufficient evidence to reject H_0: $\mu_D \leq 0$. We therefore conclude that it is not plausible that $\mu_D > 0$.

<u>Example 7.10</u> The use of computerized data reduction for handling paired comparisons is illustrated below for Example 6.6. The SAS program in this case is as follows.

```
(put initial JCL cards/statements here)
DATA;
INPUT TOP BOTTOM;
D = TOP - BOTTOM;
CARDS;
(data are entered here, one pair per line with the values separated
  by 1 space)
PROC PRINT;
PROC MEANS MEAN STD STDERR T PRT;
VAR D;
TITLE 'PAIRED COMPARISON FOR WEATHERING DATA';
(put final JCL cards/statements here)
```

The output from this program consists of two parts: the results of the PRINT procedure and those from the MEANS procedure.

SAS

OBS	TOP	BOTTOM	D
1	1.46	1.59	-0.13
2	1.53	1.27	0.26
3	1.30	1.48	-0.18
4	1.37	1.40	-0.03
5	1.57	1.45	0.12

PAIRED COMPARISON OF WEATHERING DATA

VARIABLE	MEAN	STANDARD DEVIATION	STD ERROR OF MEAN	T	PR>|T|
D	0.00800000	0.18157643	0.08120345	0.10	0.9263

From PROC MEANS, we see that the calculated value of T is 0.10. The probability 0.9248 means that T = 0.10 is not significantly different from 0 so that H_0: $\mu_D = 0$ is not rejected.

7.8 TESTING A PROPORTION P

To test the hypothesis H_0: $P = P_0$ involving a proportion P which is the parameter of the binomial distribution we will use the statistic

$$Z = \frac{\hat{P} - P_0}{\sqrt{P_0 Q_0 / n}} \tag{7.8}$$

which has an approximate standard normal distribution. This statistic was discussed in Section 6.12 where confidence intervals on P were presented. The hypothesis H_0: $P = P_0$ is rejected in favor of H_A: $P \neq P_0$ if either

$$Z > z_{1-\alpha/2} \text{ or } Z < -z_{1-\alpha/2}$$

where α is the level of significance. This and other possible situations are summarized in Table 7.6.

TABLE 7.6
Testing a Proportion P

$$\text{Statistic:} \quad Z = \frac{\hat{P} - P_0}{\sqrt{P_0(1-P_0)/n}}$$

Null hypothesis	Alternative	Rejection region
H_0: $P = P$	H_A: $P \neq P_0$	$Z > z_{1-\alpha/2}$ or $Z < -z_{1-\alpha/2}$
H_0: $P \leq P_0$	H_A: $P > P_0$	$Z > z_{1-\alpha}$
H_0: $P \geq P_0$	H_A: $P < P_0$	$Z < -z_{1-\alpha}$

Example 7.11 It is desired to know if the proportion of all batches of a particular chemical formulation which are on-grade is larger than 90%. That is, if a batch is chosen, is the probability P that it is on-grade larger than 0.9? Choose $\alpha = 0.05$.

A sample of 100 batches of chemical are chosen and 95 are on grade: $\hat{P} = 0.95$. Furthermore

$$Z = \frac{\hat{P} - P_0}{\sqrt{P_0(1 - P_0)/n}} = \frac{0.95 - 0.90}{\sqrt{\dfrac{0.90(0.10)}{100}}} = \frac{0.05}{0.03} = 1.667 \quad .$$

The hypothesis is H_0: $P \leq 0.9$ and the alternative is H_A: $P > 0.9$. From Table 7.6 the rejection region is

$$Z > z_{1-\alpha} = z_{0.95} = 1.645 \quad .$$

Since $1.667 > 1.645$, we reject H_0 and conclude that $P > 0.90$. The probability that we have reached a wrong conclusion is $\alpha = 0.05$.

7.9 TESTING THE DIFFERENCE OF TWO PROPORTIONS

To test the hypothesis H_0: $P_1 = P_2$ of equality of two proportions corresponding to two binomial populations we use the statistic

$$Z = \frac{\hat{P}_1 - \hat{P}_2 - (P_1 - P_2)}{\sqrt{\dfrac{\hat{P}_1(1 - \hat{P}_1)}{n_1} + \dfrac{\hat{P}_2(1 - \hat{P}_2)}{n_2}}} \tag{7.9}$$

where n_1 and n_2 are the sample sizes and $\hat{P}_1$ and $\hat{P}_2$ are the estimators of P_1 and P_2. The statistic Z was discussed in Section 6.13 and has an approximate normal distribution.

The results pertaining to the use of Eq. (7.9) are contained in Table 7.7.

TABLE 7.7

Testing $P_1 = P_2$

Statistic: $Z = (\hat{P}_1 - \hat{P}_2) \Big/ \sqrt{\dfrac{\hat{P}_1(1 - \hat{P}_1)}{n_1} + \dfrac{\hat{P}_2(1 - \hat{P}_2)}{n_2}}$

Null hypothesis	Alternative	Rejection region
H_0: $P_1 = P_2$	H_A: $P_1 \neq P_2$	$Z > z_{1-\alpha/2}$ or $Z < -z_{1-\alpha/2}$
H_0: $P_1 \leq P_2$	H_A: $P_1 > P_2$	$Z > z_{1-\alpha}$
H_0: $P_1 \geq P_2$	H_A: $P_1 < P_2$	$Z < -z_{1-\alpha}$

Example 7.12 Two different methods of manufacturing compression rings for 1/4-in. copper tubing are being used. It is desired to know if the proportions P_1 and P_2 of rings that do not seat properly are the same for the two methods. Can we say that $P_1 - P_2 \neq 0$? Use $\alpha = 0.01$.

One hundred rings are sampled from each method and it is found that 10 rings using method 1 and 3 rings using method 2 do not seat properly. The null hypothesis and alternative are H_0: $P_1 = P_2$ and H_A: $P_1 \neq P_2$. From Table 7.8 the critical region is

$$\frac{\hat{P}_1 - \hat{P}_2}{\sqrt{\dfrac{\hat{P}_1(1 - \hat{P}_2)}{n_1} + \dfrac{\hat{P}_2(1 - \hat{P}_2)}{n_2}}} > z_{0.995} \text{ or } < -z_{0.995} \quad .$$

Now $n_1 = n_2 = 100$, $\hat{P}_1 = 0.10$, $\hat{P}_2 = 0.03$, and the computed value for Z (under H_0) is

$$\frac{0.10 - 0.03}{\sqrt{\dfrac{0.10(0.90)}{100} + \dfrac{0.03(0.97)}{100}}} = \frac{0.07}{0.0345} = 2.03 \quad .$$

Since $z_{0.995} = 2.58$, we have $-2.58 < 2.03 < 2.58$ so we cannot reject H_0: $P_1 = P_2$. From Example 6.13, a 99% confidence interval is (-0.01, 0.15), which contains $P_1 - P_2 = 0$ and leads us to the same conclusion.

7.10 TESTS CONCERNING THE VARIANCE

Several hypotheses about a single variance are possible. If we wish, for example, to test whether a random sample is drawn from a population of a specific known variance σ_0^2, we have a two-sided hypothesis

$$H_0: \sigma^2 = \sigma_0^2$$

against the alternative

$$H_A: \sigma^2 \neq \sigma_0^2 \ .$$

The alternative could be of the form H_A: $\sigma^2 > \sigma_0^2$ or H_A: $\sigma^2 < \sigma_0^2$. Assuming H_0 is correct, the test statistic used is that given by the random variable

$$\chi^2 = (n - 1)S^2/\sigma_0^2 \ ,$$

which has a χ^2-distribution with n - 1 degrees of freedom. The critical region consists of the values in the two tails of the chi-square distribution when H_0: $\sigma^2 = \sigma_0^2$ is true. The hypothesis H_0: $\sigma^2 = \sigma_0^2$ will therefore be rejected if either $\chi^2 = (n - 1)S^2/\sigma_0^2 \leq \chi^2_{n - 1,\alpha/2}$ or $\chi^2 \geq \chi^2_{n - 1,1-\alpha}$, which represents

the conditions where the calculated χ^2 value $(n - 1)s^2/\sigma_0^2$ is too large or too small under the assumption that $\sigma^2 = \sigma_0^2$. The significance level is determined from

$$P(\chi^2_{n - 1,\alpha/2} \leq \chi^2 \leq \chi^2_{n - 1,1-\alpha/2}) = 1 - \alpha.$$

If we wish to test whether the variance of the population exceeds a given value, we have a one-sided hypothesis H_0: $\sigma^2 \le \sigma_0^2$ against H_A: $\sigma^2 > \sigma_0^2$. The corresponding one-sided critical region for which H_0 will be rejected is $\chi^2 = (n - 1)S^2/\sigma_0^2 \ge \chi^2_{n - 1,1-\alpha}$.

For the other possible one-tailed test, we have H_0: $\sigma^2 \ge \sigma_0^2$ as opposed to H_A: $\sigma^2 < \sigma_0^2$. The corresponding rejection region for this H_0 is $\chi^2 = (n - 1)S^2/\sigma_0^2 \le \chi^2_{n-1,\alpha}$. Table 7.8 summarizes the critical regions needed for each of the possible alternatives.

TABLE 7.8

Testing the Variance σ^2

Statistic: $\chi^2 = (n - 1)S^2/\sigma_0^2$, n - 1 d.f.

Hypothesis	Alternative	Rejection region
H_0: $\sigma^2 = \sigma_0^2$	H_A: $\sigma^2 \neq \sigma_0^2$	$\chi^2 > \chi^2_{n-1,1-\alpha/2}$ or $\chi^2 < \chi^2_{n-1,\alpha/2}$
H_0: $\sigma^2 \le \sigma_0^2$	H_A: $\sigma^2 > \sigma_0^2$	$\chi^2 > \chi^2_{n-1,1-\alpha}$
H_0: $\sigma^2 > \sigma_0^2$	H_A: $\sigma^2 < \sigma_0^2$	$\chi^2 < \chi^2_{n-1,\alpha}$

Example 7.13 Five similar determinations of the cold water flow rate to a heat exchanger were, in GPM, 5.84, 5.76, 6.03, 5.90, and 5.87. It is of importance that the imprecision (on a per measurement basis) does not greatly affect this measuring operation. Can we be reasonably sure that the variance per measurement is less than 0.01?

1. Assume a normal population.
2. The hypothesis is H_0: $\sigma^2 \geq 0.01$ against H_A: $\sigma^2 < 0.01$.
3. The test statistic from Table 7.8 is

$$\chi^2 = (n - 1)S^2/\sigma_0^2 .$$

4. Let $\alpha = 0.025$.
5. The test statistic in (3) has a χ^2-distribution with 4 d.f. The critical region from Table 7.8 is $\chi^2 \leq \chi^2_{4,0.025} = 0.484$.
6. From the data we obtain

$$\chi^2 = \frac{4(0.00975)}{0.01} = 3.900.$$

The sample variance $s^2 = 0.00975$ as was given in example 6.8.

7. Since $\chi^2 = 3.9 > 0.484$ we do not reject H_0: $\sigma^2 \geq 0.01$ and thus conclude that it is not reasonable to believe that $\sigma^2 < 0.01$.

7.11 TESTING THE EQUALITY OF VARIANCES

Suppose that σ_1^2 and σ_2^2 are the variances of two normal populations. The problem now is that of testing the hypothesis H_0: $\sigma_1^2 = \sigma_2^2$ against one of the three alternatives H_A: $\sigma_1^2 \neq \sigma_2^2$, H_A: $\sigma_1^2 > \sigma_2^2$, or H_A: $\sigma_1^2 < \sigma_2^2$. The hypotheses can be stated H_0: $\sigma_1^2/\sigma_2^2 = 1$ against either H_A: $\sigma_1^2/\sigma_2^2 \neq 1$, H_A: $\sigma_1^2/\sigma_2^2 > 1$, or H_A: $\sigma_1^2/\sigma_2^2 < 1$.

If $X_1, X_2, \ldots, X_{n_1}$ and $Y_1, Y_2, \ldots, Y_{n_2}$ denote random samples from two normal populations, the statistic to be used in testing the hypotheses concerning variances is

$$\Gamma = \frac{S_1^2/\sigma_1^2}{S_2^2/\sigma_2^2} . \tag{7.10}$$

Under H_0: $\sigma_1^2/\sigma_2^2 = 1$, the statistic in (7.10) becomes

$$F = S_1^2/S_2^2 ,$$

which has an F-distribution with n_1-1 and n_2-1 degrees of freedom if H_0: $\sigma_1^2/\sigma_2^2 = 1$ is true. Thus H_0 is rejected if S_1^2/S_2^2 is sufficiently different from 1. That is, α, the significance level, is such that

$$P(F_{n_1-1,n_2-1,\alpha/2} \le F \le F_{n_1-1,n_2-1,1-\alpha/2}) = 1 - \alpha ,$$

and H_0 is rejected in favor of H_A: $\sigma_1^2 \neq \sigma_2^2$ if the calculated value $f = s_1^2/s_2^2$ falls in either tail of the F-distribution.

These hypotheses and their critical regions are presented in Table 7.9

TABLE 7.9

Testing $\sigma_1^2 = \sigma_2^2$

Statistic: $F = S_1^2/S_2^2$; $\nu_1 = n_1 - 1$, $\nu_2 = n_2 - 1$ d.f.

Null hypothesis	Alternative	Rejection region
H_0: $\sigma_1^2 = \sigma_2^2$	H_A: $\sigma_1^2 \neq \sigma_2^2$	$F > F_{\nu_1,\nu_2,1-\alpha/2}$ or
		$F < F_{\nu_1,\nu_2,\alpha/2}$
H_0: $\sigma_1^2 \le \sigma_2^2$	H_A: $\sigma_1^2 > \sigma_2^2$	$F > F_{\nu_1,\nu_2,1-\alpha}$
H_0: $\sigma_1^2 \ge \sigma_2^2$	H_A: $\sigma_1^2 < \sigma_2^2$	$F < F_{\nu_1,\nu_2,\alpha}$

Example 7.14 Reaction temperature in degrees centigrade (measured on two different days) for two catalyst concentrations is given in

Example 6.9. There are 13 observations corresponding to each catalyst, i.e., $n_1 = n_2 = 13$. Is there any significant difference in the temperature variations corresponding to the two catalysts? Choose $\alpha = 0.02$.

From Example 6.9 we obtain the following statistics:

$$\bar{x}_1 = 310.8423^{\circ}\text{C} \qquad \bar{x}_2 = 310.5246^{\circ}\text{C}$$

$$s_1^2 = 1.1867 \qquad s_2^2 = 1.5757 \quad .$$

The null hypothesis is H_0: $\sigma_1^2 = \sigma_2^2$ against H_A: $\sigma_1^2 \neq \sigma_2^2$. From Table 7.9 we see that the statistic to test H_0 is $F = S_1^2/S_2^2$ with critical region $F > F_{12,12,0.99} = 4.16$ and $F < F_{12,12,0.01} = 0.241$. Now $f = s_1^2/s_2^2 = 1.1867/1.5757 = 0.753$. Since $f = 0.753$ is between 0.241 and 4.16, the null hypothesis cannot be rejected. We conclude that there is no significant difference at the 2% level of significance.

This type of analysis might precede using the t-test to test $\mu_1 - \mu_2 = 0$. If it is concluded that $\sigma_1^2 = \sigma_2^2$, then s_1^2 and s_2^2 can be pooled to estimate $\sigma_1^2 = \sigma_2^2$ and Case 2 in Section 7.6 is in effect. In fact, for the data of this example

$$T = \frac{\bar{x}_1 - \bar{x}_2}{s_p\sqrt{\frac{1}{13} + \frac{1}{13}}} = \frac{0.3117}{0.46095} = 0.689$$

where

$$s_p^2 = \frac{12s_1^2 + 12s_2^2}{24} = \frac{12(1.1867) + 12(1.5757)}{24} = 1.3812 \quad .$$

Since $t_{24,0.025} = -2.064$ and $t_{24,0.975} = 2.064$, we see that $-2.064 < 0.689 < 2.064$ and we accept H_0: $\mu_1 = \mu_2$.

7.12 OTHER χ^2 TESTS

The chi-square test was developed to facilitate testing the significance of data in light of their observed scatter. Indeed, this is the fundamental principle underlying all tests of significance. The differences studied may be those between comparable sets of data or between theoretically proposed and experimentally observed distributions and their attributes. One of the most frequent uses of the χ^2-test is in the comparison of observed frequencies (percentages cannot be used unless the sample size is exactly 100) and the frequencies we might expect from a given theoretical explanation of the phenomenon under investigation.

The χ^2 statistic used in this case is defined by

$$\chi^2 = \sum_{i=1}^{k} \frac{(O_i - E_i)^2}{E_i} \tag{7.11}$$

where O_i = observed frequency, E_i = expected frequency, and k is the number of classes. The random variable defined by Eq. (7.11) has an approximate χ^2-distribution with k-1 degrees of freedom.

Example 7.15 In the manufacture of compression rings it is desired to test whether the proportion P of rings that do not seat properly is significantly different from some specified number P_0. If n rings are chosen, then under the hypothesis H_0: $P = P_0$ against H_A: $P \neq P_0$ we would expect $E_1 = nP_0$ not to seat properly and $E_2 = n(1-P_0)$ to seat properly. However, given the sample, we observe $O_1 = s$ that do not seat properly and $O_2 = n-s$ that do seat properly. In this case we have k = 2 classes and the statistic

$$\chi^2 = \frac{(O_1 - E_1)^2}{E_1} + \frac{(O_2 - E_2)^2}{E_2}$$

will be used to determine whether what we observe is significantly different from what we expect. Suppose n = 100 with s = 10. Under H_0: P = 0.08 we obtain

$$E_1 = 100(0.08) = 8,\ O_1 = 10 \quad ,$$

$$E_2 = 100(0.92) = 92,\ O_2 = 90 \quad .$$

Thus the computed χ^2-value is

$$\begin{aligned}\chi^2 &= \frac{(10 - 8)^2}{8} + \frac{(90 - 92)^2}{92} \\ &= \frac{4}{8} + \frac{4}{92} \\ &= \frac{1}{2} + \frac{1}{23} \\ &= \frac{25}{46} \\ &= 0.54 \quad .\end{aligned}$$

We will reject H_0: P = .08, at the 10% level, if $\chi^2 > \chi^2_{1,\ 0.95}$ or $\chi^2 < \chi^2_{1,0.05}$. Since $\chi^2_{1,0.95} = 3.841$ and $\chi^2_{1,0.05} = 0.0039$, we accept H_0: P = 0.08 since 0.0039 < 0.54 < 3.841.

It is of interest to note that in this example

$$\begin{aligned}\chi^2 &= \frac{(s - n\,P_0)^2}{n\,P_0} + \frac{[(n - s) - n(1 - P_0)]^2}{n(1 - P_0)} \\ &= \frac{(s - n\,P_0)^2}{n - P_0} + \frac{(s - n\,P_0)^2}{n(1 - P_0)} \\ &= (s - n\,P_0)^2\left(\frac{n}{n^2 P_0(1 - P_0)}\right) \\ &= \frac{(s - n\,P_0)^2}{n\,P_0(1 - P_0)} = \frac{(\hat{P} - P_0)^2}{P_0(1 - P_0)/n} \quad . \qquad (7.12)\end{aligned}$$

Thus in this example the statistic χ^2 is just the square of the statistic

$$(\hat{P} - P_0) \Bigg/ \sqrt{\frac{P_0 (1 - P_0)}{n}}$$

which has an approximate standard normal distribution. Using the technique of Section 7.10 to test H_0: $P = 0.08$ would yield the same conclusion.

The technique of this section, however, which uses χ^2 in (7.11) is more general. It can be used for example to test that the data in Table 3.2 of Chapter 3 came from a normal population. The O_i in this case are the observed frequencies in each of the classes of the empirical distribution in Table 3.5, and E_i are the expected frequencies in each class using the theoretical normal curve with $\mu = \bar{x}$ and $\sigma^2 = s^2$ computed from the data. Then

$$\chi^2 = \sum_{i=1}^{7} \frac{(O_i - E_i)^2}{E_i}$$

is computed and compared with the appropriate tabular χ^2 value. The test in this case is called a <u>goodness of fit test.</u>

<u>Example 7.16</u> Test the melting point data in Table 3.2 to determine whether it came from a normal distribution. Use the empirical distribution in Table 3.5. The sample mean and standard deviation for that data are $\bar{x} = 108.05$ and $s = 0.960$. The class boundaries have to be standardized according to $x = \bar{x} + zs = 108.05 + 0.960z$ or $z = (x - \bar{x})/s$. The following table contains the needed information.

Class (z)	Observed (O_i)	Expected (E_i)
$-3.646 < z \leq -2.604$	1	0.644
$-2.604 < z \leq -1.563$	12	7.837
$-1.563 < z \leq -0.521$	33	35.929
$-0.521 < z \leq 0.521$	57	56.200
$0.521 < z \leq 1.563$	35	35.929
$1.563 < z \leq 2.604$	4	7.837
$2.604 < z \leq 3.646$	2	0.644

The expected numbers are computed by first finding the probabilities of the 7 classes using the standard normal distribution table. For example the first class has the probability $P(-3.646 \leq z \leq -2.604) = P(z \leq -2.604) - P(z \leq -3.646) = 0.00460 - 0.00013 = 0.00447$. The expected number is then $E_1 = 144(.00447) = 0.644$. The computed χ^2 value is

$$\chi^2 = \sum_{i=1}^{n} \frac{(O_i - E_i)^2}{E_i} = 7.416$$

Since $\chi^2_{4,0.95} = 9.49$ we do not reject the hypothesis that the underlying distribution of the melting point data is normal.

In Example 7.16, the degrees of freedom for the tabular χ^2 value were $(7 - 1) - 2 = 4$. The reduction from 6 degrees of freedom to 4 degrees of freedom was used because of having to estimate the two parameters μ and σ^2 in the hypothesized distribution. In using the χ^2 goodness of fit test to test that the sample came from a pecified distribution, a standard procedure is to decrease the egrees of freedom, k - 1, by c, where c is the number of paraeters estimated from the data.

If some of the expected numbers are too small, the validity of ne χ^2 test of the type discussed in this section may be questionable. If the expected numbers are all greater than or equal to 5, then the χ^2 test is valid. However, according to some prominent researchers, some of the expected numbers may be less than 5 and the χ^2 test can still be valid. For an excellent exposition on χ^2 tests the reader is referred to Conover ([1], Chapter 6).

In example 7.16, categories 1 and 2 can be combined as can categories 6 and 7. This yields 5 categories, with all categories having expected numbers greater than 5. See Problem 7.28.

7.13 CONTINGENCY TESTS

Another use of the χ^2 statistic is in contingency testing where n randomly selected items are classified according to two different criteria. The most common example is in the testing of materials subjected to a specific physical test. Here it is desired to determine whether some protective measure or sample preparation technique has been effective or not. Other examples are in quality control where it is desired to determine whether all inspectors are equally stringent. For testing the hypothesis that the outcomes are significant from such a series of binomial (yes or no) populations, the χ^2 test is a very popular procedure. The comments made in section 7.12, on the approximate nature of the χ^2 test and how the performance of the test is affected if some of the expected numbers are small, are applicable here also.

Example 7.17 Consider the case where pressure gauges are being hydraulically tested by 3 inspectors prior to shipment. It has been noted that their acceptances and rejections for some period of time have been as follows:

	Inspector			
	A	B	C	Totals
Passed	300	100	200	600
Failed	40	20	40	100
Totals	340	120	240	700

In order to test the hypothesis H_0: all inspectors are equally stringent, the test statistic used is

$$\chi^2 = \sum_{i=1}^{2} \sum_{j=1}^{3} (O_{ij} - E_{ij})^2/E_{ij} \quad , \tag{7.13}$$

where O_{ij} are the number observed to be passed by the jth inspector and E_{ij} are the number expected to be passed by the jth inspector. From the data in the table we see that

$$O_{11} = 300,\ O_{12} = 100,\ O_{13} = 200 \quad ,$$

$$O_{21} = 40,\ O_{22} = 20,\ O_{23} = 40 \quad ,$$

$$\sum_{j=1}^{3} O_{1j} = 600,\quad \sum_{j=1}^{3} O_{2j} = 100 \quad ,$$

$$\sum_{i=1}^{2} O_{i1} = 340,\quad \sum_{i=1}^{2} O_{i2} = 120,\quad \sum_{i=1}^{2} O_{i3} = 240 \quad ,$$

and

$$\sum_{i=1}^{3} \sum_{j=1}^{2} O_{ij} = n = 700 \quad .$$

In general a contingency table will have r rows and c columns. The statistic is then

$$\chi^2 = \sum_{i=1}^{r} \sum_{j=1}^{c} \frac{(O_{ij} - E_{ij})^2}{E_{ij}} \quad , \tag{7.14}$$

which has an approximate χ^2-distribution with (r-1)(c-1) d.f. The statistic is used to test whether one attribute (corresponding to row) is contingent on the other attribute (corresponding to column) or whether the two attributes are independent. Under H_0 the attributes are independent and the probability p_{ij} that an item selected at random falls in the class corresponding to the

i^{th} row and j^{th} column is $p_{i.}p_{.j}$, where $p_{i.}$ = probability that the item falls in the ith row, $p_{.j}$ = probability that the item falls in the jth column. The hypothesis can then be written as H_0: $P_{ij} = p_{i.}p_{.j}$ for $i = 1,2,\ldots,r$; $j = 1,2,\ldots,c$. The expected numbers E_{ij} under H_0 are $E_{ij} = np_{ij} = np_{i.}p_{.j}$. Now the probabilities $p_{i.}$ and $p_{.j}$ are not specified in H_0. Therefore we estimate them with

$$E_{ij} = np_{i.}p_{.j} = n\left(\frac{\sum_{j=1}^{c} O_{ij}}{n}\right)\left(\frac{\sum_{i=1}^{r} O_{ij}}{n}\right)$$

$$= \left(\sum_{j=1}^{c} O_{ij}\right)\left(\sum_{i=1}^{r} O_{ij}\right)\Big/ n \quad ,$$

and these are the E's that go into Eq. (7.14). In other words, to find the expected number for a given cell, simply multiply the corresponding row and column totals together and divide by the total number of items selected.

Thus the expected number in each case for this example is found from

$$E_{ij} = \sum_i(\text{passed})_i \sum_j(\text{inspector})_j / \sum_{ij}\sum(\text{total})$$

or

$$E_{ij} = \left(\sum_{j=1}^{3} O_{ij}\right)\left(\sum_{i=1}^{2} O_{ij}\right)\Big/ \sum_{i=1}^{2}\sum_{j=1}^{3} O_{ij} \quad .$$

We have

$$E_{11} = 600(340)/700 = 291.428 \quad ,$$

$$E_{12} = 600(120)/700 = 102.857 \quad ,$$

$$E_{13} = 600(240)/700 = 205.714 \quad ,$$

$$E_{21} = 100(340)/700 = 48.571 \quad ,$$

$$E_{22} = 100(120)/700 = 17.143 \quad ,$$

$$E_{23} = 100(240)/700 = 34.286 \quad .$$

Thus the computed χ^2 value is

$$\chi^2 = \frac{(O_{11} - E_{11})^2}{E_{11}} + \frac{(O_{12} - E_{12})^2}{E_{12}} + \frac{(O_{13} - E_{13})^2}{E_{13}}$$

$$+ \frac{(O_{21} - E_{21})^2}{E_{21}} + \frac{(O_{22} - E_{22})^2}{E_{22}} + \frac{(O_{23} - E_{23})^2}{E_{23}}$$

$$= \frac{(300 - 291.428)^2}{291.428} + \frac{(100 - 102.875)^2}{102.875} + \frac{(200 - 205.714)^2}{205.714}$$

$$+ \frac{(40 - 48.571)^2}{48.571} + \frac{(20 - 17.143)^2}{17.143} + \frac{(40 - 34.286)^2}{34.286}$$

$$= 0.4916 \quad .$$

This value of χ^2 is compared with the tabular value $\chi^2_{(r-1)(c-1),1-\alpha}$ which, as r = number of inspectors = 3 and c = number of outcomes = 2, is, by Table V, $\chi^2_{2,0.95} = 5.991$. Since $\chi^2 < \chi^2_{2,0.95}$, the hypothesis is accepted and we may say that no one inspector is more demanding than the other two.

PROBLEMS

7.1 Random samples from two batches of material were taken and analyzed in the same laboratory. Twenty analyses of samples from batch I averaged 46.0% component A with a variance s^2 of 120. Eighteen analyses from batch II had a mean value of 39.1% compound A and a variance of 180. There is no reason to expect the variances to be significantly different for the two batches. Is batch No. I significantly (at the 5% significance level) higher in component A than batch II?

7.2 Using the data presented in Problem 6.9, is there adequate evidence (you choose a satisfactory risk level) that the new rosin source is better than the old one from a product color standpoint?

7.3 The chief engineer wants you to find out if the company's "current ratio" (now 3.0) for the year 1966 is equal to, greater, or less than the mean value of the other 129 leading companies in the chemical process industries. Current ratio is the ratio obtained by dividing current assets by current liabilities. The 95% confidence interval is required. Available* current ratio data are:

2.3	2.4	3.6	1.9	2.6	2.4	1.6
1.4	2.7	2.0	2.4	2.7	2.3	2.8
3.5	4.1	2.3	2.2	2.4	2.5	2.8
2.4	1.7	1.9	2.3	6.2	2.0	2.8
2.8	4.6	2.3	3.0	2.2	2.3	4.3
2.3	2.6	4.7	1.9	4.0	2.4	2.5
2.7	2.4	1.6	2.4	5.1	2.5	2.5
3.6	2.8	1.9	2.2	1.9	2.6	4.0
3.1	1.9	2.7	2.5	3.7	2.2	2.3
1.7	2.4	2.3	1.9	2.4	2.7	2.6
1.8	1.6	2.4	3.0	2.5	3.2	2.8
1.4	2.1	2.2	4.6	1.8	2.4	1.8
2.9	3.2	2.3	3.3	2.0	2.7	2.2
4.6	1.4	2.0	3.4	1.4	2.4	1.7
2.9	2.5	2.7	2.1	2.2	2.8	2.4
2.6	2.7	2.3	3.6	2.7	3.7	2.2
1.6	1.9	3.0	1.9	2.9	3.1	3.9
2.9	2.1	2.1	1.8	3.0	2.8	3.0
2.6	2.5	2.6	2.6	2.1	3.3	2.5
2.1	2.3	2.1	1.2	2.6	3.9	

7.4 A manufacturer of light bulbs claims that on the average less than 1% of all light bulbs manufactured by his firm are defective.

*Data reprinted with permission from Chem. and Eng. News, 45, No. 34, 33A-47A (Sept. 4, 1967). Copyright 1967 by the American Chemical Society.

A random sample of 400 light bulbs contained 12 defective. On the evidence of this sample, do you believe the manufacturer's claims? Assume that the maximum risk you wish to run of falsely rejecting the manufacturer's claim has been set at 2%.

7.5 A lot of rolls of paper is acceptable for making bags for grocery stores if its mean breaking strength on a standard sample is not less than 40 lb. A random sample has a mean breaking strength of 39 lb and a standard deviation s of 2.4 lb. Twenty samples were used.

(a) Should the lot be rejected with a predetermined acceptable chance of rejecting good lots of 0.010?

(b) What is the chance of nonrejection of a lot which has true mean breaking strength of 39 lb?

7.6 Using the silicon analyses in Problem 6.11, use $\alpha = 0.02$ to test the hypothesis that the silicon content is at least as good as that when using the old supplier's raw material.

7.7 Among the design criteria for a machine producing molding powder are that it will produce cylindrical pellets by extrusion through 1/8-in. spinneret holes. These pellets are produced by cutting the strings of polymer so produced into pieces 3/16 in. long. To determine whether the unit was performing properly, a sample of 100 pellets was selected and examined. The mean length determined was 0.197 in. The standard deviation was 0.005 in. Test the hypothesis that the unit was in proper working order using a two-tailed t-test at the 0.05 and 0.01 significance levels.

7.8 For the fiber data of Problem 6.17, test the hypothesis (at the $\alpha = 0.02$ level) that the mean breaking loads are the same for the two types of fibers.

7.9 A manufacturer of electrical cable ties claims to have developed a new material with a mean breaking strength of 25 lb_f. From a sample of 50 ties tested, the mean breaking strength was found to be 24 lb_f with a standard deviation of 0.82 lb_f. State and test the appropriate hypotheses at the 2% significance level.

7.10 In the determination of carbon in benzoic acid, ten technicians each analyzed a pair of samples with the following results*

69.03	68.96
69.18	69.22
69.58	69.43
68.79	68.98
69.23	69.17
69.14	69.42
68.86	68.73
68.86	68.81
68.80	68.83
69.14	69.24

The theoretical amount of carbon is 68.84%. What is the highest significance level at which you can say that there is no statistical difference between the theoretical value and the average value determined in this test? What is the standard error of the analysis for all technicians?

7.11 A large textile mill recently installed improved air handling and cleaning equipment in an attempt to meet the OSHA standard of 200 $\mu g/m^3$. The results of 7 before and after tests are shown below. Was the expenditure for the new dust control system justified by the results?

---

*Data reprinted with permission from W. J. Youden, Anal. Chem., 20, 1136-1140 (1948). Copyright 1948 by the American Chemical Society.

Original	Modified
738.6	129.9
839.9	173.9
729.7	159.6
230.3	173.9
308.5	188.5
356.7	167.1
339.1	147.4
398.1	177.6
524.0	168.0
513.1	175.7
469.7	182.2
275.9	146.9
359.5	158.9

7.12 The unit for producing the protective atmosphere for a steel heat-treating process consists of a combustion chamber which is fed with a fuel gas and a limited amount of air. The fuel gas may be assumed to be nitrogen-free. In making material balances about the unit, the air rate is calculated from a nitrogen balance and a measurement of the product gas rate and composition. For this particular situation the nitrogen balance equation reduces to

$$Ax_a = Px_p \quad ,$$

where A and P are air and product rates in mole/hr and x_a and x_p are mole fractions nitrogen in the two streams. The mole fraction of nitrogen in the product gas is determined from an Orsat analysis by difference. For a given test period the following data were obtained on the product gas:

Volumetric analysis, dry basis

	x_i	σ_{x_i}
CO	0.122	0.002
CO_2	0.011	0.002
O_2	0.022	0.001
N_2	0.865 (by difference)	

$P = 84.2$ lb moles/hr $\qquad \sigma_p = 2.2$ lb moles/hr

The air composition is assumed to be 79.0% nitrogen but there is some variability with $\sigma_{x_{N_2}}$ in the air of 0.0025 mole fraction. Estimate the air rate and 95% confidence limits of that flow rate. Twenty samples were used.

7.13 Twelve observations were made for the flow rate in liters /min for two cotton dust samplers. The results are shown below. Are these two samplers equivalent? Use the 98% confidence interval.

# 1			# 2		
7.49	7.49	7.48	7.55	7.51	7.54
7.46	7.47	7.48	7.49	7.52	7.51
7.51	7.46	7.47	7.52	7.52	7.52
7.50	7.48	7.49	7.56	7.48	7.57

7.14 Fluid flow through a packed bed was examined on a pilot unit using 2-mm glass beads in a plastic column (i.d. = 1.51 in.). A test was made to find the pressure drop to be expected from a bed length of 60 $\frac{5}{8}$ in. and a flow rate of approximately 13.5 ml water/sec. Two pressure readings were made for each flow rate measurement. Determine the mean value, sample variance, and 95% confidence interval (using Student's t-test) for both the pressure drop and the flow rate. Does the experimental $-\Delta p$ differ significantly from the theoretical value of 60.0 cm oil?

$-\Delta p$ (cm oil, Sp.gr. = 1.75)		Flow (ml/sec)
58.0	60.5	13.60
57.5	59.5	13.45
57.0	59.0	13.59
58.0	60.0	13.52
57.0	60.5	13.45
58.0	61.0	13.52

7.15 In a routine evaluation of laboratory procedures, a revised method (#2) was checked against the current method (#1). The results are below.

Is the accuracy the same for both methods? Is the precision, as measured by the standard deviation, of the revised method as good as that of the current method?

Revised Method	Current Method
0.789	0.784
0.790	0.792
0.785	0.743
0.786	0.734
0.789	0.732
0.790	0.738
0.800	0.743
0.775	0.772
	0.748
	0.736

7.16 A group of seniors did a series of experiments to determine the overall heat transfer coefficient for the vacuum drying of sand. The group worked two successive afternoons, making five trials per day. Their results are presented below:

First day (Btu/hr ft^2°F)	Second day (Btu/hr ft^2°F)
4.68	4.70
4.73	4.67
4.65	4.63
4.69	4.58
4.70	4.57

During analysis of the data, however, the group suddenly makes the startling discovery that nobody remembered to check the pressure gauge periodically during the experiment. Hence the steam may have varied somewhat from the desired value of 20 psig. Must the group discard their second day's work? They must be 95% sure that the steam pressure did not drift during their work. Reference to old unit operations reports shows that the average variance for all prior trials is 0.20 (Btu/hr ft^{2}°F)^{2}.

7.17 The average wastage per day from a particular synthetic fiber process is 2650 lb. The standard deviation for the amount wasted on a daily basis as measured over the past 9 months is 140 lb. A new mixer has been installed in the solvent-monomer blending system in an attempt to reduce the amount of scrap material produced. It is fairly safe to assume that the daily variation in material wasted will not be significantly affected by this change. The following data were obtained from a 2-week trial of a new mixer (values are in pounds per day):

2540	2390	2530	2250	2170
2380	2400	2160	2250	2430
2310	2490	2200	2610	

(a) Was the modification successful? Use the 5% significance level to support your answer statistically.

(b) Calculate the 98% confidence interval for the daily amount of waste to be expected when using the new mixer.

Hint: Code the data by first subtracting 2400 and then dividing by 10 for ease in calculations.

7.18 Uniformity of abrasive coverage* within and between samples of coated abrasives can be measured by a uniformity index K. For the data below, assuming the samples to have been randomly selected, what is the 98% confidence interval on the difference in grit count?

---

*Bickering, C. A., New Angles on Old Problems of Measurement and Data Analysis. Ind. Qual. Control, 22, 510-512 (1966). Copyright American Society for Quality Control. Reprinted by permission.

Grit Count on Two Samples

Sample number 1	Number of grits	Sample number 2	Number of grits
1	22	1	33
2	27	2	21
3	31	3	31
4	21	4	27
5	18	5	38
6	24	6	25
7	18	7	24
8	35	8	35
9	24	9	26
10	18	10	32

7.19 A certain heat exchanger which had been performing poorly was taken out of service and cleaned thoroughly. In order to test the effectiveness of the cleaning, measurements were made before and after so as to determine the heat transfer coefficient. These were the results:

Run No.	Before	After
1	90.5	93.4
2	87.6	90.4
3	91.3	99.6
4	93.2	93.7
5	85.7	89.6
6	89.3	88.1
7	92.4	96.7
8	95.3	94.2
9	90.1	98.6
10	83.2	91.1

Did the cleaning of the heat exchanger significantly affect the heat transfer coefficient?

7.20 A random sample of 2640 students who live in the dormitories were asked about their daily eating habits by means of questionnaire X. Similar information was obtained from another sample of 2640 students living in the dormitories by means of questionnaire Y. The results were as follows:

Class interval	Number of students	
	X	Y
Miss no meals per week	385	316
Miss 1-4 meals per week	1490	1581
Miss 5-8 meals per week	728	719
Miss 9 or more meals per week	37	24

Can the difference in these distributions be purely due to chance? Support your answer from a statistical point of view?

7.21 In the manufacture of a synthetic fiber, the material, still in the form of a continuous flow of monofilaments, is subjected to high temperatures to improve its shrinkage properties. The shrinkage test results of fibers from the same source treated at two different temperatures are given below:

Percent shrinkage	
140°C	120°C
3.45	3.72
3.64	4.03
3.57	3.60
3.62	4.01
3.56	3.40
3.44	3.76
3.60	3.54
3.56	3.96
3.49	3.91
3.53	3.67
3.43	

Is the shrinkage at 140°C significantly less (at $\alpha = 0.01$) than that for material treated at 120°C? For data coded by the transformation $U = 100\ (x - 3.40)$,

$$\bar{U}_1 = 13.54\ , \qquad \bar{U}_2 = 36.0$$

$$S^2_{U_1} = 90.3\ , \qquad S^2_{U_2} = 456.89\ .$$

7.22 Pilot plant runs were made on two variations of a process to produce crude naphthalene. Product purities for the several runs are given below. In each series, all conditions were controlled in the normal manner and there is no evidence from the log sheets of bad runs.

Product purity (% naphthalene)	
Conditions A	76.0, 77.5, 77.0, 75.5, 75.0
Conditions B	80.0, 76.0, 80.5, 75.5, 78.5, 79.0, 78.5

(a) The development engineer reports that on the basis of these data conditions B give better purities but that control is poorer at these conditions. Do you agree with him? What are the chances that he is wrong in each of these conclusions?

(b) For a single run under conditions B, what is the probability that the product purity will be greater than 76.0%?

(c) What are the 95% confidence limits of the variance of the condition B population based on these data?

7.23 The percent conversion data below were obtained with two different catalysts used for the oxidation of organic materials containing N, S, H, O, and C atoms in spacecraft simulators to CO_2, H_2O, NO_2, and SO_2.

MnO_2: 55, 62, 64, 63, 58, 61, 60, 62, 64
CuO: 50, 57, 52, 55, 57, 54, 56, 51, 55

As MnO_2 is more expensive than CuO, it will be selected only if its efficiency is clearly superior to that of CuO. It has been decided that superiority can be adequately demonstrated if the conversion when using MnO_2 is at least 4% higher than that attainable with CuO. A significance level of 0.01 is required.

(a) Are the data groups above subject to equal variability?
(b) Should MnO_2 be specified for the catalytic oxidizers?
(c) What is the 99% confidence interval for the catalysts' mean percent conversion?

$\bar{x}_1 = 61$ $\Sigma x_1^2 = 33{,}559$ $\Sigma x_2^2 = 26{,}405$

$\bar{x}_2 = 54.111$ $(\Sigma x_2)^2 = 237{,}169$ $(\Sigma x_2)^2 = 237{,}169.$

7.24 The silica contents for two different cement samples were determined in triplicate by numerous laboratories in a cooperative testing program.* Selected results are presented below. From these data, was there any difference in the silica content of the samples? What are the 98% confidence intervals for mean silica content for the samples?

	Sample 1		
Test no.	X_1	X_2	X_3
5	9.96	10.12	10.32
6	10.26	10.28	10.36
7	10.22	10.11	- -
8	10.31	10.32	10.32
9	9.88	9.97	10.32
10	10.03	10.06	10.12
11	9.83	10.08	10.10
12	9.56	9.58	9.59
13	9.74	10.01	10.02
14	10.20	10.01	10.09

*Data reprinted with permission from P. J. Elving and M. G. Mellon, Anal. Chem., 20, 1140-1143 (1948).

		Sample 2	
Test no.	X_1	X_2	X_3
5	9.34	10.17	10.26
6	9.45	10.03	10.23
7	9.98	9.99	10.02
8	10.06	10.06	10.08
9	10.37	10.13	10.10
10	9.76	10.02	10.11
11	10.04	10.20	10.00
12	10.05	10.19	10.28
13	10.05	10.09	10.14
14	10.08	10.10	10.26

7.25 Two different methods for predicting the pressure drop through gate valves are to be evaluated with regard to relative variability. The X-data were calculated from the classic method; the Y-data were calculated from a supposedly improved method. The classic method uses only one flow system parameter; the improved method uses two. Does the improved method provide results which are less variable than those from the classic method? Note that the design equations for using the X- and Y-data are different so that you cannot calculate and compare the mean pressure drops. Use an F-test to evaluate the null hypothesis.

X	Y
0.363	0.240
0.277	0.208
0.536	0.201
0.634	0.278
0.549	0.371
0.714	
0.326	
$\Sigma X = 3.399$	$\Sigma Y = 1.298$
$\Sigma X^2 = 1.815223$	$\Sigma Y^2 = 0.35619$

7.26 Random number generators are essential in simulation studies. A generator that is to be used yielded n = 100 values which are tabulated as follows:

Interval	Frequency
$0 \leq x < 0.25$	18
$0.25 \leq x < 0.5$	20
$0.5 \leq x < 0.75$	40
$0.75 \leq x < 1$	22

Test the hypothesis that these data conform to a uniform distribution in the interval (0, 1). Use $\alpha = .05$.

7.27 An electronic component is under study. A sample of 100 components yields the following empirical distribution of life lengths (in 100 hrs.), where x denotes a typical life length:

Class	Frequency (O_i)
$0 < x \leq 2$	45
$2 < x \leq 4$	16
$4 < x \leq 6$	10
$6 < x \leq 8$	12
$8 < x \leq 10$	7
$10 < x \leq 12$	4
$x > 12$	6

It is desired to test the hypothesis that these data came from an exponential distribution with mean $\mu = 4$; that is, $f(x) = (1/4)e^{-x/4}$, $x > 0$. Carry out a goodness-of-fit χ^2-test for this hypothesis.

7.28 Perform a χ^2 goodness-of-fit test on the data in Example 7.16, after combining the first 2 categories and categories 6 and 7, thus yielding 5 categories. Use $\alpha = .05$.

7.29 Samples of a synthetic lipid were applied to soil samples to determine their effectiveness in reducing evaporative water losses in small scale tests. All the soil samples were made up from the same batch of well-mixed materials. Twelve soil samples were available. Half were sprayed with a wetting agent prior to application of the lipid; the other six samples were not. The results obtained in grams of water lost per square decimeter per unit time for a particular set of temperature and humidity conditions were as follows:

Sample	1	2	3	4	5	6
Lipid	11.5	13.4	14.0	13.6	11.6	14.6
Wetting agent + lipid	10.8	10.8	12.5	12.1	12.1	13.5

Did the inclusion of the wetting agent significantly affect the water loss rate? Use the 5% significance level and work this problem by the t-test and by the χ^2-test.

7.30 Regional differences in the resistance of mosquitoes to DDT are believed to be one cause of the variation in the severity of encephalitis outbreaks in the Southwest during the summer of 1968. Samples of 200 mosquitoes were randomly collected in four regions. These samples were exposed to a standard DDT level with the following results:

Area	Died	Lived
1	132	68
2	110	90
3	124	76
4	114	86

(a) State and test the null hypothesis.

(b) What inferences are possible from this data and the outcome in (a)?

7.31 An engineer has developed a correlation for the prediction of property A of a certain product line from information which is gathered routinely. He proposes that measurement of the property be discontinued since it can be predicted accurately by his correlation method and the factors on which his correlation is based are determined with high precision. The quality control engineer is not convinced and has tests made on 18 samples of material in order to compare measured and predicted values. The results are as follows:

	Property A	
Sample No.	Measured	Predicted
1	78	74
2	59	71
3	56	52
4	94	68
5	84	68
6	81	85
7	66	79
8	78	70
9	59	64
10	56	39

Sample No.	Measured	Predicted
11	88	77
12	88	83
13	75	62
14	75	74
15	72	74
16	81	83
17	84	78
18	73	70

Do you recommend discontinuance of the test? Set up a test procedure in detail showing exactly what will be calculated and how the decision is made. Do not perform calculations.

7.32 The observed values and those predicted by the Euler formula for the Prandtl number, N_{Pr},

$$N_{Pr} = \frac{C_p \mu}{k}$$

of gases at 1 atm. and 273.2°K are given below.* Are there any differences at the 95% confidence level between the predicted and observed values for the gases listed?

Gas	Predicted	Observed
Ne	0.67	0.66
Ar	0.67	0.67
H_2	0.73	0.70
N_2	0.74	0.73

*Bird, R. B., Stewart, W. E., and E. N. Lightfoot, Transport Phenomena, pp. 256-257. Copyright 1960 by John Wiley & Sons, Inc. Data reprinted by permission of John Wiley & Sons, Inc.

Gas	Predicted	Observed
O_2	0.74	0.74
SO_2	0.79	0.86
CO_2	0.78	0.78

7.33 White River Lake was recently stocked with bass. As one of the most avid fishermen in this department, you wanted to know how many bass are caught each week and what proportion of these bass are at or over the legal "keep" limit for length before you knock off working on the design project to go fishing. The boat patrol has the following information available from a sample of 5 fishermen out of the 100 at the lake week before last:

Fisherman (arbitrary numbering)	Bass caught	"Keepers"
1	0	0
2	5	4
3	2	2
4	3	2
5	5	2

From this information, what are the answers to your questions? You actually went fishing at the lake yesterday and caught three bass, two of which were of legal size. Are you an average fisherman? Better than average? Worse than average? Support your answers to these last three questions by testing the appropriate hypotheses.

7.34 A quarter is flipped 100 times. Heads turn up 65 times. Can we justifiably state that the coin is biased at the 1% significance level?

7.35 In a chemical process, the amount of a particular impurity is directly dependent on temperature as given by the relation $y = 0.0310x$ [illegible] 2.44 where y = % impurity and x = temperature $^{\circ}C$.

The impurity concentration can be uniformly determined with a standard deviation of 0.002%. What variation in impurity concentration can be expected at 86°C?

7.36 In the development of a fluorescent tracer material for cotton ginning, 2 samples of leaf were crushed. Both samples came from the same field. One sample was used as the control for the experiment. It was subjected to the full dyeing regimen but omitted the dye. The other sample was dyed by the standard procedure. The results, in terms of friability (ease of crushing) in each particle size range are given below. Did the dye itself affect the friability at the 5% level?

Size	Control	Dyed
12	123	132
20	509	477
40	312	323
100	44	61
140	3	4
270	3	2
pan	6	1

7.37 In the experimental evaluation of a manufacturing facility, the observed and predicted noise levels are shown below in decibels (A scale). Was the noise suppression system properly installed?

Area	Predicted	Measured
1	70	82
2	54	62
3	80	90
4	60	75
5	60	82
6	60	68

7.38 The following data* were obtained for the actual and predicted weight percent of the total stream for C_3 and lighter components in a hydrocarbon cracking unit. Are the predicted values acceptable estimates of system performance?

Run	Actual	Predicted
1	6.5	5.6
2	6.3	5.7
3	5.0	4.6
4	4.9	4.7
5	6.5	5.8
6	6.5	6.0

Work this problem by both the F- and χ^2-tests.

---

*White, P. J., How Cracker Feed Influences Yield. Hydrocarb. Proc., 47 (5), 103-107 (1968). Reproduced with permission from HYDROCARBON PROCESSING, May, 1968.

8
ANALYSIS OF VARIANCE

8.1 INTRODUCTION

The technique known as analysis of variance (or AOV) employs tests based on variance ratios to determine whether or not significant differences exist among the means of several groups of observations, where each group follows a normal distribution. The analysis of variance technique extends the t-test (Section 7.6) used to determine whether or not two means differ to the case where there are three or more means.

Analysis of variance is particularly useful when the basic differences between the groups cannot be stated quantitatively. A <u>one-way analysis of variance</u> is used to determine the effect of one independent variable on the dependent variable. A two-way analysis of variance is used to determine the effects of two independent variables on a dependent variable, and so on. As the number of independent variables increases, the calculations become much more complex and are best carried out on a high speed digital computer. The term independent variable used here is what is also referred to as a factor or treatment. We now develop some useful and common notation.

Suppose we have the array of data in Table 8.1. I and J denote the number of rows and colunms, respectively. Each column of observations can be thought of as a random sample of size I from a normal

TABLE 8.1

Data Array

Column →		1	2	...	j	...	J
Row ↓	1	Y_{11}	Y_{12}	...	Y_{1j}	...	Y_{1J}
	2	Y_{21}	Y_{22}	...	Y_{2j}	...	Y_{2J}
	⋮	⋮	⋮		⋮		⋮
	i	Y_{i1}	Y_{i2}	...	Y_{ij}	...	Y_{iJ}
	⋮	⋮	⋮		⋮		⋮
	I	Y_{I1}	Y_{I2}	...	Y_{Ij}	...	Y_{IJ}

population. Each data point is subscripted first to identify its row location and second to identify its column location. Thus, $_{ij}$ is the ith entry in the jth column. The sum of the observations in the ith row is designated by $Y_{i.}$ defined by

$$Y_{i.} = \sum_{j=1}^{J} Y_{ij}, \tag{8.1}$$

where the dot refers to the variable that has been summed. The mean of the values in the ith row is then

$$\overline{Y}_{i.} = Y_{i.}/J \quad . \tag{8.2}$$

Similarly, the sum of the jth column is

$$Y_{.j} = \sum_{i=1}^{I} Y_{ij} \tag{8.3}$$

and the mean is

$$\overline{Y}_{.j} = Y_{.j}/I \quad . \tag{8.4}$$

The sum of all the values in the array (matrix) is designated by $Y_{..}$ where

$$Y_{..} = \sum_{i=1}^{I} \sum_{j=1}^{J} Y_{ij} = \sum_{i=1}^{I} Y_{i.} = \sum_{j=1}^{J} Y_{.j} \quad . \tag{8.5}$$

The mean of all the values in the array is called the grand mean $\overline{Y}_{..}$ where

$$\overline{Y}_{..} = Y_{..}/IJ \quad . \tag{8.6}$$

From here on, to simplify the equations we will designate

$$\sum_{i=1}^{I} \text{ by } \sum_{i} \text{ and } \sum_{j=1}^{J} \text{ by } \sum_{j} \quad . \tag{8.7}$$

Furthermore, as before, capital letters denote random variables and lower case letters denote particular values of the corresponding random variables.

8.2. GENERAL LINEAR MODEL

The general linear model is described by means of an equation which gives the relationship between a set of independent variables and parameters and a dependent (or response) variable Y. The term linear signifies that the expression containing the independent variables and the parameters is <u>linear</u> in the <u>parameters</u>. Symbolically the model may be written as

$$Y = f(X_1, X_2, \ldots, X_p; \beta_1, \beta_2, \ldots, \beta_k) + \varepsilon$$

where $X_1, X_2, \ldots, X_p$ denote the independent variables; $\beta_1, \beta_2, \ldots, \beta_k$ denote the parameters; ε denotes a random error; and Y denotes the dependent variable. The function f is linear in $\beta_1, \beta_2, \ldots, \beta_k$. As an example, the regression model

$$Y = \beta_0 + \beta_1 X_1 + \beta_2 X_2 + \beta_3/\sqrt{X_3} + \varepsilon \tag{8.8}$$

is linear because the parameters β_i all appear to the first power. The fact that some of the independent variables have exponents other than 1 is immaterial. On the other hand, the relation

$$Y = \beta_0\beta_1^X + \varepsilon \tag{8.9}$$

is non-linear as one of the parameters, β_1, does not appear to the first power. In both cases, ε is the experimental error term.

In some situations the independent variables are not continuous, but may take on the value 0 or 1. In such cases the general linear model is termed an analysis of variance (or experimental design) model. This type of model will be discussed in this chapter and in Chapter 11, and may include terms for the effects of different "factors" and their interactions (similar to synergism) on the dependent variable Y. Terms for different sources of error (variation) are included as appropriate. The number of independent variables and the design used to perform the experiment (more of this in Chapter 11) are used to label specific models. If only one independent variable β_j is involved, the result is a one-way analysis of variance model:

$$Y_{ij} = \mu + \beta_j + \varepsilon_{ij}. \tag{8.10}$$

If two parameters α_i and β_j are involved, the model is

$$Y_{ij} = \mu + \alpha_i + \beta_j + \varepsilon_{ij} \quad . \tag{8.11}$$

When more than one sample is taken for some experimental conditions, the model is unaffected: the experimental results should become more precise. If, however, the experiment results in samples which can be divided (subsampled), then a sampling error η_{ijk} exists and must be evaluated separately from the experimental error. An example of the model for one-way analysis of variance when subsampling is involved is

$$Y_{ijk} = \mu + \beta_j + \varepsilon_{ij} + \eta_{ijk} \quad . \qquad (8.12)$$

Suitable limits on the subscripts are given where each of these models is discussed in detail.

8.3 ONE-WAY ANALYSIS OF VARIANCE

One-way analysis of variance is used when we wish to test the equality of J population means. The procedure is based on the assumptions that each of J groups of observation is a random sample from a normal distribution and that the population variance $\sigma_Y^2 = \sigma^2$ is constant among the groups. There are two possible ways to estimate the population variance: pooled estimates and calculation of the variance of group means around the grand mean. We shall see that these two ways of estimating σ^2 will lead to a test of the equality of the J population means.

8.3.1 Pooled Variance Estimates

In the pooled estimate the sample variance for each group (each column of data in a one-way analysis) is calculated. These estimates are then weighted by the corresponding degrees of freedom to obtain a pooled sample variance. For any column j the sample variance is

$$S_j^2 = \frac{\sum_i (Y_{ij} - \overline{Y}_{\cdot j})^2}{I - 1} \quad . \qquad (8.13)$$

From Section 5.4 we know that $E(S_j^2) = \sigma^2$ for every j. We are assuming, for the sake of simplicity, that each column contains the same number of values. We then have J estimates in the form of Eq. (8.13). To pool them we weight each by its degrees of freedom (I-1), add them up, and divide by the total degrees of freedom which is J(I-1). The pooled sample variance is then

$$S_p^2 = \frac{\sum_j [(I-1)S_j^2]}{J(I-1)} \tag{8.14}$$

which may be combined with Eq. (8.13) to give

$$S_p^2 = \frac{\sum_j [\sum_i (Y_{ij} - \overline{Y}_{\cdot j})^2]}{J(I-1)} = \frac{SS_W}{J(I-1)} = MS_W \quad . \tag{8.15}$$

In this equation, the term SS_W is referred to as the sum of squares within groups or error sum of squares. The quantity SS_W when divided by the appropriate degrees of freedom J(I-1) is referred to as the mean square for within groups (or error mean square) and is denoted by MS_W.

As Eq. (8.15) is not particularly convenient for calculation purposes, it can be presented in the more usable form

$$S_p^2 = \frac{\sum_{ij}\sum (Y_{ij})^2 - (\sum_j Y_{\cdot j}^2)/I}{J(I-1)} = \frac{SS_W}{J(I-1)} = MS_W \quad . \tag{8.16}$$

The pooled estimator of the population variance, S_p^2, is an unbiased estimator for σ^2 regardless of whether the population means $\mu_1, \mu_2, \ldots, \mu_J$ are equal or not, because it takes into account deviations from each group mean $\overline{Y}_{\cdot j}$, j = 1,2,...,J. Unbiasedness follows from Eq. (8.15) since

$$\begin{aligned} E(S_p^2) &= \sum_j \{E[\sum_i (Y_{ij} - \overline{Y}_{\cdot j})^2/(I-1)]\}/J \\ &= \sum_j E(S_j^2)/J \\ &= \sigma^2 \quad . \end{aligned} \tag{8.17}$$

8.3.2 Variance of Group Means

A second method of estimating the population variance σ^2 is to calculate the sample variance of the group means around the grand mean by use of

$$S_{\bar{Y}}^2 = \frac{\sum_j (\bar{Y}_{\cdot j} - \bar{Y}_{\cdot\cdot})^2}{J - 1} \quad . \tag{8.18}$$

If the group population means are <u>all</u> equal, then $S_{\bar{Y}}^2$ is an unbiased estimate of the variance of the population mean $\sigma_{\bar{Y}}^2$. To obtain an estimator of the population variance σ^2 recall that

$$S_{\bar{Y}}^2 = S_Y^2/I \quad . \tag{8.19}$$

Combining Eqs. (8.18) and (8.19) we have our second estimator of the population variance σ^2:

$$S_Y^2 = \frac{I\sum_j (\bar{Y}_{\cdot j} - \bar{Y}_{\cdot\cdot})^2}{J - 1} = \frac{SS_B}{J - 1} = MS_B \quad . \tag{8.20a}$$

If there are unequal numbers of observations for the groups, Eq. (8.20a) may be written as

$$S_Y^2 = \sum_j I_j(\bar{Y}_{\cdot j} - \bar{Y}_{\cdot\cdot})^2/(J - 1) = \frac{SS_B}{J - 1} = MS_B \quad . \tag{8.20b}$$

The quantities SS_B and MS_B are usually referred to as the <u>between-groups sum of squares</u> and <u>mean square between groups</u>, respectively. The estimator S_Y^2 can be written as

$$S_Y^2 = \frac{\sum_j Y_{\cdot j}^2/I - Y_{\cdot\cdot}^2/IJ}{J - 1} = \frac{SS_B}{J - 1} = MS_B \quad , \tag{8.21}$$

which is a more useful form for calculation purposes.

The estimator S_Y^2 given in (8.20a or b) or (8.21) is an unbiased estimator of σ^2 only when the group population means are equal.

If the population means $\mu_1, \mu_2, \ldots, \mu_J$ are not all equal then the estimator S_Y^2 overestimates σ^2, that is, $E(S_Y^2) > \sigma^2$. The estimators S_Y^2 and S_p^2 are linked by a very important identity given by

$$\sum_j \sum_i (Y_{ij} - \overline{Y}_{..})^2 = I \sum_j (\overline{Y}_{.j} - \overline{Y}_{..})^2 + \sum_j \sum_i (Y_{ij} - \overline{Y}_{.j})^2 \quad . \qquad (8.22)$$

The left-hand side of Eq. (8.22) is usually referred to as the total sum of squares corrected for the mean and is denoted by SS_{TC}. To verify (8.22) we write

$$\begin{aligned}
\sum_j \sum_i (Y_{ij} - \overline{Y}_{..})^2 &= \sum_j \sum_i [(Y_{ij} - \overline{Y}_{.j}) + (\overline{Y}_{.j} - \overline{Y}_{..})]^2 \\
&= \sum_j \sum_i [(Y_{ij} - \overline{Y}_{.j})^2 + (\overline{Y}_{.j} - \overline{Y}_{..})^2 \\
&\qquad + 2(Y_{ij} - \overline{Y}_{.j})(\overline{Y}_{.j} - \overline{Y}_{..})] \\
&= \sum_j \sum_i (Y_{ij} - \overline{Y}_{.j})^2 + \sum_j \sum_i (\overline{Y}_{.j} - \overline{Y}_{..})^2 \quad . \qquad (8.23)
\end{aligned}$$

The cross-product term vanishes since

$$\sum_j \sum_i 2(Y_{ij} - \overline{Y}_{.j})(\overline{Y}_{.j} - \overline{Y}_{..}) = \sum_j 2(\overline{Y}_{.j} - \overline{Y}_{..}) \sum_i (Y_{ij} - \overline{Y}_{.j})$$

and $\sum_i (Y_{ij} - \overline{Y}_{.j}) = 0$. Equation (8.22) thus follows from (8.23). The identity given by Eq. (8.22) can be written as $SS_{TC} = SS_B + SS_W$. In other words the total variation is partitioned into two components, a component SS_B which reflects variation among groups and a component SS_W which reflects experimental error or sampling variation. The degrees of freedom associated with SS_{TC} are also partitioned into the degrees of freedom associated with SS_B and SS_W, i.e., $IJ - 1 = (J - 1) + J(I - 1)$.

If the means $\mu_1, \mu_2, \ldots, \mu_j$ are all equal then S_p^2 and S_Y^2 are independent so that the random variable

$$F = \frac{S_Y^2}{S_p^2} = \frac{SS_B/(J - 1)}{SS_W/J(I - 1)} = \frac{MS_B}{MS_W} \tag{8.24}$$

has an F-distribution with J - 1 and J(I - 1) degrees of freedom. Thus under H_0: $\mu_1 = \mu_2 = \ldots \mu_J$ we would expect the value of F to be close to 1. If H_0 is not true then the value of S_Y^2 would tend to be larger than S_p^2 which would force F to be larger than 1. Consequently, based on the data, the hypothesis H_0 would be rejected if the computed F-value, $f = s_Y^2/s_p^2$, is too large. That is, the rejection region is of the form $F > F_{\nu_1,\nu_2,1-\alpha}$, where $\nu_1 = J - 1$ and $\nu_2 = J(I - 1)$.

8.3.3 Model for One-way Analysis of Variance

Let us consider the model for a one-way analysis of variance. Here it is assumed that the columns of data are J random samples from J independent normal populations with means $\mu_1, \mu_2, \ldots, \mu_J$ and common variance σ^2. The one-way analysis of variance technique will give us a procedure for testing the hypothesis H_0: $\mu_1 = \mu_2 = \ldots = \mu_J$ against the alternative H_A: at least two μ_j not equal. The statistical model gives us the structure of each observation in the I by J array in Table 8.1. The model is

$$Y_{ij} = \mu_j + \varepsilon_{ij} \quad . \tag{8.25}$$

This model says that the dependent variable Y_{ij} is made up of two parts: the first part μ_j which is the mean of the population corresponding to the jth column (population) and the second part, ε_{ij}, the random experimental error which is taken to have mean 0, i.e., $E(\varepsilon_{ij}) = 0$. This must be the case since $E(Y_{ij}) = \mu_j$. The model in Eq. (8.25) can be reduced to Eq. (8.10):

$$\begin{aligned} Y_{ij} &= \mu_j + \mu - \mu + \varepsilon_{ij} \\ &= \mu + (\mu_j - \mu) + \varepsilon_{ij} \\ &= \mu + \beta_j + \varepsilon_{ij} , \end{aligned} \tag{8.10}$$

where $\mu = \bar{\mu}_{\cdot} = \sum_j \mu_j/J$ is called the grand population mean and $\beta_j = \mu_j - \bar{\mu}_{\cdot}$ is called the effect of the jth population. Equation (8.10) states that any experimental value is the sum of a term representing the general location of the grand population mean plus a term β_j showing the displacement of a given population from the general location, plus a term giving the random experimental error ε_{ij} of the particular observation. The ε_{ij} are independent and normally distributed with mean 0 and variance σ^2.

If all the column means are equal, $\mu = \mu_j$ and $\beta_j = 0$. Therefore, the hypothesis H_0: $\mu_1 = \mu_2 = \ldots = \mu_J$ is the same as the hypothesis H_0: $\beta_j = 0$ for all j.

One is generally interested in the hypothesis H_0: $\beta_1 = \beta_2 = \ldots = \beta_J = 0$, which states that there is no population or column effect. This means that the variation in $\bar{y}_{\cdot 1}, \bar{y}_{\cdot 2}, \ldots, \bar{y}_{\cdot J}$ is due to experimental error and not to any difference in population means. To test the hypothesis H_0 we use the F-statistic given in Eq. (8.24),

$$F = \frac{SS_B/(J - 1)}{SS_W/J(I - 1)} = \frac{MS_B}{MS_W} .$$

The rejection region at significance level α is $F > F_{J-1,J(I-1),1-\alpha}$.

Recall from Eq. (8.17) that $E(S_p^2) = E(MS_W) = \sigma^2$. Furthermore, if H_0: $\beta_1 = \beta_2 = \ldots = \beta_J = 0$ is true, then $E(S_Y^2) = E(MS_B) = \sigma^2$. However, if H_0 is not true then $E(MS_B) \neq \sigma^2$. To show this, consider

$$E(SS_B) = E[\sum_{ij}\sum(\overline{Y}_{\cdot j} - \overline{Y}_{\cdot\cdot})^2]$$

$$= I\sum_j E(\overline{Y}_{\cdot j} - \overline{Y}_{\cdot\cdot})^2.$$

Now, $Y_{ij} = \mu + \beta_j + \varepsilon_{ij}$. Therefore,

$$\overline{Y}_{\cdot j} = \sum_i Y_{ij}/I = \sum_i(\mu + \beta_j + \varepsilon_{ij})/I = \mu + \beta_j + \overline{\varepsilon}_{\cdot j}$$

and

$$\overline{Y}_{\cdot\cdot} = \sum_j \overline{Y}_{\cdot j}/J = \sum_j(\mu + \beta_j + \overline{\varepsilon}_{\cdot j})/J = \mu + \overline{\varepsilon}_{\cdot\cdot} ,$$

since

$$\sum_j \beta_j = \sum_j(\mu_j - \mu) = 0 .$$

Thus,

$$E(SS_B) = I \cdot E\{\sum_j[\beta_j + (\overline{\varepsilon}_{\cdot j} - \overline{\varepsilon}_{\cdot\cdot})]^2\}$$

$$= I \cdot E\{\sum_j[\beta_j^2 + (\overline{\varepsilon}_{\cdot j} - \overline{\varepsilon}_{\cdot\cdot})^2]\}$$

since

$$E[2\beta_j(\overline{\varepsilon}_{\cdot j} - \overline{\varepsilon}_{\cdot\cdot})] = 2\beta_j E(\overline{\varepsilon}_{\cdot j} - \overline{\varepsilon}_{\cdot\cdot}) = 0 .$$

Consequently,

$$E(SS_B) = I\sum_j \beta_j^2 + I \cdot E[\sum_j(\overline{\varepsilon}_{\cdot j} - \overline{\varepsilon}_{\cdot\cdot})^2] ,$$

and since $\overline{\varepsilon}_{\cdot 1}, \overline{\varepsilon}_{\cdot 2}, \ldots, \overline{\varepsilon}_{\cdot J}$ can be considered to be a random sample of size J from a normal population with mean 0 and variance

σ^2/I, it follows that $E\left[\sum_j(\bar{\varepsilon}_{\cdot j} - \bar{\varepsilon}_{\cdot\cdot})^2/(J-1)\right] = \sigma^2/I$. Thus

$$E(SS_B) = I\sum_j \beta_j^2 + I(J-1)(\sigma^2/I)$$

$$= I\sum_j \beta_j^2 + (J-1)\sigma^2$$

and

$$E(S_Y^2) = E(MS_B)$$

$$= E\left[SS_B/(J-1)\right]$$

$$= \sigma^2 + I\sum_j \beta_j^2/(J-1) .$$

The quantities $E(MS_W)$ and $E(MS_B)$ are called <u>expected mean squares.</u> The expected mean squares suggest what ratio to use in testing H_0. That is, since $E(MS_B) = \sigma^2$ when H_0: $\beta_1 = \beta_2 = \dots = \beta_J = 0$ is true and $E(MS_B) > \sigma^2$ otherwise, and since MS_B/MS_W has an F-distribution when H_0 is true, then H_0 will have the rejection region $F > F_{I-1,J(I-1),1-\alpha}$. For the F-test we use the unbiased estimate of σ^2 in the denominator of the F-ratio and the biased estimate in the numerator.

The results needed to test H_0: $\beta_1 = \beta_2 = \dots = \beta_J = 0$ are usually summarized in the form of Table 8.2. It should be noted that $SS_T = SS_M + SS_B + SS_W$ and that the degrees of freedom add up similarly; that is, $IJ = 1 + (J-1) + J(I-1)$.

TABLE 8.2
One-Way Analysis of Variance

Source of variation	Degrees of Freedom	Sum of Squares	Mean square
Mean	1	$SS_M = y_{..}^2/IJ$	
Between columns	J-1	$SS_B = \sum_j y_{.j}^2/I - y_{..}^2/IJ$	$MS_B = \frac{SS_B}{J-1}$
Within columns (error)	J(I-1)	$SS_W = \sum_{ij}\sum y_{ij}^2 - \sum_j y_{.j}^2/I$	$MS_W = \frac{SS_W}{J(I-1)}$
Total	IJ	$SS_T = \sum_{ij}\sum y_{ij}^2$	

Example 8.1 The amount of fluoride in the local water supply was determined by the strontium chloranilate (A), ferric thiocyanate (B), alizarin lanthanum complexan (C), and eriochrome cyanine RC (D) colorimetric methods in a comparative study. Five replications were made for each test. To preclude bias from variations in the sample over the time required for the analyses, all samples were taken from a single 10-gal carboy of water. The results in ppm are

A	B	C	D
2	5	1	2
3	4	3	1
6	4	2	1
5	2	4	2
4	3	4	1

(a) Are the methods equivalent? Use the 5% significance level.

(b) What are the 95% confidence limits on the values obtained from each method?

(a) We first calculate the required sums and squares:

$$y_{\bullet A} = 20 \qquad \sum_i y_{iA}^2 = 90 \qquad \bar{y}_{\bullet A} = 4.0 \qquad \sum_j y_{\bullet j}^2 = 969$$

$$y_{\bullet B} = 18 \qquad \sum_i y_{iB}^2 = 70 \qquad \bar{y}_{\bullet B} = 3.6$$

$$y_{\bullet C} = 14 \qquad \sum_i y_{iC}^2 = 46 \qquad \bar{y}_{\bullet C} = 2.8$$

$$y_{\bullet D} = 7 \qquad \sum_i y_{iD}^2 = 11 \qquad \bar{y}_{\bullet D} = 1.4 \ .$$

From Eq. (8.16) we calculate the pooled sample variance:

$$s_p^2 = \left[\sum_j \sum_i y_{ij}^2 - \left(\sum_j y_{\bullet j}^2\right)/I\right]/J(I-1) = \frac{217 - 193.8}{16} = 1.45 = MS_W \ .$$

We then calculate the variance of the group means around the overall mean using Eq. (8.21):

$$s_Y^2 = \left[\sum_j y_{\bullet j}^2/I - y_{\bullet\bullet}^2/IJ\right]/(J-1) = \frac{969/5 - (59)^2/20}{3} = 6.58 = MS_B \ .$$

We next compare the equality of group means by calculating $f = s_Y^2/s_p^2 = 6.58/1.45 = 4.54$ according to Eq. (8.24). To test the null hypothesis H_0: $\beta_1 = \beta_2 = \beta_3 = \beta_4 = 0$, the calculated value of F is compared to $F_{3,16,0.95} = 3.25$. Since $f = 4.54 > 3.25$, we reject H_0 and conclude that there is a significant difference among the four methods. The AOV results are summarized below.

AOV Table

Source	d.f	SS	MS	F
Mean	1	174.05	--	--
Between	3	19.75	6.58	4.54
Error	16	23.20	1.45	--
Total	20	217.00		

(b) For 16 degrees of freedom, at the 95% confidence level, $t_{16,0.975} = 2.120$ and $t_{16,0.025} = -2.120$. The standard error is found as before from $s_{\bar{p}} = s_p/\sqrt{I} = \sqrt{1.45/5} = 0.5385$. The confidence limits for the means corresponding to the four colorimetric methods are [according to Section (6.8)]

$$\mu_A: \quad 4.0 \pm 2.12(0.5385) = 2.86 \text{ to } 5.14$$

$$\mu_B: \quad 3.6 \pm 2.12(0.5385) = 2.46 \text{ to } 4.74$$

$$\mu_C: \quad 2.8 \pm 2.12(0.5385) = 1.66 \text{ to } 3.94$$

$$\mu_D: \quad 1.4 \pm 2.12(0.5385) = 0.26 \text{ to } 2.54 \;..$$

Example 8.2 Evaluate the coke yield data in volume percent given in Problem 8.2. Using the following SAS program (SAS for Linear Models) for the analysis of variance when equal numbers of observations are present for all variables, we prepare the program below.

```
(put initial JCL cards/statements here)
DATA;
INPUT FEED $ 1 YIELD 3-6;
CARDS;
(yield data inserted here according to INPUT format)
PROC PRINT;
PROC ANOVA;
CLASS FEED;
MODEL YIELD = FEED;
TITLE 'COMPARISON OF HYDROGENATION LEVEL IN FEED ON COKE YIELD';
TITLE 'NOTE: A = 0, B = 300, C = 500';
(put final JCL cards/statements here)
```

The results below were obtained from this program.

SAS

COMPARISON OF HYDROGENATION LEVEL IN FEED ON COKE YIELD

ANALYSIS OF VARIANCE PROCEDURE

CLASS LEVEL INFORMATION

CLASS	LEVELS	VALUES
FEED	3	A B C

NUMBER OF OBSERVATIONS IN DATA SET = 18

NOTE: A = 0, B = 300, C = 500

OBS	FEED	YIELD
1	A	4.2
2	A	6.8
3	A	5.9
4	A	11.0
5	A	16.0
6	A	8.0
7	B	4.1
8	B	4.8
9	B	6.2
10	B	8.0
11	B	14.0
12	B	4.0
13	C	4.0
14	C	4.4
15	C	8.1
16	C	8.3
17	C	11.9
18	C	6.1

COMPARISON OF HYDROGENATION LEVEL IN FEED ON COKE YIELD

ANALYSIS OF VARIANCE PROCEDURE

DEPENDENT VARIABLE: YIELD

SOURCE	DF	SUM OF SQUARES	MEAN SQUARE	F VALUE
MODEL	2	11.24111111	5.62055556	0.41
ERROR	15	206.88333333	13.79222222	PR > F
CORRECTED TOTAL	17	218.12444444		0.6724

SOURCE	DF	ANOVA SS	F VALUE	PR > F
FEED	2	11.24111111	0.41	0.6724

The null hypothesis, as in all one-way analyses of variance, was that the β_j (treatment = hydrogenation level) do not affect the coke yield. From the calculated value of f = 0.41 and the 67.24% probability of finding a larger value, there is no reason to reject the hypothesis: it is thus accepted as probably true at the $\alpha = 0.05$ level.

If an unequal number of observations is available for the independent variables, the SAS program must be modified. PROC GLM (for general linear model) must be used instead of PROC ANOVA. This approach (SAS for Linear Models) is illustrated in the following example. The calculational routine is presented in Table 11.4.

Example 8.3 A sample of the coeds taking English 231 in one of our classrooms last summer was classified according to hair color (blond, brunette, redhead). Their heights in inches were then measured. What statistical model would be appropriate for analyzing these data? Indicate what hypotheses and tests would be useful in analyzing these data. Perform the tests of the hypotheses.

Blondes	Brunettes	Redheads
60	70	65
62	61	63
68	64	61
65	67	
63	64	
	66	
	63	

The SAS program is:

```
(put JCL cards/statements here)
DATA;
INPUT COLOR $ 1-2 HEIGHT 4-5;
CARDS;
(type in the data here in format specified by the INPUT statement
  using BL, BR, and RH for blonde, brunette, redhead)
PROC GLM;
CLASS COLOR;
MODEL HEIGHT = COLOR;
TITLE 'EFFECT OF HAIR COLOR ON HEIGHT OF FEMALE STUDENTS';
(put final JCL cards/statements here).
```

The SAS output obtained is shown below.

```
                              SAS
          EFFECT OF HAIR COLOR ON HEIGHT OF FEMALE STUDENTS
                  GENERAL LINEAR MODELS PROCEDURE
                     CLASS LEVEL INFORMATION
                  CLASS          LEVELS          VALUES
                  COLOR            3            BL BR RH
              NUMBER OF OBSERVATIONS IN DATA SET = 15

DEPENDENT VARIABLE: HEIGHT
SOURCE          DF   SUM OF SQUARES    MEAN SQUARE     F VALUE
MODEL            2      10.53333333     5.26666667        0.65
ERROR           12      97.20000000     8.10000000      PR > F
CORRECTED       14     107.73333333                     0.5394
 TOTAL
SOURCE          DF        TYPE I SS    F VALUE     PR > F
COLOR            2      10.53333333       0.65     0.5394

SOURCE          DF      TYPE III SS    F VALUE     PR > F
COLOR            2      10.53333333       0.65     0.5394
```

The model sum of squares appears three times in the output: in the AOV table and as separate entries below that table where it is labelled both as Type I and as Type III. The Type I sum of squares is a result of adding each source of variation sequentially to the model in the order listed in the MODEL statement. Type III sums of squares are used when it is necessary to adjust the effect of each independent variable for the presence of the others in order to eliminate interaction terms. As we have a one-way AOV model in this example, all of the sums of squares should be identical and are.

The interpretation of the results is straightforward. As the calculated value of $f = 0.65$ and there is a 53.94% probability of obtaining a greater value, the null hypothesis that hair color does not affect height is accepted as probably true.

8.4 TWO-WAY ANALYSIS OF VARIANCE

If we desire to study the effects of two independent variables (factors) on one dependent variable, we will have to use a _two-way analysis of variance._ For this case the columns represent various values or levels of one independent variable and the rows represent levels or values of the other independent variable. Each entry in the array in Table 8.1 of data points then represents one of the possible combinations of the two independent variables and how it affects the dependent variable. Here, we will consider the case of only one observation per data point.

8.4.1 Model for Two-Way Analysis of Variance

For a two-way analysis of variance the assumed model is

$$Y_{ij} = \mu_{ij} + \varepsilon_{ij} \quad .$$

In this case Y_{ij} is assumed to come from a normal population with mean μ_{ij} and variance σ^2. Furthermore, the Y_{ij} are independently distributed. The two-way model can be written as

$$Y_{ij} = \mu + \alpha_i + \beta_j + \varepsilon_{ij} \tag{8.26}$$

where

$\mu = \bar{\mu}_{..} = \sum_{ij}\sum \mu_{ij}/IJ$, $\alpha_i = \bar{\mu}_{i.} - \bar{\mu}_{..}$, $\beta_j = \bar{\mu}_{.j} - \bar{\mu}_{..}$, $\bar{\mu}_{i.} = \sum_j \mu_{ij}/J$,

and $\bar{\mu}_{.j} = \sum_i \mu_{ij}/I$. The parameter μ is the contribution of the grand mean, α_i is the contribution of the ith level of the row variable, β_j is the contribution of the jth level of the column variable, and ε_{ij} is the random experimental error. The model in Eq. (8.26) does not contain what is usually referred to as a row-column interaction; that is, the row and column effects are additive. The idea of interaction will be discussed briefly in Section 8.5 and in more detail in Chapter 11.

Two restrictions (or assumptions) related to the model in Eq. (8.26) are $\sum_i \alpha_i = 0$ and $\sum_j \beta_j = 0$. These follow from $\alpha_i = \bar{\mu}_{i.} - \bar{\mu}_{..}$ and $\beta_j = \bar{\mu}_{.j} - \bar{\mu}_{..}$.

Two hypotheses related to the model in (8.26) are

$$H_0: \mu_{1.} = \mu_{2.} = \dots = \mu_{I.}$$

and

$$H_0: \mu_{.1} = \mu_{.2} = \dots = \mu_{.J} \quad .$$

The first hypothesis says that there is no row effect, that is, the means across rows have the same value. Similarly, the second hypothesis says there is no column effect. The two hypotheses above can be written equivalently as

$$H_0: \alpha_1 = \alpha_2 = \dots = \alpha_I = 0 \tag{8.27}$$

and

$$H_0: \beta_1 = \beta_2 = \dots = \beta_J = 0 \quad . \tag{8.28}$$

The hypotheses relevant to no row or column effects are usually written in the form of (8.27) and (8.28). If both hypotheses are true, then $Y_{ij} = \mu + \varepsilon_{ij}$, which says that all the observations came from one fixed normal population.

It can be shown that the total variation can be partitioned according to the identity

$$\sum_{ij}\sum(Y_{ij} - \overline{Y}_{..})^2 = \sum_{ij}\sum(\overline{Y}_{.j} - \overline{Y}_{..})^2 + \sum_{ij}\sum(\overline{Y}_{i.} - \overline{Y}_{..})^2 + \sum_{ij}\sum(Y_{ij} - \overline{Y}_{.j} - \overline{Y}_{i.} + \overline{Y}_{..})^2 .$$

This identity can be written as

$$SS_{TC} = SS_C + SS_R + SS_E .$$

That is, the total variation is partitioned into variation due to columns (SS_C), variation due to rows (SS_R), and variation due to experimental error (SS_E). The degrees of freedom are likewise partitioned as

$$IJ-1 = (J-1) + (I-1) + (I-1)(J-1) .$$

The mean squares are defined by $MS_C = SS_C/(J-1)$, $MS_R = SS_R/(I-1)$, and $MS_E = SS_E/(I-1)(J-1)$. Furthermore, it can be shown (just as for the one-way model) that the expected mean squares are

$$E\left[MS_R\right] = \sigma^2 + J \sum_i \alpha_i^2/(I-1)$$

$$E\left[MS_C\right] = \sigma^2 + I \sum_j \beta_j^2/(J-1)$$

$$E\left[MS_E\right] = \sigma^2 .$$

Thus, to test H_0: all $\alpha_i = 0$, we use the F-statistic with I-1 and (I-1)(J-1) degrees of freedom,

$$F = \frac{MS_R}{MS_E} .$$

The rejection region is $F > F_{(I-1),(I-1)(J-1),1-\alpha}$.

Similarly, to test H_0: all $\beta_j = 0$, we use $F = MS_C/MS_E$ with J-1 and (I-1)(J-1) degrees of freedom. The rejection region is $F > F_{J-1,(I-1)(J-1),1-\alpha}$.

The results pertinent to the two-way model are summarized in Table 8.3.

TABLE 8.3

Two-Way Analysis of Variance

Source of variance	Degrees of freedom	Sum of squares	Mean square
Mean	1	$SS_M = y_{..}^2/IJ$	
Between columns	J-1	$SS_C = \sum_j y_{.j}^2/I - y_{..}^2/IJ$	$MS_C = \frac{SS_C}{J-1}$
Between rows	I-1	$SS_R = \sum_i y_{i.}^2/J - y_{..}^2/IJ$	$MS_R = \frac{SS_R}{I-1}$
Error	(I-1)(J-1)	$SS_E = SS_T - SS_C - SS_R - SS_M$	$MS_E = \frac{SS_E}{(I-1)(J-1)}$
Total	IJ	$SS_T = \sum_{ij}\sum y_{ij}^2$	

We note from Table 8.3 that the sum of squares due to error, SS_E, is $SS_T - SS_C - SS_R - SS_M$. The procedure is to compute SS_T, SS_M, SS_C, and SS_R and then determine SS_E according to the difference in Table 8.3. Statistical theory has shown that MS_E is an unbiased estimate of the population variance σ^2 regardless of whether the hypotheses are true or not. However, MS_R and MS_C are unbiased for σ^2 only if the corresponding hypotheses are true.

Example 8.4 In an experiment to determine the effects of varying the reflux ratio on the number of required stages Y_{ij} used in the separation of benzene and toluene, four different lab groups used the same four reflux ratios with the following results:

Reflux ratio (moles liquid per mole vapor)

Lab Group	1	2	3	4
1	11.4	9.2	7.5	6.2
2	10.7	8.6	8.3	5.9
3	11.9	8.7	9.3	5.4
4	9.9	9.0	7.1	5.6

You should realize that any differences that occur due to lab groups will affect the total for each column. However, as each group checked each reflux ratio, all column totals should be equally affected by the differences in groups. We now have identified the two major sources of variation: groups and reflux ratio. Prepare an analysis of variance table for these results. What can you conclude from the entries in this table?

We obtain

$y_{\cdot 1} = 43.9,\quad y_{1\cdot} = 34.3,\quad y_{\cdot\cdot} = 134.7,$

$y_{\cdot 2} = 35.5,\quad y_{2\cdot} = 33.5$

$y_{\cdot 3} = 32.3,\quad y_{3\cdot} = 35.3$

$y_{\cdot 4} = 23.1,\quad y_{4\cdot} = 31.6$

$$SS_T = \sum_{ij}\sum y_{ij}^2 = 1195.7 \quad ,$$

$$SS_M = y_{\cdot\cdot}^2/IJ = 18{,}144.09/16 = 1134.0056 \quad ,$$

$$SS_C = \sum_j y_{\cdot j}^2/I - SS_M = 4757.91/4 - SS_M = 55.4719 \quad ,$$

$$SS_R = \sum_i y_{i\cdot}^2/J - SS_M = 4543.39/4 - SS_M = 1.8419 \quad ,$$

$$SS_E = SS_T - SS_C - SS_R - SS_M = 3.8506 \ .$$

The analysis of variance table is:

Source	d.f.	Sum of squares	Mean square	F
Mean	1	1134.0056	---	---
Reflux ratio	3	55.4719	18.4906	43.222
Lab group	3	1.8419	0.6139	1.435
Error	9	3.8506	0.4278	---
Total	16	1195.7	---	---

When the calculated values of F are compared to $F_{3,9,0.95} = 3.86$ from Table VI, it is seen at once that the differences in the number of required stages is significantly affected by the reflux ratio. No significant differences between lab groups were found.

The use of PROC GLM for two-way AOV is illustrated by the following example.

Example 8.5 Astarita et al. [1] give the following data for non-Newtonian gravity flow along an inclined plane:

β	δ, mm at $Q = 2$ cm^3/min	δ, mm at $Q = 10$ cm^3/min
20	1.63	2.89
10	2.06	3.80
6.5	2.42	4.41
3.5	3.06	5.46
2.0	3.84	6.90

where β = Angle of inclination, degrees;

δ = Liquid layer depth, mm; and
Q = Total flow rate, cm^3/sec

Determine if the angle β and/or the flow rate Q make a significant difference in the liquid depth δ at a 95% significance level. The SAS program is as follows.

```
(put initial JCL cards/statements here)
DATA;
INPUT DELTA 1-4 BETA 6-9 FLOW 11-12;
CARDS;
(enter data from problem statement here using format given by
  INPUT statement)
PROC PRINT;
PROC ANOVA;
CLASSES BETA DELTA;
MODEL FLOW = BETA DELTA;
TITLE 'EFFECT OF ANGLE AND THICKNESS ON GRAVITY FLOW';
(put final JCL cards/statements here)
```

The SAS output for this experiment is shown below. (As the data have already been given, we have omitted that portion of the output.)

```
                         SAS

  EFFECT OF ANGLE AND FLOW RATE ON LAYER THICKNESS

           ANALYSIS OF VARIANCE PROCEDURE

              CLASS LEVEL INFORMATION

        CLASS      LEVELS     VALUES

        BETA            5     2.0 3.5 6.5 10 20

        FLOW            2     2 10

     NUMBER OF OBSERVATIONS IN DATA SET = 10
```

EFFECT OF ANGLE AND FLOW RATE ON LAYER THICKNESS

ANALYSIS OF VARIANCE PROCEDURE

DEPENDENT VARIABLE: DELTA

SOURCE	DF	SUM OF SQUARES	MEAN SQUARE	F VALUE
MODEL	5	22.59261000	4.51852200	19.45
ERROR	4	0.92920000	0.23230000	
CORRECTED TOTAL	9	23.52181000		

SOURCE	DF	ANOVA SS	F VALUE	PR > F
BETA	4	11.67236000	12.56	0.0155
FLOW	1	10.92025000	47.01	0.0024

From the calculated f = 19.45 for the model as a whole, and the very slight (0.66%) probability of obtaining a higher value of F, we reject the null hypothesis that the model is invalid at α = 0.05. Examining the individual model components, f_β = 12.56 with only a 1.55% probability of being exceeded. We reject the null hypothesis that angle of inclination does not affect flow rate as probably false. Similarly, f_Q = 47.01 with only a 0.24% probability of being exceeded. We thus reject as probably false the hypothesis that flow rate does not influence the liquid layer depth.

8.5 CONFIDENCE INTERVALS AND TESTS OF HYPOTHESES

In the two-way model $Y_{ij} = \mu + \alpha_i + \beta_j + \varepsilon_{ij}$ it may be of interest to compare the effects due to two rows, say α_1 and α_2 if H_0: all $\alpha_i = 0$ has been rejected. The mean for the ith row is $\overline{Y}_{i.} = \mu + \alpha_i + \overline{\beta}_. + \overline{\varepsilon}_{i.}$ where $\overline{\beta}_. = \sum_j \beta_j/J = 0$ and $\overline{\varepsilon}_{i.} = \sum_j \varepsilon_{ij}/J$. Thus $\overline{Y}_{1.} - \overline{Y}_{2.} = \alpha_1 - \alpha_2 + \overline{\varepsilon}_{1.} - \overline{\varepsilon}_{2.}$ and $E(\overline{Y}_{1.} - \overline{Y}_{2.}) = \alpha_1 - \alpha_2$ since $E(\varepsilon_{ij}) = 0$ for all i and j. Now MS_E is an unbiased estimator for σ^2 and the random variable

$$T = \frac{\overline{Y}_{1\cdot} - \overline{Y}_{2\cdot} - (\alpha_1 - \alpha_2)}{S_{\overline{Y}_1 - \overline{Y}_2}} = \frac{\overline{Y}_{1\cdot} - \overline{Y}_{2\cdot} - (\alpha_1 - \alpha_2)}{\sqrt{\hat{\sigma}^2\left(\frac{1}{J} + \frac{1}{J}\right)}}$$

$$= \frac{\overline{Y}_{1\cdot} - \overline{Y}_{2\cdot} - (\alpha_1 - \alpha_2)}{\sqrt{MS_E\left(\frac{2}{J}\right)}}$$

has a t-distribution with (I-1)(J-1) degrees of freedom. The development regarding the 2-sample t-test in Section 7.6 yields $(1-\alpha)100\%$ confidence limits for $\alpha_1 - \alpha_2$:

$$\overline{Y}_{1\cdot} - \overline{Y}_{2\cdot} \pm t_{\nu,1-\alpha/2}\left(\frac{2MS_E}{J}\right)^{1/2}, \tag{8.29}$$

where $\nu = (I-1)(J-1)$. If these confidence limits cover the value zero then we accept H_0: $\alpha_1 = \alpha_2 = 0$, that is, there is no significant difference between the effects due to rows 1 and 2. The confidence limits (8.29) can be used on any pair of row effects α_i and α_i'. They can also be used for any pair of column effects β_j and β_j' if the difference of the sample means is replaced by $\overline{y}_{\cdot j} - \overline{y}_{\cdot j}'$ and J is replaced with I.

Consider the random variable $\sum_i c_i\overline{Y}_{i\cdot}$ where $c_1, c_2, \ldots, c_I$ are any constants. It is not difficult to show that $\sum_i c_i\overline{Y}_{i\cdot}$ has mean $\sum_i c_i(\mu + \alpha_i)$ and variance $\sum_i c_i^2\sigma^2/J$. Now if we assume $\sum_i c_i = 0$ then the mean becomes $\sum_i c_i\alpha_i$, since $\sum_i c_i\mu = \mu\sum_i c_i = 0$. The linear combination $\sum_i c_i\alpha_i$ is called a <u>contrast</u> if $\sum_i c_i = 0$. Let $L = \sum_i c_i\overline{Y}_{i\cdot}$ be a contrast. Then L has a normal distribution with mean $\sum_i c_i\alpha_i$ and variance $\sigma_L^2 = \sum_i c_i^2\sigma^2/J$. Thus the random variable

$$T = \frac{L - \mu_L}{\hat{\sigma}_L} = \frac{\sum_i c_i \overline{Y}_{i\cdot} - \sum_i c_i \alpha_i}{\sqrt{\frac{MS_E}{J} \sum_i c_i^2}} \tag{8.30}$$

has a t-distribution with (I-1)(J-1) degrees of freedom. From this we obtain a (1-α)100% confidence interval for the contrast $\sum_i c_i \alpha_i$:

$$\sum_i c_i \overline{Y}_{i\cdot} - t_{\nu,1-\alpha/2} \left(\frac{MS_E}{J} \sum_i c_i^2\right)^{1/2} < \sum_i c_i \alpha_i < \sum_i c_i \overline{Y}_{i\cdot} + t_{\nu,1-\alpha/2} \left(\frac{MS_E}{J} \sum_i c_i^2\right)^{1/2} \tag{8.31}$$

where $\nu = (I-1)(J-1)$. Similarly, for the contrast $\sum_j a_j \beta_j$,

$$\sum_j a_j \overline{Y}_{\cdot j} - t_{\nu,1-\alpha/2} \left(\frac{MS_E}{I} \sum_j a_j^2\right)^{1/2} < \sum_j a_j \beta_j < \sum_j a_j \overline{Y}_{\cdot j} + t_{\nu,1-\alpha/2} \left(\frac{MS_E}{I} \sum_j a_j^2\right)^{1/2} . \tag{8.32}$$

We have considered only simple contrasts. However multiple contrasts arise quite naturally. In Example 8.4, for instance, since H_0: $\beta_1 = \beta_2 = \beta_3 = \beta_4 = 0$ was rejected we conclude that there is a significant difference due to reflux ratio. In trying to determine where the difference is occurring we may want to consider $\beta_1 - \beta_2$ or $\beta_1 - (\beta_2 + \beta_3 + \beta_4)/3$, which are contrasts. We can use the confidence intervals above to test H_0: $\beta_1 = \beta_2$ or H_0: $\beta_1 = (\beta_2 + \beta_3 + \beta_4)/3$.

<u>Example 8.6</u> In Example 8.4 there was a significant difference among the four levels of reflux ratio at the 5% level of significance. Find 95% confidence intervals for the contrasts (a) $\beta_1 - \beta_2$ and (b) $\beta_1 - (\beta_2 + \beta_3 + \beta_4)/3$.

(a) For the contrast we need to use the interval given in (8.32) with $a_1 = 1$, $a_2 = -1$, and $a_3 = a_4 = 0$. From the analysis of variance of Example 8.4 we have $MS_E = 0.4278$. Also $I = 4$ and $\nu = (I-1)(J-1) = 9$. From Table IV we find $t_{9,0.975} = 2.262$. Since $\bar{y}_{\cdot 1} = 10.975$ and $\bar{y}_{\cdot 2} = 8.875$, the 95% confidence limits are

$$\bar{y}_{\cdot 1} - \bar{y}_{\cdot 2} \pm 2.262 \left(\frac{0.4278}{4} \ (1 + 1)\right)^{1/2} ,$$

$$10.975 - 8.875 \pm 2.262(0.2139)^{1/2} ,$$

$$2.1 \pm 1.046 ,$$

$$1.054, \ 3.146 ,$$

Since the confidence interval does not contain zero we can say that $\beta_1 - \beta_2 > 0$ or $\beta_1 > \beta_2$ or that the effect from the first reflux ratio is greater than that from the second reflux ratio.

(b) For the contrast $\beta_1 - (\beta_2 + \beta_3 + \beta_4)/3$ we have $a_1 = 1$ and $a_2 = a_3 = a_4 = -1/3$. The 95± confidence limits are, by (8.32)

$$10.975 - (8.875 + 8.05 + 5.775)/3 \pm 2.262 \left(\frac{0.4278}{4} \ (1 + \frac{3}{9})\right)^{1/2} ,$$

$$10.475 - 7.717 \pm 2.262(0.1426)^{1/2} ,$$

$$3.258 \pm 0.854 ,$$

$$2.404, \ 4.112 .$$

Since both confidence limits are positive we conclude that $\beta_1 > (\beta_2 + \beta_3 + \beta_4)/3$, that is, the effect from the first reflux ratio is greater than the average effect from the last three reflux ratios.

Confidence intervals can also be constructed for contrasts in the one-way model $Y_{ij} = \mu + \beta_j + \varepsilon_{ij}$. Confidence limits on $\sum_j c_j \beta_j = \sum_j c_j(\mu_j - \mu) = \sum_j c_j \mu_j$ are

$$\sum_j c_j \bar{y}_{\cdot j} \pm t_{J(I-1),1-\alpha/2} \left(\frac{SS_E}{J(I-1)} \sum c_j^2 / I \right)^{1/2} \quad . \tag{8.33}$$

Recall that MS_E from Table 8.2 has $J(I-1)$ degrees of freedom associated with it.

Example 8.7 For the fluoride data of Example 8.1 determine a 95% confidence interval on $\beta_A - \beta_B$.

The contrast $\beta_A - \beta_B = \mu_A - \mu - (\mu_B - \mu) = \mu_A - \mu_B$. We thus desire a 95% confidence interval on $\mu_A - \mu_B$, the difference of mean responses of ferric thiocyanate and alizarin lanthanum complexan methods. From the results of Example 8.1 we have $\bar{y}_{\cdot A} = 4.0$. $\bar{y}_{\cdot B} = 3.6$, $J(I-1) = 4(4) = 16$, and $MS_E = 1.45$. Also $c_A = 1$, $c_B = -1$, and $t_{16,0.975} = 2.120$. The 95% confidence limits on $\mu_A - \mu_B$ are, from (8.33),

$$4.0 - 3.6 \pm 2.120 \left[\frac{23.2}{16} \left(\frac{1+1}{5} \right) \right]^{1/2} ,$$

$$0.4 \pm 2.120(0.58)^{1/2} ,$$

$$0.4 \pm 2.120(0.7616) ,$$

$$-1.2146, \ 2.0146 \quad .$$

Since the interval contains zero we conclude that there is no difference between methods A and B.

The information from Tables 8.2 and 8.3 for the one-way and two-way models, respectively, can be used to construct a confidence interval on σ^2 or to test hypotheses concerning σ^2. This can be done by observing that MS_E/σ^2 has a χ^2-distribution with the same degrees of freedom as MS_E. The details then follow along the lines of Sections 7.10 and 7.11 and will be omitted.

One can also test any hypothesis concerning contrasts $\sum_i c_i \alpha_i$ or $\sum_j a_j \beta_j$ for both the one-way and two-way models. The confidence intervals presented in this section can of course be used to test a corresponding hypothesis. If the hypothesis is one-sided then one can find the critical regions just as was done in Chapter 7.

8.6. MULTIPLE COMPARISONS AMONG TREATMENT MEANS

In comparing treatment means, e.g. in a one-way AOV, the comparisons to be considered should be selected prior to any analysis of the data. Otherwise, the analysis of contrasts such as that discussed in the previous section would not in general be valid. However, there are situations in which the experimenter wishes to gain some insight into the data which has been collected. No clues may be available regarding which contrasts should be examined at the time the experiment is planned. That is, the experiment may be of an exploratory nature and designed to gain more information than a mere statement that the treatment means are (or are not) significantly different.

The method discussed in the previous section of comparing treatment means, or combinations of treatment means, by way of contrasts can be applied to individual contrasts, one at a time. However, if one is interested in studying two or more contrasts <u>simultaneously</u>, then the method of the previous section is not recommended, especially if the number of treatments is high. If there are I treatments then there are $I(I - 1)/2$ treatment differences which might be of interest.

One method for making all possible comparisons among treatment means is due to Scheffé [12]. In a one-way AOV, the probability is $1 - \alpha$ that all possible contrasts will be contained in the intervals given by

$$\hat{L} - S\hat{\sigma}_{\hat{L}} \leq L \leq \hat{L} + S\hat{\sigma}_{\hat{L}} \qquad (8.34)$$

where

$$L = \sum_{i=1}^{I} c_i\alpha_i = \sum_{i=1}^{I} c_i\mu_i ,$$

$$\hat{L} = \sum_{i=1}^{I} c_i\overline{Y}_{.i} ,$$

$$S^2 = (I - 1)F_{I-1,J(I-1),1-\alpha},$$

and

$$\hat{\sigma}^2_{\hat{L}} = \frac{MS_E}{J} \sum_{i=1}^{I} c_i^2 \quad .$$

In a two-way analysis of variance, the intervals needed to construct Scheffé type multiple confidence intervals are given as follows: Comparisons among $\alpha_1,\ldots,\alpha_I$ require

$$\hat{L} = \sum_{i=1}^{I} c_i Y_{i\cdot} \quad ,$$

$$\hat{\sigma}^2_{\hat{L}} = \frac{MS_E}{J} \sum_{i=1}^{I} c_i^2 \quad ,$$

and

$$S^2 = (I-1)F_{I-1,(I-1)(J-1),1-\alpha} \quad .$$

Comparisons among $\beta_1,\ldots,\beta_J$ require

$$\hat{L} = \sum_{j=1}^{J} a_j \overline{Y}_{\cdot j} \quad ,$$

$$\hat{\sigma}^2_{\hat{L}} = \frac{MS_E}{I} \sum_{j=1}^{J} a_j^2 \quad ,$$

and

$$S^2 = (J-1)F_{J-1,(I-1)(J-1),1-\alpha} \quad .$$

It should be noted that if the AOV F-test rejects the hypothesis of equal means, then at least one of the contrasts will be judged different from zero by the Scheffé method. Thus this method allows more conclusions than merely stating that the treatments are not all the same.

Example 8.8 For the fluoride data of Example 8.1, determine 95% multiple confidence intervals on $\alpha_A - \alpha_B$, $\alpha_A - \alpha_C$, and $\alpha_A - \alpha_D$. The contrasts are:

$$L_1 = \alpha_A - \alpha_B = \mu_A - \mu_B \quad ,$$
$$L_2 = \alpha_A - \alpha_C = \mu_A - \mu_C \quad ,$$
$$L_3 = \alpha_A - \alpha_D = \mu_A - \mu_D \quad .$$

From Example 8.1 we have

$$\hat{L}_1 = \overline{Y}_{\cdot A} - \overline{Y}_{\cdot B} = 4.0 - 3.6 = 0.4$$

$$\hat{L}_2 = \overline{Y}_{\cdot A} - \overline{Y}_{\cdot C} = 4.0 - 2.8 = 1.2$$

$$\hat{L}_3 = \overline{Y}_{\cdot A} - \overline{Y}_{\cdot D} = 4.0 - 1.4 = 2.6 \quad .$$

Also, S^2 and $\hat{\sigma}^2_{\hat{L}}$ for each contrast are

$$S^2 = 3F_{3,16,.95} = 3(3.24) = 9.72 \quad ,$$

$$\hat{\sigma}^2_{\hat{L}} = \frac{MS_E}{5} \sum_{i=1}^{5} c_i^2 = \frac{1.45}{5}(2) = 0.58 \quad ,$$

and

$$S\hat{\sigma}_{\hat{L}} = [9.72(0.58)]^{1/2} = 2.3744 \quad .$$

Thus the multiple confidence intervals are:

$$0.4 - 2.3744 \leq L_1 \leq 0.4 + 2.3744$$

$$1.2 - 2.3744 \leq L_2 \leq 1.2 + 2.3744$$

$$2.6 - 2.3744 \leq L_3 \leq 2.6 + 2.3744$$

or

$$-1.97 \leq \alpha_A - \alpha_B \leq 2.77$$

$$-1.17 \leq \alpha_A - \alpha_C \leq 3.57$$

$$0.23 \leq \alpha_A - \alpha_D \leq 4.97 \quad .$$

Based on these confidence intervals, there is probably a difference between methods A and D but not between A and B or between A and C as those last two confidence intervals contain zero.

In Example 8.4, the 95% confidence interval on $\alpha_A - \alpha_B$ is (-1.2146, 2.0146), based on use of the t-distribution and Eq. (8.33). This interval is shorter than the interval (-1.97, 2.77) obtained in Example 8.7 by using the Scheffé method. The result from Eq. (8.33) can be applied to only one contrast using a 95% confidence level. The Scheffé method can be applied to all possible contrasts, with an overall confidence level of 95%. The price we pay in obtaining multiple confidence intervals is that of having longer intervals. This is to be expected from the nature of the two methods.

There are several other multiple comparison methods which are used. For a discussion and comparison of such methods the reader is referred to Bancroft [3].

Some existing computer statistical packages which provide procedures to perform analyses of variance also have the feature of computing multiple confidence intervals as an option. SAS is one of these packages (SAS User's Guide: Statistics, 1982 Edition, SAS Institute Inc., Cary, NC: SAS Institute Inc. 1982. 584 pp.). We now consider the analysis of the data in Example 8.1 by means of PROC ANOVA or PROC GLM. This procedure has several multiple comparisons methods as options. For references and documentation on the various multiple comparisons methods used the reader is referred to the SAS manual referenced above.

Example 8.9 The data of Example 8.1 was processed using the ANOVA procedure in the SAS system to obtain another multiple comparison between the treatment means by the methods of Duncan's multiple range test and Scheffé's method. The program is as follows below.

```
(put initial JCL cards/statements here)
DATA FLUORIDE;
INPUT METHOD $ 1 PPM 3;
CARDS;
(data here according to INPUT format)
PROC ANOVA;
CLASS METHOD;
MODEL PPM = METHOD;
TITLE 'ONE-WAY ANALYSIS OF VARIANCE';
MEANS METHOD/DUNCAN;
```

```
MEANS METHOD/SCHEFFE;
(put final JCL cards/statements here)
```

The output contains the AOV table and the results of the multiple comparisons requested, namely, those using the Duncan method and the Scheffé method. The output is summarized below:

SAS

ONE-WAY ANALYSIS OF VARIANCE

ANALYSIS OF VARIANCE PROCEDURE

DEPENDENT VARIABLE: PPM

SOURCE	DF	SUM OF SQUARES	MEAN SQUARE	F VALUE
MODEL	3	19.75000000	6.58333333	4.54
ERROR	16	23.20000000	1.45000000	PR > F
CORRECTED TOTAL	19	42.95000000		0.0174

SOURCE	DF	ANOVA SS	F VALUE	PR > F
METHOD	3	19.75000000	4.54	0.0174

The SAS output was condensed to yield the table above. It gives the corrected total rather than the uncorrected total. The number 0.0174 under PR > F is $P(F \geq 4.54)$, where F has an F distribution with 3 and 16 d.f. That is, 0.0174 is the probability of observing a F ratio larger than or equal to 4.54, given that H_0: $\beta_1 = \beta_2 = \beta_3 = \beta_4 = 0$ is true. The SAS output can be compared with that in Example 8.1.

The multiple comparisons results are shown below. Using Duncan's multiple range test, the results for the four means are:

```
                              SAS

              ONE-WAY ANALYSIS OF VARIANCE

             ANALYSIS OF VARIANCE PROCEDURE

DUNCAN'S MULTIPLE RANGE TEST FOR VARIABLE: PPM
NOTE: THIS TEST CONTROLS THE TYPE I COMPARISONWISE ERROR RATE,
      NOT THE EXPERIMENTWISE ERROR RATE.

ALPHA=0.05   DF=16   MSE=1.45

MEANS WITH THE SAME LETTER ARE NOT SIGNIFICANTLY DIFFERENT.

DUNCAN    GROUPING              MEAN      N  METHOD

                 A            4.0000      5  A
                 A
                 A            3.6000      5  B
                 A
       B         A            2.8000      5  C
       B
       B                      1.4000      5  D
```

These results indicate that the means $\bar{y}_{.A} = 4.0$, $\bar{y}_{.B} = 3.6$, $\bar{y}_{.C} = 2.8$ are not significantly different. Also, the means $\bar{y}_{.C} = 2.8$ and $\bar{y}_{.D} = 1.4$ are not significantly different. There are two "homogeneous" groups of means, that is, (A, B, C) and (C, D). The letters A, B in the SAS output are not to be confused with the letters A and B in Example 8.1.

The results of the Scheffé option are as follows:

```
                              SAS

                 ONE-WAY ANALYSIS OF VARIANCE

                ANALYSIS OF VARIANCE PROCEDURE

SCHEFFE'S TEST FOR VARIABLE: PPM
NOTE: THIS TEST CONTROLS THE TYPE I EXPERIMENTWISE ERROR RATE
      BUT GENERALLY HAS A HIGHER TYPE II ERROR RATE THAN REGWF
      FOR ALL PAIRWISE COMPARISONS.

ALPHA=0.05  DF=16  MSE=1.45
CRITICAL VALUE OF T=1.79969
MINIMUM SIGNIFICANT DIFFERENCE=2.37395

MEANS WITH THE SAME LETTER ARE NOT SIGNIFICANTLY DIFFERENT.

SCHEFFE  GROUPING              MEAN      N  METHOD

              A              4.0000      5  A
              A
       B      A              3.6000      5  B
       B      A
       B      A              2.8000      5  C
       B
       B                     1.4000      5  D
```

There are two groups deemed "homogeneous" by the Scheffé method; namely, (A, B, C) and (B, C, D). The minimum significant difference in the previous example is $\hat{s\sigma}_{\hat{L}} = [9.75(0.58)]^{1/2} = 2.378$, as compared with 2.37395 from the AOV results. From the previous example, we can see that $\bar{y}_{.B} - \bar{y}_{.D} = 2.2$ and $\bar{y}_{.C} - \bar{y}_{.D} = 1.4$ even though the Scheffé method indicates that neither B and D nor C and D are different. The Duncan method does not place B in the same group with C and D. One can infer from the Duncan method that D is different from A and B; whereas, from the Scheffé method one can only say that D is different from A.

It has been established that the Scheffé method is quite conservative in the number of comparisons declared significant. It is more conservative than the Duncan method as can be seen from this example. For further details on multiple comparisons see Bancroft [3] or the references on the ANOVA and GLM procedures in the SAS publication previously referenced in this section.

8.7 BARTLETT'S TEST FOR EQUALITY OF VARIANCES

The analysis of variance technique for testing equality of means is a rather robust procedure. That is, when the assumption of normality and homogeneity of variances is "slightly" violated the F-test remains a good procedure to use. In the one-way model, for example, with an equal number of observations per column it has been exhibited that the F-test is not significantly effected. However, if the sample size varies across columns, then the validity of the F-test can be greatly affected. There are various techniques for testing the equality of k variances $\sigma_1^2, \sigma_2^2, \ldots, \sigma_k^2$. We discuss here a very widely used technique, Bartlett's χ^2-test [4] for examining homogeneity of variances which is valid when the sample populations are normal.

Let $S_1^2, S_2^2, \ldots, S_k^2$ be k independent sample variances corresponding to k normal populations with means μ_i and variance σ_i^2, $i = 1,2,\ldots,k$. Suppose $n_1-1, n_2-1, \ldots, n_k-1$, are the respective degrees of freedom. Bartlett [3] proposed the statistic

$$\chi^2 = \left((\ln V) \sum_{i=1}^{k} (n_i-1) - \sum_{i=1}^{k} (n_i-1) \ln S_i^2 \right) \Big/ \ell \quad , \qquad (8.35)$$

where

$$V = \sum_{i=1}^{k} (n_i-1) S_i^2 \Big/ \sum_{i=1}^{k} (n_i-1)$$

and $\ln$ denotes the natural logarithm. The denominator in Eq. (8.35) is defined by

$$\ell = 1 + \frac{1}{3(k-1)}\left(\sum_{i=1}^{k} \frac{1}{n_i - 1} - \frac{1}{\sum_{i=1}^{k} (n_i - 1)}\right) .$$

It can be shown that the statistic in Eq. (8.35) has an approximate χ^2-distribution with k-1 degrees of freedom when used as a test statistic for H_0: $\sigma_1^2 = \ldots = \sigma_k^2$. Given k random samples of sizes $n_1, n_2, \ldots, n_k$, from k independent normal populations the statistic χ^2 in (8.35) can be used to test H_0. Recall that a sample variance S^2 is

$$S^2 = \left(\sum_{i=1}^{n} Y_i^2 - n\bar{Y}^2\right) \Big/ (n-1) .$$

If all the samples are of the same size, say n, then Bartlett's statistic in Eq. (8.35) becomes

$$\chi^2 = (n-1)\left(k \ln V - \sum_{i=1}^{k} \ln S_i^2\right) \Big/ \ell , \tag{8.36}$$

where $\ell = 1 + \frac{(k+1)}{3k(n-1)}$. This statistic can be written in terms of common logarithms by observing that $\ln a = 2.3026 \log a$, where a is any positive real number. The statistic in Eq. (8.36) becomes

$$\chi^2 = 2.3026(n-1)\left(k \log V - \sum_{i=1}^{k} \log S_i^2\right) \Big/ \ell . \tag{8.37}$$

If $n_i = n$, $i = 1,2,\ldots,k$, then

$$V = \sum_{i=1}^{k} S_i^2/k.$$

The rejection region for testing H_0: $\sigma_1^2 = \sigma_2^2 = \ldots = \sigma_k^2$ is $\chi^2 > \chi^2_{(k-1),1-\alpha}$.

Example 8.10 In an experiment to determine the effects of sample size and amount of liquid phase on the height equivalent to a theoretical plate in gas chromatography, it was necessary to utilize solid support material from different batches. It was therefore imperative that the resulting data be checked for homogeneity prior to attempting to develop any quantitative expressions regarding the effects of these variables on HETP. Several sets of data points were selected at random and examined using Bartlett's test.

In particular, a set of four HETP values obtained for cyclohexane for a 4-$\mu\ell$ sample injected into a 40 wt% β,β'-oxydipropionitrile column were

$$y_1 = 0.44 \qquad y_1^2 = 0.1936 \qquad \bar{y} = \Sigma y_i/n = 1.71/4 = 0.4275$$

$$y_2 = 0.44 \qquad y_2^2 = 0.1936 \qquad \bar{y}^2 = (0.4275)^2 = 0.18275625$$

$$y_3 = 0.40 \qquad y_3^2 = 0.1600 \qquad n\bar{y}^2 = 0.731025$$

$$y_4 = 0.43 \qquad y_4^2 = 0.1849$$

$$\Sigma y_i = 1.71 \qquad \Sigma y_i^2 = 0.7321$$

$$s^2 = \frac{\sum_{i=1}^{n} y_i^2 - n\bar{y}^2}{n-1} = \frac{0.7321-0.731025}{3} = \frac{0.001075}{3} = 3.583 \times 10^{-4}.$$

The variance of 10 cyclohexane data sets, each consisting of 4 observations, are thus calculated and presented below. In this case $n_i = 4$, $i = 1,2,\ldots,10$, and $k = 10$.

Liquid phase β,β'-oxydipropionitrile	s_i^2	$\log s_i^2$
40% - 4 μℓ sample	0.0003583	-3.44575
30% - 8 μℓ sample	0.0002250	-3.64782
20% -10 μℓ sample	0.0002250	-3.64782
20% - 4 μℓ sample	0.0000916	-4.03810
10% - 4 μℓ sample	0.0000916	-4.03810
5% -10 μℓ sample	0.0003000	-3.52288
5% - 2 μℓ sample	0.0002250	-3.64782
3% - 8 μℓ sample	0.0002250	-3.64782
3% - 6 μℓ sample	0.0003000	-3.52288
10% - 2 μℓ sample	0.0002250	-3.64782
	0.0022665	-36.80681

The computations needed to calculate χ^2 according to Eq. (8.37) are

$$v = 0.0022665/10 = 0.00022665$$

$$\log v = -3.64464$$

$$k \log v = 10(-3.64464) = -36.4464$$

$$\ell = 1 + 11/30(3) = 1.1222$$

$$\chi^2 = 2.3026(n - 1)(k \log v - \Sigma \log s_i^2)/\ell$$

$$= 2.3026(3)\ [-36.4464 - (-36.8068)]\ /1.1222$$

$$= 6.9078(0.3604)/1.1222 = 2.218 \quad .$$

Reference to a χ^2 table (Table V) shows $\chi^2_{9,0.975} = 19.0$ and $\chi^2_{9,0.99} = 21.7$, which says that the computed χ^2 value is not significantly large and therefore we do not reject the hypothesis of homogeneous variances. These data were therefore subsequently used for quantitative effects determination by the method of least squares (to be discussed in Chapter 9).

The F-test for testing the equality of two variances is quite sensitive to non-normality. Bartlett's test suffers from the same undesirable characteristic. If non-normality is suspected, one should resort to other methods such as those discussed in Conover [5].

PROBLEMS

8.1 The following data give the yields of a product that resulted from trying catalysts from four different suppliers in a process.

Catalyst			
I	II	III	IV
36	35	35	34
33	37	39	31
35	36	37	35
34	35	38	32
32	37	39	34
34	36	38	33

(1) Are yields influenced by catalysts?

(2) What are your recommendations in the selection of a catalyst? Assume that economics dictate 95% probability of being right on a decision. Hint: reexamine the catalyst group means to find the corresponding confidence intervals.

8.2 Pilot plant runs were made to determine the effect of percent conversion (defined as volume converted per 100 volumes fed) on coke yields in catalytic cracking of cracked cycle gas oil, hydrogenated to varying degrees. It was also desired to determine whether hydrogenation has any effect on the coke-yield relationship and if so to estimate its effect. Three samples of gas oil were subjected to cracking runs, each made at different severity to vary conversion. Coke yields were determined by measuring the CO_2 in fuel gas from burning off coke from the cracking catalyst following each run. Feed A is untreated; feed B is A hydrogenated to a hydrogen consumption of 300 std cu ft per barrel of feed;

feed C is A hydrogenated to 500 std cu ft per barrel. The following data were obtained.

Feed A		Feed B		Feed C	
Conv. vol%	Coke yield wt%	Conv. vol%	Coke yield wt%	Conv. vol%	Coke yield wt%
20.1	4.2	20.3	4.1	25.8	4.0
32.2	6.8	27.6	4.8	31.6	4.4
24.7	5.9	35.4	6.2	38.8	8.1
41.2	11.0	38.0	8.0	45.0	8.3
47.0	16.0	49.0	14.0	50.2	11.9
32.7	8.0	24.2	4.0	36.1	6.1

Correlate the data, getting as much information as possible out of it.

8.3 Random samples of size ten were drawn from normal populations A, B, C, and D. The measurements of property x are given in the table below along with totals and sums of squares in columns.

	y_A	y_B	y_C	y_D	
	3.355	0.273	3.539	3.074	
	1.086	2.155	2.929	3.103	
	2.367	1.725	3.025	2.389	
	0.248	0.949	4.097	4.766	
	1.694	0.458	2.236	2.553	
	1.546	1.455	3.256	3.821	
	1.266	2.289	3.374	1.905	
	0.713	2.673	1.781	2.350	
	0.000	1.800	2.566	1.161	
	3.406	2.407	2.510	2.122	
$\Sigma y_{i\cdot}$	15.681	16.184	29.313	27.244	$\Sigma y_{\cdot\cdot} = 88.422$
$\bar{y}$	1.5681	1.6184	2.9313	2.7244	
$\sum_j y_{ij}^2$	37.071327	32.339668	90.081121	83.620342	$\sum_{ij}\Sigma y_{ij}^2 = 243.112458$

(a) Are there significant differences between the population means? If so, which are different?

(b) An independent estimate of a population variance σ^2 is 1.0 (property units)^2. Does the variation in these data differ from what would be expected on the basis of the independent estimate?

(c) Compare each group with every other group by appropriate t-tests. Discuss the advantages of using the F-test for group comparisons as a composite t-test.

8.4 Six vertical elutriators were calibrated after 100 hr. service intervals. The flow rates in liters/min are below. Is there a difference in flow rates? If so, does the change depend on cumulative time in service?

Time, hrs.	VE_1	VE_2	VE_3	VE_4	VE_5	VE_6
0	7.51	7.71	7.45	7.44	7.35	7.59
100	7.58	7.72	7.48	7.48	7.38	7.63
200	7.57	7.71	7.43	7.44	7.34	7.60
300	7.60	7.70	7.44	7.48	7.32	7.61
400	7.60	7.71	7.50	7.53	7.37	7.65
500	7.67	7.67	7.55	7.58	7.40	7.64

8.5 You work for one of two companies that manufactures thermocouples for caustic service. The rival company advertises that its new thermocouple lasts longer than any other one on the market. From a great many tests you know that the estimated mean length of life for your company's unit is 400 days.

The rival's advertising campaign is damaging to your company but if the claims are publicly disputed and the rival's thermocouples are shown to last as long or longer than yours, the resulting publicity will be even more damaging to your company than the rival's advertising campaign.

(a) Explain how you would test the rival's claim and give the null and alternative hypotheses for your experiment.

(b) In view of the above, explain if you would dispute the claim or not and explain why in each of the following situations:

(1) The mean of your sample of the rival's thermocouples is 4250 hr and the estimated variance of the estimated mean is 6250.

(2) The mean of the sample is 4250 hr and the estimated variance of the estimated mean is 25000.

(3) The sample mean is 3500 hr and the estimated variance of the estimated mean is 640.

8.6 The additives shown below were introduced into a trioxane system [9] in hopes of strengthening its mechanical properties without altering its physical properties as estimated by the following intrinsic viscosities:

None	Formic acid	H_2O	Methanol	HCHO	Acetic anhydride
2.1	2.3	2.2	2.2	1.9	1.9
1.9	1.6	1.9	2.6	1.9	2.1
2.1	1.5	1.6	2.0	1.9	2.4

Do any of the additives effect intrinsic viscosity significantly? Use the $\alpha = 0.01$ significance level.

8.7 Samples of steels from 4 different batches were analyzed for carbon content. The results are shown below for quadruplicate determinations by the same analyst. Are the carbon contents (given in weight percent) of these batches the same? What are the 99% confidence limits on the average carbon content of each?

1	2	3	4
0.39	0.36	0.32	0.43
0.41	0.35	0.36	0.39
0.36	0.35	0.42	0.38
0.38	0.37	0.40	0.41

The following additional information is available.

$$y_{\cdot 1} = 1.54 \qquad \sum_i y_{i1}^2 = 0.5942 \qquad \bar{y}_{\cdot\cdot} = 0.3850 \qquad \bar{y}_{\cdot 1} = 0.3850$$

$$y_{\cdot 2} = 1.43 \qquad \sum_i y_{i2}^2 = 0.5115 \qquad \sum_i\sum_j y_{ij}^2 = 2.3236 \qquad \bar{y}_{\cdot 2} = 0.3575$$

$$y_{\cdot 3} = 1.50 \qquad \sum_i y_{i3}^2 = 0.5684 \qquad y_{\cdot\cdot}^2 = 36.9664 \qquad \bar{y}_{\cdot 3} = 0.3750$$

$$y_{\cdot 4} = 1.61 \qquad \sum_i y_{i4}^2 = 0.6495 \qquad \sum_j y_{\cdot j}^2 = 9.2586 \qquad \bar{y}_{\cdot 4} = 0.4025 .$$

8.8 Auerbach et al [2] have performed an exhaustive check of wavelength fidelity in the ultraviolet and visible wavelength portion of spectrophotometers. A holmium glass filter was used for this purpose. The wavelengths corresponding to the maximum absorption peaks for this filter were recorded at 241, 361, 446, 454, and 460.5 mμ by two different operators on each of 27 instruments tested. This study was performed to see if spectral findings in different laboratories on different instruments could be used interchangeably. The results are below for 361 and 454 mμ. With what confidence can you state that the absorption maxima at these two wavelengths will deviate no more than $\pm$ 0.3 mμ from the true value?

λ = 361		λ = 454		λ = 361		λ = 454	
1	2	1	2	1	2	1	2
361	361	453	453	361.3	361	453.5	452.8
361	361	453	453	---	---	454	454
361	361	454	453.5	---	---	454	454
361	361	454	453.5	362.5	362	455	454
361	360	454	454	362	361	455	454
360.8	360	454.2	455	361.5	361.5	455.5	455.5
360.7	361	453.5	454	361.5	361.5	456	455
360.7	361	453.5	454	361	361.5	455.5	455
367	367	458	458	361	361.5	455	455
367	367	458	458	360	361	454	454
360.8	361	453	453	360	361	454	454
360.8	360.5	453	453	361.6	362	455.8	455.5
361	361	454	454	361.6	361.8	456	456
361	360.8	454	454	361	360.5	454	454
361	361	453	453	360.5	360.5	454	454
361	361	453	453	360	360	453	454
360.5	---	453	---	360	360	452.5	454
360.4	---	452.2	---	363	---	460	---
362	362	456	456	360.5	---	453.5	---
362	362	456	456	360.5	---	453.5	---
360	360	---	---	361	361	454	---
360	360	---	---	361	361	454	---
359.5	359.7	451.5	454.2	361	361	454	454
359.5	359.1	451	450.8	361	361	454	454
361	361.1	---	455	360.3	---	452.8	---
361	251.1	---	454.7	360.4	---	453.0	---

8.9 Harper and El Sahrigi [7] give the following data for the measured thermal conductivity of a quick-frozen pear containing the following gases in the pores.

$$K_e = \frac{\text{Btu}}{\text{hr ft}^2\ {}^{\circ}\text{F/ft}} \times 10^3$$

F-12	CO_2	N_2	Ne	He
12.6	12.7	12.6	12.6	12.6
12.6	12.7	12.6	12.9	12.6
12.8	13.0	12.6	13.7	12.6
13.5	13.4	12.6	14.2	12.8

Do the different gases make a significant difference in the thermal conductivity of the pear at the 95% confidence level?

8.10 A solution of HPC (hot potassium carbonate) is used to scrub CO_2 from gas feed stock in the production of nylon intermediate materials. Pilot plant data for effect of HPC flow rate on CO_2 leakage are given in Table I below. Leakage is expressed in weight percent.

Prepare an anlysis of variance for these results. Does flow rate affect CO_2 leakage?

TABLE I

HPC Flow Rate, gal/hr

0.5	0.7	0.9	1.1
2.00	1.90	1.65	1.31
2.15	1.87	1.54	1.29
2.21	1.75	1.49	1.18
2.12	1.82	1.60	1.20
2.35	1.78	1.52	1.09

8.11 Four groups in Physical Chemistry lab are told to find the molecular weight of dichloroethane by the method of Dumas using the same equipment over a period of 4 weeks. Test the data below to see if there is a significant difference in the lab technique of the groups.

Group (M.W. calc.)

1	2	3	4
120.0	105.8	130.8	116.2
127.0	116.7	123.6	108.8
118.0	104.8	104.2	111.3
121.6	109.7	100.3	115.3
110.8	110.4	102.8	107.4

8.12 Suppose that the data given in Problem 8.7 had been obtained by four different technicians, each working with one sample from each batch of steel. Are there significant differences between analysts or is the major source of variation in wt% C due to the differences between steel batches? The following calculations have been made:

$y_{1\cdot} = 1.50$ $\qquad \sum_j y_{1j}^2 = 0.5690$ $\qquad \sum_i y_{i\cdot}^2 = 9.2438$

$y_{2\cdot} = 1.51$ $\qquad \sum_j y_{2j}^2 = 0.5723$

$y_{3\cdot} = 1.51$ $\qquad \sum_j y_{3j}^2 = 0.5729$

$y_{4\cdot} = 1.56$ $\qquad \sum_j y_{4j}^2 = 0.6094$

8.13 The 6 ft × 3 in. packed gas absorber in the unit operations laboratory was used by 4 different groups who each found the concentration (in ppm) of ammonia in outlet air at 3 different inlet NH_3 concentrations. A constant value of moles liquid/mole total gas was used. The following data was obtained:

Group	Inlet Concentration, %		
	1	3	5
1	49.8	92.1	138.4
2	52.6	89.3	135.9
3	47.9	93.4	136.2
4	48.7	90.6	139.9

Do either of the variables affect the outlet ammonia concentration?

8.14 In an experiment to determine the effects of varying inlet water flow rate on the HTU (in) for the cross-flow, pilot-scale cooling tower, different lab groups used the same water flow rates to yield the following results:

Group	Flow Rate, GPM				
	1	2	3	4	5
1	12.6	6.2	4.8	3.5	2.9
2	13.1	7.0	5.1	3.8	3.0
3	12.7	6.7	4.8	3.6	2.7

Is the difference in HTU due to group differences or flow rates?

8.15 The torque output for several pneumatic actuators at several different supply pressures are given in the table below.

Model	Torque (in. lb.)		
	60 psi	80 psi	100 psi
A	205	270	340
B	515	700	880
C	1775	2450	3100
D	7200	9600	12100

(1) Do the different supply pressures significantly affect the output?

(2) Does the output vary for the different models?

8.16 In the colorimetric analysis of chloride ion in aqueous solution, the following data were obtained:

Modified Iwasaki method
wavelength: 465 mμ

Sample	% Transmittance	
	Trial 1	Trial 2
1	100	99.8
2	98.5	99.0
3	94.0	98.0
4	84.0	97.0
5	88.0	96.0
6	90.0	95.0
7	91.0	92.2
8	89.0	92.0
9	85.0	88.5
10	84.0	70.5

Mercuric chloranilate method
wavelength: 485 mμ

Sample	% Transmittance	
	Trial 1	Trial 2
1	96.4	99.8
2	99.0	97.2
3	94.5	92.5
4	91.3	96.8
5	99.2	97.0
6	92.8	92.8
7	94.5	93.2
8	90.2	92.8
9	92.4	91.2
10	95.5	94.3

(a) Show statistically that the procedures yield different results.

(b) When down-graded for the inconsistency of the results in the pairs of trials, the students who obtained these data said that the results of these trials were obtained with reagents prepared from two different water sources: trial 1 using laboratory distilled water and trial 2 using commercial distilled water. They contended that the difference in water sources is the major source of error. Perform suitable calculations to make a definite comment on their contention.

8.17 In unit operations lab, 3 different people read the outlet water temperature for 4 different inlet hot water flow rates. The data are shown below. Was there a difference in temperature due to flow rate or are the results due to the observer or both?

	Flow Rate, gal/min			
Observer	1	2	3	4
1	67	72	77	80
2	71	75	80	81
3	69	71	76	80

8.18 It is proposed that the addition of an amine to the carbonate solution in Problem 8.10 will reduce CO_2 leakage in the purified or scrubbed gas stream. Pilot plant data are given in Table II. Does the additive have the desired effect?

TABLE II

HPC Flow Rate (gal/hr)

		0.5	0.7	0.9	1.1
Amine	0.5	1.64	1.32	0.76	0.62
Conc.,	1.0	1.31	1.09	0.47	0.38
Wt%	3.0	0.99	0.81	0.15	0.08
	6.0	0.61	0.45	0.04	0.01

8.19 The data below show the relationship between the temperature of molasses to its velocity through several pipe diameters. The velocity is due to gravity at sea level.

	Velocity (ft/min)		
Smooth pipe diameter, in.	$70^{\circ}F$	$100^{\circ}F$	$212^{\circ}F$
1	2.0	15	20
3	4.4	17	21
5	5.1	21	21
6	5.2	21	24

(a) Are the flows influenced by the temperature?

(b) Are the flows influenced by the pipe diameter?

8.20 The following data show the total conversion for the catalytic cracking of acetic acid to acetic anhydride at $750^{\circ}C$ with different concentrations of triethyl phosphate (TEP) as the catalyst. The feed rate was also changed for each run.

Feed rate (gal/hr)	% Conversion		
	0.5% TEP	0.3% TEP	0.1% TEP
200.0	77.85	76.03	73.87
147.0	89.12	88.94	87.65
98.5	99.09	97.14	91.78
61.0	99.55	99.51	97.60

(1) Are yields influenced by the amount of catalyst?

(2) Are yields influenced by feed rate?

8.21 In an air-water contact utilizing a packed tower, the following [8] gas film heat transfer coefficient data were recorded in Btu/hr ft^2 °F where L = liquid rate, $\frac{lb}{hr\ ft^2}$, G = gas rate, $\frac{lb}{hr\ ft^2}$ and H = heat transfer coefficient, $\frac{Btu}{hr\ ft^2\ °F}$

Gas rate (G)	Liquid rate (L)			
	190	250	300	400
200	200	226	240	261
400	278	312	330	381
700	369	416	462	517
1100	500	575	645	733

(a) Perform the indicated two-way AOV for these data. Use the 0.01 significance level.

(b) Which variable has the greater effect on the heat transfer coefficient?

8.22 Dana et al. [6] have presented data on the lightness (L), greenness (a), and yellowness (b) of a semimatte opaque glaze for use on ceramic tile as a function of the amounts of blue and yellow colorants added in formulation. Their data are given below.

	Blue colorant			
	1.4%	2.0%	2.6%	3.2%
Yellow Colorant				
	L 78.8	L 77.6	L 76.5	L 75.3
1.1%	a 11.5	a 12.4	a 13.4	a 14.3
	b 5.9	b 10.0	b 2.5	b 0.8
	L 78.6	L 77.3	L 75.9	L 74.7
1.5%	a 12.0	a 12.9	a 13.9	a 14.8
	b 8.8	b 4.2	b 5.3	b 3.5
	L 78.3	L 76.8	L 75.4	L 74.1
1.9%	a 12.5	a 13.4	a 14.3	a 15.2
	b 11.8	b 7.1	b 8.1	b 6.3

For the data presented above, determine whether the blue or yellow colorant has the more pronounced effect on the lightness of the glaze. Present the complete AOV table as part of your answer.

8.23 Repeat Problem 8.22 for the effect of both colorants on greenness (a). Give the appropriate AOV table. State and test the appropriate null hypothesis. Show the model involved and fully define all its terms.

8.24 Repeat Problem 8.22 for the effect of both colorants on yellowness (b). Give the model involved and state the null hypothesis. Test it, interpret the results at the 5% significance level, and present a complete AOV table.

8.25 Lemus [10] presented the force required to cause an electrical connection to separate as a function of angle of pull. The data are given below. Each entry is in pounds-force.

Angle degrees	Connectors 1	2	3	4	5
0	45.3	42.2	39.6	36.8	45.8
2	44.1	44.1	38.4	38.0	47.2
4	42.7	42.7	42.6	42.2	48.9
6	43.5	45.8	47.9	37.9	56.4

What conclusions are possible from these data? Support your answer by stating and testing the appropriate hypothesis. Show a complete analysis of variance.

8.26 Mouradian [11] has presented the results of a study to determine the pull strength of adhesive systems on primed and unprimed surfaces. The data for test specimens of two thicknesses are given below.

	Adhesive systems I	II	III	IV
With primer	60.0*	57.0*	19.8*	52.0*
	73.0	52.0	32.0	77.0
	63.0*	52.0*	19.5*	53.0*
	79.0	56.0	33.0	78.0
	57.0*	55.0*	19.7*	44.0*
	70.0	57.0	32.0	70.0
	53.0*	59.0*	21.6*	48.0*
	69.0	58.0	34.0	74.0
	56.0*	56.0*	21.1*	48.0*
	78.0	52.0	31.0	74.0
	57.0*	54.0*	19.3*	53.0*
	74.0	53.0	27.3	81.0

	Adhesive systems			
	I	II	III	IV
Without primer	59.0*	51.0*	29.4*	49.0*
	78.0	52.0	37.8	77.0
	48.0*	44.0*	32.2*	59.0*
	72.0	42.0	36.7	76.0
	51.0*	42.0*	37.1*	55.0*
	72.0	51.0	35.4	79.0
	49.0*	54.0*	31.5*	54.0*
	75.0	47.0	40.2	78.0
	45.0*	47.0*	31.3*	49.0*
	71.0	57.0	40.7	79.0
	48.0*	56.0*	33.0*	58.0*
	72.0	45.0	42.6	79.0

Perform the analysis of variance to determine whether the major source of variation is due to surface pretreatment or adhesive type. Use only the data marked with an asterisk.

8.27 Repeat Problem 8.26 using the unmarked data.

8.28 A psychology class performed an experiment to see if there is any correlation between amount of sleep and test-taking ability. Four students participated in the experiment. Each student varied his normal sleeping time from 4 hours to 7 hours and took a simple test the next day. The test occurred once a week for 4 weeks. Corresponding grades (out of 100%) are shown below. Does sleep time or student affect the grades?

Student	Sleep Time, Hours			
	4	5	6	7
1	78	79	84	89
2	84	82	84	91
3	80	84	86	83
4	71	73	73	77

8.29 A sophomore quantitative analysis student standardized three sulfide solutions by each of 3 methods: A (sodium nitroprusside), B (lead acetate), and the reference (modified calcium hydroxide). The data are below. Were the solutions different? Did the analysis procedure affect the results? Support your answer by a suitable AOV table and interpret the results.

Method	Solution		
	1	2	3
A	5.21	5.18	5.22
B	5.30	5.23	5.12
C	5.18	5.38	5.09

8.30 Relative effectiveness has been measured for a catalyst at 5 different space velocities (V) for 2 different surface areas (A). The data are below.

		V, sec^{-1}						
		1.5	2.4	3.2	4.6	7.3	$y_{i.}$	$\sum_j y_{ij}^2$
$A, m^2/gm$	1	3.84	3.06	2.42	2.06	1.63	13.01	36.8661
	3	6.90	5.46	4.41	3.80	2.89	23.46	119.6618
	$y_{.j}$	10.74	8.52	6.83	5.86	4.52		
	$\sum_i y_{ij}^2$	62.3556	39.1752	25.3045	18.6836	11.009		

Is catalyst effectiveness influenced by either space velocity or surface area? "Prove" your answer by stating and testing the appropriate null hypotheses at the $\alpha = 0.05$ level. Note that although the independent variables were arranged in log-spacing to facilitate regression analysis (if warranted), that has nothing to do with the solution to this problem.

8.31 Four vertical elutriators were used to obtain samples of the concentration of cotton dust in the open-end spinning room of a large textile mill. The results, in micrograms per cubic meter, are shown below. Each elutriator is located in a different part of the room; all have been calibrated to the OSHA standard. Is there any difference ($\alpha = 0.05$) between areas with regard to dust concentration?

VE_1	VE_2	VE_3	VE_4
182.6	174.3	182.0	181.7
173.4	178.5	182.1	183.4
190.1	180.0	184.6	180.6
178.6		180.9	
188.2			

REFERENCES

1 Astarita, G., Marrocci, G., and Palumbo, G., Non-Newtonian Flow Along Inclined Plane Surfaces, I&EC Fundamentals, 3, 333 (1964). Copyright 1964 by the American Chemical Society. Data reprinted by permission of the copyright owner.

2 Auerback, M. E., Bauer, E. L., and Naehod, F. C., Spectrophotometer Wavelength Reliability. Ind. Qual. Control, 20, 45-48 (1964). Copyright American Society for Quality Control, Inc., Reprinted by permission.

3 Bancroft, T. A.; Topics in Intermediate Statistical Methods, Volume 1, Chapter 8, Iowa State University Press, Ames (1968).

4 Bartlett, M. S., Proc. Roy. Soc. (London), A901, 160, 273-275 (1964).

5 Conover, W. J.; Practical Nonparametric Statistics, Ch. 5, 2nd ed., John Wiley and Sons, Inc., New York (1980).

6 Dana, R., Bayer, H. S., and McElrath, G. W., Color and Shade Control of Ceramic Tile. Ind. Qual. Control, 21, 609-614 (1965). Copyright American Society for Quality Control, Inc. Reprinted by permission.

7 Harper, J. C., and El Sahrigi, A. F., Thermal Conductivities of Gas-Filled Porous Solids, I&EC Fundamentals, 3, 318 (1964). Copyright 1964 by the American Chemical Society. Data reprinted by permission of the copyright owner.

8 Hensel, S. L., and Treybal, R. E., Air-Water Contact. Adiabatic Humidification of Air with Water in a Packed Tower. Chem. Eng. Prog., 48, 362-370 (1952). Copyright 1952 by the American Institute of Chemical Engineers. Data reprinted by permission of the copyright owner.

9 Ito, A. and Hayashi, K., Polyoxymethylene via Radiation. Hydrocarb. Proc., 47 (11), 197-202 (1968). Reproduced with permission from HYDROCARBON PROCESSING, Nov., 1968.

10 Lemus, F., A Mixed Model Factorial Experiment in Testing Electrical Connectors. Ind. Qual. Control, 17 (6), 12-16 (1960). Copyright American Society for Quality Control, Inc. Reprinted by permission.

11 Mouradian, G., A Statistical Approach to Design Review. Ind. Qual. Control, 22, 516-520 (1966). Copyright American Society for Quality Control, Inc. Reprinted by permission.

12 Scheffé, H.; A Method for Judging All Contrasts in the Analysis of Variance, Biometrika, 40: 87-104 (1953).

9
REGRESSION ANALYSIS

9.1 INTRODUCTION

Regression is a highly useful statistical technique for developing a quantitative relationship between a dependent variable and one or more independent variables. It utilizes experimental data on the pertinent variables to develop a numerical relationship showing the influence of the independent variables on a dependent variable of the system.

Throughout engineering, regression may be applied to correlating data in a wide variety of problems ranging from the simple correlation of physical properties to the analysis of a complex industrial system. If nothing is known from theory about the relationship among the pertinent variables, a function may be assumed and fitted to experimental data on the system.

Frequently a linear function is assumed. In other cases where a linear function does not fit the experimental data properly, the engineer might try a polynomial or exponential function.

9.2 SIMPLE LINEAR REGRESSION

In the simplest case the proposed functional relationship between two variables is

$$Y = \beta_0 + \beta_1 X + \varepsilon \quad . \tag{9.1}$$

In this model Y is the dependent variable, X is the independent variable, and ε is a random error (or residual) which is the amount of variation in Y not accounted for by the linear relationship. The parameters β_0 and β_1 are called the regression coefficients which are unknown and are to be estimated. The variable X is not a random variable but takes on fixed values. It will be assumed that the errors ε are independent and have a normal distribution with mean 0 and variance σ^2, regardless of what fixed value of X is being considered. Taking the expectation of both sides of Eq. (9.1), we have

$$E(Y) = \beta_0 + \beta_1 X \quad , \tag{9.2}$$

where we note that the expected value of the errors is zero.

In the simple linear regression model, the variable X can be taken to be a random variable in which case Eq. (9.2) is written as

$$E(Y|X) = \beta_0 + \beta_1 X \quad . \tag{9.3}$$

In this representation $E(Y|X)$ is the mean, or expected value of Y, given a fixed value of X. The mean $E(Y|X)$ is a conditional mean, that is, the mean of Y given X. This conditional mean can be written as $E(Y|X) = \mu_{Y|X}$. Equation (9.3) is called the regression of Y on X.

Under equation (9.1) the ε are normally distributed and the random variable Y has a normal distribution with mean $\beta_0 + \beta_1 X$ and variance σ^2. Under equation (9.3), the random variable Y, at a given value of X, has a normal distribution with mean $\beta_0 + \beta_1 X$ and variance σ^2. The main distinction between (9.1) and (9.3) is that under (9.1) the X values are fixed (nonrandom) and repeated values of Y can often be obtained for some X values, whereas, under (9.3) X and Y have a joint distribution, and if X has a continuous distribution (such as the normal) then repeated Y values for a given X will not be available in the sample. The model in (9.1) is termed a simple linear model and the one in (9.3) is called a simple linear regression model. The estimation of β_0 and β_1 is the same under (9.1) and (9.3). Thus we will use the two terms for models (9.1) and (9.3) interchangeably. However, in discussing the correlation between X and Y in section 9.7 we assume model (9.3), that is, X and Y have a joint distribution.

In order to estimate the relationship between Y and X suppose we have n observations on Y and X, denoted by $(X_1,Y_1),(X_2,Y_2),\ldots,(X_n,Y_n)$. By Eqs. (9.1) and (9.3) we can write the assumed relationship between Y and X as

$$Y = E(Y|X) + \varepsilon \quad , \tag{9.4}$$

where by Y on the left-hand side of Eq. (9.4) is meant Y, given X. The aim here is to estimate β_0 and β_1 and thus $E(Y|X)$ or Y in terms of the n observations. If $\hat{\beta}_0$ and $\hat{\beta}_1$ denote estimates of β_0 and β_1, then an estimate of E(Y) is denoted by $\hat{Y} = \hat{E}(Y) = \hat{\beta}_0 + \hat{\beta}_1 X$. Thus each observed Y_i can be written as

$$Y_i = \hat{Y}_i + e_i, \quad i = 1,2,\ldots,n \quad ,$$

where $\hat{Y}_i$ is the estimate of $E(Y_i)$ and e_i is the estimate of ε_i. Therefore, if E(Y) is a linear relationship,

$$Y_i = \beta_0 + \beta_1 X_i + \varepsilon_i = \hat{\beta}_0 + \hat{\beta}_1 X_i + e_i, \quad i = 1,2,\ldots,n \ . \tag{9.5}$$

The two equations (9.5) are illustrated in Fig. 9.1. The point (X_i,Y_i) denotes the ith observation. The "true" error or residual is $Y_i - (\beta_0 + \beta_1 X_i)$, the difference between the observed Y_i and the true unknown value $\beta_0 + \beta_1 X_i$. The observed residual e_i is $Y_i - (\hat{\beta}_0 + \hat{\beta}_1 X_i) = Y_i - \hat{Y}_i$ which is the difference between the observed Y_i and the estimated $\hat{Y}_i = \hat{\beta}_0 + \hat{\beta}_1 X_i$. The quantity $\hat{Y} = \hat{\beta}_0 + \hat{\beta}_1 X$ is

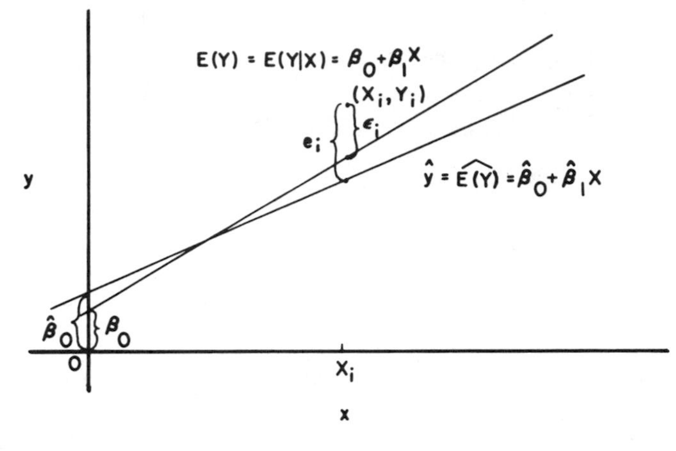

Fig. 9.1 True and estimated regression lines.

commonly called the predicted value of Y resulting from the estimated regression line.

The problem is now to obtain estimates $\hat{\beta}_0$ and $\hat{\beta}_1$ from the sample for the unknown parameters β_0 and β_1. This can best be done by the method of least squares. This method minimizes the sum of squares, $\sum_{i=1}^{n} e_i^2 = SS_E$, of the differences between the predicted values and the experimental values for the dependent variable. The method is based on the principle that the best estimators of β_0 and β_1 are those that minimize the sum of squares due to error, SS_E. The error sum of squares is

$$SS_E = \sum_{i=1}^{n} e_i^2 = \sum_{i=1}^{n} (Y_i - \hat{Y}_i)^2 \tag{9.6}$$

$$SS_E = \sum_{i=1}^{n} (Y_i - \hat{\beta}_0 - \hat{\beta}_1 X_i)^2 \quad . \tag{9.7}$$

To determine the minimum, the partial derivative of the error sum of squares with respect to each constant ($\hat{\beta}_0$ and $\hat{\beta}_1$ for this model) is set equal to zero to yield

$$\frac{\partial(SS_E)}{\partial\hat{\beta}_0} = \frac{\partial}{\partial\hat{\beta}_0}\left(\sum_{i=1}^{n} (Y_i - \hat{\beta}_0 - \hat{\beta}_1 X_i)^2\right) = 0 \quad , \tag{9.8}$$

$$\frac{\partial(SS_E)}{\partial\hat{\beta}_1} = \frac{\partial}{\partial\hat{\beta}_1}\left(\sum_{i=1}^{n} (Y_i - \hat{\beta}_0 - \hat{\beta}_1 X_i)^2\right) = 0 \quad . \tag{9.9}$$

These equations are called <u>normal equations</u>, indicating that they were obtained from the least-squares differentiation. Carrying out the differentiation, we obtain

$$n\hat{\beta}_0 + \hat{\beta}_1 \sum_i X_i = \sum_i Y_i \quad , \tag{9.10}$$

$$\hat{\beta}_0 \sum_i X_i + \hat{\beta}_1 \sum_i X_i^2 = \sum_i X_i Y_i \quad , \tag{9.11}$$

where all the summations go from i = 1 to i = n. The solutions to these normal equations are

$$\hat{\beta}_0 = \overline{Y} - \hat{\beta}_1 \overline{X} \quad , \tag{9.12}$$

$$\hat{\beta}_1 = \sum_i (X_i - \overline{X})(Y_i - \overline{Y}) / \sum_i (X_i - \overline{X})^2 \quad . \tag{9.13}$$

This solution for estimating β_0 and β_1 is called the least-squares solution. These estimates are used to give the regression equation

$$\hat{Y} = \hat{\beta}_0 + \hat{\beta}_1 X \tag{9.14}$$

which is an estimate of the true linear relationship between Y and X.

The estimator $\hat{\beta}_1$ can also be written in the form

$$\hat{\beta}_1 = (\sum_i X_i Y_i - n\overline{X}\overline{Y}) / (\sum_i X_i^2 - n\overline{X}^2) \quad ,$$

which is often more useful for computation. It can be shown that $\hat{\beta}_0$ and $\hat{\beta}_1$ are unbiased for β_0 and β_1, that is, $E(\hat{\beta}_0) = \beta_0$ and $E(\hat{\beta}_1) = \beta_1$.

The error sum of squares can be written as

$$SS_E = \sum_i (Y_i - \hat{\beta}_0 - \hat{\beta}_1 X_i)^2$$

$$= \sum_i [(Y_i - \overline{Y}) - \hat{\beta}_1 (X_i - \overline{X})]^2$$

$$= \sum_i (Y_i - \overline{Y})^2 - 2\hat{\beta}_1 \sum_i (Y_i - \overline{Y})(X_i - \overline{X}) + \hat{\beta}_1^2 \sum_i (X_i - \overline{X})^2 \quad .$$

The middle term above becomes $-2\hat{\beta}_1^2 \sum_i (X_i - \overline{X})^2$ on multiplying and dividing it by $\sum_i (X_i - \overline{X})^2$. Thus

$$SS_E = \sum_i (Y_i - \overline{Y})^2 - \hat{\beta}_1^2 \sum_i (X_i - \overline{X})^2 \quad . \tag{9.15}$$

The first term on the right-hand side of Eq. (9.15) is the total corrected sum of squares, SS_{TC}, of the Y's. The linear relationship between X and Y accounts for a reduction of $\hat{\beta}_1^2 \sum_i (X_i - \overline{X})^2$ in SS_{TC}. That is, if there is no linear relationship between X and Y (i.e., $\beta_1 = 0$) then $\sum_i (Y_i - \overline{Y})^2 = SS_E$. If there is a linear relationship between X and Y, then SS_E (or SS_{TC}) is reduced by an amount $\hat{\beta}_1^2 \sum_i (X_i - \overline{X})^2$, which is called the sum of squares due to regression and is denoted by SS_R. Equation (9.15) can be written as

$$SS_E = SS_{TC} - SS_R$$

or

$$SS_T \equiv \sum_i Y_i^2 = n\overline{Y}^{\,2} + SS_E + SS_R$$

$$= SS_M + SS_E + SS_R \quad .$$

Thus regression analysis may be looked on as the process of partitioning the total sum of squares, SS_T, into three parts: (1) the sum of squares due to the mean, SS_M, plus (2) the sum of squares due to error, SS_E (or deviations about the regression line), plus (3) the sum of squares due to regression, SS_R. Another way of stating this result is that each Y value is made up of three parts (or partitioned into three segments), each one leading to the corresponding sum of squares. That is,

$$Y_i = \overline{Y} + (\hat{Y}_i - \overline{Y}) + (Y_i - \hat{Y}_i), \qquad i = 1,2,\ldots,n \quad .$$

Figure 9.2 shows the partition of Y in graphical form. It should be noted that the estimated regression line always passes through the point $(\overline{X},\overline{Y})$. This is obvious from $\hat{Y} = \hat{\beta}_0 + \hat{\beta}_1 X = \overline{Y} - \hat{\beta}_1\overline{X} + \hat{\beta}_1 X$.

It can be shown that $SS_E/(n-2)$ is an unbiased estimator of σ^2. Furthermore SS_E/σ^2 has a χ^2-distribution with n - 2 degrees of freedom.

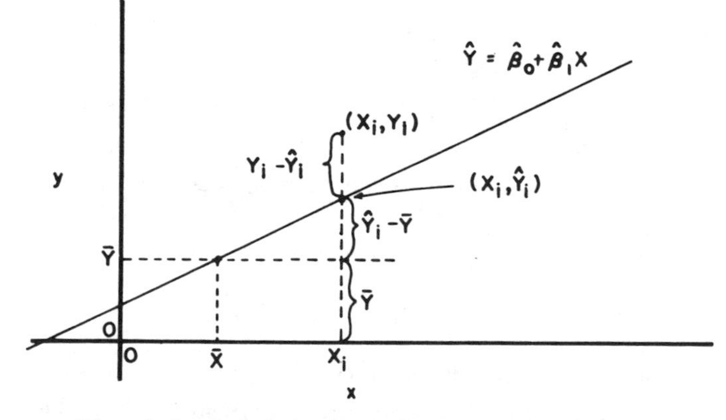

Fig. 9.2 Partitioning of total sum of squares in simple linear regression.

There are various hypotheses that are of interest in relation to the simple linear regression model. One of these is H_0: $\beta_1 = \beta_1'$. The special case H_0: $\beta_1 = 0$ (i.e., $\beta_1' = 0$) is perhaps the most important since if $\beta_1 = 0$ then there is no linear relationship between X and Y.

To make inferences regarding the parameter β_1 we need to know the distribution of its estimator $\hat{\beta}_1$. The estimator $\hat{\beta}_1$ can be written as a linear combination of $Y_1, Y_2, \ldots, Y_n$. That is

$$\hat{\beta}_1 = \sum_i \left(\frac{X_i - \bar{X}}{\sum_j (X_j - \bar{X})^2} \right) Y_i \quad .$$

From this it can be shown that β_1 has a normal distribution with mean β_1 and variance $\sigma^2_{\hat{\beta}_1} = \sigma^2 / \sum_i (X_i - \bar{X})^2$, keeping in mind that the X's are not random variables. From this it follows (see Section 6.6) that the random variable

$$T = \frac{\hat{\beta}_1 - \beta_1}{\sigma_{\hat{\beta}_1}} = \frac{\hat{\beta}_1 - \beta_1}{s_{\hat{\beta}_1}} = \frac{\hat{\beta}_1 - \beta_1}{\left[\left(\frac{SS_E}{n-2} \right) \Big/ \sum_i (X_i - \bar{X})^2 \right]^{1/2}} \qquad (9.16)$$

has a t-distribution with n - 2 degrees of freedom, which can be used to make inferences relative to β_1. Note in Eq. (9.16) that σ^2 is estimated by $MS_E = SS_E/(n-2)$. The term $s_{\hat{\beta}_1}$ is the estimate of the standard deviation of $\hat{\beta}_1$, the slope of the "fitted" regression line.

Example 9.1 An intermediate step in a reaction process is A → B. While this reaction is carried out at atmospheric pressure, the temperature varies from 1° to 10°C. As a beginning step in the optimization of this process, the relation between conversion of A and temperature must be obtained. Pilot plant studies have provided the following data (X = temperature, Y = yield):

x	y
1	3
2	5
3	7
4	10
5	11
6	14
7	15
8	17
9	20
10	21

The proposed model is $Y = \beta_0 + \beta_1 X + \varepsilon$. From these data we obtain the following:

$$\Sigma x_i = 55.0 \qquad \Sigma y_i = 123.0, \qquad \Sigma x_i y_i = 844.0,$$

$$\Sigma x_i^2 = 385.0, \qquad \Sigma y_i^2 = 1855.0, \qquad (\Sigma x_i)(\Sigma y_i) = 6765.0,$$

$$(\Sigma x_i)^2 = 3025.0, \qquad (\Sigma y_i)^2 = 15129.0,$$

$$\bar{x} = \frac{\Sigma x_i}{n} = 5.5; \qquad \bar{y} = \frac{\Sigma y_i}{n} = 12.3 \quad ,$$

$$\Sigma(x_i - \bar{x})(y_i - \bar{y}) = \Sigma x_i y_i - \frac{\Sigma x_i \; \Sigma y_i}{n} = 167.5 \quad ,$$

$$\Sigma(x_i - \bar{x})^2 = \Sigma x_i^2 - \frac{(\Sigma x_i)^2}{n} = 82.5 \quad ,$$

$$\Sigma(y_i - \bar{y})^2 = \Sigma y_i^2 - \frac{(\Sigma y_i)^2}{n} = 342.1 \quad ,$$

$$\hat{\beta}_1 = \frac{\Sigma(x_i - \bar{x})(y_i - \bar{y})}{\Sigma(x_i - \bar{x})^2} = 2.0303 \quad ,$$

$$\hat{\beta}_0 = \bar{y} - \beta_1 \bar{x} = 1.1333 \quad .$$

(Remember that lower case letters denote values of random variables, hence lower case y's in the above computations. The variable X is not a random variable, but we use the same convention for X also.)

To test the usefulness of the regression line, we must find the standard deviation of the line. This is done in the following manner. Our regression model is

$$Y_i = \beta_0 + \beta_1 X_i + \varepsilon_i, \quad i = 1,2,\ldots,n \quad , \tag{9.17}$$

where the ε_i are normally and independently distributed with mean 0 and variance $\sigma_\varepsilon^2 = \sigma^2$ which is the same for each X. The subscript "ε" is used because σ_ε^2 is the variance of the errors. If we assume that the failure of the assumed model to fit is solely a function of errors, the mean square for deviation about regression (the residual mean square) can be used as an estimate of σ^2. The estimator for σ^2 is $\hat{\sigma}^2 = MS_E = SS_E/(n-2)$ or

$$s^2 = \hat{\sigma}^2 = \sum_i \frac{(Y_i - \hat{Y}_i)^2}{n-2} = \frac{[\sum_i (Y_i - \overline{Y})^2 - \hat{\beta}_1 \sum_i (X_i - \overline{X})(Y_i - \overline{Y})]}{n-2} \quad . \tag{9.18}$$

Having determined the variance estimate $\hat{\sigma}^2$ it is a simple matter to obtain an estimate of the variance of the statistic $\hat{\beta}_1$, calculated in the regression analysis. The estimated variance of the regression coefficient $\hat{\beta}_1$ is

$$s_{\hat{\beta}_1}^2 = s^2 / \sum_{i=1}^{n} (X_i - \overline{X})^2 \quad , \tag{9.19}$$

since

$$\sigma_{\hat{\beta}_1}^2 = \sigma^2 / \sum_{i=1}^{n} (X_i - \overline{X})^2 \quad .$$

The sum of squares of deviations from the regression line is, according to Eq. (9.15),

$$\sum_i e_i^2 = \sum_i (y_i - \overline{y})^2 - \hat{\beta}_1^2 \sum_i (x_i - \overline{x})^2 = 2.025 \quad .$$

The mean square deviation is s^2 given by

$$\hat{\sigma}^2 = s^2 = \sum_i e_i^2/(n-2) = 0.2531 \quad .$$

Note that we have reduced the degrees of freedom for s^2 by 2 to account for fitting the mean and the regression coefficient $\hat{\beta}_1$ for these data. The standard deviation of experimental data from the line is s, where

$$\hat{\sigma} = s = \sqrt{s^2} = 0.5031 \quad .$$

The standard deviation of the slope of the regression line is

$$s_{\hat{\beta}_1} = s/[\sum_i (x_i - \bar{x})^2]^{1/2} = 0.05539 \quad .$$

To test the regression of Y on X, that is, H_0: $\beta_1 = 0$, we use the t-statistic in Eq. (9.16) with $\beta_1 = 0$. The computed t-value is

$$t = \frac{\hat{\beta}_1}{s_{\hat{\beta}_1}} = 36.655 \quad .$$

From Table IV, t = 2.306 with 8 degrees of freedom at a 0.05 significance level. Since the calculated t-value falls above the tabular value $t_{8,0.975} = 2.306$, the hypothesis that $\beta_1 = 0$ is rejected.

Suppose that theoretical considerations lead us to believe that $\beta_1 = 2.000$. In order to test whether or not the proposed β_1 is valid, we use the t-statistic in Eq. (9.16) with $\beta_1 = 2.0$ which is computed as

$$t = \frac{\hat{\beta}_1 - \beta_1}{s_{\hat{\beta}_1}} = \frac{2.0303 - 2.000}{0.05539} = 0.5470 \quad .$$

Since this value of "t" is less than the tabular value of $t_{8,0.975} = 2.306$, the hypothesis that $\beta_1 = 2.000$ is accepted.

9.2.1 Interval Estimation in Simple Linear Regression

Prior to determining confidence intervals or determining test procedures to be used, we recall three assumptions in the model $Y = \beta_0 + \beta_1 X$: (1) The independent variable X is a fixed variable whose values can be observed without error, (2) for any given value of X, Y is normally distributed with mean $\mu_{Y|X} = \beta_0 + \beta_1 X$ and variance $\sigma^2_{Y|X} = \sigma^2$, and (3) that the variance can be represented as $\sigma^2_\varepsilon = \sigma^2$ which is the same for each X. The estimator for σ^2, as previously mentioned, is $\hat{\sigma}^2 = MS_E = SS_E/(n-2)$.

Estimation and testing of the parameter β_1 was discussed in the previous section. Three other parameters of interest are the intercept β_0, the true mean, $\mu_{Y|X} = E(Y)$, of Y given a value of X, and the true predicted value, Y_i, corresponding to a fixed value of X. The variances of the estimators of these three parameters can be shown to be

$$\sigma^2_{\hat{\beta}_0} = \left(\frac{1}{n} + \frac{\overline{X}^2}{\sum_i (X_i - \overline{X})^2}\right)\sigma^2 , \tag{9.20}$$

$$\sigma^2_{\hat{\mu}_{Y|X}} = \left(\frac{1}{n} + \frac{(X - \overline{X})^2}{\sum_i (X_i - \overline{X})^2}\right)\sigma^2 , \tag{9.21}$$

and

$$\sigma^2_{\hat{Y}} = \left(1 + \frac{1}{n} + \frac{(X - \overline{X})^2}{\sum_i (X_i - \overline{X})^2}\right)\sigma^2 . \tag{9.22}$$

The estimators of the variances in (9.20), (9.21), and (9.22) are obtained by replacing σ^2 with $\hat{\sigma}^2 = SS_E/(n-2)$ as defined by Eq. (9.18) to give

$$s^2_{\hat{\beta}_0} = \left(\frac{1}{n} + \frac{\overline{X}^2}{\sum_i (X_i - \overline{X})^2} \right) \hat{\sigma}^2 , \tag{9.23}$$

$$s^2_{\hat{\mu}_{Y|X}} = \left(\frac{1}{n} + \frac{(X - \overline{X})^2}{\sum_i (X_i - \overline{X})^2} \right) \hat{\sigma}^2 , \tag{9.24}$$

and

$$s^2_{\hat{Y}} = \left(1 + \frac{1}{n} + \frac{(X - \overline{X})^2}{\sum_i (X_i - \overline{X})^2} \right) \hat{\sigma}^2 \tag{9.25}$$

These variance estimators may be used in testing hypotheses about the unknown parameters or they may be used to provide interval estimates of the same unknown parameters. As noted earlier, in linear regression problems the estimator of greatest interest is the slope $\hat{\beta}_1$. This, of course, is an estimator of β_1. To provide a $(1-\alpha)100\%$ confidence interval for β_1, we compute

$$\left. \begin{matrix} L \\ U \end{matrix} \right\} = \hat{\beta}_1 \mp s_{\hat{\beta}_1} t_{n-2,1-\alpha/2} \tag{9.26}$$

for a given significance level α, where $s_{\hat{\beta}_1}$ is given in Eq. (9.19) and L and U represent the lower and upper confidence limits, respectively. The confidence limits follow quite easily by considering the random variable given in Eq. (9.16).

If a confidence interval on β_0 is needed, we have only to compute

$$\left. \begin{matrix} L \\ U \end{matrix} \right\} = \hat{\beta}_0 \mp s_{\hat{\beta}_0} t_{n-2,1-\alpha/2} , \tag{9.27}$$

where $s^2_{\hat{\beta}_0}$ is given in Eq. (9.23).

A (1-α)100% confidence interval on $\mu_{Y|X} = \beta_0 + \beta_1 X$ is given by

$$\left.\begin{matrix} L \\ U \end{matrix}\right\} = \hat{Y} \mp s_{\hat{\mu}_{Y|X}} t_{n-2,1-\alpha/2} , \tag{9.28}$$

where $s^2_{\hat{\mu}_{Y|X}}$ is given in Eq. (9.24) and $\hat{Y} = \hat{\beta}_0 + \hat{\beta}_1 X$. The limits L, U yield a (1-α)100% confidence interval on the mean value of Y given a value of X.

Occasionally it is desired to predict an individual Y value, Y', associated with a given X value. This can be done by using $\hat{Y}$ to predict an individual value other than a mean value. In this case the prediction interval is provided by

$$\left.\begin{matrix} L \\ U \end{matrix}\right\} = \hat{Y} \mp s_{\hat{Y}} t_{n-2,1-\alpha/2} , \tag{9.29}$$

where $s^2_{\hat{Y}}$ is given in Eq. (9.25).

It should be noted that the confidence interval on $\mu_{Y|X}$ is shorter than the corresponding interval on a predicted Y since the latter takes into account the variability of individual Y's. This comes from the fact that $s^2_{\hat{Y}} = s^2_{\hat{\mu}_{Y|X}} + \hat{\sigma}^2$ from which we see that $s^2_{\hat{Y}} > s^2_{\hat{\mu}_{Y|X}}$.
Furthermore, the intervals on $\mu_{Y|X}$ and Y' are shortest when $X = \bar{X}$. This follows from considering Eq. (9.24) and (9.25).

It is quite conceivable that we may at some time wish to determine a confidence region for the simultaneous estimation of β_0 and β_1. To do this we use the fact that the random variable

$$F = \frac{n(\hat{\beta}_0 - \beta_0)^2 + 2n\bar{X}(\hat{\beta}_0 - \beta_0)(\hat{\beta}_1 - \beta_1) + (\hat{\beta}_1 - \beta_1)^2 \sum_i X_i^2}{2\hat{\sigma}^2} \tag{9.30}$$

has an F-distribution with $\nu_1 = 2$ and $\nu_2 = n-2$ degrees of freedom Thus

$$P(F \leq F_{\nu_1,\nu_2,1-\alpha}) = 1 - \alpha \quad ,$$

and using the definition of F in Eq. (9.30) we obtain the two-dimensional confidence region given by the interior of the ellipse

$$\frac{n(\hat{\beta}_0 - \beta_0)^2 + 2n\bar{X}(\hat{\beta}_0 - \beta_0)(\hat{\beta}_1 - \beta_1) + (\hat{\beta}_1 - \beta_1)^2 \sum_i X_i^2}{2\hat{\sigma}^2} = F_{\nu_1,\nu_2,1-\alpha}. \tag{9.31}$$

Equation (9.31) is a second-degree equation in β_0 and β_1. The hypothesis H_0: $\beta_0 = \beta_0'$ and $\beta_1 = \beta_1'$ is rejected if the left-hand side of Eq. (9.31) evaluated at $\beta_0 = \beta'_0$ and $\beta_1 = \beta_1'$ exceeds $F_{\nu_1,\nu_2,1-\alpha}$.

Example 9.2 For the data in Example 9.1, construct 95% confidence intervals on (a) β_0, (b) β_1, (c) $\mu_{Y|X=4}$, and (d) predicted value, Y', when X = 4.

(a) The standard error of $\hat{\beta}_0$ is, by Eq. (9.23),

$$s_{\hat{\beta}_0} = \left(\frac{1}{10} + \frac{(5.5)^2}{82.5}\right)^{1/2} (0.5031)$$

$$= (0.10 + 0.3667)^{1/2} (0.5031)$$

$$= 0.3437 \quad .$$

Thus, 95% confidence limits on β_0 are, by Eq. (9.27),

$$1.1333 \pm t_{8,0.975} s_{\hat{\beta}_0} \quad ,$$

$$1.1333 \pm 2.306(0.3437) \quad ,$$

$$1.1333 \pm 0.7926$$

$$0.3407, \quad 1.9259 \quad .$$

(b) The standard error of $\hat{\beta}_1$ is $s_{\hat{\beta}_1} = 0.05539$, from Example 9.1. Thus by Eq. (9.26) the 95% confidence limits on β_1 are

$$2.0303 \pm 2.306(0.05539) \quad ,$$

$$2.0303 \pm 0.1277 \quad ,$$

$$1.9026, \quad 2.1580 \quad .$$

(c) The standard error of $\hat{\mu}_{Y|X}$ is, obtained from Eq. (9.24),

$$s_{\hat{\mu}_{Y|X=4}} = \left(\frac{1}{10} + \frac{(4 - 5.5)^2}{82.5}\right)^{1/2} (0.5031)$$

$$= (0.1 + 0.0273)^{1/2} (0.5031)$$

$$= 0.3568(0.5031)$$

$$= 0.1795 \quad ,$$

and, by Eq. (9.28), the 95% confidence limits for $\mu_{Y|X}$ are

$$[1.1333 + 2.0303(4)] \mp 0.1795(2.306) \quad ,$$

$$9.2545 \mp 0.4139 \quad ,$$

$$8.8406, \quad 9.6684 \quad .$$

(d) From Eq. (9.25) we have

$$s_{\hat{Y}} = (1 + 0.1273)^{1/2} (0.5031)$$

$$= 1.0617(0.5031)$$

$$= 0.5341$$

and the 95% confidence limits on Y' are, from Eq. (9.29),

$$9.2545 \mp 2.306(0.5341) \quad ,$$

$$9.2545 \mp 1.2316 \quad ,$$

$$8.0229, \quad 10.4861 \quad .$$

9.2.2 Hypothesis Testing in Simple Linear Regression

Sometimes it is desired to determine whether the estimated slope $\hat{\beta}_1$ is significantly different from some hypothetical value of β_1, say β_1'. In this case the hypothesis to be tested is H_0: $\beta_1 = \beta_1'$. The appropriate test statistic, as illustrated in Example 9.1, is

$$T = (\hat{\beta}_1 - \beta_1')/s_{\hat{\beta}_1} \tag{9.32}$$

A popular value for β_1' is 0, as this reflects the hypothesis that Y is independent of X. The critical value for testing H_0: $\beta_1 = \beta_1'$ against H_A: $\beta_1 \neq \beta_1'$ is

$$T = \left| (\hat{\beta}_1 - \beta_1')/s_{\hat{\beta}_1} \right| > t_{n-2,1-\alpha/2} \quad .$$

Critical regions for one-sided alternatives can easily be obtained.

Other test procedures in simple linear regression are concerned with such hypotheses as (1) H_0: $\beta_0 = \beta_0'$, (2) H_0: $\mu_{Y|X=X_0} = \mu_0$, and (3) H_0: $\beta_0 = 0$. Appropriate test procedures, statistics, and rejection regions are given in Table 9.1 below.

Example 9.3 For the data in Example 9.1, test the hypotheses (a) H_0: $\beta_0 = 1$, (b) H_0: $\beta_1 \leq 1.8$ against H_A: $\beta_1 > 1.8$, (c) H_0: $\mu_{Y|X=4} = 9$, (d) H_0: $\beta_0 = 1$ and $\beta_1 = 1.8$. Use $\alpha = 0.05$.

(a) From Table 9.1 we see that we need to compute $T = (\hat{\beta}_0 - 1)/s_{\hat{\beta}_0}$ to test H_0: $\beta_0 = 1$. We have

$$\begin{aligned} t &= (1.1333 - 1)/0.3437 \\ &= 0.3878 \quad . \end{aligned}$$

Since $0.3878 < t_{8,0.975} = 2.306$ we accept H_0: $\beta_0 = 1$. This conclu-

TABLE 9.1

Test Procedures in Simple Linear Regression

H_0	Statistic	Rejection region
$\beta_0 = \beta_0'$	$T = \dfrac{\hat{\beta}_0 - \beta_0'}{s_{\hat{\beta}_0}}$	$T \geq t_{n-2,1-\alpha/2}$ or $T \leq t_{n-2,\alpha/2}$
$\beta_1 = \beta_1'$	$T = \dfrac{\hat{\beta}_1 - \beta_1'}{s_{\hat{\beta}_1}}$	
$\mu_{Y\|X=X_0} = \mu_0$	$T = \dfrac{\hat{\beta}_0 + \hat{\beta}_1 X_0 - \mu_0}{s_{\hat{y}}}$	
$\beta_0 = \beta_0'$ and $\beta_1 = \beta_1'$	$F = [n(\hat{\beta}_0-\beta_0')^2+2n\bar{X}(\hat{\beta}_0-\beta_0')(\hat{\beta}_1-\beta_1') + (\hat{\beta}_1-\beta_1')^2\Sigma X^2]/2\hat{\sigma}^2$	$F \geq F_{2,n-2,1-\alpha}$
$\beta_1 = 0$	$F = MS_R/MS_E$	$F \geq F_{2,n-2,1-\alpha}$

sion is also obtained from the confidence interval in Example 9.2(a) since $\beta_0' = 1$ is contained in that confidence interval.

(b) The hypothesis H_0: $\beta_1 \leq 1.8$ is one-sided and the pertinent statistic is $T = (\hat{\beta}_1 - 1.8)/s_{\hat{\beta}_1}$. We have

$$t = (2.0303 - 1.8)/0.05539 = 4.1578 .$$

Since $4.1578 > t_{8,0.95} = 1.860$ we reject H_0: $\beta_1 \leq 1.8$.

(c) The confidence interval in Example 9.2(c) contains $\mu_{Y|X=4} = 9$, therefore we accept H_0: $\mu_{Y|X=4} = 9$.

(d) To test H_0: $\beta_0 = 1$ and $\beta_1 = 1.8$ we need to evaluate the F-statistic defined by Eq. (9.30) which is also in Table 9.1. Since n = 10, $\hat{\beta}_0 = 1.1333$, $\beta_0' = 1$, $\bar{x} = 5.5$, $\hat{\beta}_1 = 2.0303$, $\beta_1' = 1.8$, $\Sigma x^2 = 385.0$, and $\hat{\sigma}^2 = 0.2351$, we have

$$f = \frac{10(0.1333)^2 + 20(5.5)(0.1333)(0.2303) + (0.2303)^2(385.0)}{2(0.2531)}$$

$$= \frac{0.1777 + 3.3769 + 20.4197}{0.5062}$$

$$= 47.36 \quad .$$

Since $47.36 > F_{2,8,0.95} = 4.46$ we reject the joint hypothesis that $\beta_0 = 1$ and $\beta_1 = 1.8$.

9.2.3 Inverse Prediction in Simple Linear Regression

The equation $Y = \hat{\beta}_0 + \hat{\beta}_1 X$ may sometimes be used to estimate the unknown value of X associated with an observed value of Y. The procedure is as follows. We compute

$$\hat{X} = (Y_0 - \hat{\beta}_0)/\hat{\beta}_1 \tag{9.33}$$

for which Y_0 is the observed value of Y from which we desire to estimate the associated value of X. The confidence interval for the true but still unknown X value is obtained from

$$\left.\begin{matrix} L \\ \\ U \end{matrix}\right\} = \bar{x} + \frac{\hat{\beta}_1(y_0 - \bar{y})}{\lambda} \mp \frac{t\hat{\sigma}}{\lambda}\left[c(y_0 - \bar{y})^2 + \lambda\left(\frac{n+1}{n}\right)\right]^{1/2} \tag{9.34}$$

where

$$c = \frac{1}{\sum_i (X_i - \bar{X})^2} \tag{9.35}$$

$$\lambda = \hat{\beta}_1^2 - ct^2\hat{\sigma}^2 = \hat{\beta}_1^2 - t^2 s_{\hat{\beta}_1}^2 \tag{9.36}$$

and

$$t = t_{n-2,\alpha/2} \quad . \tag{9.37}$$

If as frequently occurs, several, say m, values of Y are associated with the same unknown value of X, Eqs. (9.33) and (9.34) are modified as follows:

$$\hat{x} = (\bar{y}_0 - \hat{\beta}_0)/\hat{\beta}_1 , \tag{9.38}$$

$$\left.\begin{matrix} L \\ U \end{matrix}\right\} = \bar{x} + \frac{\hat{\beta}_1(\bar{y}_0 - \bar{y})}{\lambda} \mp \frac{t}{\lambda}\hat{\sigma}' \left[c(\bar{y}_0 - \bar{y})^2 + \lambda\left(\frac{n+m}{nm}\right)\right]^{1/2} , \tag{9.39}$$

where

$$\bar{y}_0 = \sum_{i=1}^{m} y_{0i}/m , \tag{9.40}$$

$$(\hat{\sigma}')^2 = \frac{(n-2)\hat{\sigma}^2 + \sum_{i=1}^{m} (y_{0i} - \bar{y}_0)^2}{n + m - 3} , \tag{9.41}$$

$$t = t_{n+m-3,\alpha/2} , \tag{9.42}$$

and λ and c are as defined by Eqs. (9.35) and (9.36), respectively. If the number of Y values m associated with the unknown X is small by comparison to the total number of observations, an approximate solution may be obtained by using $\hat{\sigma}^2$ rather than $(\hat{\sigma}')^2$ in Eq. (9.39).

Example 9.4 Using the data in Example 9.1, construct a 99% confidence interval for the true value of X corresponding to the observed value $Y = Y_0 = 12$.

The confidence limits are given in Eq. (9.34). The required quantities are $\bar{x} = 5.5$, $\hat{\beta}_1 = 2.0303$, $y_0 = 12$, $\bar{y} = 12.3$, $\Sigma x_i^2 = 385.0$, $t = t_{8,0.995} = 3.355$, and $\hat{\sigma}^2 = 0.2531$. From these quantities we obtain $\lambda = 4.1147$. Thus the 99% confidence limits for X are

$$5.5 - 0.1480 \mp 0.4102(4.5264)^{1/2} \quad ,$$

$$5.352 \mp 0.8727 \quad ,$$

$$4.4793,\ 6.2247 \quad .$$

9.2.4 Analysis of Variance in Simple Linear Regression

The mean square due to error, $MS_E = SS_E/(n-2)$, is an unbiased estimator for σ^2 because $E(MS_E) = \sigma^2$ regardless of whether or not the hypothesis H_0: $\beta_1 = 0$ is true. It can be shown that the expected value of the mean square due to regression, $MS_R = SS_R/(1)$, is a biased estimator for σ^2 unless $\beta_1 = 0$. This can be shown by

$$\begin{aligned} E(MS_R) &= E(SS_R/1) \\ &= \sigma^2 + \beta_1^2 \Sigma(X_i - \overline{X})^2 > \sigma^2 \quad . \end{aligned}$$

These two expected mean squares suggest the use of the ratio

$$F = MS_R/MS_E$$

in testing H_0: $\beta_1 = 0$. This ratio has an F-distribution with $\nu_1 = 1$ and $\nu_2 = n-2$ degrees of freedom if H_0: $\beta_1 = 0$ is true. Thus to test H_0: $\beta_1 = 0$ one rejects H_0 if $F > F_{1,n-2,1-\alpha}$. Since MS_R and MS_E both estimate σ^2 under H_0: $\beta_1 = 0$, one rejects H_0 if F is significantly larger than 1 since $E(MS_R) > \sigma^2$ when H_0: $\beta_1 = 0$ is not true. One should observe that H_0: $\beta_1 = 0$ can also be tested by using $T = (\hat{\beta}_1 - 0)/s_{\hat{\beta}_1}$ and the t-statistic with n-2 degrees of freedom. In fact it can be shown that T^2 is an F-statistic with 1 and n-2 degrees of freedom so that using F is equivalent to using T or T^2. The F-test for H_0: $\hat{\beta}_1 = 0$ is the bottom entry in Table 9.1.

The analysis of variance in simple linear regression is summarized in Table 9.2.

TABLE 9.2

Analysis of Variance for Simple Linear Regression

Source	d.f.	SS	MS	EMS
Due to mean	1	$n\bar{Y}^2$	---	---
Due to regression (β_1)	1	$\hat{\beta}_1\Sigma(X_i-\bar{X})(Y_i-\bar{Y}) = SS_R$	$MS_R = SS_R$	$\sigma^2+\beta_1^2\sum_i(X_i-\bar{X})^2$
Error or residual	n-2	$\sum_i(Y_i-\hat{Y}_i)^2 = SS_E$	$MS_E = SS_E/(n-2)$	σ^2
Total	n	$\sum_i Y_i^2 = SS_T$		

The AOV Table for the data in Example 9.1 is given below.

Source	d.f	SS	MS	EMS
Mean	1	1512.900	---	---
β_1	1	340.075	340.1	$\sigma^2 + 82.5\beta_1^2$
Error	8	2.025	0.2531	σ^2
Total	10	1855.000	---	---

The F-value for testing H_0: $\beta_1 = 0$ is f = 340.075/0.2531 = 1343.6. Since $F_{1,8,0.95} = 5.32$ the hypothesis is rejected. The computed T-value in Example 9.1 for testing the same hypothesis was t = 36.654. The two tests are equivalent with $t^2 = (36.654)^2 = 1343.5 \doteq f = 1343.6$, the difference being due to rounding error.

The use of the SAS System for simple linear regression is quite easy. (This program, as all the others for regression models, have been taken with permission from SAS for Linear Models (SAS Institute, Inc. Cary, NC: SAS Institute Inc. 1981. 231 pp.)). In this procedure, we introduce several new uses of the SAS general linear models routine.

```
(put initial JCL cards/statements here)
DATA;
INPUT depvar LOCATION indepvar LOCATION;
CARDS;
(put data here according to INPUT format)
PROC PRINT;
PROC PLOT;
PLOT depvar * indepvar;
PROC GLM;
MODEL depvar = indepvar;
OUTPUT OUT = NEW P = YHAT R = RESID;
PROC PLOT;
PLOT depvar * indepvar YHAT * indepvar = 'P'/OVERLAY;
PLOT RESID * indepvar/VREF = 0;
TITLE 'SIMPLE LINEAR REGRESSION OF indepvar ON depvar';
(put final JCL cards/statements here)
```

Some comments are in order. As usual, the dependent (depvar) variable must be listed first in the PROC PLOT command. This is also true for the MODEL statement. There, as elsewhere, "indepvar" refers to the independent variable. The MODEL statement instructs the SAS System to apply the least squares routine to the INPUT data set. The OUTPUT statement tells the SAS System to create a new data set containing the independent variable and the predicted values of the dependent variable. The name appearing after OUT = is the name you have chosen for the new data set (NEW for this example). The name (YHAT in this example) appearing after P = in the OUTPUT statement names the predicted variable in the new set. The predicted values (YHAT) are found by inserting the values of the independent variable into the regression equation. The name appearing after R = in the OUTPUT statement assigns a name to the variable which contains the residuals in the new data set. A residual is the original value of the dependent variable minus its value as predicted corresponding to some particular value of the independent variable. If an observation is missing, no residual will be calculated but the YHAT will contain the value projected by the regression model for the corresponding value of the independent variable.

The second PLOT statement asks for the original and new data sets to be plotted on the same graph by use of the /OVERLAY instruction. The predicted (YHAT) values are labeled P by enclosing them in apostrophes so you can distinguish them from the original data. (See Figure 9.3 in Example 9.5 below.) The last PLOT statement tells the SAS System to plot the residual (RESID) vs. the independent variable. The statement also instructs the SAS System to draw a horizontal line at 0 on the vertical scale (/VREF = 0). (See Figure 9.4 in Example 9.5 below.) This PLOT will show you whether you had the correct regression model. The values plotted should be randomly scattered about the reference line. If they show any recognizable trend, the mathematical expression for that shape should be added to a new trial model.

The use of the SAS program where "depvar" = CV and "indepvar" = NRE is shown below in Example 9.5.

<u>Example 9.5</u> The discharge coefficient C_V of venturi flowmeters in the turbulent regime should be constant at 0.98. Data as shown in the SAS output below were obtained for $18{,}600 < N_{Re} < 39{,}130$. Does C_V vary with the Reynolds number, N_{RE}? Using the program above with the variables as defined, the following output was obtained.

SAS

OBS	CV	NRE
1	1.190	18600
2	0.849	24120
3	0.804	26130
4	0.945	29480
5	0.856	32170
6	1.030	33510
7	1.010	39130

SAS

GENERAL LINEAR MODELS PROCEDURE

DEPENDENT VARIABLE: CV

SOURCE	DF	SUM OF SQUARES	MEAN SQUARE	F VALUE
MODEL	1	0.00307222	0.00307222	0.15
ERROR	5	0.10474064	0.02094813	PR > F
CORRECTED TOTAL	6	0.10781286		0.7175

SOURCE	DF	TYPE I SS
NRE	1	0.00307222

SOURCE

NRE

PARAMETER	ESTIMATE	T FOR H0: PARAMETER=0	PR > |T|	STD ERROR OF ESTIMATE
INTERCEPT	1.05212932	4.05	0.0098	0.25982499
NRE	-3.3519014E-06	-0.38	0.7175	0.00000875

From the value of T_{β_0}, we conclude that the intercept is non-zero.

For H_0: $\beta_0 = 0.98$, we have

$$T_{\beta_0} = \frac{\beta_0 - \hat{\beta}_0}{s_{\hat{\beta}_0}} = \frac{0.98 - 1.0521293}{0.25982499} = -0.278$$

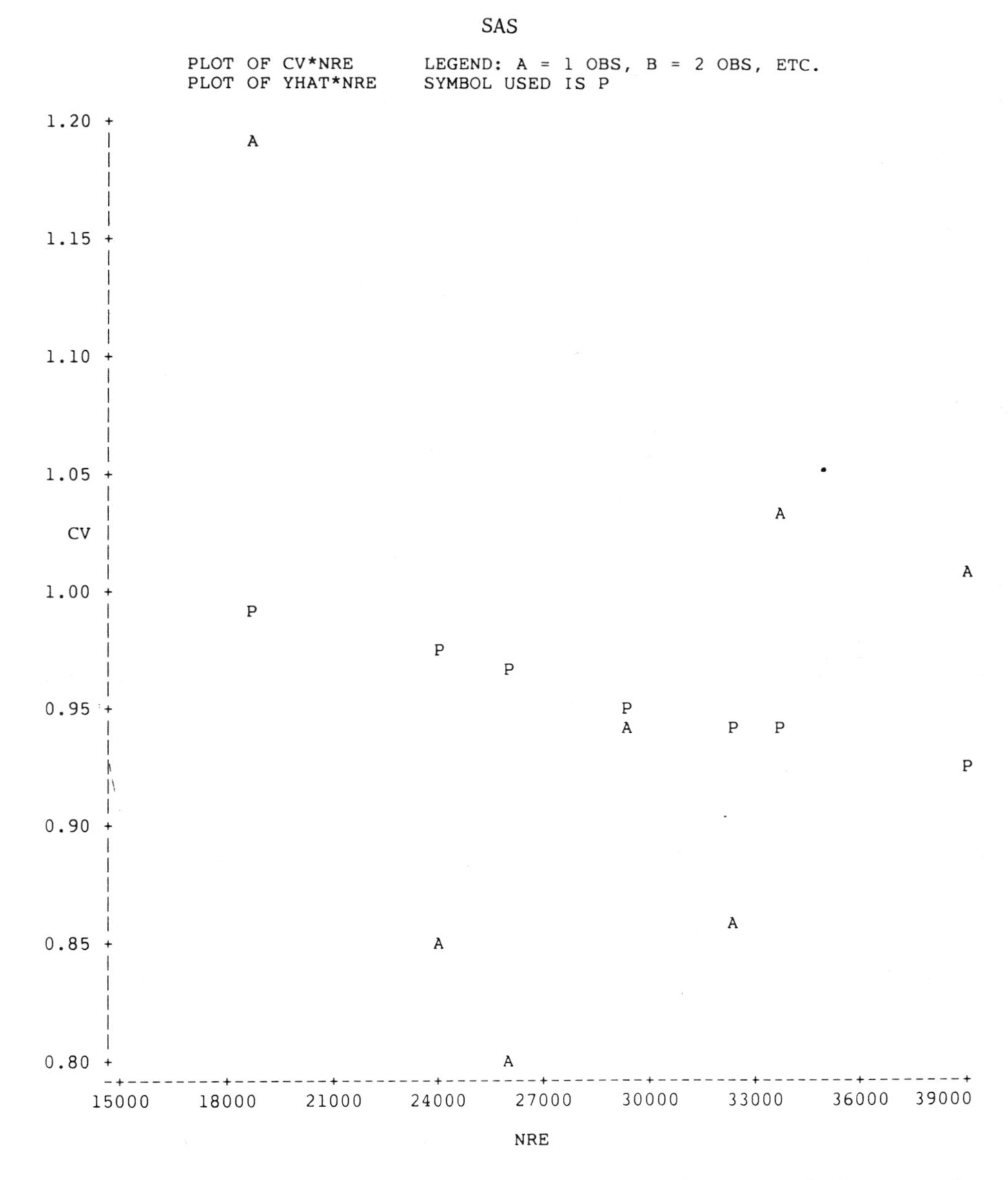

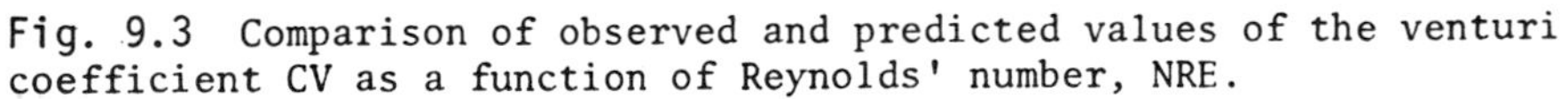
Fig. 9.3 Comparison of observed and predicted values of the venturi coefficient CV as a function of Reynolds' number, NRE.

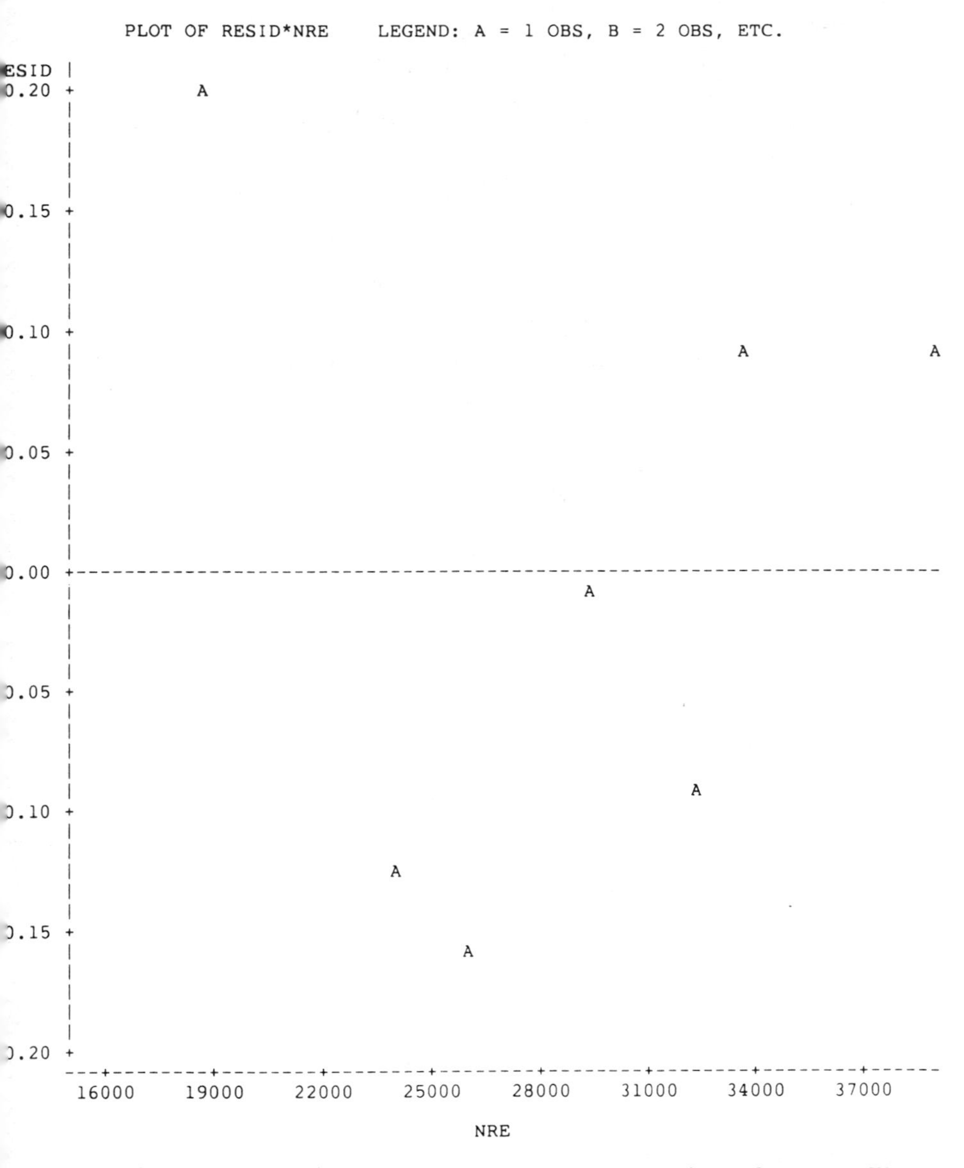

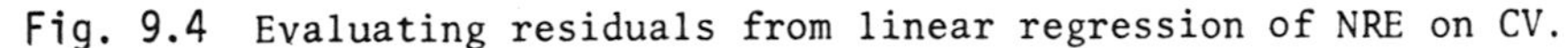
Fig. 9.4 Evaluating residuals from linear regression of NRE on CV.

so we accept H_0 and tentatively conclude that the experiment verifies theory as C_v is statistically ($\alpha = 0.05$) = 0.98.

To complete this example, we must show that C_v does not vary with N_{RE}. To do this, note that $T_{\beta_1} = -0.38$ for H_0: $\beta_1 = 0$. We thus accept this hypothesis also and conclude that for the flow rate range examined, C_v does not vary with N_{RE}.

9.2.5 Lack of Fit

The test for lack of fit of the regression model breaks up the residual sum of squares into a sum of squares for lack of fit and an experimental error sum of squares. This can be done only if we have some values of X for which we have more than one value for Y.

Suppose that of the n X's there are k distinct X's, where $k < n$, which occur with frequencies $n_1, n_2, \ldots, n_k$, where $n_1 + n_2 + \cdots + n_k = n$. The sum of squares of the n_i Y's corresponding to an X_i,

$$\sum_{j=1}^{n_i} (Y_j - \overline{Y})^2 = \sum_{j=1}^{n_i} Y_j^2 - n_i\overline{Y}^2 = (SS_E)_i \quad ,$$

is a sum of squares due to experimental error. The pooled sum of squares due to experimental error is denoted by

$$SS_E = \sum_{i=1}^{k} (SS_E)_i \tag{9.43}$$

with $\sum_{i=1}^{k} (n_i - 1) = n - k$ degrees of freedom. The mean square due to experimental error is $MS_E = SS_E/(n-k)$. The usual residual sum of squares is $SS_{Res.} = \sum_{i=1}^{n} (Y_i - \hat{Y}_i)^2$ with n-2 degrees of freedom. The lack of fit sum of squares is $SS_{L.F.} = SS_{Res.} - SS_E$ with

$n - 2 - (n-k) = k - 2$ degrees of freedom. The critical region in testing H_0: Lack of Fit is $F > F_{k-2,n-k,1-\alpha}$ where

$$F = \frac{SS_{L.F.}/(k-2)}{SS_E/(n-k)} = \frac{MS_{L.F.}}{MS_E}$$

has an F-distribution with k-2 and n-k degrees of freedom.

9.2.6 Regression Through a Point

The least squares regression technique will not necessarily produce an equation which will pass through some pre-determined point. Such an occurrence is a necessity in many physical situations: at zero flow, the float in a rotameter should be at zero and the pressure drop across an orifice or other flowmeter should likewise be zero. If no liquid is flowing down a gas absorber, its efficiency cannot be anything other than zero. Given a particular point (X_p, Y_p) through which a line must pass, how can we find the coefficients which will cause that to occur? The model in question is now

$$Y_i - Y_p = \beta(X_i - X_p) + \varepsilon_i, \quad i = 1,2,\ldots n$$

for which the normal equations must be solved as before. For the case of a linear model, the fitted equation must pass through (X_p, Y_p) instead of $(\overline{X}, \overline{Y})$. The normal equations for this case yield

$$\hat{\beta} = \frac{\sum_i (X_i - X_p)(Y_i - Y_p)}{\sum_i (X_i - X_p)^2} .$$

Example 9.6 The calibration data for the feed rotameter of the pilot plant distillation column are as follows:

Y = Reading	0	10	20	30	40	50	60	70	80	90	100
X = Flow rate, ml/min	0	22	40	61	86	103	120	143	164	186	210

$\Sigma X_i Y_i = 79{,}510$ $\quad \Sigma Y_i = 550$ $\quad \bar{Y} = 50$

$\Sigma X_i = 1135$ $\quad \Sigma X_i^2 = 164{,}251$ $\quad \bar{X} = 103.181818$

$X_p = 0,\ Y_p = 0$

$\hat{\beta} = 0.484076$

The corresponding standard least squares fit of this data yields $\hat{\beta}_1 = 0.482821$, $\hat{\beta}_0 = 0.181665$. The difference of course is due to the fact that in the first case, the line was forced through the origin.

9.3 REGRESSION USING MATRICES*

A convenient technique for general linear regression analysis involves presentation of the data in matrix notation. Consider the simple linear regression model

$$Y = \beta_0 + \beta_1 X + \varepsilon \tag{9.44}$$

which, if n observations are made on X and Y, can be written in matrix notation as $\underline{Y} = \underline{X\beta} + \underline{\varepsilon}$ where

$$\underline{Y} = \begin{pmatrix} Y_1 \\ Y_2 \\ \vdots \\ Y_n \end{pmatrix}, \quad \underline{X} = \begin{pmatrix} 1 & X_1 \\ 1 & X_2 \\ \vdots & \vdots \\ 1 & X_n \end{pmatrix}, \quad \underline{\beta} = \begin{pmatrix} \beta_0 \\ \beta_1 \end{pmatrix}, \quad \underline{\varepsilon} = \begin{pmatrix} \varepsilon_1 \\ \varepsilon_2 \\ \vdots \\ \varepsilon_n \end{pmatrix}.$$

The sum of squares due to error, $SS_E = \sum_i (Y_i - \beta_0 - \beta_1 X_i)^2$, can be written as the quadratic form

*Those students who are unfamiliar with matrix operations should see the Appendix for a brief introduction to this topic.

$$Q = (\underline{Y} - \underline{X\beta})'(\underline{Y} - \underline{X\beta}) = \underline{Y}'\underline{Y} - \underline{\beta}'\underline{X}'\underline{Y} - \underline{Y}'\underline{X\beta} + \underline{\beta}'\underline{X}'\underline{X\beta} \quad . \tag{9.45}$$

If as before we proceed to take the partial derivative of Q with respect to the parameters β_0 and β_1, we have

$$\frac{\partial Q}{\partial \underline{\beta}} = - 2\underline{X}'\underline{Y} + 2\underline{X}'\underline{X\beta} \tag{9.46}$$

where

$$\frac{\partial Q}{\partial \underline{\beta}} = \begin{pmatrix} \frac{\partial Q}{\partial \beta_0} \\ \frac{\partial Q}{\partial \beta_1} \end{pmatrix} \quad .$$

Setting $\frac{\partial Q}{\partial \underline{\beta}}$ equal to zero leads to

$$\underline{X}'\underline{X\beta} = \underline{X}'\underline{Y} \quad . \tag{9.47}$$

If we now let $\underline{S} = \underline{X}'\underline{X}$, we have the form

$$\underline{S\beta} = \underline{X}'\underline{Y} \quad , \tag{9.48}$$

from which we have the solution $\underline{\hat{\beta}}$ given by

$$\underline{\hat{\beta}} = \underline{S}^{-1}\underline{X}'\underline{Y} \quad . \tag{9.49}$$

To use these relationships, write the $\underline{S}$ matrix as

$$\underline{S} = \underline{X}'\underline{X} = \begin{pmatrix} n & \sum_i X_i \\ \sum_i X_i & \sum_i X_i^2 \end{pmatrix} \tag{9.50}$$

and the cross product matrix $\underline{X}'\underline{Y}$ as

$$\underline{X}'\underline{Y} = \begin{pmatrix} \sum_i Y_i \\ \sum_i X_i Y_i \end{pmatrix} \tag{9.51}$$

and observe that the following forms are immediately obtained when Eq. (9.48) is expanded by using the definition of $\underline{\beta}$ and Eqs. (9.50) and (9.51). That is, Eq. (9.48) in expanded form is

$$\begin{pmatrix} n & \sum_i X_i \\ \sum_i X_i & \sum_i X_i^2 \end{pmatrix} \begin{pmatrix} \hat{\beta}_0 \\ \hat{\beta}_1 \end{pmatrix} = \begin{pmatrix} \sum_i Y_i \\ \sum_i X_i Y_i \end{pmatrix}$$

which becomes

$$n\hat{\beta}_0 + \hat{\beta}_1 \sum_i X_i = \sum_i Y_i \quad , \tag{9.52}$$

$$\hat{\beta}_0 \sum_i X_i + \hat{\beta}_1 \sum_i X_i^2 = \sum_i X_i Y_i \quad . \tag{9.53}$$

Note that $\hat{\beta}_i$ are used as estimators of the components of $\underline{\beta}$. These are the normal equations previously obtained as Eqs. (9.10) and (9.11).

9.4 MULTIPLE LINEAR REGRESSION

Matrix algebra is readily adaptable to multiple linear regression where the model to be fitted is

$$Y = \beta_0 + \beta_1 X_1 + \beta_2 X_2 + \dots + \beta_p X_p + \varepsilon \quad . \tag{9.54}$$

The assumptions are the same as for the simple linear model except that now we have p independent variables. The same discussion of equation (9.1) and (9.3) in simple linear regression is pertinent to the model (9.54). That is, we can consider (9.54) or the model $E(Y|X_1,\dots,X_p) = \beta_0 + \beta_1 X_1 + \beta_2 X_2 + \dots + \beta_p X_p$, which considers

$\beta_0 + \beta_1X_1 + \beta_2X_2 + \ldots + \beta_pX_p$ to be the conditional mean of Y, given $X_1, X_2, \ldots, X_p$. The assumptions are the same as for the simple linear model (9.1) and the simple linear regression model (9.3) except that now we have p independent variables.

To obtain the least squares estimates for the β_i, we must again minimize the error sum of squares. As with simple linear regression, we have n observations on Y, $X_1, X_2, \ldots, X_p$, and the error sum of squares is

$$SS_E = \sum_i e_i^2 = \sum_i (Y_i - \hat{Y}_i)^2$$

$$= \sum_i (Y_i - \hat{\beta}_0 - \hat{\beta}_1X_{1i} - \hat{\beta}_2X_{2i} - \ldots - \hat{\beta}_pX_{pi})^2 , \tag{9.55}$$

which is minimized by setting $\partial(SS_E)/\partial\hat{\beta}_i = 0$ to get the system of normal equations (9.56) as follows:

$$\begin{aligned}
n\hat{\beta}_0 + \hat{\beta}_1 \Sigma X_{1i} + \hat{\beta}_2 \Sigma X_{2i} + \ldots + \hat{\beta}_p \Sigma X_{pi} &= \Sigma Y_i , \\
\hat{\beta}_0 \Sigma X_{1i} + \hat{\beta}_1 \Sigma X_{1i}^2 + \hat{\beta}_2 \Sigma X_{1i}X_{2i} + \ldots + \hat{\beta}_p \Sigma X_{1i}X_{pi} &= \Sigma X_{1i}Y_i , \\
\hat{\beta}_0 \Sigma X_{2i} + \hat{\beta}_1 \Sigma X_{1i}X_{2i} + \hat{\beta}_2 \Sigma X_{2i}^2 + \ldots + \hat{\beta}_p \Sigma X_{2i}X_{pi} &= \Sigma X_{2i}Y_i , \\
\vdots \qquad \vdots \qquad \vdots \qquad \vdots \\
\hat{\beta}_0 \Sigma X_{pi} + \hat{\beta}_1 \Sigma X_{1i}X_{pi} + \hat{\beta}_2 \Sigma X_{2i}X_{pi} + \ldots + \hat{\beta}_p \Sigma X_{pi}^2 &= \Sigma X_{pi}Y_i ,
\end{aligned} \tag{9.56}$$

where all the summations go from i = 1 to i = n. To obtain the estimates $\hat{\beta}_0, \hat{\beta}_1, \ldots, \hat{\beta}_p$ one needs to solve the system (9.56) of p + 1 linear equations, in the unknowns $\hat{\beta}_0, \hat{\beta}_1, \ldots, \hat{\beta}_p$. In the simple linear case we had two equations in two unknowns. A much easier approach to the normal equations is found from matrix algebra where the β_i are estimated by solving $\underline{S}\underline{\hat{\beta}} = \underline{X}'\underline{Y}$ which is actually the system (9.56) in matrix form.

In the case of p independent variables the sum of squares due to error is $SS_E = \sum_i (Y_i - \hat{Y}_i)^2$, where $\hat{Y}_i = \hat{\beta}_0 + \hat{\beta}_1 X_1 + \ldots + \hat{\beta}_p X_p$.
It can be shown that SS_E can be written as

$$SS_E = \sum_i (Y_i - \hat{Y}_i)^2$$

$$= \sum_i (Y_i - \overline{Y})^2 - [\hat{\beta}_1 \sum_i (X_{1i} - \overline{X}_1)(Y_i - \overline{Y}) + \ldots + \hat{\beta}_p \sum_i (X_{pi} - \overline{X}_p)(Y_i - \overline{Y})]$$

where $\overline{X}_j = \sum_{i=1}^{n} X_{ji}/n$. The sum enclosed within brackets is the sum of squares due to regression and thus

$$SS_E = SS_{TC} - SS_R$$

or

$$SS_{TC} = SS_E + SS_R$$

where

$$SS_R = \hat{\beta}_1 \sum_i (X_{1i} - \overline{X}_1)(Y_i - \overline{Y}) + \ldots + \hat{\beta}_p \sum_i (X_{pi} - \overline{X}_p)(Y_i - \overline{Y}). \tag{9.57}$$

A hypothesis of great interest is H_0: $\beta_1 = \beta_2 = \ldots = \beta_p = 0$. To test this hypothesis the AOV Table 9.3 is prepared. An unbiased estimate of σ^2 is furnished by $MS_E = \sum_i (Y_i - \hat{Y})^2/(n-p-1)$.

The hypothesis is tested by calculating $f = \frac{MS_R}{MS_E}$ and comparing it with $F_{p,(n-p-1),1-\alpha}$. The critical region is $F > F_{p,n-p-1,1-\alpha}$.

TABLE 9.3

Analysis of Variance for Multiple Linear Regression

Source	df	SS	MS
Regression	p	SS_R	$MS_R = SS_R/p$
Error	n-p-1	SS_E	$MS_E = SS_E/(n-p-1)$
Total	n-1	SS_{TC}	

Another hypothesis of interest is H_0: $\beta_k = 0$ where β_k is a particular β. To test this hypothesis we need to know the distribution of the statistic $(\hat{\beta}_k - \beta_k)/s_{\hat{\beta}_k}$ where $s^2_{\hat{\beta}_k}$ denotes an estimate of the variance of $\hat{\beta}_k$. It can be shown from the theory regarding multiple regression (see Ref. [10]) that $\hat{\beta}_{k-1}$ is unbiased for β_{k-1} and furthermore that the variance of $\hat{\beta}_{k-1}$ is given by $c_{kk}\sigma^2$ where c_{kk} is the kk^{th} element of the matrix $\underline{S}^{-1} = \underline{C}$. (It would prove instructive for the reader to check this out for the simple regression model.) In the same reference [10] it is shown that $c_{kk}\sigma^2$ is an unbiased estimator for $\sigma^2_{\hat{\beta}_{k-1}}$. Finally, the random variable $(\hat{\beta}_{k-1} - \beta_{k-1})/\hat{\sigma}\sqrt{c_{kk}}$ has a t-distribution with (n-p-1) degrees of freedom. Thus to test H_0: $\beta_{k-1} = 0$, we use the statistic $T = \hat{\beta}_{k-1}/\hat{\sigma}\sqrt{c_{kk}}$ and the rule is to reject H_0 if the computed T-value is such that $T > t_{n-p-1,1-\alpha/2}$ or $T < t_{n-p-1,\alpha/2}$.

Example 9.7 It is necessary to relate the percent gas absorbed in a tower, Y, to the gas temperature X_1 and the vapor pressure of the absorbing liquid, X_2. The postulated model is

$$Y = \beta_0 + \beta_1 X_1 + \beta_2 X_2 + \varepsilon \quad .$$

The following data are available:

X_1, $^{\circ}$F	X_2, mm Hg	Y, % absorbed
78.0	1.0	1.5
113.5	3.2	6.0
130.0	4.8	10.0
154.0	8.4	20.0
169.0	12.0	30.0
187.0	18.5	50.0
206.0	27.5	80.0
214.0	32.0	100.0

From these data the following values are calculated:

$\Sigma x_{1i} = 1{,}251.5$ $\quad \Sigma x_{2i} = 107.4$ $\quad \Sigma x_{1i}x_{2i} = 20{,}359.29$

$\Sigma x_{1i}^2 = 211{,}344.25$ $\quad \Sigma x_{2i}^2 = 2{,}371.339$ $\quad \Sigma x_{1i}\Sigma x_{2i} = 134{,}411.0$

$(\Sigma x_{1i})^2 = 1{,}566{,}252.25$ $\quad (\Sigma x_{2i})^2 = 11{,}534.75$ $\quad \Sigma y_i = 297.5$

$\Sigma x_{1i}y_i = 57{,}478.0$ $\quad \Sigma x_{2i}y_i = 6{,}921.70$ $\quad \Sigma y_i^2 = 20338.25$

$\Sigma x_{1i}\Sigma y_i = 372{,}321.25$ $\quad \Sigma x_{2i}\Sigma y_i = 31{,}951.49$

$\bar{x}_1 = 156.4375$ $\quad \bar{x}_2 = 13.4249$ $\quad \bar{y} = 37.1875$

The corresponding corrected sums of squares are

$$\sum_i (x_{1i} - \bar{x}_1)^2 = 15{,}562.75 \quad ,$$

$$\sum_i (x_{2i} - \bar{x}_2)^2 = 929.4951 \quad ,$$

$$\sum_i (x_{1i} - \bar{x}_1)(x_{2i} - \bar{x}_2) = 3557.913 \quad ,$$

$$\sum_i (x_{1i} - \bar{x}_1)(y_i - \bar{y}) = 10{,}937.844 \quad ,$$

$$\sum_i (x_{2i} - \bar{x}_2)(y_i - \bar{y}) = 2927.763 \quad ,$$

$$\sum_i (y_i - \bar{y})^2 = 9274.969 \quad .$$

The regression coefficients may now be found by solving the system

$$8\hat{\beta}_0 + 1251.5\hat{\beta}_1 + 107.4\hat{\beta}_2 = 297.5 \ ,$$

$$1251.5\hat{\beta}_0 + 211{,}344.25\hat{\beta}_1 + 20{,}359.29\hat{\beta}_2 = 57{,}478.0 \ ,$$

$$107.4\hat{\beta}_0 + 20{,}359.29\hat{\beta}_1 + 2{,}371.339\hat{\beta}_2 = 6{,}921.7 \ .$$

The solution is $\hat{\beta}_1 = -0.13840$, $\hat{\beta}_2 = 3.6796$, and $\hat{\beta}_0 = 9.4398$. It can be shown that in general

$$\hat{\beta}_1 = (s_{22}s_{1y} - s_{12}s_{2y})/(s_{11}s_{22} - s_{12}^2) \ ,$$

$$\hat{\beta}_2 = (s_{11}s_{2y} - s_{12}s_{1y})/(s_{11}s_{22} - s_{12}^2) \ ,$$

and

$$\hat{\beta}_0 = \overline{Y} - \hat{\beta}_1\overline{X}_1 - \hat{\beta}_2\overline{X}_2 \ ,$$

where $s_{22} = \Sigma(X_{2i} - \overline{X}_2)^2$, $s_{1y} = \Sigma(X_{1i} - \overline{X}_1)(Y_i - \overline{Y})$, $s_{12} = \Sigma(X_{1i} - \overline{X}_1)(X_{2i} - \overline{X}_2)$ and similarly for s_{11} and s_{2y}.

The resulting regression equation is

$$\hat{Y} = 9.4398 - 0.13840X_1 + 3.67961X_2 \ .$$

In this example the sum of squares due to regression is

$$SS_R = \hat{\beta}_1 \ \Sigma(x_{1i} - \overline{x}_1)(y_i - \overline{y}) + \hat{\beta}_2 \ \sum_i(x_{2i} - \overline{x}_2)(y_i - \overline{y})$$

$$= -0.13840(10{,}937.84) + 3.67961(2927.762)$$

$$= 9259.22194 \ .$$

The sum of squares due to error is thus

$$SS_E = \sum_i(y_i - \hat{y})^2$$

$$= \sum_i(y_i - \overline{y})^2 - SS_R$$

$$= 9274.96875 - 9259.22194$$

$$= 15.7468 \ .$$

The hypothesis H_0: $\beta_1 = \beta_2 = 0$ is tested by using the F-ratio $F = MS_R/MS_E$. The computed F-value is $f = 4629.6107/3.1494 = 1470.02$ which is highly significant, so H_0: $\beta_1 = \beta_2 = 0$ is rejected. The AOV table summarizing our results is given below.

Source	d.f.	SS	MS	f
Regression	2	9259.22194	4629.6107	1470.02
Error	5	15.74681	3.1494	
Total	7	9274.96875		

If one wanted to test hypotheses involving β_1 or β_2 one would use the t-statistics given by

$$T = \frac{\hat{\beta}_i - \beta_i}{s_{\hat{\beta}_i}}, \quad i = 1,2 \quad .$$

As mentioned previously, estimates of $\sigma^2_{\hat{\beta}_1}$ and $\sigma^2_{\hat{\beta}_2}$ are obtained by finding the inverse of the coefficient matrix on the left-hand side of system (9.56) which in our example is

$$8\hat{\beta}_0 + \hat{\beta}_1 \sum_i X_{1i} + \hat{\beta}_2 \sum_i X_{2i} = \sum_i Y_i \quad ,$$

$$\hat{\beta}_0 \sum_i X_{1i} + \hat{\beta}_1 \sum_i X^2_{1i} + \hat{\beta}_2 \sum_i X_{1i}X_{2i} = \sum_i X_{1i}Y_i \quad ,$$

$$\hat{\beta}_0 \sum_i X_{2i} + \hat{\beta}_1 \sum_i X_{1i}X_{2i} + \hat{\beta}_2 \sum_i X^2_{2i} = \sum_i X_{2i}Y_i \quad .$$

The estimates of $\sigma^2_{\hat{\beta}_1}$ and $\sigma^2_{\hat{\beta}_2}$ are then given by $s^2_{\hat{\beta}_1} = c_{22}\hat{\sigma}^2$ and $s^2_{\hat{\beta}_2} = c_{33}\hat{\sigma}^2$ where c_{22} and c_{33} are the second and third diagonal elements in the inverse of

$$\begin{pmatrix} 8 & 1,251.5 & 107.4 \\ 1251.5 & 211,344.25 & 20,359.29 \\ 107.4 & 20,359.29 & 2,371.339 \end{pmatrix} = S$$

In our example $c_{22} = 5.1446 \times 10^{-4}$ and $c_{33} = 8.6137 \times 10^{-3}$. Thus, $s^2_{\hat{\beta}_1} = 5.1446 \times 10^{-4} \cdot \sigma^2 = 5.1446 \times 10^{-4}(3.154) = 0.001623$. Similarly $s^2_{\hat{\beta}_2} = 0.02717$. To test the hypothesis H_0: $\beta_1 = 0$ the computed t-value is $t = \hat{\beta}_1/s_{\hat{\beta}_1} = -0.1384/0.0402 = -3.443$. Since $-3.433 < t_{5,0.025} = -2.571$, H_0: $\beta_1 = 0$ is rejected at the 5% level of significance. To test H_0: $\beta_2 = 0$ the computed t-value is $t = \hat{\beta}_2/s_{\hat{\beta}_2} = 3.67961/0.1648 = 22.328$. Since $22.328 > 2.571$, H_0: $\beta_2 = 0$ is likewise rejected at the 5% level of significance.

The use of the SAS System for multiple linear regression can be illustrated by the following example.

Example 9.8 In the production of ethylene glycol from ethylene oxide, the conversion of ethylene to ethylene oxide, Y, is a function of the activity X_1 of the silver catalyst and the residence time X_2. The following coded data are available:

Y:	12.1	11.9	10.2	8.0	7.7	5.3	7.9	7.8	5.5	2.6
X_1:	0	1	2	3	4	5	6	7	8	9
X_2:	7	4	4	6	4	2	1	1	1	0

(a) Calculate the regression coefficients.

The program is as follows.

```
(put initial JCL cards/statements here)
DATA;
INPUT X1  1  X2  3  Y  5-8;
CARDS;
(data here according to INPUT format)
PROC PRINT;
PROC GLM;
MODEL Y = X1  X2;
OUTPUT OUT = NEW P = YHAT R = RESID;
PROC GLM;
MODEL YHAT = Y;
PROC PLOT;
PLOT YHAT * Y;
TITLE 'NOTE: IF MLR OK, GRAPH SLOPE = 1, INTERCEPT = 0';
(put final JCL cards/statements here)
```

Note that the INPUT statement can be adjusted to accommodate any number of variables of any size. The MODEL statement instructs the program to perform a multiple regression by the method of least squares, using in this case two independent variables. The PLOT statement in this program will result in a graph of YHAT, the predicted value, vs. Y, the original data (Figure 9.5). If the regression has been significant, the resulting line should pass through the origin with slope of 1.0. To find out whether it does, we can look at f for the first MODEL and use PROC GLM for the YHAT vs. Y data to get the slope and intercept and their corresponding standard errors of estimate. The results are below.

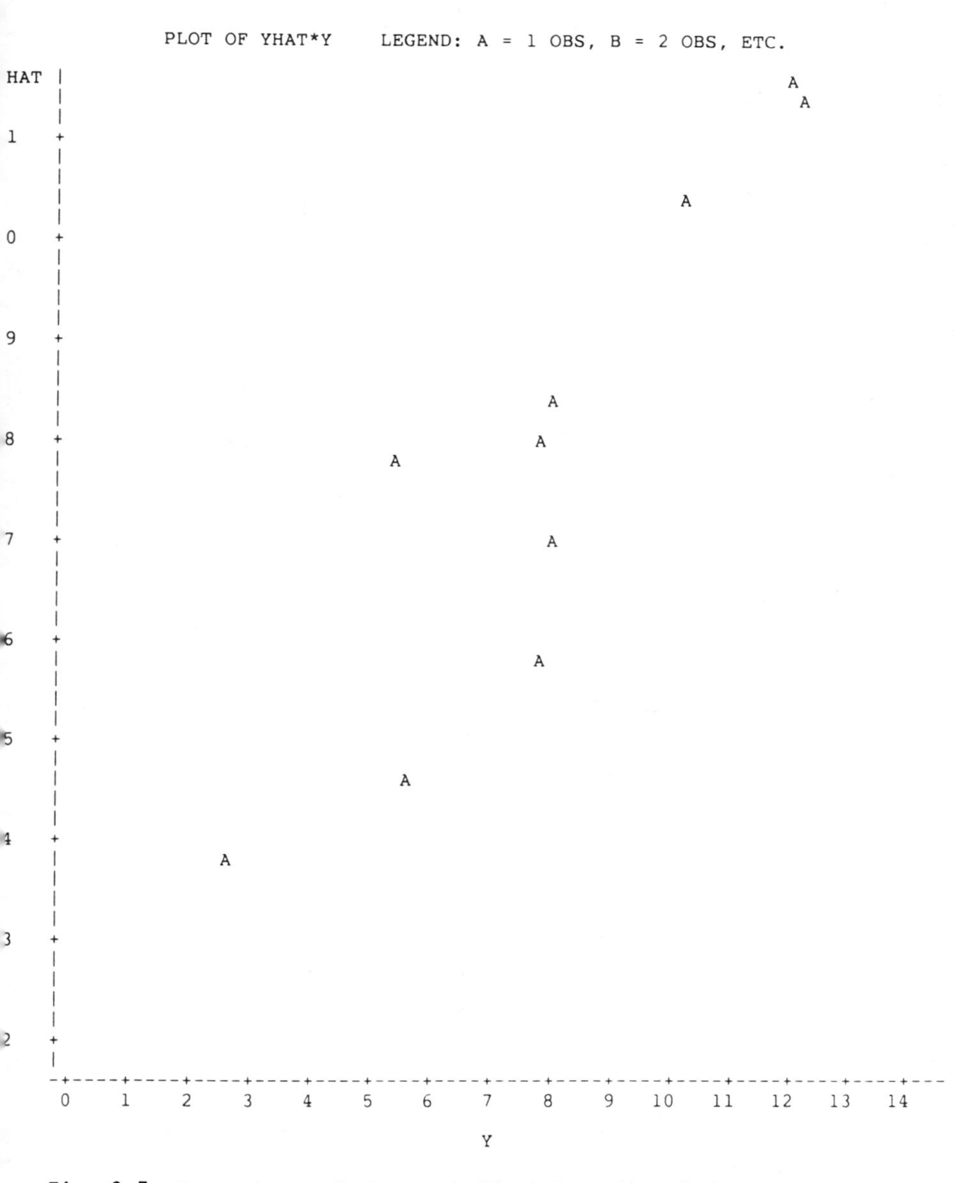

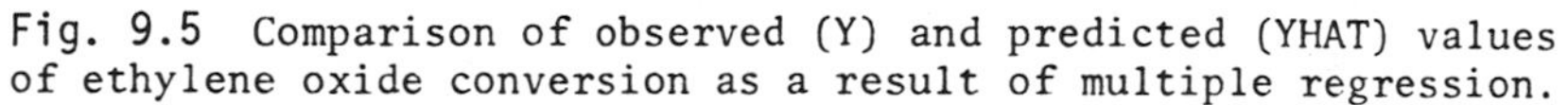
Fig. 9.5 Comparison of observed (Y) and predicted (YHAT) values of ethylene oxide conversion as a result of multiple regression.

SAS

GENERAL LINEAR MODELS PROCEDURE

DEPENDENT VARIABLE: Y

SOURCE	DF	SUM OF SQUARES	MEAN SQUARE	F VALUE
MODEL	2	65.68846912	32.84423456	16.53
ERROR	7	13.91153088	1.98736155	PR > F
CORRECTED TOTAL	9	79.60000000		0.0022

SOURCE	DF	TYPE I SS	F VALUE	PR > F
X1	1	63.71212121	32.06	0.0008
X2	1	1.97634791	0.99	0.3519

PARAMETER	ESTIMATE	T FOR H0: PARAMETER=0	PR > |T|	STD ERROR OF ESTIMATE
INTERCEPT	14.70755585	4.94	0.0017	2.97851835
X1	-1.20420499	-3.33	0.0125	0.36135254
X2	-0.46287779	-1.00	0.3519	0.46416575

From these results, we see that the null hypothesis $H: \beta_1 = \beta_2 = 0$ is rejected at $\alpha = 0.05$. It appears, however, from the T_{β_1} values that β_2 is probably 0. From plotting the residuals (Y_i - $YHAT_i$) vs. each of the independent variables, it appears that X_2^2 should have been included in the model rather than the X_2 term.

Continuing, let us examine the result of comparing the predicted ($YHAT_i$) and original (Y_i) conversion data.

SAS

GENERAL LINEAR MODELS PROCEDURE

EPENDENT VARIABLE: YHAT

OURCE	DF	SUM OF SQUARES	MEAN SQUARE	F VALUE
ODEL	1	54.20822833	54.20822833	37.77
RROR	8	11.48024079	1.43503010	PR > F
ORRECTED TOTAL	9	65.68846912		0.0003

OURCE	DF	TYPE I SS	F VALUE	PR > F
	1	54.20822833	37.77	0.0003

ARAMETER	ESTIMATE	T FOR H0: PARAMETER=0	PR > |T|	STD ERROR OF ESTIMATE
NTERCEPT	1.38066701	1.23	0.2551	1.12633543
	0.82523202	6.15	0.0003	0.13426845

From these results, we see that we would accept H_0: $\beta_0 = 0$ and reject H_0: $\beta_1 = 0$ at $\alpha = 0.05$. From the standard error, we calculate

$$T_{\beta_1} = \frac{1 - 0.82523202}{0.13426845} = 1.302$$

We thus accept the hypothesis that $\beta_1 = 1$ and conclude that the regression is probably satisfactory, even if the data do appear badly scattered.

9.5 POLYNOMIAL REGRESSION

In the case of polynomial or curvilinear regression, as given by the model

$$Y = \beta_0 + \beta_1 X + \beta_2 X^2 + \dots + \beta_p X^p + \varepsilon \quad , \tag{9.58}$$

there is only one independent variable: X. However, p - 1 other independent variables are defined as powers of X. The powers of X can be considered as $W_1 = X$, $W_2 = X^2, \ldots, W_p = X^p$ and the model (9.58) is reduced to the multiple linear regression model given by Eq. (9.54) with independent variables $W_1, W_2, \ldots, W_p$. Thus the determination of the $\hat{\beta}_i$ in model (9.58) is done in the fashion of Section 9.4, the only difference being that to determine $W_1, W_2, \ldots, W_p$ one needs to know only X.

As an example consider the quadratic model

$$Y = \beta_0 + \beta_1 X + \beta_2 X^2 + \varepsilon \quad .$$

In this case $W_1 = X$ and $W_2 = X^2$ and the normal equations are, given n observations on X and Y,

$$n\hat{\beta}_0 + \hat{\beta}_1 \sum_i W_{1i} + \hat{\beta}_2 \sum_i W_{2i} = \sum_i Y_i \quad ,$$

$$\hat{\beta}_0 \sum_i W_{1i} + \hat{\beta}_1 \sum_i W_{1i}^2 + \hat{\beta}_2 \sum_i W_{1i} W_{2i} = \sum_i W_{1i} Y_i \quad ,$$

$$\hat{\beta}_0 \sum_i W_{2i} + \hat{\beta}_1 \sum_i W_{1i} W_{2i} + \hat{\beta}_2 \sum_i W_{2i}^2 = \sum_i W_{2i} Y_i \quad .$$

However, $W_{1i} = X_i$ and $W_{2i} = X_i^2$. Thus the normal equations become

$$\begin{aligned} n\hat{\beta}_0 + \hat{\beta}_1 \sum_i X_i + \hat{\beta}_2 \sum_i X_i^2 &= \sum_i Y_i \quad , \\ \hat{\beta}_0 \sum_i X_i + \hat{\beta}_1 \sum_i X_i^2 + \hat{\beta}_2 \sum_i X_i^3 &= \sum_i X_i Y_i \quad , \qquad (9.59) \\ \hat{\beta}_0 \sum_i X_i^2 + \hat{\beta}_1 \sum_i X_i^3 + \hat{\beta}_2 \sum_i X_i^4 &= \sum_i X_i^2 Y_i \quad . \end{aligned}$$

The equations (9.59) can be solved for $\hat{\beta}_0$, $\hat{\beta}_1$, and $\hat{\beta}_2$. Extensions to polynomials of higher degree are obvious and the solution follows in the same manner.

It should be pointed out that when one speaks of a <u>linear</u> model in regression the term linear means linear in the parameters $\beta_0, \beta_1, \ldots, \beta_p$

and not in the independent variable X. Thus the polynomial regression model in Eq. (9.58) is a linear model. However, the model $Y = \beta_0 e^{\beta_1 X}$ is not since Y is not linear in β_0 and β_1. The nonlinear type model will be discussed in the next section.

Other examples of linear models (linear in the parameters) are

$$(1)\quad Y = \beta_0 + \beta_1 \log X + \beta_2 X^2 + \varepsilon \ ,$$

$$(2)\quad Y = \beta_0 + \beta_1 e^{-X} + \beta_2 X^{1/2} + \varepsilon \ ,$$

$$(3)\quad Y = \beta_0 + \beta_1 e^{-X_1} + \beta_2 X_2^2 + \beta_3 X_3 + \varepsilon \ .$$

Models (1) and (2) involve one variable X but two independent variables defined in terms of X. Model (3) contains three independent variables and could be written as $Y = \beta_0 + \beta_1 W_1 + \beta_2 W_2 + \beta_3 W_3 + \varepsilon$ where $W_1 = e^{-X_1}$, $W_2 = X_2^2$, and $W_3 = X_3$.

Example 9.9 It is believed that the effect of temperature on catalyst activity is quadratic. Eight different temperatures (coded X data below) were used. The resulting activities are given as Y.

The proposed model is

$$Y = \beta_0 + \beta_1 X + \beta_2 X^2 + \varepsilon \ .$$

The following data are available:

X	Y
2	0.846
4	0.573
6	0.401
8	0.288
10	0.209
12	0.153
14	0.111
16	0.078

If we let $W_1 = X$ and $W_2 = X^2$, the model reduces to the form

$$Y = \beta_0 + \beta_1 W_1 + \beta_2 W_2 + \varepsilon \quad .$$

Following the same procedure as for multiple linear regression, the values of the β_i are estimated as

$$\hat{\beta}_0 = 1.05652 \quad ,$$

$$\hat{\beta}_1 = -0.13114 \quad ,$$

$$\hat{\beta}_2 = 0.00447 \quad .$$

The resulting regression equation is then

$$Y = 1.05652 - 0.13114X + 0.00447X^2$$

In studies that involve only one independent variable it is generally advisable to consider a plot of the data. This is an easy task and it often suggests the type of model to be used. For example, the plot may suggest an absence of linearity in X in favor of a quadratic regression model. Or it may suggest an exponential model $Y = \beta_0 e^{-\beta_1 X}$ which will be discussed in the next section.

The use of SAS procedures for quadratic regression follows the same pattern as for multiple linear regression with one change: a new variable, X^2, must be generated from the original data. Instructions for this must be placed between the INPUT and CARDS statements. The following example illustrates the procedure.

<u>Example 9.10</u> Magnetic taconite is separated mechanically from a crushed ore slurry and rolled into marble-sized balls before charging to a pelletizing furnace. After burning in the furnace, the pellets must have a crushing strength of 8 to 10 lb. A binder, natural peat, is added during the balling process to increase the strength of the pellets. From the data below, propose an equation for the effect of peat concentration on crushing strength of taconite balls [18].

Strength of taconite balls, lb	Peat content, lb/ton of taconite
3.6	0.0
9.8	4.0
14.7	8.0
16.2	12.0
16.0	16.0
15.5	20.0

```
(put initial JCL cards/statements here)
DATA;
INPUT STR 1-4  PEAT 6-9;
PEATSQ = PEAT * PEAT;
CARDS;
(put data here according to INPUT format)
PROC GLM;
MODEL STR = PEAT  PEATSQ;
OUTPUT OUT = NEW P = YHAT R = RESID;
PROC PRINT;
PROC PLOT;
PLOT STR * PEAT YHAT * PEAT = 'P'/OVERLAY;
PLOT RESID * PEAT/VREF = 0;
TITLE 'QUAD REGR OF PEAT CONC ON STRENGTH OF TACONITE';
PROC GLM;
MODEL YHAT = STR;
PROC PLOT;
PLOT YHAT * STR;
TITLE 'NOTE:  IF QUAD REGR OK, SLOPE = 1 AND INTERCEPT = 0';
(put final JCL cards/statements here)
```

Note that in all MODEL and PLOT statements, the dependent variable must be listed first. The form of the first PROC GLM and MODEL statements instruct the program to perform a quadratic least squares regression. The second PROC GLM and MODEL statements produce a linear regression of the original data (STR_i) on the predicted ($YHAT_i$) values. The / OVERLAY command causes the original and predicted strength values to be plotted vs. peat concentration on the same graph. The last plot statement produces a graph of the residuals ($RESID_i = STR_i - YHAT_i$) vs. the peat concentration (PEAT). If the quadratic model was appropriate for the data set, the residuals should be randomly scattered about the reference line (/VREF = 0). The output is below.

SAS

DEP VARIABLE: STR

SOURCE	DF	SUM OF SQUARES	MEAN SQUARE	F VALUE	PROB>F
MODEL	2	125.102	62.551202	175.225	0.0008
ERROR	3	1.070929	0.356976		
C TOTAL	5	126.173			

DEP MEAN	12.633333	ADJ R-SQ	0.9859
C.V.	4.729352		

VARIABLE	DF	PARAMETER ESTIMATE	STANDARD ERROR	T FOR H0: PARAMETER=0	PROB > |T|
INTERCEP	1	3.739286	0.541508	6.905	0.0062
PEAT	1	1.771696	0.127339	13.913	0.0008
PEATSQ	1	-0.060156	0.006111548	-9.843	0.0022

From the F-test and the t-tests of the individual coefficients we accept the regression as significant. The correlation index R^2 seems to confirm this.

Fig. 9.6 shows the original and predicted strength values as a function of peat concentration. From this graph, the regression looks even more reasonable. Fig. 9.7 shows the residuals ($STR_i - YHAT_i$) vs. peat concentration. As the residuals are randomly scat-

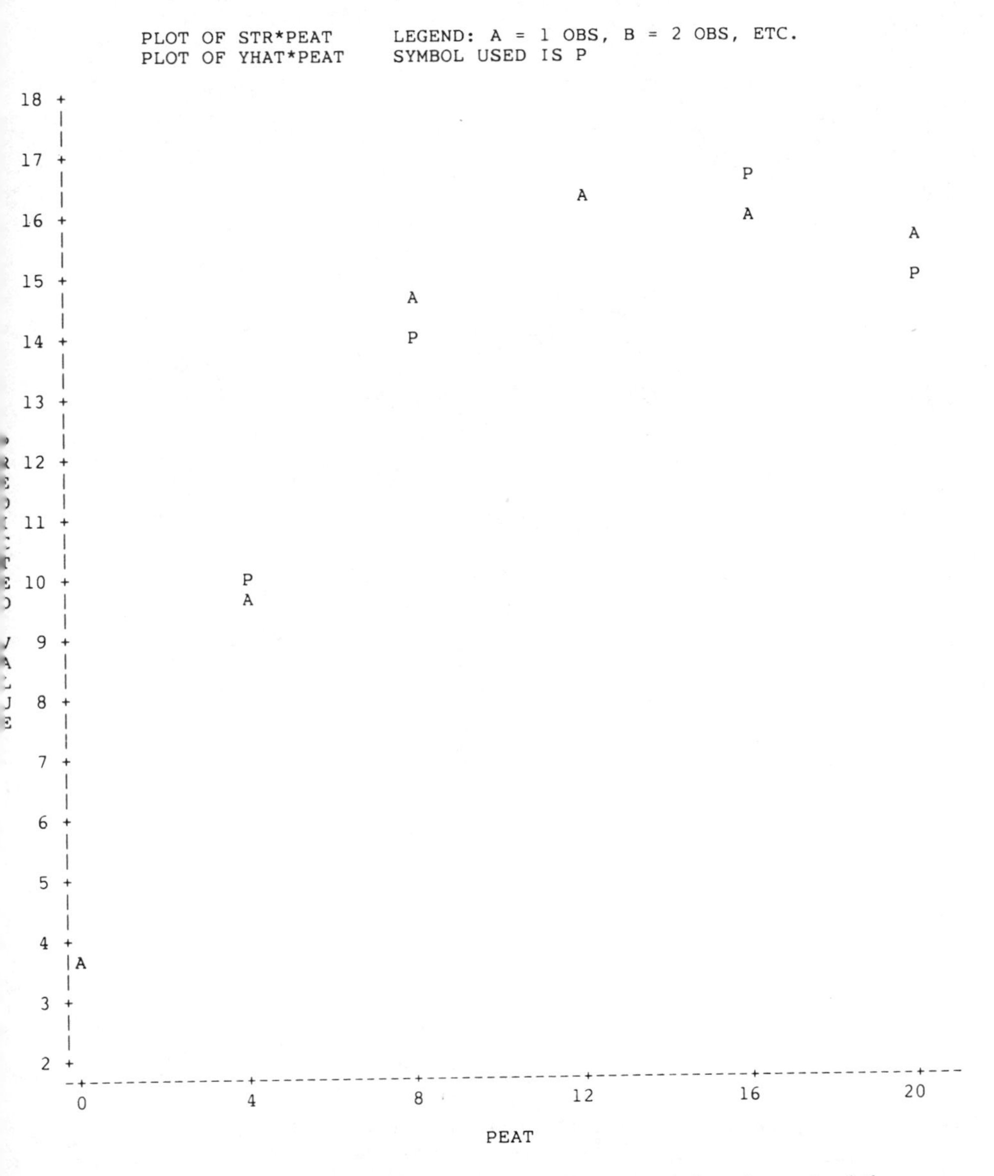

Fig. 9.6 Comparison of original crushing strength values A with those predicted P by quadratic regression of peat concentration.

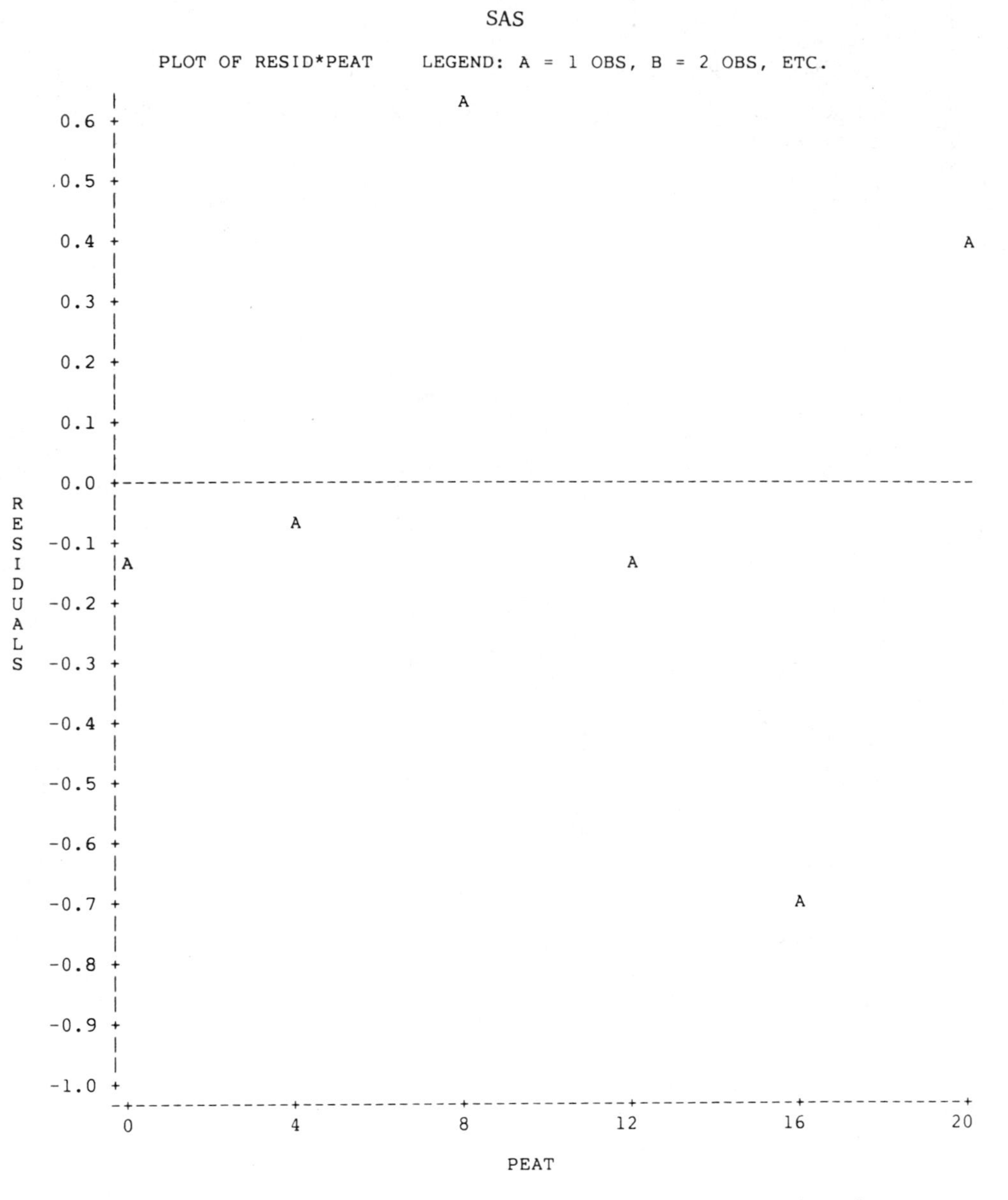

Fig. 9.7 Evaluation of residuals from quadratic regression for random scatter.

tered about the reference line, this is even more convincing evidence that the regression model used was the correct one.

Finally, we can use the $YHAT_i$ values generated by the OUTPUT statement in yet another way: to find the regression of strength (STR_i) on $YHAT_i$. As shown in the results from that PROC GLM routine, a linear model is appropriate for comparing the observed and predicted strength values. From the t-tests, $T_{\beta_0} = 0.17$ which leads us to accept the hypothesis that this model does pass through the origin. Using the standard error of estimate $s_{\hat{\beta}_1}$ to find $T_{\beta_1} = 0.185$, we similarly accept the null hypothesis that the slope is 1.0 as indicated in Figure 9.8.

SAS

QUAD REGR OF PEAT CONC ON STRENGTH OF TACONITE

EP VARIABLE: YHAT PREDICTED VALUE

OURCE	DF	SUM OF SQUARES	MEAN SQUARE	F VALUE	PROB>F
ODEL	1	124.041	124.041	467.267	0.0001
RROR	4	1.061839	0.265460		
TOTAL	5	125.102			

ARIABLE	DF	PARAMETER ESTIMATE	STANDARD ERROR	T FOR H0: PARAMETER=0	PROB > |T|
NTERCEP	1	0.107229	0.616468	0.174	0.8704
TR	1	0.991512	0.045869	21.616	0.0001

9.6 NONLINEAR REGRESSION

A nonlinear model which occurs quite frequently is

$$Y = \beta_0 e^{\beta_1 X} \quad . \tag{9.60}$$

If β_1 is positive, Eq. (9.60) is termed an exponential growth

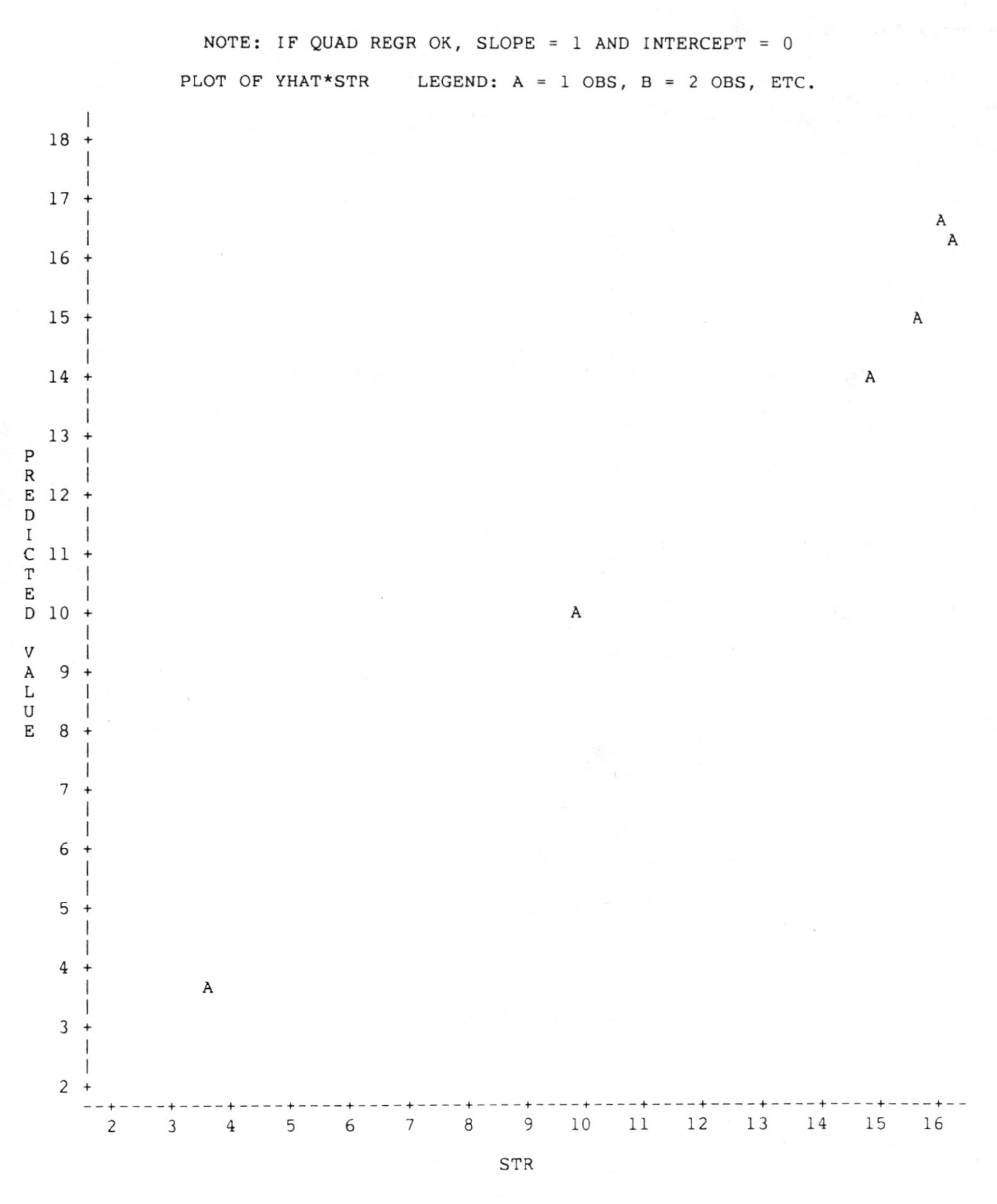

Fig. 9.8 Comparison of predicted (YHAT) and observed (Y) values of crushing strength.

curve; if negative, an exponential decay curve. Both have many applications in engineering and the basic sciences.

The problem in fitting a non-linear model is that the resulting normal equations are not linear in the $\hat{\beta}$'s and consequently there is no clear-cut solution. There do exist iterative procedures for solving such systems of equations, however, we will not discuss that approach here. The interested reader is referred to Ref. [8].

The model $Y = \beta_0 e^{\beta_1 X} + \varepsilon' = E(Y) + \varepsilon'$ is usually handled by using the log transformation and "linearizing" the model to yield the model

$$\ln Y = \ln \beta_0 + \beta_1 X + \varepsilon .$$

Letting $Z = \ln Y$, $\alpha_0 = \ln \beta_0$, and $\alpha_1 = \beta_1$, the model thus reduces to the linear model

$$Z = \alpha_0 + \alpha_1 X + \varepsilon . \tag{9.61}$$

Using the linear model (9.61), estimates $\hat{\alpha}_0$ and $\hat{\alpha}_1$ are obtained. From these one obtains the estimates $e^{\hat{\alpha}_0}$ and $\hat{\alpha}_1$. Now $\hat{\alpha}_0$ and $\hat{\alpha}_1$ are the least squares estimates of α_0 and α_1, however, $e^{\hat{\alpha}_0}$ and $\hat{\alpha}_1$ are not necessarily the least squares estimates of β_0 and β_1, the original parameters. Nevertheless, fitting the exponential model is commonly done by first linearizing and then doing the least squares estimation.

Another nonlinear model which often occurs is the so-called simple exponential model

$$y = \beta_0 \beta_1^X . \tag{9.62}$$

Using the natural logarithm transformation we have the model

$$Z = \alpha_0 + \alpha_1 X + \varepsilon \tag{9.63}$$

where $Z = \ln Y$, $\alpha_0 = \ln \beta_0$, and $\alpha_1 = \ln \beta_1$. Estimates of β_0 and β_1 can be obtained by fitting the linearized model (9.63).

One should be careful in using transformations such as the above, since if it is assumed that the original variable is normally distributed, then the transformed variable may not be. The homogeneity of variance property may be likewise violated. Frequently, however, the original assumption of normality may not be justified and the transformed variables may have a distribution closer to normal. Other non-linear models which can be linearized are shown below with their corresponding linearized forms.

(1) Multiplicative: $$Y_i = \alpha X_{1i}^{\beta} X_{2i}^{\gamma} X_{3i}^{\delta} + \varepsilon_i' \quad (9.64)$$

$$Z_i = \ln Y_i = \ln \alpha + \beta \ln X_{1i} + \gamma \ln X_{2i} + \delta \ln X_{3i} + \varepsilon_i \quad (9.65)$$

(2) Exponential: $$Y_i = \exp(\beta_0 + \beta_1 X_{1i} + \beta_2 X_{2i}) + \varepsilon_i' \quad (9.66)$$

$$Z_i = \ln Y_i = \beta_0 + \beta_1 X_{1i} + \beta_2 X_{2i} + \varepsilon_i \quad (9.67)$$

(3) Reciprocal: $$Y_i = 1/(\beta_0 + \beta_1 X_{1i} + \beta_2 X_{2i}) + \varepsilon_i' \quad (9.68)$$

$$Z_i = 1/Y_i = \beta_0 + \beta_1 X_{1i} + \beta_2 X_{2i} + \varepsilon_i \quad (9.69)$$

(4) $$Y_i = 1/(1 + \exp(\beta_0 + \beta_1 X_{1i} + \beta_2 X_{2i})) + \varepsilon_i' \quad (9.70)$$

$$Z_i = \ln(1/Y_i = 1) = \beta_0 + \beta_1 X_{1i} + \beta_2 X_{2i} + \varepsilon_i' \quad (9.71)$$

When using a transformed model that is linear in the "new parameters" the assumptions pertinent to the new model need to be checked very carefully. This is especially true if any inferences are to be made on the parameters in the transformed model. If the non-linear model, such as Eq. (9.60), is used directly, then the problem is to determine the values of $\hat{\beta}_0$ and $\hat{\beta}_1$ that minimize

$$SSE = \sum_{i=1}^{n} [Y_i - \hat{\beta}_0 e^{\hat{\beta}_1 X_i}]^2 .$$

This process is termed nonlinear least squares. If the derivatives $\partial(SSE)/\hat{\beta}_0$ and $\partial(SSE)/\hat{\beta}_1$ are set equal to zero, then non-linear equations in $\hat{\beta}_0$ and $\hat{\beta}_1$ result. Such equations are not solved as easily as the linear forms [Eq. (9.10) and (9.11)]. However, computer software is available that can be used to solve such equations (SAS for Linear Models. SAS Institute Inc. Cary, NC: SAS Institute Inc. 1981. 231 pp.). It should also be noted that the estimates obtained from the transformed model will not be the same as those obtained from the original model by nonlinear least squares.

Other frequently encountered linear (in the parameters β_i) models are the asymptotic growth curve expressed by

$$Y_i = \beta_0 - \beta_1 e^{-X_i} + \varepsilon_i' \tag{9.72}$$

and the sigmoid or logistic growth curve so common in biological and medical research:

$$Y_i = \beta_0/(1 + \beta_1 e^{X_i}) + \varepsilon_i' \quad . \tag{9.73}$$

Other non-linear models are:

$$Y_i = \beta_0 + \beta_1 e^{-\beta_2 X_i} + \varepsilon_i' \tag{9.74}$$

and

$$Y_i = \beta_0 + \beta_1 X_i + \beta_2 \beta_3^{X_i} + \varepsilon_i' \quad . \tag{9.75}$$

Unfortunately, these models cannot be linearized by a transformation of variables. Least squares fits may be obtained by employing the appropriate software to yield estimates of the parameters by nonlinear least squares.

Example 9.11 The Gurney-Lurie charts give the relations between fractional temperature change accomplished (dimensionless)

$(T_s - T)/(T_s - T_0)$ and the dimensionless group $kt/\rho C_p r^2$ for different values of the reciprocal Nusselt number k/hr and the fractional radius (dimensionless) x/r for cylinders in unsteady state conduction. These relations consist of families of essentially straight lines for varying k/hr values for set levels of x/r when $(T_s - T)/(T_s - T_0)$ is plotted vs $kt/\rho C_p r^2$ on semilog paper. For the case of x/r = 0.6 and k/hr = 1, find the correct mathematical expression for the following data:

$(T_s - T)/(T_s - T_0)$	$kt/\rho C_p r^2$
0.002	3.82
0.005	3.36
0.01	2.91
0.02	2.49
0.05	1.92
0.1	1.49
0.2	1.05
0.5	0.67

For relations such as this, which are straight or essentially so on semilog paper, a model of the form $Y = \beta_0 \beta_1^X$ is the proper starting point. The substitutions $Z = \log Y$, $\alpha_0 = \log \beta_0$, and $\alpha_1 = \log \beta_1$ are made to give a revised simple linear model $Z = \alpha_0 + \alpha_1 X$. The following quantities are calculated:

$$\bar{z} = -1.500000 \qquad \bar{x} = 2.213749$$

$$\Sigma z = -12.000000 \qquad \Sigma x = 17.70990$$

$$\Sigma zx = -33.015480 \qquad \Sigma x^2 = 48.008040$$

$$\Sigma z^2 = 22.737470 \qquad (\Sigma x)^2 = 313.64350$$

$$\Sigma z \Sigma x = -212.519800 \qquad (\Sigma z)^2 = 143.99980 \quad .$$

Solution of the normal equations for the revised model gives the expression

$$Z = 0.122229 - 0.73279X$$

which is then converted to the original form to obtain the final result as

$$Y = 1.3250(0.18501)^X$$

Example 9.12 An orifice meter has been calibrated by varying the water flow rate over the range of 1-10 ft/sec. The data are below where U = flow rate, ft/sec and DELTAP = pressure drop, inches of mercury. The theoretical relationship is

$$U_i = C_0 \sqrt{2g_c/(1 - \beta^4)\rho}\ \sqrt{-\Delta P_i} + \varepsilon_i$$

$$= 0.61(2.49690198) \sqrt{-\Delta P_i} + \varepsilon_i$$

where $-\Delta P_i$ = in. water and the parenthetical term is the first square root term and the unit conversion terms as evaluated for water (68°F) flowing through a 0.625 inch orifice in a 1.025 inch ID pipe.

The model may be approximated as

$$\log_{10} U_i = 0.18273133 + 0.5 \log_{10}(\text{DELTAP}_i)$$

Are the experimental data consistent with theory? We will use the SAS program for simple linear regression after transformation of variables to evalute the data. The SAS program is below.

```
(put initial JCL cards/statements here)
DATA;
INPUT U 1-2  DELTAP 4-5;
Z = LOG10(U);
W = LOG10(DELTAP);
CARDS;
 1   5
 2   8
 3  10
 4  12
 5  14
 6  16
 7  18
 8  19
 9  23
10  24
PROC PLOT;
PLOT U * DELTAP;
TITLE 'EFFECT OF PRESSURE DROP ON FLOW RATE';
PROC GLM;
MODEL Z = W;
OUTPUT OUT = NEW P = YHAT R = RESID;
PROC PRINT;
PROC PLOT;
PLOT Z * W YHAT * W = 'P'/OVERLAY;
TITLE 'COMPARISON OF ORIGINAL AND PREDICTED DATA';
PLOT RESID * W;
TITLE 'CHECKING RESIDUALS FOR RANDOM SCATTER';
(put final JCL cards/statements here)
```

The results are shown below.

SAS

EFFECT OF PRESSURE DROP ON FLOW RATE

GENERAL LINEAR MODELS PROCEDURE

EPENDENT VARIABLE: Z

OURCE	DF	SUM OF SQUARES	MEAN SQUARE	F VALUE
ODEL	1	0.90685273	0.90685273	1379.88
RROR	8	0.00525757	0.00065720	PR > F
ORRECTED TOTAL	9	0.91211030		0.0001

OURCE	DF	TYPE I SS	F VALUE	PR > F
	1	0.90685273	1379.88	0.0001

ARAMETER	ESTIMATE	T FOR H0: PARAMETER=0	PR > |T|	STD ERROR OF ESTIMATE
NTERCEPT	-1.01226115	-22.18	0.0001	0.04563522
	1.47534307	37.15	0.0001	0.03971662

From the F-test ($\alpha = 0.05$) we accept the regression as significant. Examining the T-values for testing the β_i values, we conclude that both are non-zero. We also calculate

$$T_{\alpha_0} = \frac{-1.01226115 - 0.18273133}{0.04563522} = -18.177$$

and

$$T_{\alpha_1} = \frac{1.47534307 - 0.5}{0.03971662} = 24.557 \quad .$$

From the calculated T-values, we see that we must reject H_0: $\alpha_0 = 0.18273133$ and H_0: $\alpha_1 = 0.5$. The first result leads us to the conclusion that, as g_c and β are constants, the orifice discharge coefficient C_o is not 0.61 for this experiment. The second result leads us to doubt the well-accepted square-root dependency of flow rate on pressure drop. Note that all tests of hypotheses have been made in the "log domain" as that is how the data were handled to obtain the regression coefficients. In this case, the high

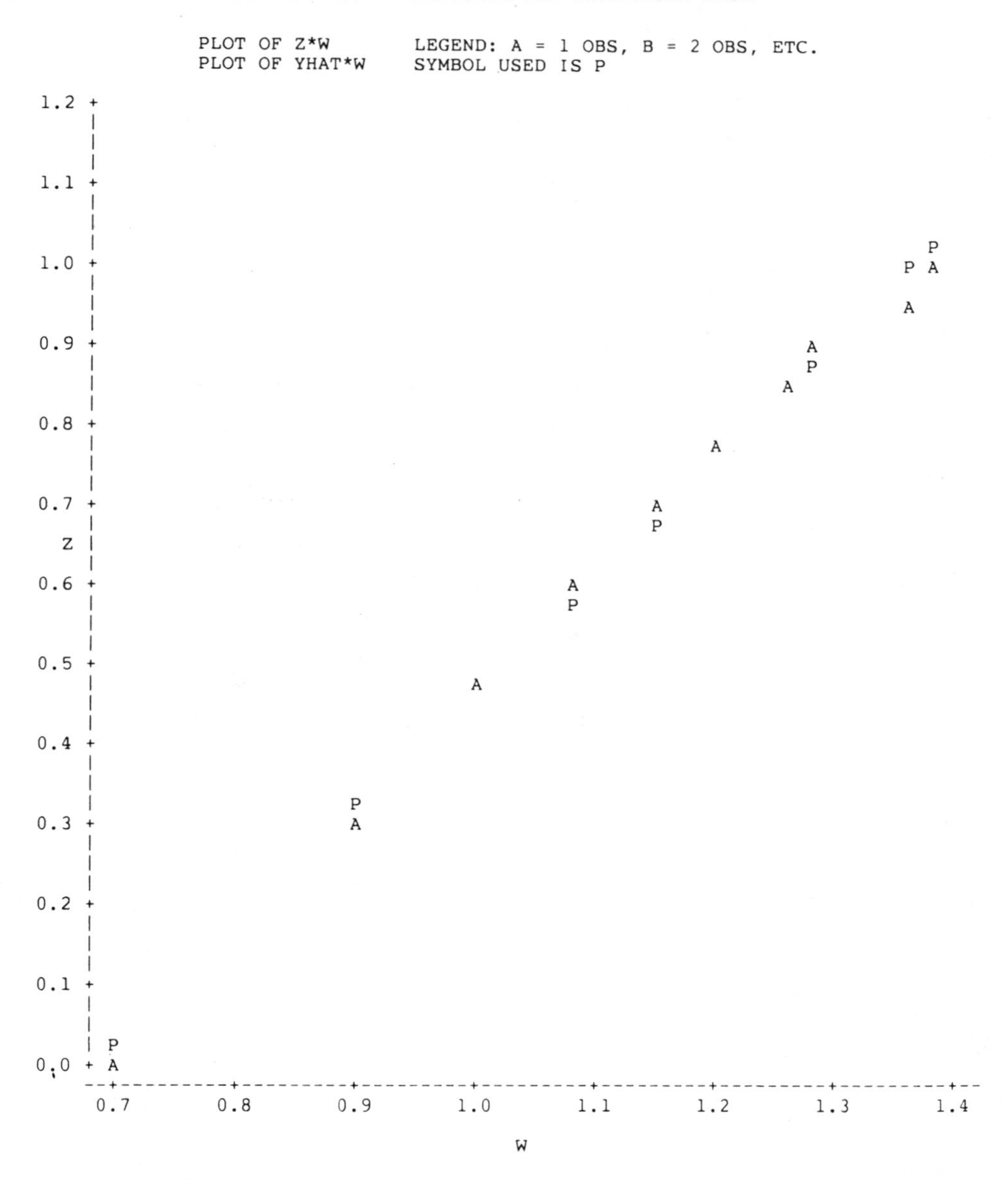

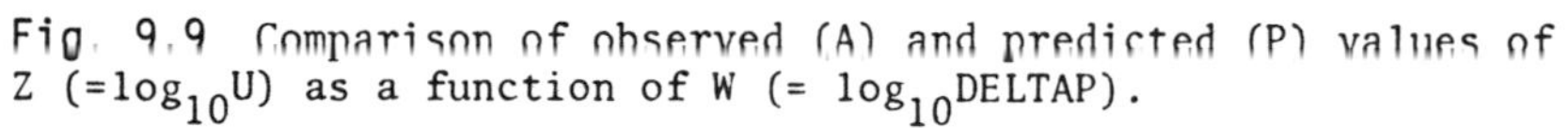

Fig. 9.9 Comparison of observed (A) and predicted (P) values of Z (=$\log_{10}$U) as a function of W (= $\log_{10}$DELTAP).

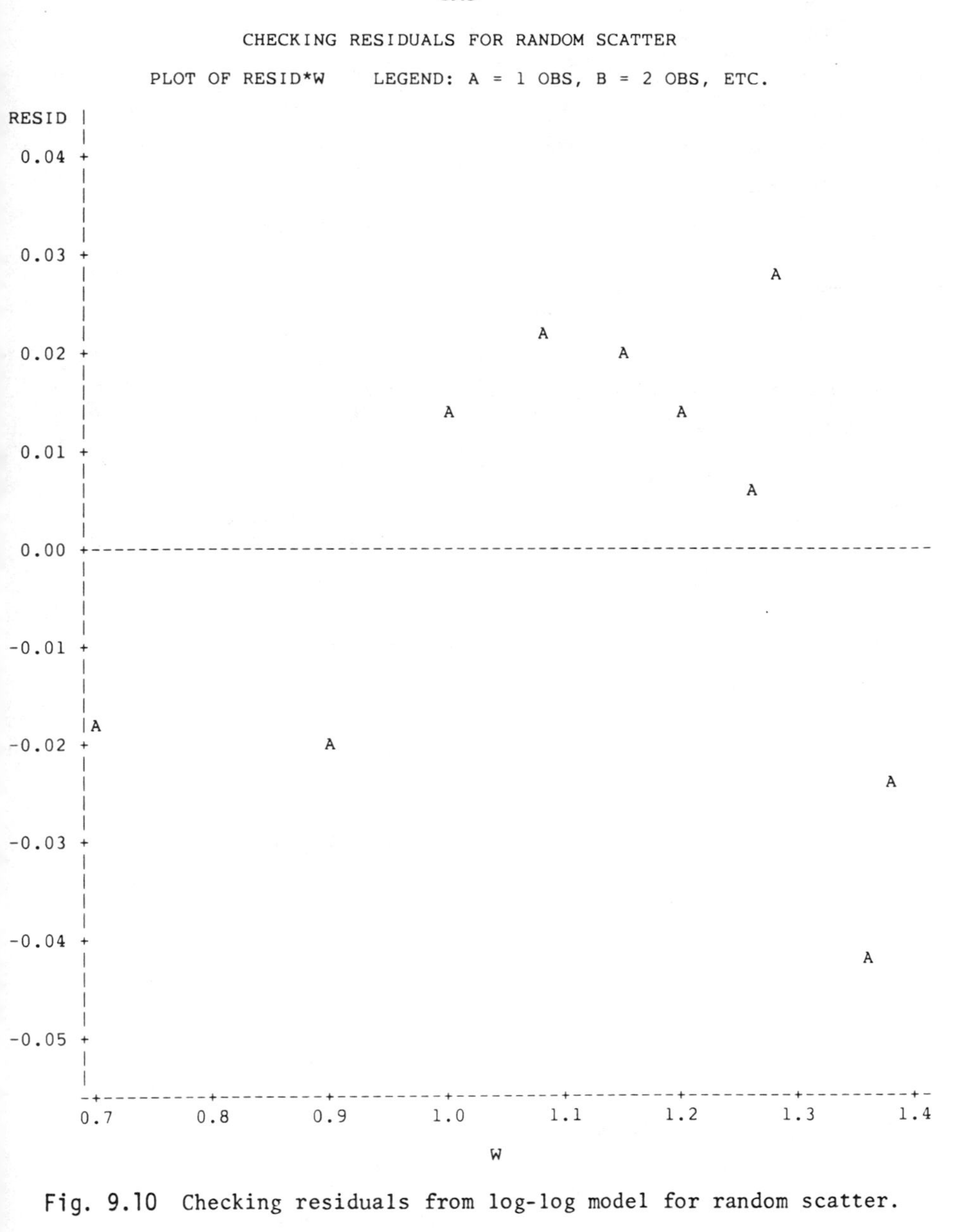

Fig. 9.10 Checking residuals from log-log model for random scatter.

(0.994^+) value of R^2 obtained by the methods described in Section 9.7 is totally misleading: the experimental data do not confirm the expected values. Figure 9.9 shows the observed (A) and predicted (P) values of $\log_{10} U_i$ or Z_i vs. $\log_{10}(DELTAP_i)$ or W_i. The agreement appears quite good. Do not be misled by the apparent shape of the residuals plot, Figure 9.10, as the vertical scale there is expanded about 8-fold compared to that in Figure 9.9.

Example 9.13 The vapor pressure of water absorbed on silica gel can be expressed as a function of the vapor pressure of pure water for various gel loadings in spacecraft humidity water recovering systems. For the water loading of 0.1 lb water/lb dry silica gel, the following data were obtained:

p^*, absorbed H_2O in. Hg,abs.	p^*, pure H_2O in. Hg,abs.
0.038	0.2
0.080	0.4
0.174	0.8
0.448	2.0
1.43	6.0
5.13	20.
9.47	35.

A plot of the p data on log-log paper yields a straight line so an equation of the form $Y = \beta_0 X^{\beta_1}$ is assumed as a model. If we let $Z = \log Y$, $\alpha_0 = \log \beta_0$, and $W = \log X$, we can reduce this to the simple linear case, $Z = \alpha_0 + \alpha_1 W$. The following quantities are then calculated:

$\bar{z} = -0.254785$ $\qquad$ $\bar{w} = 0.390065$

$\Sigma z = -1.783492$ $\qquad$ $\Sigma w = 2.730458$

$\Sigma z^2 = 5.400247$ $\qquad$ $\Sigma w^2 = 5.429265$

$\Sigma zw = 3.950120$ $\qquad$ $(\Sigma w)^2 = 7.455401$

$(\Sigma z)^2 = 3.3180843$ $\qquad$ $\Sigma z \Sigma w = -4.869750$.

Solution of the normal equations for $\hat{\alpha}_0$ and $\hat{\alpha}_1$ gives

$$Z = -0.670018 + 1.06452W$$

which, on conversion to the original form, gives the desired result as

$$Y = 0.21379 \; X^{1.06452} \quad .$$

9.7 CORRELATION ANALYSIS

Having determined that a relationship exists between variables, the next question which arises is that of how closely the variables are associated. The statistical techniques which have been developed to measure the amount of association between variables are called correlation methods. A statistical analysis performed to determine the degree of correlation is called a correlation analysis. The term used to measure correlation is referred to as a correlation coefficient. The correlation coefficient measures how well the regression equation fits the experimental data. As such, it is closely related to the standard error of estimate, $\hat{\sigma}$.

The correlation coefficient R should exhibit two characteristics:

(1) It should be large when the variables are closely associated and small when there is little association.

(2) It must be independent of the units used to measure the variables.

An effective correlation coefficient which exhibits these two characteristics is the square root of the fraction of the sum of squares of deviations of the original data from the regression curve that has been accounted for by the regression. This is a justifiable definition since the closeness of the regression curve to the data points is reflected in how much of the total corrected sum of squares, SS_{TC}, is accounted for by the sum of squares due to regression, SS_R. From Eq. (9.15) we have the identity $SS_{TC} = SS_E + SS_R$. If $SS_E = 0$

or $SS_R = SS_{TC}$, then the data points fall on the regression curve which fits the data perfectly. On the other hand, if $SS_R = 0$, then there is no reduction in SS_{TC} due to the linear relation between X and Y, or we say there is no relation (i.e., zero correlation) between X and Y. In view of this we define the correlation coefficient in terms of the proportional reduction in the error sum of squares accounted for by the regression of Y on X. The precise definition is

$$R^2 = SS_R/SS_{TC} = (SS_{TC} - SS_E)/SS_{TC} = 1 - SS_E/SS_{TC} \quad .$$

As the $SS_E \leq SS_{TC}$, R^2 will lie between 0 and 1. If the regression curve is a poor fit of the experimental data, R^2 will be close to zero.

9.7.1 Correlation in Simple Linear Regression

For the simple linear regression model, $Y = \beta_0 + \beta_1 X + \varepsilon$, the sum of squares due to regression is

$$SS_R = \hat{\beta}_1^2 \sum_i (X_i - \overline{X})^2 \quad .$$

Thus for the simple linear model we have

$$\begin{aligned} R^2 &= SS_R/SS_{TC} \\ &= \hat{\beta}_1^2 \sum_i (X_i - \overline{X})^2 / \sum_i (Y_i - \overline{Y})^2 \\ &= \left[\sum_i (X_i - \overline{X})(Y_i - \overline{Y})\right]^2 / \sum_i (X_i - \overline{X})^2 \sum_i (Y_i - \overline{Y})^2 \quad , \end{aligned} \tag{9.76}$$

since $\hat{\beta}_1 = \sum_i (X_i - \overline{X})(Y_i - \overline{Y}) / \sum_i (X_i - \overline{X})^2$. From Eq. (9.76) we have

$$R = \sum_i (X_i - \overline{X})(Y_i - \overline{Y}) / \left[\sum_i (X_i - \overline{X})^2 \sum_i (Y_i - \overline{Y})^2\right]^{1/2} \quad .$$

The correlation index for the simple linear model is usually denoted by r^2, rather than R^2. However, we shall use R^2 reserving r^2 to denote a particular value of the statistic R^2. From Eq. (9.76) it is obvious that the correlation coefficient associated with simple linear regression may be readily obtained from the statistics already calculated in the regression analysis. The values of R^2 lie between 0 and 1 and hence $-1 \leq R \leq 1$. A value of $R = 0$ means that there is no correlation between the variables and a value of either +1 or -1 implies perfect correlation. We say there is perfect positive or negative correlation according as $r = +$ or -1.

Example 9.14 Referring to the data of Example 9.1 and using Eq. (9.76) we calculate the simple linear correlation index as

$$r^2 = \frac{(167.5)^2}{82.5(342.1)} = 0.994 \quad ,$$

indicating that the regression equation accounts for 99.4% of the variability of the data about $\bar{x}$. Since $\sum_i (x_i - \bar{x})(y_i - \bar{y}) = 167.5$, $r = +\sqrt{0.994} = 0.996$. This means that X and Y are positively correlated. That means that as X increases or decreases the corresponding values of Y increase or decrease, accordingly. This also implies that the slope of the regression line is positive. In this example, the value of the correlation coefficient is quite high, $r = 0.996$, indicating a "strong" linear relationship.

From the foregoing discussion it is seen that r^2 (or R^2) is just the proportion of the variation in $\sum_i (Y_i - \bar{Y})^2$ accounted for by the linear relationship between Y and X (or the X's). The simple correlation coefficient is defined as $r = \sqrt{r^2}$, $-1 \leq r \leq 1$. This term implies that X and Y have a joint distribution, that is, we are assuming model (9.3). In this case we can use the term correlation coefficient between X and Y and covariance of X and Y, since X is also a random variable.

If we denote the sample covariance of X and Y by s_{XY}, β_1 can be found from

$$\hat{\beta}_1 = s_{XY}/s_X^2 \tag{9.77}$$

and r can be found from

$$r = s_{XY}/s_X s_Y \quad . \tag{9.78}$$

The quantity $100r^2$ is the percentage of the corrected sum of squares that is accounted for by the simple linear regression $\hat{Y} = \hat{\beta}_0 + \hat{\beta}_1 X$. If this model does not account for enough of the variation in the data, a new regression equation should be used. Frequently the form of the regression equation which will prove most effective in handling the data can be determined from the shape of the curve obtained from an appropriate plot of the data.

We have discussed linear regression assuming that the dependent variable Y, given a value of X, has a normal distribution with mean $\mu_{Y|X}$ and variance σ^2. Suppose that X is also a random variable. Then we have a two-dimensional random variable (X,Y) which has a bivariate probability density function f(x,y). If it is assumed that f(x,y) is the bivariate normal density function, then the simple correlation coefficient r is an estimator for the population correlation coefficient ρ, which is defined by

$$\begin{aligned}\rho &= E[(X - \mu_X)(Y - \mu_Y)/\sigma_X\sigma_Y] \\ &= \sigma_{XY}/\sigma_X\sigma_Y \quad ,\end{aligned}$$

where $\sigma_{XY} = E[(X - \mu_X)(Y - \mu_Y)]$ is the population covariance of X and Y.

If the variable X is also a random variable, then it has some variance which implies that the values of X are not necessarily observed or measured without error. The probability that X is not measured without error must be taken into consideration. The problem of estimating the parameters involved under these circumstances has not been resolved in general and so we will not pursue it here. Thus we will assume that X can be observed or measured without error.

In sampling from a bivariate normal population it is of interest to test the hypothesis H_0: $\rho = 0$, that is, that there is no linear relation between X and Y. It can be shown that, if $\rho = 0$, then

$$\frac{R - \rho}{s_R} = \frac{R}{s_R} = \frac{\hat{\beta}_1}{s_{\hat{\beta}_1}} . \tag{9.79}$$

We already know [Eq. (9.16)] that $T = \hat{\beta}_1/s_{\hat{\beta}_1}$ has a t-distribution with (n-2) degrees of freedom. Thus, in testing H_0: $\rho = 0$ against H_A: $\rho \neq 0$, the rejection region is $T \leq t_{n-2,\alpha/2}$ or $T \geq t_{n-2,1-\alpha/2}$, where $T = R/s_R$. Actually

$$s_R = \sqrt{(1-r^2)/(n-2)} .$$

It should be noted that testing H_0: $\rho = 0$ is the same as testing H_0: $\beta_1 = 0$.

If H_0: $\rho = 0$ is rejected, then the question of how good the linear relation is remains. Thus one may want to test a hypothesis of the type H_0: $\rho = \rho_0$ where $\rho_0 \neq 0$. To this end one can use the statistic

$$Z_r = \frac{1}{2} \ell n \left(\frac{1 + R}{1 - R}\right)$$

which has been shown to have an approximate normal distribution with mean $\mu_{Z_r} = (1/2)\ \ell n[(1 + \rho)/(1 - \rho)]$ and variance $\sigma^2_{Z_r} = 1/(n-3)$. Thus the critical region in testing H_0: $\rho = \rho_0$ against H_0: $\rho \neq \rho$ is

$$Z > z_{1-\alpha/2} \quad \text{or} \quad Z < z_{\alpha/2}$$

where $Z = (Z_r - \mu_{Z_r})/\sigma_{Z_r}$ is the standard normal random variable (deviate) as discussed in Section 3.3.6.

Example 9.15 For the data in Example 9.1, test the hypothesis H_0: $\rho \geq 0.95$ against H_A: $\rho < 0.95$. Use $\alpha = 0.01$.

From Example 9.14 the sample correlation coefficient is r = 0.997. The statistic $Z_r = (1/2)\,\ell n[(1 + R)/(1 - R)]$ when computed is

$$\begin{aligned} z_r &= 0.5\,\ell n(1.997/0.003) \\ &= 0.5(6.501) \\ &= 3.250 \quad . \end{aligned}$$

If $\rho = \rho_0 = 0.95$, then $\mu_{Z_r} = 0.5\,\ell n[1.95/0.05] \doteq 1.832$ and $\sigma^2_{Z_r} = 1/(n-3) = 1/7 = 0.1429$. Thus the computed Z-value is $z = (3.250 - 1.832)/0.1429 = 9.923$, and since $z_{0.01} = -2.33$ we accept H_0.

9.7.2 Correlation in Multiple Linear Regression

In multiple linear regression where the model is

$$Y = \beta_0 + \beta_1 X_1 + \beta_2 X_2 + \ldots + \beta_p X_p + \varepsilon \quad , \tag{9.80}$$

the correlation index R^2 is, according to Eq. (9.57),

$$\begin{aligned} R^2 &= SS_R/SS_{TC} \\ &= \frac{\hat{\beta}_1 \sum_i (X_{1i} - \overline{X}_1)(Y_i - \overline{Y}) + \ldots + \hat{\beta}_p \sum_i (X_{pi} - \overline{X}_p)(Y_i - \overline{Y})}{\sum_i (Y_i - \overline{Y})^2} \quad . \end{aligned} \tag{9.81}$$

Equation (9.81) is analogous to Eq. (9.76). The coefficient R as defined by Eq. (9.81) is called the multiple correlation coefficient and can be viewed as being the simple correlation coefficient between Y and $\hat{Y}$, where $\hat{Y} = \hat{\beta}_0 + \hat{\beta}_1 X_1 + \ldots + \hat{\beta}_p X_p$.

One of the most common faults in regression analysis is failure to plot the data before postulating a model. Consider the data below obtained from an experiment designed to measure the velocity profile in a tube. The location X is measured in the radial direction; the flow rate Y is measured in the longitudinal direction.

X:	-2	-1	0	1	2
Y:	4.2	0.9	0	1.1	4.0

Some students actually fitted this data to a simple linear model. Needless to say, they were quite surprised to find $r^2 = 0.00027$. The data actually described a nearly perfect parabola. Had the data been plotted first, the proper model would have been obvious. The data would have then been fitted with a quadratic model and the resulting regression equation would have been

$$Y = -0.01714 - 0.02000X + 1.02857X^2 \quad .$$

R^2 for the quadratic model is 0.99754, that is, the simple correlation between Y and $\hat{Y}$ can be viewed as being 0.99754.

Example 9.16 The correlation index R^2 may be obtained for the data of Example 9.5 as a means of determining the "goodness-of-fit" of the regression equation already estimated. Equation (9.70) is used to give

$$R^2 = \frac{\hat{\beta}_1 \sum_i (X_{1i} - \overline{X}_1)(Y_i - \overline{Y}) + \hat{\beta}_2 \sum_i (X_{2i} - \overline{X}_2)(Y_i - \overline{Y})}{\sum_i (Y_i - \overline{Y})^2}$$

where

$$\sum_i (y_i - \overline{y})^2 = \sum_i y_i^2 - (\sum_i y_i)^2/n = 9274.97 \quad .$$

Therefore,

$$r^2 = \frac{-0.13840(10{,}937.84) + 3.67961(2927.762)}{9274.97}$$

$$= 0.9983 \quad ,$$

indicating that the proposed model fits the data exceptionally well and the regression equation in fact accounts for 99.9% of the variability of the experimental data about the mean $\overline{y}$. If the correlation coefficient between any two of the independent variables is high, say at least 0.8, then generally no great increase in information can be obtained by using multiple correlation for both variables as opposed to simple linear correlation for using only one of them. For example, using the data of Example 9.5, the simple correlation coefficients are $r^2_{X_1X_2} = 0.8712$, $r^2_{X_1Y} = 0.8288$, and $r^2_{X_2Y} = 0.9956$. Since R^2, as shown by Example 9.11, is 0.9991, very little was gained by adding X_1 to the correlation. However, using Y and both X_1 and X_2 increases the correlation from 0.8288 to 0.9991 over using Y and X_1. In Example 9.5 it was seen that H_0: $\beta_1 = 0$ was rejected at the 5% significance level. The hypothesis H_0: $\beta_2 = 0$ was also rejected at the 5% significance level.

One warning should be repeated here: High correlation coefficients can be obtained in models which leave very little or no physical meaning. It is good practice to plot the data, whenever possible, and select trial models according to the results of the plot. Frequently the data will have to be plotted in several ways before a good, i.e., reasonable and usable, model is apparent. Consider the example below.

Example 9.17 All too often we encounter data which are nonlinear on semilog plots. One good example of this is the O'Connell [14] correlation between overall column efficiency E vs the product of relative volatility α and liquid viscosity μ for multicomponent distillation systems. From the points below (estimated from O'Connell's original curve) find a mathematical relationship between E and $\alpha\mu$.

E	$\alpha\mu$
100	0.05
90	0.082
80	0.13
70	0.24
60	0.46
50	0.91
40	2.19
30	7.64

When these data were plotted on rectangular coordinate paper, a quadratic model was suggested, which yielded

$$E = 85.6529 - 31.7893(\alpha\mu) + 3.21634(\alpha\mu)^2 , \qquad (A)$$

for which $R^2 = 0.8404$. As the model did not account for a large enough portion of the data variation, the data were plotted on semi-log paper. From that, two models seemed obvious. The fitted results were

$$E = 52.6465 - 32.4268 \log (\alpha\mu) \qquad (B)$$

for which $R^2 = 0.9743$ and

$$E = 47.1390 - 28.8853 \log (\alpha\mu) + 9.54988 [\log(\alpha\mu)]^2 \qquad (C)$$

for which $R^2 = 0.9916$. Although inclusion of the quadratic term in log $\alpha\mu$ raised R^2, there are no known instances of $[\log(X)]^2$ having any real significance in a physical model. Therefore equation (C) was discarded as useless.

One student suggested reversing the axes. This was done and gave

$$\log E = 1.86758 - 0.0582711(\alpha\mu) \qquad (D)$$

for which $R^2 = 0.7023$. As we seemed to be headed in the wrong direction, the data were plotted on log-log paper and a nearly straight line was obtained. The data were then correlated using an appropriate model which gave the result

$$\log E = 1.69014 - 0.242014 \log(\alpha\mu) \qquad (E)$$

for which $R^2 = 0.9996$. At this point, the question was raised: "Which relation should we use, (B) or (E)?" This is almost like asking, "What is the best way to give change for a dollar?" While (E) may be slightly better than (B), it is not as easy to use because it involves log E rather than E itself. In this example, one would choose model (B) since it appears to be more useful than model (E). In similar circumstances, a choice is made based on usefulness. If the choice were between (A) and either (B) or (E), (A) would not be selected because of the lower R^2 value.

An alternate approach would be to break the curve into a series of segments which can be approximated as linear and using the techniques of Example 9.4. When this was done in two sections, $\alpha\mu \leq 2$ and $\alpha\mu > 2$, the results were, respectively,

$$E = 0.5117 - 0.14751 \log(\alpha\mu) \quad \text{for } \alpha\mu \leq 2$$

and

$$E = 0.44916 - 0.07220 \log(\alpha\mu) \quad \text{for } \alpha\mu > 2 \quad .$$

Potentially the errors encountered can be severe when taking this approach. At the break point in this case, the error is a maximum. The error excursions in the center 60% of each segment are fairly small. The errors in the outer 20% of the range of each line are so large that the lines, in the vicinity of the break point, are virtually useless for estimation purposes in the initial stages of bubble cap tower design. This illustrates an all too common failing on the part of the beginning statistics practitioner: laziness. Granted it is easier to segment the curve and solve the resulting pair of simple linear regressions. As we see in this example, the predictions of efficiency as it depends on the $\alpha\mu$ product are worthless. It is far better to expend the additional effort required to obtain a continuous curve which permits adequate predictions.

9.8 STEPWISE REGRESSION

The situation occasionally arises when several independent variables affect the outcome of an experiment. In that case, you will want to find an empirical equation relating them that simultaneously satisfies two criteria: has a high value of R^2 and uses only the most important independent variables. This can be handled in two ways by use of PROC STEPWISE or PROC RSQUARE (SAS User's Guide: Statistics, 1982 Edition. SAS Institute Inc., Cary, NC: SAS Institute Inc. 1979. 494 pp.). PROC STEPWISE has 5 methods of retaining individual terms in the model: forward selection, backward elimination, etc. More discussion of these techniques can be found in Draper and Smith [8] and Daniel and Wood [7]. Stepwise regression requires the assumption that there is only one best equation and that the procedure will find it. This of course presupposes that the "right" independent variables were selected for the model. Another problem arises when there is a high correlation between independent variables: difficulty in interpreting the results. The general forward selection procedure begins by using the independent variables one at a time, comparing the corresponding reduction in the error sum of squares with some pre-set criterion (α = 0.15 is the default for PROC STEPWISE) and then either retaining or rejecting the term. PROC STEPWISE is a modified forward selection method in which the f-value for each term in the model is calculated, compared to the corresponding tabular value of F, and rejected if it is not significant at the pre-set significance level. Then the next term is added to the model and the process is repeated.

Backward elimination, as the name implies, starts with the full model and eliminates the terms one at a time based on their lack of contribution to the reduction in the error sum of squares. It should be noted that all independent variables not "read in" by INPUT and CARDS statements must be generated in the program. Such generations (X^2, $\sqrt{X}$, log X, etc.) must be placed between the INPUT and CARDS statements.

PROC RSQUARE constructs regression analyses for all possible combinations of the independent variables. The output includes the variables used and their associated values of R^2 for all possible combinations of the independent variables in the MODEL statement. The seemingly "best" models based on the R^2-values and compatibility with physical theory can hopefully be identified for further scrutiny or experimentation.

Example 9.18 This example adapted from data given by Ostle* demonstrates the use of PROC STEPWISE and PROC RSQUARE.

The size (Y values measured in microns) of bainite grains was determined at different time intervals (X values, seconds) for a particular steel alloy undergoing isothermal transformation to bainite. Examine these data to find the most reasonable regression model.

---

*Data reprinted from Ostle, B. and R. W. Mensing, Statistics in Research, 3rd ed., copyright 1975 by the Iowa State University Press, Ames, Iowa, by permission.

X, sec	Y, μm	X, sec	Y, μm
1	17	8	64
2	21	9	80
3	22	10	86
4	27	11	88
5	36	12	92
6	49	13	94
7	56		

```
(put initial JCL cards/statements here)
DATA TIME;
INPUT X1  1-2  Y  4-5;
X2 = X1*X1;
X3 = X1*X1*X1;
CARDS;
(enter data here according to INPUT format)
PROC PRINT;
PROC PLOT;
PLOT Y*X1;
TITLE 'EFFECT OF ISOTHERMAL TRANSFORMATION TIME ON GRAIN SIZE';
PROC STEPWISE DATA = TIME;
MODEL Y = X1  X2  X3;
TITLE 'EFFECTIVENESS OF STEPWISE REGRESSION';
(put final JCL cards/statements here)
```

Note that data set should be named in both the DATA and PROC STEPWISE statements.

The results of PROC STEPWISE are shown below.

SAS

STEPWISE REGRESSION OF TRANSFORMATION TIME IN GRAIN SIZE

STEPWISE REGRESSION PROCEDURE FOR DEPENDENT VARIABLE Y

STEP 1 VARIABLE X1 ENTERED R SQUARE = 0.96977720

	DF	SUM OF SQUARES	MEAN SQUARE	F	PROB>F
REGRESSION	1	10177.58791209	10177.58791209	352.96	0.0001
ERROR	11	317.18131868	28.83466533		
TOTAL	12	10494.76923077			

	B VALUE	STD ERROR	TYPE II SS	F	PROB>F
INTERCEPT	3.96153846				
X1	7.47802198	0.39803546	10177.58791209	352.96	0.0001

From the analysis of variance, we see that only the linear term was significant in the regression with the pre-set $\alpha = 0.15$ (default level if another option is not specified).

To continue this example, replace the PROC STEPWISE statement by PROC RSQUARE. Again, the data set is named in the DATA and PROC RSQUARE statements. The results are shown below.

USE OF PROC RSQUARE ON TTT - GRAIN SIZE DATA

N= 13 REGRESSION MODELS FOR DEPENDENT VARIABLE Y

NUMBER IN MODEL	R-SQUARE	VARIABLES IN MODEL
1	0.80322270	X3
1	0.91122065	X2
1	0.96977720	X1
2	0.97003083	X1 X2
2	0.97135430	X1 X3
2	0.99130555	X2 X3
3	0.99523836	X1 X2 X3

This shows that a cubic model may be warranted for the data. These results do not conflict with those from PROC STEPWISE because of the default significance level for entering a new variable. We recommend plotting the data first, then follow with PROC RSQUARE, and finally with the appropriate regression model using PROC GLM. The results from PROC GLM for this data set are shown below.

SAS

P VARIABLE: Y

URCE	DF	SUM OF SQUARES	MEAN SQUARE	F VALUE	PROB>F
DEL	3	10444.797	3481.599	627.035	0.0001
ROR	9	49.972278	5.552475		
TOTAL	12	10494.769			

RIABLE	DF	PARAMETER ESTIMATE	STANDARD ERROR	T FOR H0: PARAMETER=0	PROB > |T|
TERCEP	1	21.727273	3.595816	6.042	0.0002
	1	-5.839577	2.141837	-2.726	0.0234
	1	2.343781	0.348835	6.719	0.0001
	1	-0.113345	0.016421	-6.903	0.0001

From the AOV for the entire model, the calculated f is much greater than $F_{3,9,0.95}$. We also note, however, that all the terms in the model are significant based on the t-tests.

Figures 9.11 and 9.12, respectively, illustrate another way of examining the regression. The predicted values ($YHAT_i$) should fall on or very near the observed Y_i values on an OVERLAY plot. Such is the case here. The residuals ($YHAT_i - Y_i$) should be randomly scattered about the reference line and they are. If the regression is valid, a plot of YHAT vs. Y should have a slope of 1.0 and pass through the origin. Figure 9.13 shows that this is so for this example.

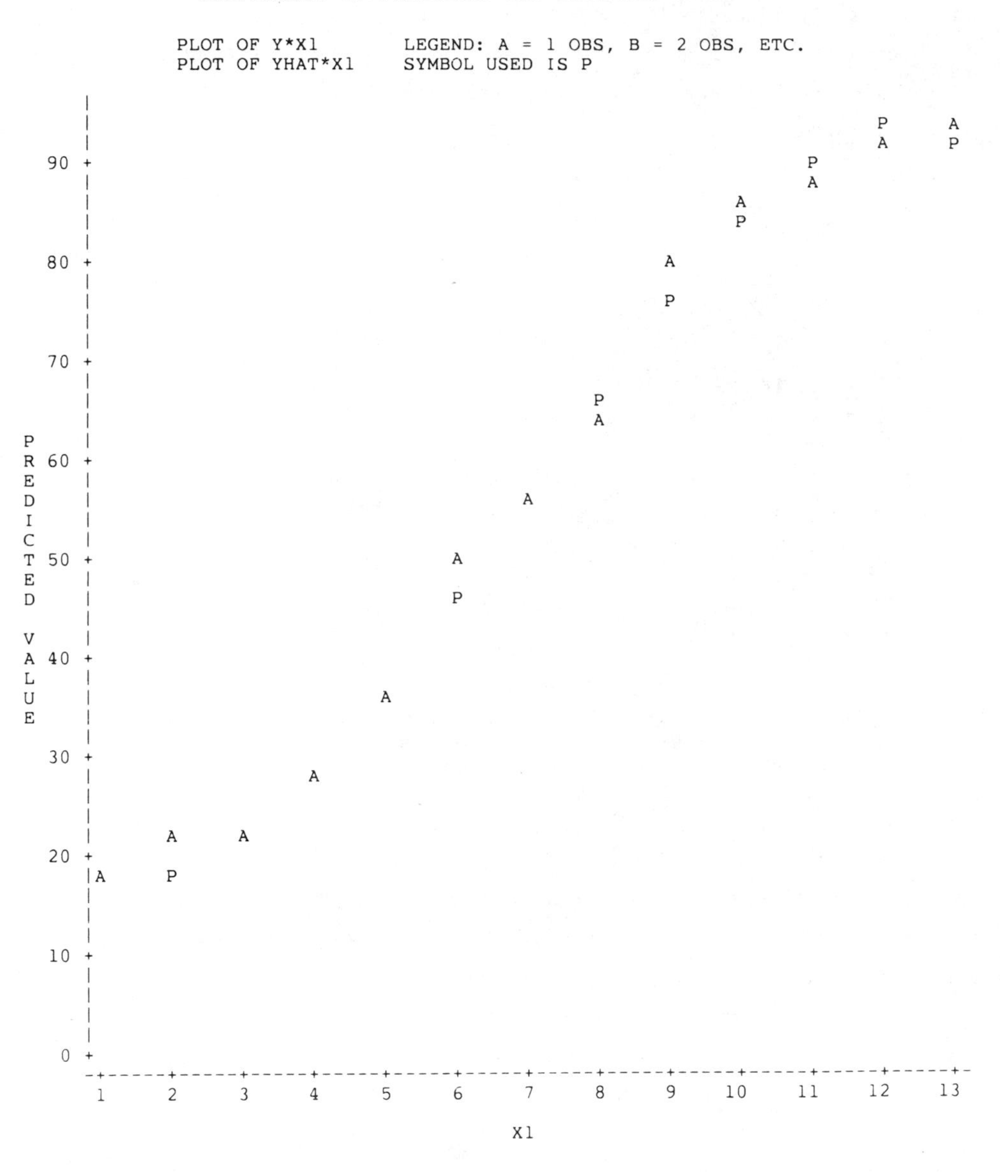

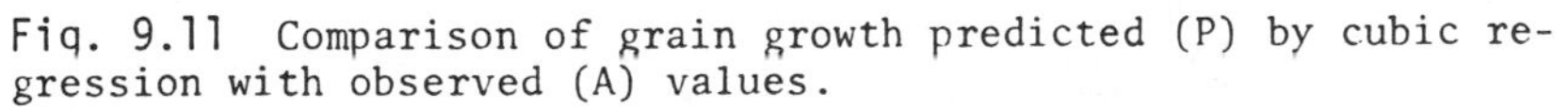
Fig. 9.11 Comparison of grain growth predicted (P) by cubic regression with observed (A) values.

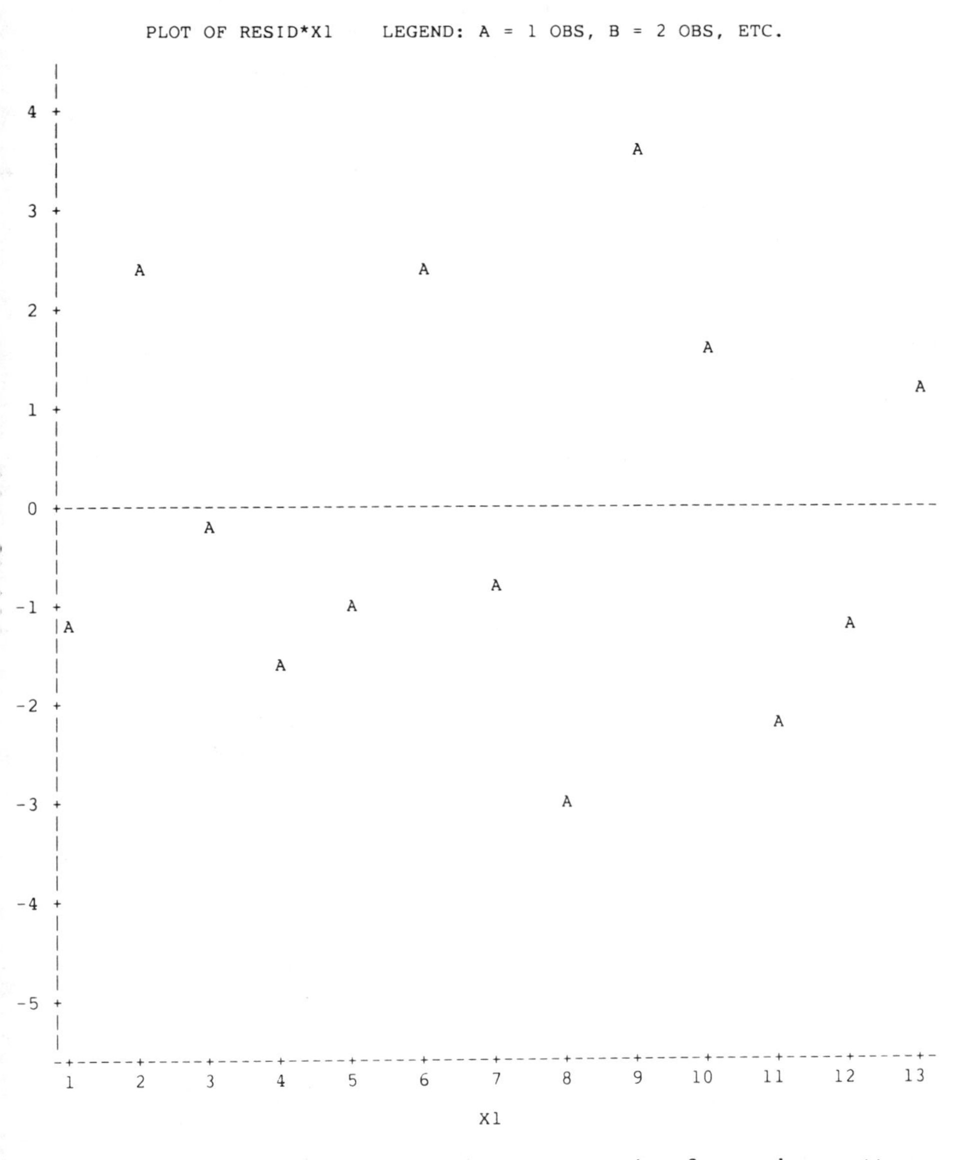

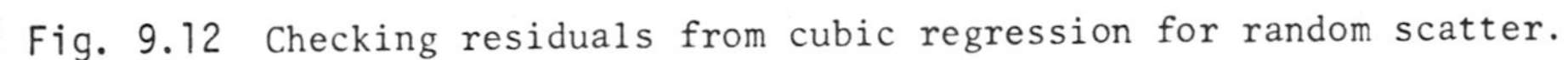
Fig. 9.12 Checking residuals from cubic regression for random scatter.

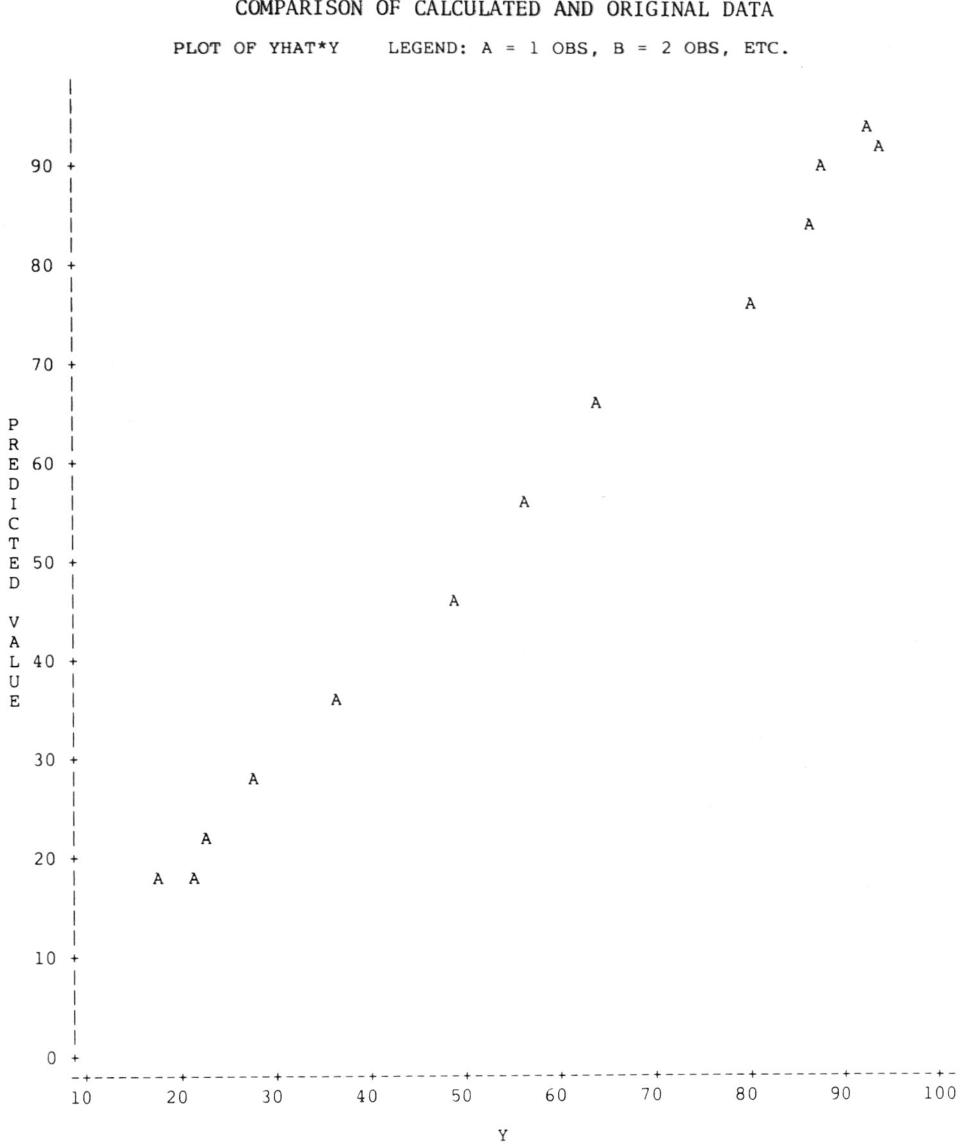

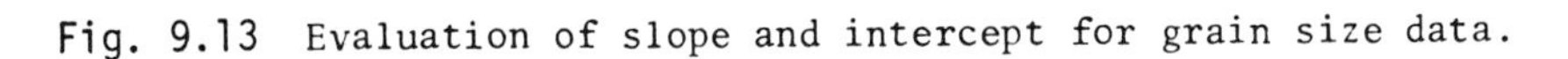
Fig. 9.13 Evaluation of slope and intercept for grain size data.

Further examination of the results using

```
PROC GLM;
MODEL Y = X1  X2  X3;
OUTPUT OUT = NEW P = YHAT R = RESID;
PROC GLM;
MODEL YHAT = Y;
```

illustrates another way to test the effectiveness of the cubic regression: the intercept and slope should be 0 and 1, respectively. The results are below.

SAS

EFFECTIVENESS OF CUBIC REGRESSION OF TIME ON GRAIN SIZE

DEP VARIABLE: YHAT PREDICTED VALUE

SOURCE	DF	SUM OF SQUARES	MEAN SQUARE	F VALUE	PROB>F
MODEL	1	10395.063	10395.063	2299.130	0.0001
ERROR	11	49.734328	4.521303		
C TOTAL	12	10444.797			

VARIABLE	DF	PARAMETER ESTIMATE	STANDARD ERROR	T FOR H0: PARAMETER=0	PROB > |T|
INTERCEP	1	0.268117	1.309089	0.205	0.8415
	1	0.995238	0.020756	47.949	0.0001

From the T value for β_0 (intercept), we accept the null hypothesis that the intercept of the $\hat{Y}$ vs. Y model is zero. Calculating $T_{\beta_1} = (\beta_1 - \hat{\beta}_1)/S_{\hat{\beta}_1}$ which is

$$T_{\hat{\beta}_1} = \frac{1 - 0.99523836}{0.02075608} = 0.229$$

and comparing it with $t_{11,0.975} = 2.201$, we see that there is no reason to believe that the slope is not 1 as seen in Figure 9.13.

The use of R^2 as a criterion for evaluating empirical models is illustrated by the example below.

Example 9.19 The activity coefficients γ for toluene (T) and n-octane (0) vs. their corresponding mole fractions X in a binary mixture are shown below.

γ_T	γ_0	X
1.1757	1.2972	0.07
1.1492	1.2647	0.10
1.1168	1.2411	0.15
1.0965	1.2162	0.20
1.0944	1.1513	0.30
1.0522	1.0972	0.40
1.0387	1.0593	0.50
1.0221	1.0290	0.60
1.0116	1.0116	0.70
1.0039	1.0053	0.80
1.0028	1.0023	0.90
1.0000	1.0000	1.00

Find the most promising empirical models by use of PROC RSQUARE. Using the usual linearizing transformations to logarithms we create

```
U = LOG10(GAMMAT)
W = LOG10(XT)   .
```

The results of using PROC RSQUARE are below.

SAS

```
N=      12     REGRESSION MODELS FOR DEPENDENT VARIABLE GAMMAT

NUMBER IN      R-SQUARE      VARIABLES IN MODEL
  MODEL

    1         0.70560351     XTSQ
    1         0.88134528     XT
    1         0.98552199     W
----------------------------
    2         0.97805571     XT XTSQ
    2         0.98562837     XT W
    2         0.98584669     XTSQ W
----------------------------------
    3         0.98738233     XT XTSQ W
-------------------------------------
```

Figure 9.14 shows the original data. Initial trial models come immediately to mind: quadratic in XT, $\log_{10}\gamma_T$ vs. XT, $\log_{10}\gamma_T$ vs. $\log_{10}$XT. As seen in Fig. 9.15, the log-log model produced an approximately linear relation. When we look at the OVERLAY plot, Fig. 9.16, we see that the linearizing transformation used was appropriate: the predicted values γ_T differ only slightly from the original values. This is even more obvious when we look at the residuals plot, Fig. 9.17: the lack of a regular shape for the graph indicates again that our selection of model was appropriate. The analysis of variance below shows that the log-log model is significant ($f \cong 659.7$) at $\alpha = 0.05$. Checking the residuals and comparing the predicted values to the original data is the only

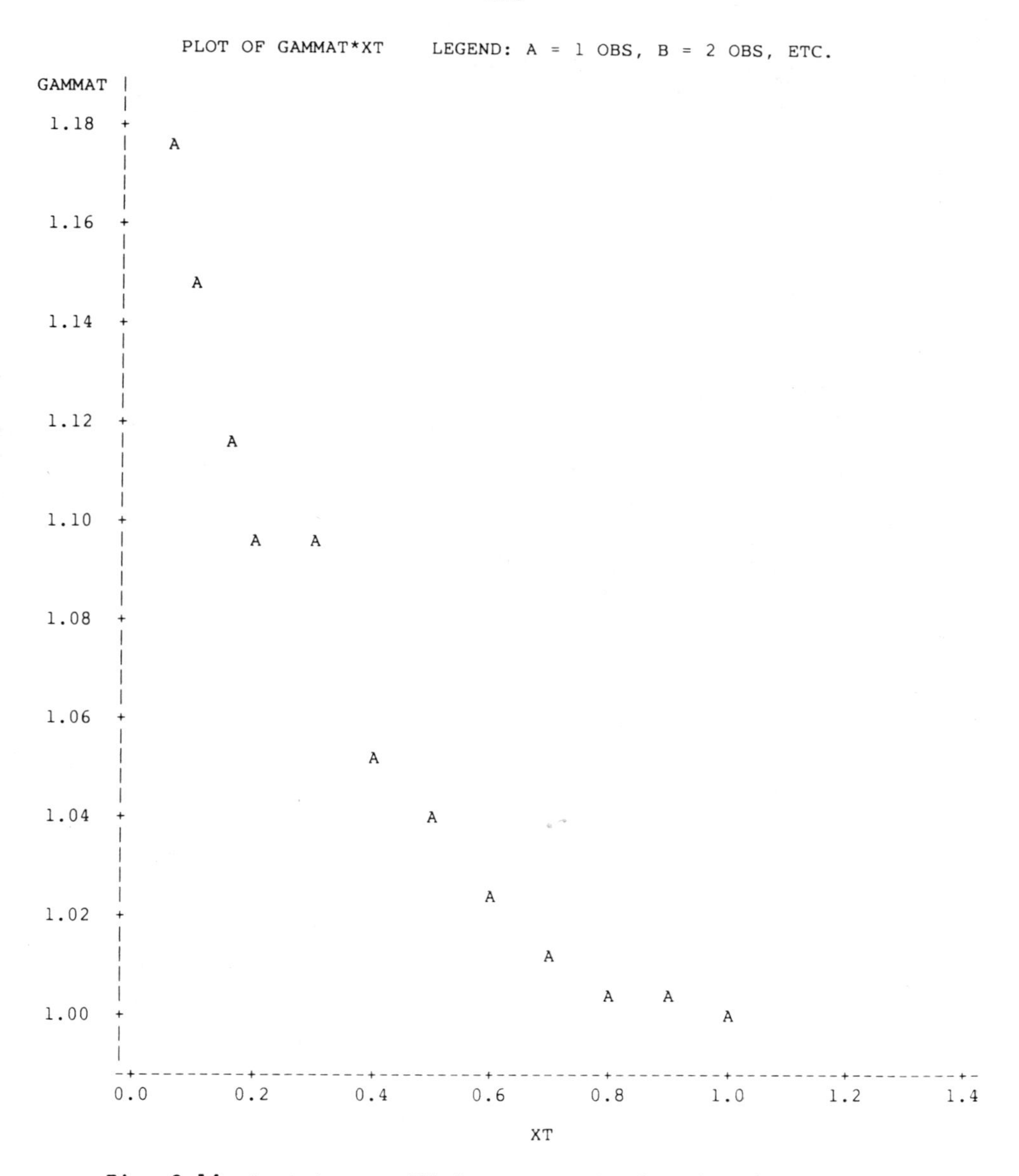

Fig. 9.14 Activity coefficient vs. mole fraction for toluene.

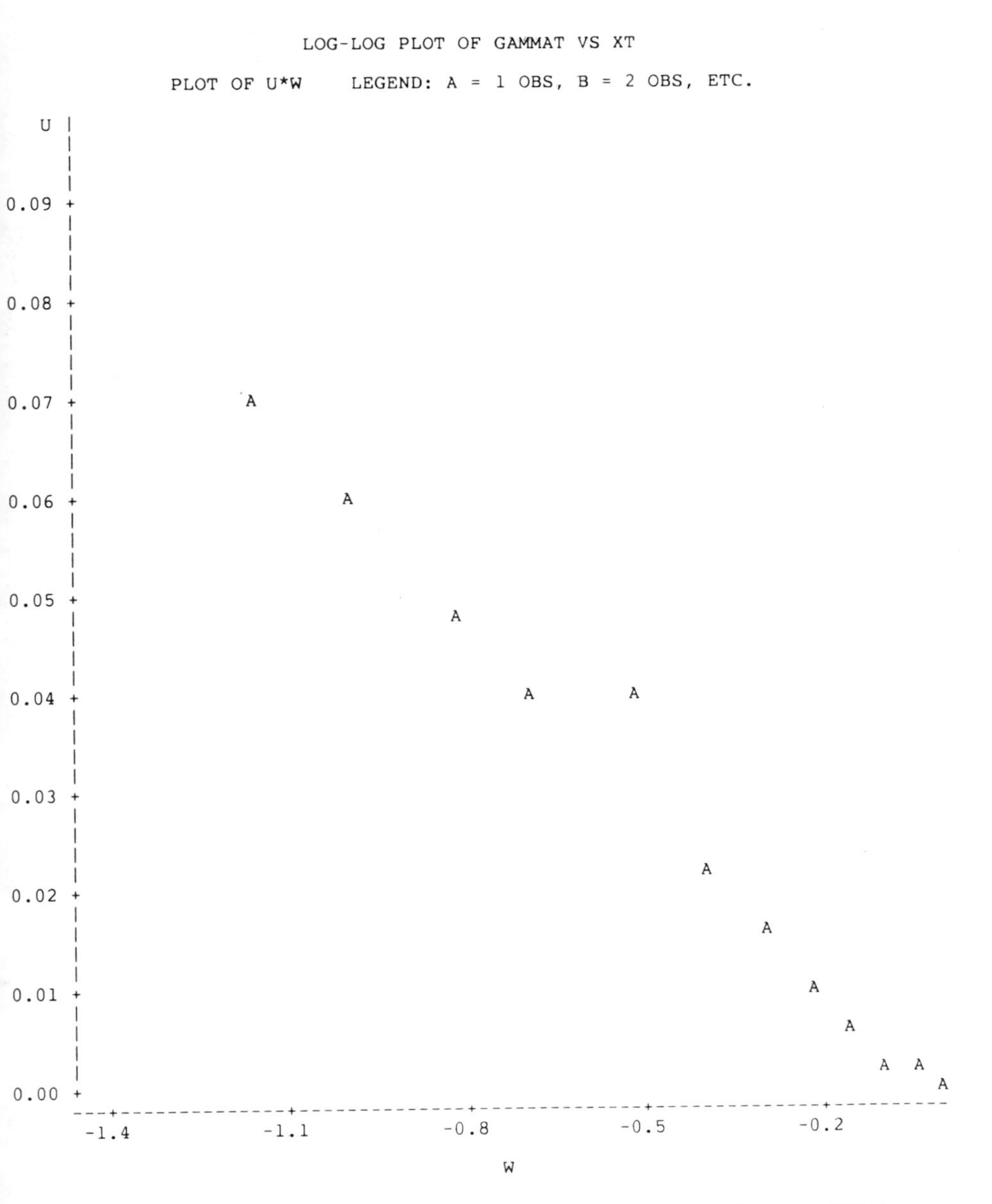

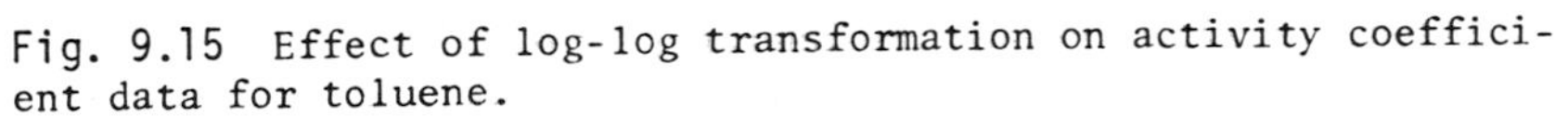
Fig. 9.15 Effect of log-log transformation on activity coefficient data for toluene.

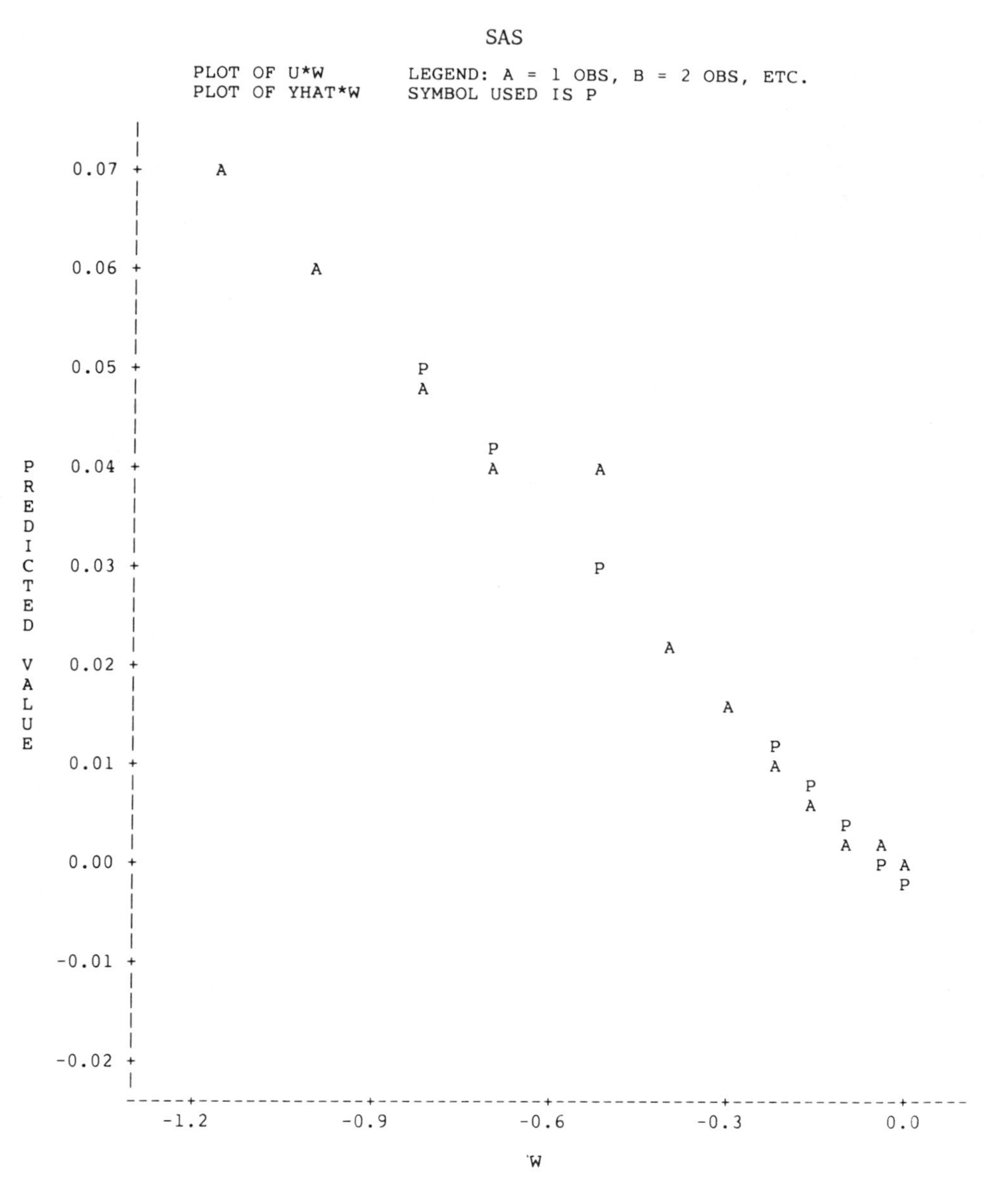

Fig. 9.16 Comparison of observed (A) and predicted (P) toluene activity coefficients after log-log transformation.

SAS

CHECKING RESIDUALS FOR RANDOM SCATTER

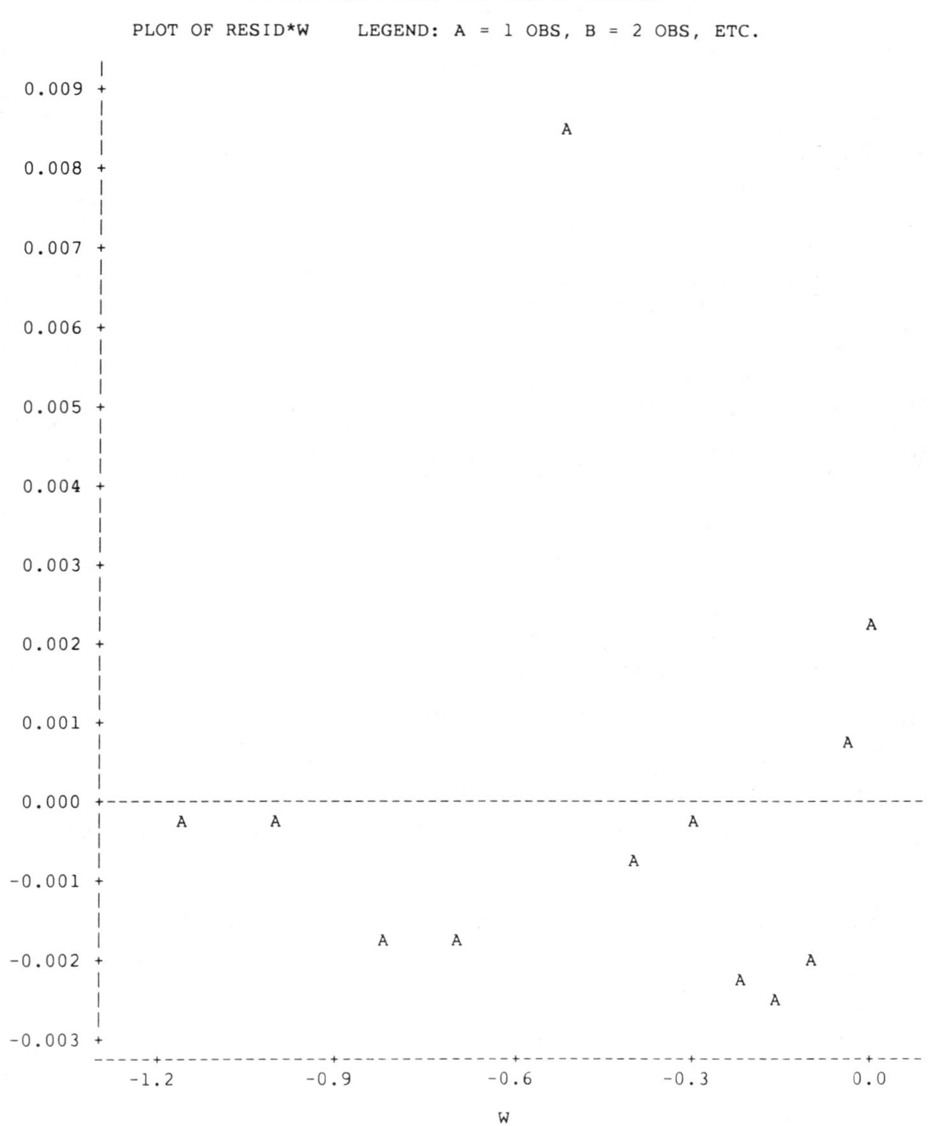

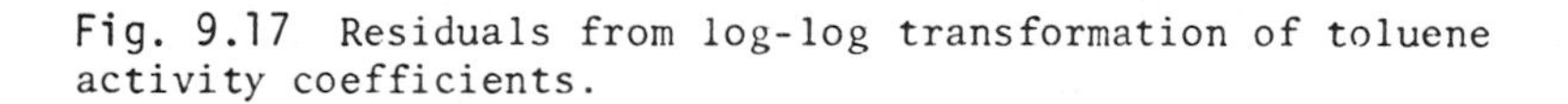
Fig. 9.17 Residuals from log-log transformation of toluene activity coefficients.

reasonable way to evaluate transformed models after obtaining favorable AOV results for them.

SAS

DEP VARIABLE: U

SOURCE	DF	SUM OF SQUARES	MEAN SQUARE	F VALUE	PROB>F
MODEL	1	0.006597756	0.006597756	659.698	0.0001
ERROR	10	0.0001000118	.00001000118		
C TOTAL	11	0.006697767			

VARIABLE	DF	PARAMETER ESTIMATE	STANDARD ERROR	T FOR H0: PARAMETER=0	PROB > |T|
INTERCEP	1	-0.00230662	0.001435714	-1.607	0.1392
W	1	-0.063023	0.002453747	-25.685	0.0001

9.9 TESTING EQUALITY OF SLOPES

Suppose a series of n observations of pairs (X,Y) can be partitioned into k groups with n_i pairs in the ith group such that $n_1 + n_2 + \ldots + n_k = n$. Assume a simple linear regression model for each group of observations with a common error variance for all groups. In this section we will describe a procedure for testing the hypothesis H_0: $\beta_1 = \beta_2 = \ldots = \beta_k$, the equality of the k slopes of k regression lines.

Consider first the case k = 2, that is two groups (X_{11}, Y_{11}), $(X_{12}, Y_{12}), \ldots, (X_{1n_1}, Y_{1n_1})$ and (X_{21}, Y_{21}), $(X_{22}, Y_{22}), \ldots, (X_{2n_2}, Y_{2n_2})$. The estimates of β_1 and β_2 are, from Eq. (9.13),

$$\hat{\beta}_i = \sum_{j=1}^{n_i} (X_{ij} - \bar{X}_i)(Y_{ij} - \bar{Y}_i) / \sum_{j=1}^{n_i} (X_{ij} - \bar{X}_i)^2, \quad i = 1,2 \quad ,$$

where $\overline{X}_i$ and $\overline{Y}_i$ are the means for the ith group. The variances of $\hat{\beta}_1$ and $\hat{\beta}_2$ are

$$s^2_{\hat{\beta}_1} = \frac{\hat{\sigma}^2}{\sum_j (X_{1j}-\overline{X}_1)^2} \quad \text{and} \quad s^2_{\hat{\beta}_2} = \frac{\hat{\sigma}^2}{\sum_j (X_{2j}-\overline{X}_2)^2} .$$

The variance of $\hat{\beta}_1 - \hat{\beta}_2$ is $s^2_{\hat{\beta}_1} + s^2_{\hat{\beta}_2}$. Since $E(\hat{\beta}_1 - \hat{\beta}_2) = \beta_1 - \beta_2 = 0$ under H_0: $\beta_1 = \beta_2$ and since $\hat{\beta}_1 - \hat{\beta}_2$ has a normal distribution with mean $\beta_1 - \beta_2$ and variance $\sigma^2_{\hat{\beta}_1} + \sigma^2_{\hat{\beta}_2}$, then the statistic

$$T = \frac{\hat{\beta}_1 - \hat{\beta}_2}{\hat{\sigma}\sqrt{\dfrac{1}{\sum_j (X_{1j}-\overline{X}_1)^2} + \dfrac{1}{\sum_j (X_{2j}-\overline{X}_2)^2}}} \qquad (9.82)$$

has a t-distribution with $n_1 - 2 + n_2 - 2 = n_1 + n_2 - 4$ degrees of freedom, where

$$\hat{\sigma}^2 = \frac{(SS_E)_1 + (SS_E)_2}{n_1 + n_2 - 4} .$$

The terms $(SS_E)_1$ and $(SS_E)_2$ denote the error sums of squares corresponding to the two sets of data, that is,

$$(SS_E)_i = \sum_j (Y_{ij} - \overline{Y}_i)^2 - \hat{\beta}_i^2 \sum_j (X_{ij} - \overline{X}_i)^2, \quad i = 1,2 ,$$

according to Eq. (9.15). To test the hypothesis H_0: $\beta_1 = \beta_2$ the procedure is to use the statistic in (9.71) with rejection region $T > t_{\nu, 1-\alpha/2}$ or $T < t_{\nu, \alpha/2}$ where $\nu = n_1 + n_2 - 4$ degrees of freedom.

Example 9.20 Using the results of Problem 9.8, compare the slopes of the two calibration curves. The slopes, $\hat{\beta}_{1,i}$ and their standard errors $s_{\hat{\beta}_{1,i}}$ are:

$$\hat{\beta}_{1,1} = -0.13081677 \qquad s_{\hat{\beta}_{1,1}} = 0.00536172$$

$$\hat{\beta}_{1,2} = -0.13072794 \qquad s_{\hat{\beta}_{1,2}} = 0.00483616$$

From Eq. (9.71), which can be written as

$$T = \frac{\hat{\beta}_{1,1} - \hat{\beta}_{1,2} - (\beta_{1,1} - \beta_{1,2})}{\sqrt{s^2_{\hat{\beta}_{1,1}} + s^2_{\hat{\beta}_{1,2}}}} ,$$

we calculate T = -1.703 with $n_1 + n_2 - 4 = 18$ degrees of freedom for this example. As $t_{18,0.975} = 2.101$, the null hypothesis of equal slopes is accepted.

In the case of k groups let

$$(X_{i1}, Y_{i1}), (X_{i2}, Y_{i2}), \ldots, (X_{in_i}, Y_{in_i})$$

represent the data for the ith group. The estimates of the slopes $\beta_1, \beta_2, \ldots, \beta_k$ are given by

$$\hat{\beta}_i = \sum_j (X_{ij} - \overline{X}_i)(Y_{ij} - \overline{Y}_i) / \sum_j (X_{ij} - \overline{X}_i)^2, \quad i = 1,2,\ldots,k \quad . \tag{9.83}$$

The corresponding slope obtained by treating all the individual groups as one large group is $\hat{\beta}_p$ where

$$\hat{\beta}_p = \sum_{i=1}^{k} \sum_{j=1}^{n_i} (X_{ij} - \overline{X})(Y_{ij} - \overline{Y}) / \sum_{i=1}^{k} \sum_{j=1}^{n_i} (X_{ij} - \overline{X})^2 , \tag{9.84}$$

where $\overline{X}$ and $\overline{Y}$ denote the overall means using all $n = \sum_{i=1}^{k} n_i$

values. The error sum of squares for the ith group is

$$(SS_E)_i = \sum_j (Y_{ij} - \overline{Y}_i)^2 - \hat{\beta}_i^2 \sum_j (X_{ij} - \overline{X}_i)^2 \quad , \qquad (9.85)$$

and the pooled error sum of squares, denoted by S_1, is

$$S_1 = \sum_{i=1}^{k} (SS_E)_i$$

$$= \sum_{i=1}^{k} [\sum_j (Y_{ij} - \overline{Y}_i)^2 - \hat{\beta}_i^2 \sum_j (X_{ij} - \overline{X}_i)^2] \quad , \qquad (9.86)$$

with $\sum_{i=1}^{k} (n_i - 2) = \sum_i n_i - 2k$ degrees of freedom. An unbiased estimate of σ^2 is furnished by

$$\hat{\sigma}^2 = S_1/(\sum_i n_i - 2k) \quad .$$

The total sum of squares based on all k data sets collectively can be partitioned as

$$\sum_{i=1}^{k} \sum_{j=1}^{n_i} (Y_{ij} - \overline{Y})^2 = S_1 + \sum_{i=1}^{k} \sum_{j=1}^{n_i} (\hat{\beta}_i - \hat{\beta}_p)^2 (X_{ij} - \overline{X}_i)^2$$
$$+ \hat{\beta}_p^2 \sum_{i=1}^{k} \sum_{j=1}^{n_i} (X_{ij} - \overline{X}_i)^2 \quad . \qquad (9.87)$$

The term S_1 is the pooled error sum of squares given by Eq. (9.86) and the second term, denoted by S_2, is the sum of squares due to differences between group slopes. The statistic

$$F = \frac{S_2/(k-1)}{S_1/(\sum_{i=1}^{k} n_i - 2k)}$$

has an F-distribution with (k-1) and $(\sum_{i=1}^{k} n_i - 2k)$ degrees of freedom if $\beta_1 = \beta_2 = \ldots = \beta_k = \beta$. Thus the critical region for H_0: $\beta_1 = \beta_2 = \ldots = \beta_k = \beta$ is $F > F_{\nu_1,\nu_2,1-\alpha}$, where $\nu_1 = k-1$ and $\nu_2 = \sum_{i=1}^{k} n_i - 2k$. The sum S_2 can be written as

$$S_2 = \sum_i \sum_j (\hat{\beta}_i - \hat{\beta}_p)^2 (X_{ij} - \overline{X})^2 \tag{9.87a}$$

$$= \sum_{i=1}^{k} \hat{\beta}_i^2 A_i^2 - \hat{\beta}_p^2 \sum_{i=1}^{k} A_i \quad , \tag{9.87b}$$

where

$$A_i = \sum_{j=1}^{n_i} (X_{ij} - \overline{X}_i)^2 \quad .$$

9.10 TRANSFORMATION OF DATA IN REGRESSION ANALYSIS

The basic assumptions in regression analysis are that the errors are independently and normally distributed with mean 0 and variance σ^2 and, moreover, that the errors are homogeneous over the region of interest. If the error distribution is not normal or if the error variance is not homogeneous, it may be necessary to transform the dependent variable to attain desirable properties. When taking a transformation to linearize a given nonlinear model it may also be useful to determine the variance of the transformed variables. These ideas will be briefly discussed.

9.10.1 Propagation of Error

We have seen that if $Y = a_1X_1 + a_2X_2 + \ldots + a_nX_n$, where the a's are constants and $X_1,X_2,\ldots,X_n$ are independent random variables, then the variance of Y is $\sigma^2 = a_1^2\sigma_{X_1}^2 + a_2^2\sigma_{X_2}^2 + \ldots + a_n^2\sigma_{X_n}^2$.

More generally, let $Y = f(X_1, X_2, \ldots, X_n)$ be some function of the independent random variables $X_1, X_2, \ldots, X_n$. The variance of Y is given, approximately, by

$$\sigma_Y^2 \cong \left(\frac{\partial f}{\partial X_1}\right)^2 \sigma_{X_1}^2 + \left(\frac{\partial f}{\partial X_2}\right)^2 \sigma_{X_2}^2 + \cdots + \left(\frac{\partial f}{\partial X_n}\right)^2 \sigma_{X_n}^2 , \qquad (9.88)$$

where the partial derivatives are evaluated at $X_1 = \mu_1, \ldots, X_n = \mu_n$. Equation (9.88) can be verified very easily by using the first (n+1) terms of the Taylor expansion of the function f of $X = (X_1, \ldots, X_n)$ about the point $\mu = (\mu_1, \mu_2, \ldots, \mu_n)$ where $\mu_i = E(X_i)$. We have

$$f(X_1, X_2, \ldots, X_n) \cong f(\mu_1, \mu_2, \ldots, \mu_n) + \sum_{i=1}^{n} \left(\left.\frac{\partial f}{\partial X_i}\right|_{X_i = \mu_i} \right) (X_i - \mu_i)$$

Since the X_i are independent, the variance of the right-hand side yields Eq. (9.88) as required.

Equation (9.88) is an approximation. In many cases it adequately describes the total variance (precision) of the function f and furthermore it dictates which variable makes the largest contribution to the variance of the function. The partial derivatives in Eq. (9.88) are evaluated at $X_i = \mu_i$, $i = 1, 2, \ldots, n$, where $\mu_i = E(X_i)$.

As a special case let $Y = f(X) = a_1X_1 + a_2X_2 + \cdots + a_nX_n$. We have $\partial f/\partial X_i = a_i$, $i = 1, 2, \ldots, n$. Thus

$$\sigma_Y^2 = a_1^2\sigma_{X_1}^2 + a_2^2\sigma_{X_2}^2 + \cdots + a_n^2\sigma_{X_n}^2$$

which we know holds exactly.

For another example suppose $Z = \ell n\, Y$. Since $\partial Z/\partial Y = 1/Y$ the variance of Z is $\sigma_Z^2 = \left(\frac{1}{\mu_Y}\right)^2 \cdot \sigma_Y^2$. If $\mu_Y = E(Y)$ is not known then

$\sigma_Z^2 = \left(\frac{1}{\overline{Y}}\right)^2 \cdot \sigma_Y^2$, and if an estimate, $\hat{\sigma}_Y^2$, of σ_Y^2 is available then

$\hat{\sigma}_Z^2 = (1/\overline{y})^2 \hat{\sigma}_Y^2$.

9.10.2 On Transforming the Data

Two reasons for transforming data are (1) lack of normality of error structure and (2) lack of homogeneity in error structure. Lack of normality could occur when the tolerance on one side of the setpoint differed from the other side in reading an instrument. In other words the error distribution could be skewed instead of bell-shaped. A heterogeneous error structure could occur in cases where the magnitude of the error depends on the region of values of the independent variable. For example, data consisting of values read from an instrument whose precision varies with the level of the reading are of this type.

Transformations to adjust for nonhomogeneity of the error structure are performed on the dependent variable. Sinibaldi et al. [17] discuss a method of Box and Cox [2] that is used in transforming the dependent variable to develop models for engineering applications. In many cases these transformations not only simplify the model but yield data which follow the underlying basic assumptions.

In some instances there may be a significant lack of fit in the model being used (see Section 9.3). Box and Tidwell [3] present methods of transforming the independent variables so that the resulting model is in as simple a form as possible. For further details see Ref. [3].

PROBLEMS

9.1 The relation between the heat capacity of liquid sulfuric acid in cal/g $^{\circ}$C and temperature in $^{\circ}$C is as follows:

heat capacity:	0.377	0.389	0.396	0.405	0.446	0.458
°C:	50	100	150	200	250	300

The proposed model is

$$C_p = C_{p_0} + at + \varepsilon \ .$$

(a) By use of simple linear regression, evaluate C_{p_0} and a.

(b) What portion of the data does your equation explain?

(c) Are C_{p_0} and a significantly different from zero? Prove your answer.

9.2 In a study to determine the relationship between the incidence of sandspur and "goat-head" clumps on the intramural field behind our building, Y, and the number of times the grass is cut each year, X, the results of a 4-year study showed:

$\bar{x} = 20$, $\bar{y} = 22$, $\sum_i (x_i - \bar{x})^2 = 225$, $\sum_i (y_i - \bar{y})^2 = 414$, $\sum_i (x_i - \bar{x})(y_i - \bar{y}) = 180$, $n = 32$.

(a) What is the estimated population regression coefficient?

(b) Test the hypothesis that $\beta = 1$ for the model $Y_i = \alpha + \beta X_i + \varepsilon_i$.

9.3 For the data presented in Problem 8.25, determine the relationship between separation force and pulling angle. Test the resulting equation if it is believed that the true relation is

$$\text{force}(\text{lb}_F) = 40 + 0.75 \text{ (angle, in degrees)}.$$

9.4 Obtain the form of the linear relationships between blue colorant and lightness (L) for each level of yellow colorant from the data in Problem 8.22. Plot your results. Show the 98% confidence limit for each such relationship graphically.

9.5 The data for the calibration curve for the colorimetric analysis of fluoride ions in aqueous solution by the Lacroix and Labalade [12] method is as follows:

Concentration in cuvet mg $F^-/m\ell$	Percent transmittance		
	Trial 1	Trial 2	Average
0.000608	80.3	80.3	80.30
0.001216	80.5	80.4	80.45
0.001824	80.9	81.0	80.95
0.002432	81.2	81.8	81.50
0.003040	81.6	82.0	81.80
0.003648	82.9	82.5	82.70
0.004256	83.0	83.1	83.05
0.004864	83.9	84.0	83.95
0.005472	84.0	84.0	84.00
0.006080	85.0	84.8	84.90

The data can be expressed as

$$\%T = 79.46637 + 865.2737C$$

where %T = percent transmittance and C = fluoride concentration in mg $F^-/m\ell$. The equation above is difficult to use for determining the concentration of an unknown from its %T. Find, by the least-squares technique, the desired relation between C and %T, test the validity of the regression constants, and give the 95% confidence interval on concentration for an unknown sample having 82.1%T.

9.6 Use the data below from the packed 6 ft × 3 in gas absorber (water flow rate = 4 GPM) to find the loading and flooding points mathematically as a result of suitable regression analyses. Compare the results with the corresponding L/G, moles water per mole gas, vs. pressure drop ($-\Delta P$, in H_2O) results you obtain by plotting the data. Comment on any differences.

$-\Delta P$	L/G	$-\Delta P$	L/G
0.10	21.7	0.75	5.8
0.13	19.0	0.90	5.0
0.12	17.2	1.05	4.6
0.20	13.0	1.10	4.0
0.30	11.4	1.22	3.6
0.35	9.6	1.35	3.4
0.50	8.5	1.55	3.2
0.60	7.8	1.70	2.6
0.65	7.0		

9.7 A preliminary experiment has been run to check the working condition of the steam kettle in Ch.E. B-4. The experiment is carried out at atmospheric pressure with the steam pressure varying from 1 psig to 10 psig. You are asked to find the relationship between steam pressure and the temperature in the center of the kettle at t = 1 min. Let X be the steam pressure in psig and Y be the temperature in $^{\circ}$C.

X	Y
1	5
2	7
3	9
4	12
5	13
6	16
7	17
8	19
9	21
10	23

9.8 The data below were obtained two months apart for the relationship between refractive index and composition of mixtures of benzene and 1-propanol. Different lots of materials were used in each case.

(a) Determine the relationship between these variables.

(b) Test each term at the $\alpha = 0.01$ level.

Support your answer statistically.

		Refractive index	
Benzene, ml	1-Proponol, ml	Run 1	Run 2
10	1	1.4837	1.4842
10	2	1.4724	1.4716
10	4	1.4650	1.4651
10	6	1.4571	1.4555
10	8	1.4430	1.4420
10	10	1.4385	1.4368
10	14	1.4263	1.4244
10	20	1.4108	1.4100
10	30	1.3941	1.3940
10	50	1.3892	1.3888
10	100	1.3816	1.3819

9.9 In an effort to work out a program for control of stack loss of a relatively valuable chemical, data were obtained on stack losses and suspected related independent variables for 139 days. Data on stack loss, water temperature, and air temperature are given below where

$y = 10$ (stack loss in g/m^3 - 3.0)
X_w = water temperature in (^oC - 20^oC)
X_a = air temperature in (^oF - 50^oF).

Test No.	y	X_w	X_a
1	-20	-6	-11
2	-9	1	-3
3	-18	1	-2
4	-6	-1	-1
5	-3	2	-3
6	-5	-5	-10
7	-20	2	-20
8	+19	5	+14
9	-4	-1	-14
⋮	⋮	⋮	⋮
139			

$\Sigma y = 1479$ $\quad$ $\Sigma x_w y = 12{,}797$

$\Sigma x_w = 406$ $\quad$ $\Sigma y^2 = 99{,}077$

$\Sigma x_a = 1$ $\quad$ $\Sigma x_w^2 = 3{,}456$

$\Sigma x_w x_a = 3{,}533$ $\quad$ $\Sigma x_a^2 = 10{,}525$

$\Sigma x_a y = 17{,}340$ $\quad$ $n = 139$.

An independent investigation was made to determine the reproducibility of the stack loss measurement. Ten measurements at identical conditions were made and the data coded to correspond to y given above. From the replicate data [$y = 10$ (stack loss in g/m^3 - 3.0)], $\Sigma y = 42$ and $\Sigma y^2 = 1560$.

(a) Considering only the relationship between stack loss and water temperature,

1. Determine the best linear relationship between the variables and test its significance.
2. Establish 95% confidence limits of the regression equation for predicting stack loss at water temperatures of 15° to $30^{\circ}C$.

(b) Determine whether the addition of air temperature in the regression equation significantly improves the correlation.

(c) Do you feel that your best expression adequately accounts for the variation in stack losses? If not what do you suggest as the next move?

9.10 The irritant factor Y of polluted air can be determined as a function of the concentrations of SO_2 and NO_2 in the atmosphere. The following data are available where X_1 = parts NO_2 per ten million parts of air and X_2 = parts SO_2 per hundred million parts of air. Determine the irritant factor as a function of X_1 and X_2.

Y	X_1	X_2
65	10	12.5
72	12	15
82	15	18
95	16	21
110	19	26
122	21	30
125	25	35
130	28	40

9.11 Liquid Murphree plate efficiencies, E_L, in distillation are dependent upon the closeness of approach to equilibrium on that plate. Equilibrium can be upset by changes in feed location or temperature. The data below were obtained for the methanol-water system using a 3-in ID column with 6 sieve plates, a total condenser, and a partial reboiler. For a constant external reflux ratio, the following data were obtained for E_L for plate 3. The feed was the same composition and flow rate in all cases.

Feed Plate	Feed Temperature, °F			
	90	105	125	150
2	7.5	41.2	51.6	65.2
3	54.0	62.5	84.2	100
4	29.0	48.6	41.0	36.0
5	12.3	14.6	18.3	8.3

Do either of these variables affect the Murphree efficiency?

9.12 In an investigation of the effects of carrier gas velocity and support particle diameter on retention time in gas chromatography, the following data have been obtained. Using multiple linear regression, develop a relationship among these three variables. Let Y be the retention time in minutes, X_1 be the carrier gas velocity in ml/min, and X_2 be the average particle diameter in μm.

Y	X_1	X_2
6.5	76	81
8.4	92	96
9.7	106	108
11.4	120	150
13.4	136	159
15.6	159	173
18.7	200	214
20.0	218	230

9.13 Given the following data from the performance tests on a degassing tower:

X_1	X_2	Y
14	1.00	46
15	1.25	51
16	3.00	69
17	3.25	74
18	4.00	80
19	5.25	82
20	5.50	97

$$\sum_i (x_{1i} - \bar{x}_1)^2 = 28.0$$
$$\sum_i (x_{1i} - \bar{x}_1)(x_{2i} - \bar{x}_2) = 22.50$$
$$\sum_i (x_{2i} - \bar{x}_2)^2 = 18.7143$$
$$\sum_i (x_{1i} - \bar{x}_1)(y_i - \bar{y}) = 226.$$
$$\sum_i (x_{2i} - \bar{x}_2)(y_i - \bar{y}) = 183.86$$
$$\sum_i (y_i - \bar{y})^2 = 1915.43$$

where X_1 = liquor rate in hundreds of gallons per hour, X_2 = air velocity, ft/sec, and Y = % dissolved CO removed,

(a) Calculate the regression equation between these variables.

(b) Test each term for significance at the 5% level.

(c) Give the 95% confidence interval for the true value of Y estimated from your regression equation for $x_1 = 18.5$, $x_2 = 4.5$.

9.14 Consider the breakthrough curve below obtained in an investigation of the adsorbents and/or catalysts best suited for the reduction or removal of NO_2 from space cabin atmospheres. The adsorption is most likely of the Langmuir type. As such, the model describing the adsorption process is quadratic in form. Describe how you would find the regression coefficients involved for this model. Be specific and complete.

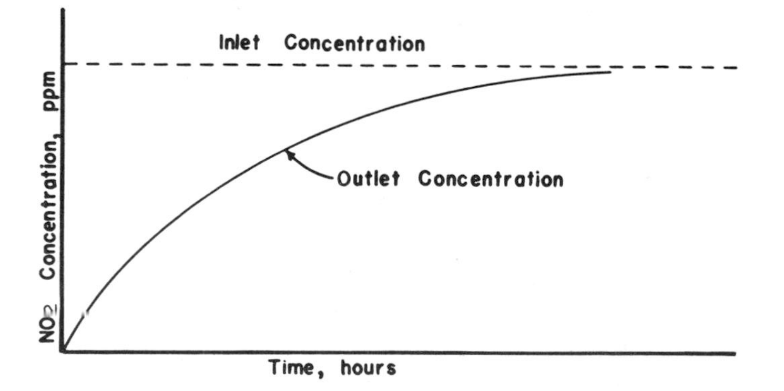

9.15 In the determination of the calibration curve for the analysis of ionic fluoride in mg F^-/mℓ of solution (the Y variable) vs %T (the X variable), the data in Problem 9.5 were refit to the simple linear model $Y = \beta_0 + \beta_1 X + \varepsilon$ from which the following calculated results are available:

$\sum_i x = 1{,}647.198$ | $(\sum_i y)^2 = 0.004473$ | $\sum_i (x_i - \bar{x})^2 = 46.8125$

$\sum_i x^2 = 135{,}710.$ | $\sum_i xy = 5.561005$ | $\sum_i (y_i - \bar{y})^2 = 0.000061$

$(\sum_i x)^2 = 2{,}713{,}264$ | $\sum_i x \sum_i y = 110.1646$ | $\sum_i (x_i - x)(y_i - y) = 0.052773$

$\bar{x} = 82.53993$ | $\bar{y} = 0.003344$

$\sum_i y = 0.066880$ | $\sum_i y^2 = 0.00285$

(a) Find the equation of the best line through these data.

(b) What portion of the variability in the data is accounted for by your equation?

(c) What is the range of fluoride concentration expected if %T = 82.0?

9.16 In a heat transfer experiment, data were obtained on the Nusselt, Prandtl, and Reynolds numbers in the form

$$Y = \ln[(N_{Nu}/10^3)/N_{Pr}^{0.4}],$$

$$X = \ln(N_{Re}/10^5)$$

by Burrows [4] in order to evaluate

$$N_{Nu} = C(N_{Re})^B (N_{Pr})^K.$$

Devise the best straight line through the points below, plot it, and show the 99% confidence limit on the line.

Y	X
-0.3929	0.5417
-0.9978	0.0060
-0.5986	0.3542
-0.8564	-0.2303
-1.6724	-1.0524
-1.7430	-1.0496
-2.1144	-1.6100
-0.5211	0.5429
-1.0167	-0.0240

9.17 For the batch saponification of ethyl acetate a second-order rate equation was assumed to describe the rate of reaction. In order to establish that the second-order equation was a correct assumption, specific reaction rate constants were calculated at various times during the reaction. They are as follows:

Time, sec	$k_2 \times 10^2$, liters/g mole-sec
69	0.7948
124	0.6572
179	0.7950
253	0.8157
305	0.9062
420	0.9249
600	0.8584
727	0.8653
902	0.8620
1201	0.7885

If the values for k_2 are nearly identical it can be assumed that the second-order equation satisfactorily explained the kinetic data. Is it reasonable to believe that the second-order equation is the correct equation?

9.18 Two supposedly identical Brooks model R-2-65-5 rotameters with 316 stainless-steel spherical floats were calibrated for helium service at 20 psig input, $74^{\circ}F$. The data are below. Let Y = flow rate and X = scale reading.

(a) Obtain the regression coefficients for the quadratic model involved for each rotameter.

(b) Calculate the r^2 for each regression equation from (a).

(c) Compare the regression equations at the 0.05 significance level by comparing the individual regression coefficients, i.e., $\hat{\beta}_0$ with $\hat{\beta}'_0$, $\hat{\beta}_1$ with $\hat{\beta}'_1$, and $\hat{\beta}_2$ with $\hat{\beta}'_2$ where the $\hat{\beta}_i$ are for tube 1 and the $\hat{\beta}'_i$ are for tube 2.

Scale reading, mm	Flow rate, mℓ He/min[a] Tube 1	Tube 2
10	9.2	8.4
10	9.5	8.6
14	---	12.6
16	15.0	---
19	---	18.6
20	21.3	18.0
25	29.4	27.0
30	41.0	36.8
30	40.5	36.3
35	54.0	49.1
35	55.0	50.0
40	68.0	64.8
40	68.8	64.8
45	86.0	81.3
45	88.1	82.0
50	103.2	100.4
50	104.6	97.0
55	124.0	114.7
55	123.1	117.0
60	144.0	134.0
60	145.3	136.0
60	145.1	134.0

[a] A dash, ---, means that no flow rate reading was taken. This does not mean that the flow rate at that scale reading was zero.

9.19 Over the Reynolds number range $15{,}000 < N_{RE} < 40{,}000$, the following data were measured (f_{exp}) and predicted solely as a function of N_{Re} by the Konakov [15] equation (f_K) for the Fanning friction factor.

N_{Re}	f_{exp}	f_K
16,500	0.0285	0.0269
17,610	0.0298	0.0265
17,890	0.0224	0.0264
18,340	0.0289	0.0262
19,400	0.0275	0.0258
21,990	0.0239	0.0251
25,700	0.0265	0.0241
27,970	0.0231	0.0236
28,940	0.0270	0.0230
31,210	0.0182	0.0230
38,230	0.0107	0.0220

From the Konakov model, we have

$$f_K = 0.0858882 - 0.0060812 \ln N_{Re}$$

Find the corresponding equation for the experimental data and compare those results with the predicted values.

9.20 In a recent experiment to determine the distillation curve for gasoline, acetone, and various mixtures of the two, the data below were obtained. The procedure of ASTM Standard D-86 was used.

	Temperature, °F		
	Gasoline	Acetone	Acetone
Initial boiling-point	95	128	133
End point	368	140	142
% Recovered			
5	110	132	133
10	117	132	133
20	134	134	134
30	158	134	134
40	182	134	134
50	204	134	135
60	218	134	136
70	247	136	138
80	274	138	138
90	321	139	140
95	368	140	142
% Recovered	95	98	97
% Total recovery	97	99	98
% Loss	3	1	2

	Temperature, °F		
	50% Gasoline 50% acetone	30% Gasoline 70% acetone	70% Gasoline 30% acetone
Initial boiling point	110	125	135
End point	345	270	319
% Recovered			
5	115	127	135
10	120	130	135
20	123	132	135
30	126	132	135
40	126	135	138

	Temperature, ^{o}F		
% Recovered	50% Gasoline 50% acetone	30% Gasoline 70% acetone	70% Gasoline 30% acetone
50	135	135	150
60	140	135	168
70	148	136	225
80	238	138	260
90	302	142	319
95	345	270	---
% Recovered	95	96	93
% Total recovery	96	97	95
% Loss	4	3	5

Using the two determinations for acetone as a measure of experimental error, statistically show how the addition of acetone to gasoline affects the ASTM distillation curve.

9.21 Calibration data for the ammonium electrode for effluent gas sample analysis are:

NH_3, ppm	MV
5	+13.8
13	0
20	-10.6
34	-20.0
50	-26.7
90	-35.0
150	-45.5
250	-56.0
500	-66.8
1000	-79.1

Find a suitable equation for the calibration curve in the form of log (NH_3 concentration) = f(millivolt reading from the electrode).

9.22 Nitric acid is utilized to dissolve uranium metal. It has been determined that impurity concentration in the metal has a direct effect on its dissolution rate in HNO_3. Data for the dissolution of two samples of uranium, one containing 400-500 ppm carbon, and one containing < 35 ppm carbon, are given below. Plot the data and determine the best equations for the resultant curves [11].

Dissolution time, hr	mg/cm^3 dissolved	
	Impurity, 400-500 Ppm	Impurity, < 35 Ppm
0.0	0.0	0.0
10.0	8.0	5.5
20.0	18.5	10.0
30.0	33.0	18.5
40.0	57.0	25.0
50.0	88.5	36.0

9.23 During air pollution studies concerning nitrogen oxides and hydrocarbon vapors, it has been determined that the use of chromite catalysts will catalytically promote the removal of NO in the presence of reducing agents such as CO. Removal is presumed to follow the reaction

$$2\ CO + 2\ NO \xrightarrow{cat} 2\ CO_2 + N_2\ .$$

Results of several runs given below for copper chromite indicate a reduction of catalyst effectiveness with time. Plot the data and determine an equation for the effect of time on % CO removal [16].

Time, min	% CO removal
4.0	99.0
8.0	100.0
12.0	99.5
16.0	98.0
20.0	95.0
24.0	82.0
28.0	51.0
32.0	23.0

9.24 For fluids being heated or cooled inside tubes the following relation has been found empirically:

$$\frac{h}{C_p G} = 0.023 \left(\frac{\mu}{DG}\right)^{0.2} \left(\frac{k}{C_p \mu}\right)^{0.7}$$

or more simply $N_{St} = 0.023\ N_{Re}^{-0.2}\ N_{Pr}^{-0.7}$. This relation may be expressed graphically by plotting $N_{St}N_{Re}^{0.2}$ vs N_{Pr} on log-log paper to yield a straight line for such fluids as water, air, acetone, ethanol, and n-butane. Data from these and many other systems have been lumped together and the best straight line "eyeballed" for most practical purposes. You, however, need to use this relation in a computer program for neopentane. Before you can do so, you need proof that neopentane behaves in a similar manner. From the data below, determine whether or not the given functional relation will hold for your system

$N_{St}N_{Re}^{0.2}$			N_{Pr}
0.007,	0.0073,	0.0077	6
0.0045,	0.0050,	0.0049	10
0.0034	0.0036,	0.0032	20
0.0024,	0.0028,	0.0027	30
0.0020,	0.0021,	0.0023	40

9.25 The overall heat-transfer coefficient U, Btu/hr ft^2 ^oF shows a log-log dependence on ν_{max}, ft/sec, in heat exchangers with parallel banks of finned tubes. Typical data are below. For these data, determine the relation between U and ν_{max}.

U			ν_{max}
8.4,	7.9		300
9.7,	9.1		500
9.7,	9.9		600
10.4,	10.1,	10.1	800
11.3,	11.6	11.4	1000
12.9,	14.0,	13.8	1200
14.8,	16.0,	15.1	1500

Also show that any constants you obtain are nonzero. What error can you expect if using your relation to predict the value of U corresponding to a ν_{max} of 900 ft/sec?

9.26 In recent research by Graham [9] on the removal of SO_2 from air by reaction with MnO_2, the following data were obtained at 650^oF. Run 12-18-1 used 30 MnO_2 pellets with initial loaded container weight of 1.4188 g and an inlet concentration of 3620 ppm SO_2. Run 12-18-2 used 20 pellets with an initial loaded container weight of 1.3935 g. Tare weight in both cases was 1.3576 g. Times are in seconds, weights are for the container + pellets.

Run 12-18-1

Time	Wt.	Time	Wt.
0	1.4188	520	1.4225
178	1.4200	588	1.4230
366	1.4210	640	1.4235
382	1.4215	716	1.4240
462	1.4220	863	1.4245

Time	Wt.	Time	Wt.
1003	1.4250	2556	1.4330
1314	1.4270	2668	1.4335
1389	1.4275	2779	1.4340
1504	1.4280	2880	1.4345
1595	1.4285	2980	1.4350
1702	1.4290	3085	1.4355
1825	1.4295	----	1.4360
1920	1.4300	3285	1.4365
2040	1.4305	3390	1.4370
2145	1.4310	3530	1.4375
2250	1.4315	3650	1.4380
2340	1.4320	3771	1.4385
2455	1.4325	3882	1.4390

Run 12-18-2

Time	Wt.	Time	Wt.
0	1.3935	2176	1.4028
150	1.3950	2336	1.4032
244	1.3955	2470	1.4035
335	1.3960	2610	1.4039
518	1.3965	2783	1.4043
635	1.3970	2905	1.4046
761	1.3975	3040	1.4049
880	1.3980	3120	1.4052
985	1.3985	3220	1.4055
1111	1.3990	3340	1.4059
1240	1.3995	3512	1.4063
1347	1.4000	3605	1.4066
1460	1.4005	3727	1.4069
1588	1.4010	3860	1.4073
1724	1.4015	3980	1.4077
1828	1.4018	4325	1.4082
1935	1.4021	4445	1.4085
2075	1.4025		

(a) Obtain the breakthrough curves for these runs by plotting % wt. gain vs time on arithmetic and semilog paper.

(b) Use nonlinear regression techniques to obtain the equations of these curves.

(c) Compare the regression coefficients at the α = 0.05 level and comment on the effect of the change in number of MnO_2 pellets.

9.27 The SO_2 removal studies in Problem 9.26 were extended to include runs at 800°F, 31 psig, 20 pellets per run. For runs 1-13-3 and 1-13-6, 20 pellets MnO_2 each were used. Tare weight was 1.3885 g for each. In Run 1-13-3, the filled container weighed 1.3918 g; for Run 1-13-6, it was 1.3938 g. The data are below:

Run 1-13-3

Time	Wt.	Time	Wt.
0	1.3885	870	1.4005
75	1.3905	960	1.4015
128	1.3915	1010	1.4020
170	1.3920	1190	1.4035
232	1.3930	1264	1.4040
281	1.3940	1345	1.4045
360	1.3950	1433	1.4050
406	1.3955	1578	1.4060
480	1.3965	1665	1.4065
530	1.3970	1904	1.4080
600	1.3975	2020	1.4085
665	1.3980	2140	1.4090
778	1.3995	2410	1.4100
		2875	1.4110

Run 1-13-6

Time	Wt.	Time	Wt.
0	1.3937	245	1.3995
150	1.3975	310	1.4005
190	1.3985	366	1.4015

Time	Wt.	Time	Wt.
461	1.4030	1300	1.4125
526	1.4040	1380	1.4130
600	1.4050	1442	1.4135
672	1.4055	1546	1.4140
850	1.4080	1620	1.4145
917	1.4085	1774	1.4150
955	1.4090	1879	1.4165
1050	1.4100	1980	1.4170
1124	1.4105	2050	1.4175
1200	1.4115	2160	1.4180

SO_2 inlet concentration was 7540 ppm for both runs.

(a) Plot the breakthrough curves for these runs and determine statistically their mathematical form.

(b) Is there a significant difference in these results and those of Run 12-18-2? How would you determine whether the difference is due to differences in MnO_2 pellets, SO_2 inlet concentration, or temperature?

9.28 Conley and Valint [6] monitored the continuous oxidative degradation of poly(ethyl acrylate) at various temperatures. They determined the activation energy of the degradation reaction(s) as a function of temperature in a novel manner. The absorbance of infrared energy at 3.35 μm was measured at each temperature as a function of time. Those data were correlated with polymer weight loss so as to eliminate differences in film thickness between samples. The rate constants for the degradation (k) were obtained from the relationship between % polymer remaining on the specimen holders and time. From the temperature dependence of (- log k), the activation energy of degradation was found to be 37 kcal/g mole. Assume that the variation in weight loss as measured by absorption at 3.35 μm is the same for all degradation temperatures. What confidence limit can be placed on their values of the rate constants? What is the expected variance in the activation energy at any temperature within the range studied?

Time, min	Absorbance	% wt. remaining
0	0.243	100
15	0.221	91
30	0.204	84
45	0.195	80
60	0.170	70
75	0.158	65
90	0.146	60
105	0.125	51
120	0.113	46
135	0.107	44

Temperature, ^{o}C	Rate constant k, hr^{-1}
180	0.64
190	1.28
200	9.3
210	11.9
220	27.4
230	221.6
240	472.2

9.29 - 9.41 Calculate the correlation coefficients for Problems 9.1 - 9.13.

9.42 The grades in Physical Chemistry 347 and Ch.E. 330 (an applied solid state chemistry course dealing with the physical properties of materials of all types) are as follows:

x_1 = Chem. 347	x_2 = Ch.E. 330
86	92
78	89
94	83
85	75
94	82
90	96
89	84
74	72

$\Sigma x_1 = 670$ $\quad \Sigma x_2^2 = 57{,}272$

$\Sigma x_2 = 674$ $\quad \Sigma x_1 x_2 = 56{,}663$

$\Sigma x_1^2 = 56{,}514$

(a) What is the correlation coefficient for grades in these two courses?

(b) Develop an equation to predict Ch.E. 330 grades from the Chemistry 347 grades above.

(c) Present and use two different techniques for testing the hypothesis that there is no relation between course grades.

9.43 Below are data obtained for the minimum bending radius for thin-wall (22 or 24 gage) aluminum tubing to avoid crimping. The data are averages of 10 determinations per tube size.

Bending radius, in.	Tube diameter, in.
0.91	1/8
1.73	3/16
2.41	1/4
3.64	5/16
3.81	3/8
4.96	1/2
6.51	5/8
7.42	3/4
8.71	7/8
9.79	1

(a) Find the appropriate relationship between bending radius and tube diameter.

(b) What is the best estimate of the change in bending radius as the tube diameter is varied from 3/16 in. to 7/8 in.?

(c) What portion of the data is accounted for by your relationship?

9.44 Bennett and Franklin [1] present data on the effect of annealing temperature on the density of a borosilicate glass with high silica content. For the data below, they obtained a cubic equation in terms of the orthogonal polynomial coefficients. Rework this problem to obtain the actual relationship between the variables involved. What part of the variation in the data does your regression equation explain?

Annealing temperature, °C	Density, g/cm^3
450	2.23644
475	2.23534
500	2.23516
525	2.23574
550	2.23513
575	2.23674
600	2.23748
625	2.23902
650	2.23985
675	2.24005
700	2.24085
725	2.24120
750	2.24218

After plotting the data and your equation it appears that the relationship involved is, at least over the range of variables studied, part of a sine wave. Using appropriate transformations, fit the data to this new model. Transform the results of the fitted curve as necessary to the form of the original sine model. What portion of the variation in data does this approach provide? Which technique would you recommend for future use?

9.45 In an experiment on quantitative analysis by gas chromatography, the data below were obtained.

Composition vol. %	EtFo pk. ht	Avg.	Bz. pk. ht.	Avg.
100 Bz	---	---	49.8	49.8
20% EtFo	28.0		45.6	
20% EtFo	28.3	28.166	46.6	46.166
20% EtFo	28.2		46.8	
40% EtFo	41.5		40.2	
40% EtFo	41.8	41.533	41.7	41.133
40% EtFo	41.3		41.5	
60% EtFo	51.5		28.2	
60% EtFo	51.5	51.900	27.0	27.733
60% EtFo	52.7		28.0	
80% EtFo	58.1		27.9	
80% EtFo	58.5	58.560	28.6	28.166
80% EtFo	59.1		28.0	
100% EtFo	63.3	63.3	----	----
Unknown	53		24.4	
Unknown	55	54.367	25.2	24.333
Unknown	55.1		23.4	

(a) Plot the two calibration curves on a single graph.

(b) Determine, using the least squares technique, the best calibration curves to use in the analysis.

(c) One group of students obtained the following relation:

$$Y_1 = 0.050 + 186.706X_1 - 313.525X_1^2 + 312.482X_1^3 - 122.475X_1^4$$

$$Y_2 = 1.761 + 136.333X_2 - 168.966X_2^2 + 81.516X_2^3$$

where

Y = peak height

X = wt. fraction

1 = ethyl formate, EtFo

2 = benzene, Bz

Comment on their solution with regard to form, model, correlation, and usefulness.

9.46 Chen et al. [5] studied heat transfer in the streamline flow of water through an annulus. For their data, fit an appropriate relation by the method of least squares. What portion of the variability in the data is accounted for by your equation?

9.47 Newton and Dodge [13] studied the temperature dependence of thermodynamic equilibrium constant K on absolute temperature T (in $^{\circ}K$), for the synthesis of methanol from CO and H_2. From their results, postulate the model involved and obtain the best possible estimates of the population parameters.

1000/T, $^{\circ}K^{-1}$	log K
1.66	-4.15
1.73	-3.75
1.72	-3.65
1.75	-3.30
1.82	-3.10
1.81	-3.20
1.82	-3.00
1.82	-2.90
1.83	-2.95
1.88	-2.60
1.91	-2.70
1.91	-3.00
1.92	-2.30
2.05	-2.30
2.05	-2.15
2.05	-2.35

9.48 In the analysis of isopropanol and methylethylketone mixtures, a gas chromatographic procedure was used. Quantitative data were obtained and are shown below where A = isopropanol and B = methylethylketone.

Weight %	Peak height			Avg.
100.0 A	69.3	70.0	71.1	70.10
89.1 A	64.9	67.0	63.6	65.67
10.9 B	9.0	9.1	8.7	8.93
78.8 A	58.8	60.4		59.60
21.2 B	19.2	19.8		19.50
69.9 A	53.0	52.6		52.80
30.1 B	28.7	28.4		28.55
60.9 A	42.8	43.6		43.20
39.1 B	35.0	35.8		35.40
51.1 A	36.5	35.5		36.00
48.9 B	45.3	44.2		44.80
41.5 A	28.5	31.4	30.0	30.00
58.5 B	52.4	57.9	55.2	55.20
32.6 A	22.9	23.1		23.00
67.4 B	63.9	64.9		64.40
24.4 A	16.4	15.4	16.0	15.70
75.6 B	70.0	66.3	68.0	68.10
14.3 A	9.5	9.5		9.50
85.7 B	78.7	79.1		78.90
100.0 B	95.9	94.1	92.6	93.20
---A	44.1	43.5		43.80
---B	35.5	35.0		35.25

(a) What was the composition of the unknown sample?

(b) Give the 98% confidence limits for your answer.

(c) How good are these data? Support your answer by showing the analysis of variance on the calibration equations involved.

9.49 - 9.55 What portion of the data variation in Problems 10.6 - 10.12 is explained by the regression equation?

9.56 - 9.61 For the data in Problems 10.13 - 10.18, calculate r^2 as a measure of the variability of the data accounted for by the regression equation.

9.62 The overall heat transfer coefficient U in evaporators is a function of the logarithm of the LMTD (labelled ΔT below). For the following data, find the relation between U and Z where $Z = \log \Delta T$. Is the semilog model reasonable?

U	ΔT	Z
151.7	75.9	1.88024
186.2	84.25	1.92557
204.3	90.41	1.95622
221.7	97.04	1.98695
231.7	101.74	2.00749

$\Sigma(U_i - \overline{U})(Z_i - \overline{Z}) = 6.3632296$ $\Sigma(U_i - \overline{U})^2 = 4013.728$

$\Sigma(Z_i - \overline{Z})^2 = 0.01016400$ $\overline{U} = 199.12$ $\overline{Z} = 1.951294$

REFERENCES

1. Bennett, C. A., and Franklin, N. L., Statistical Analysis in Chemistry and the Chemical Industry, p. 260. John Wiley and Sons, Inc., New York, 1954. Data reprinted by permission of John Wiley and Sons, Inc.

2. Box, G. E. P., and Cox, D. R., An analysis of transformations, J. Roy. Statist. Assoc., 26 (2), 211-243 (1964).

3. Box, G. E. P., and Tidwell, P. W., Transformation of the independent variables, Technometrics, 4 (4), 531-550 (1962).

4. Burrows, G. L., Interpreting straight line relationships, Ind. Qual. Control, 15 (1), 15-16 (1958). Copyright American Society for Quality Control, Inc. Reprinted by permission.

5. Chen, C. Y., Hawkins, G. A., and Solberg, H. L., Heat transfer in annuli, Trans. Amer. Soc. Mech. Engrs., 68, 99-106 (1946).

6. Conley, R. T., and Valint, P. L., Oxidative Degradation of poly(ethyl acrylate), J. Appl. Polym. Sci., 9, 785-797 (1965). Copyright 1965 by John Wiley and Sons, Inc. Data reproduced by permission of John Wiley and Sons, Inc.

7. Daniel, C. and Wood, F. S., Fitting Equations to Data, Wiley-Interscience, New York, 1971.

8. Draper, N. R., and Smith, H., Applied Regression Analysis, Chap. 10. John Wiley and Sons, Inc., New York, 1966.

9. Graham, R. R., Reactor Performance Prediction for the Reaction of Sulfur Dioxide with Manganese Dioxide, unpublished Ph.D. dissertation, Library, Texas Tech University, Lubbock, Texas, 1969, pp. 84-87.

10. Graybill, F. A., Introduction to Linear Statistical Models, Chap. 6, Vol. I, McGraw-Hill, New York, 1961.

11. Lacher, J. R., Dissolving uranium in nitric acid. Ind. Eng. Chem., 53 , 282-284 (1961). Copyright 1961 by the American Chemical Society. Data reprinted by permission of the copyright owner.

12. Lacroix, S., and Labalade, M. M., Anal. Chem. Acta., 4, 68 (1950).

13. Newton, R. H., and Dodge, B. F., The equilibrium between hydrogen, formaldehyde, and methanol. II. The reaction $CO + 2H_2 \rightleftarrows CH_3OH$. J. Amer. Chem. Soc., 56, 1287 (1934). Copyright 1934 by the American Chemical Society, Reprinted by permission of the copyright owner.

14. O'Connell, Trans. Amer. Inst. Chem. Engr., 42, 791 (1946).

15. Olujić, Ž., Compute friction factors fast for flow in pipes, Chem. Eng., 88 (25), 91-93 (1981).

16. Roth, J. F., Oxidation-reduction catalysis. Ind. Eng. Chem., 53, 293 (1961). Copyright 1961 by the American Chemical Society. Data reprinted by permission of the copyright owner.

17. Sinibaldi, F. J., Jr., Koehler, T. L., and Babis, A. H., Transformed data simplify and confirm math models, Chem. Eng., 71(10), 139-146 (1971).

18. White, R. G., Pelletizing magnetic taconite concentrate, Ind. Eng. Chem., 53, 215-216 (1961). Copyright 1961 by the American Chemical Society. Data reprinted by permission of the copyright owner.

10
ORTHOGONAL POLYNOMIALS IN POLYNOMIAL REGRESSION

10.1 INTRODUCTION

The method of orthogonal polynomials is used in the fitting of data represented by the polynomial (curvilinear) regression model

$$Y = \beta_0 + \beta_1 X + \beta_2 X^2 + \dots + \beta_k X^k + \varepsilon \, , \tag{10.1}$$

provided the values of the independent variable X are equally spaced. This is not a serious restriction for the application of this technique. If you desire to use orthogonal polynomials for model-fitting, you should make every effort to insure that your data are evenly incremented in the independent variable. The technique is generally faster than applying least squares directly to the model (10.1) and has a distinct advantage in that the $\hat{\beta}_i$ values are computed independently of each other. If the X_i data are not evenly spaced, the method of Grandage [4] should be used to obtain the orthogonal polynomial coefficients. If the treatments are also unequally replicated, the method of Carmer and Seif [2] should be used to obtain the orthogonal polynomial coefficients. If the values of X are evenly spaced but not at unit intervals, the data may be coded by dividing through by the length of the common interval. The method outlined below is also restricted to the case where an equal number of Y values are present for each X value.

10.1.1 Orthogonal Model

In using the method of orthogonal polynomials, the model of Eq. (10.1) is replaced by

$$Y = \beta'_0 + \beta'_1\xi'_1 + \beta'_2\xi'_2 + \dots + \beta'_k\xi'_k + \varepsilon \qquad (10.2)$$

where the $\xi'_i (i = 1,2,3,\dots,k)$ are orthogonal polynomials. Two polynomials ξ'_i and ξ'_j, $i \neq j$, are orthogonal if when X takes on n specified values, say $x_1, x_2, \dots, x_n$,

$$\sum_{r=1}^{n} \xi'_i(x_r)\xi'_j(x_r) = \sum_{r=1}^{n} \xi'_{ir}\xi'_{jr} = 0.$$

The main idea in using orthogonal polynomials is to replace the independent variables $X, X^2, \dots, X^k$ in Eq. (10.1) with a new set of independent variables (orthogonal polynomials) $\xi'_1\ \xi'_2, \dots, \xi'_k$ so that the resulting computations relative to the new model are greatly simplified. The polynomials $\xi'_1, \xi'_2, \dots, \xi'_k$ are of degrees $1,2,\dots,k$, respectively. For example, an equation involving $\xi'_1, \xi'_2, \dots, \xi'_{10}$ will be of degree 10. Consequently, fitting a model involving $\xi'_1, \xi'_2, \dots, \xi'_{10}$, is exactly the same as fitting a polynomial model of degree 10. The orthogonal polynomials can be chosen so that the additional condition, $\sum_{r=1}^{n} \xi'_{ir} = 0$, $i = 1,2,\dots,k$, is also satisfied.

Our orthogonal model, given n observations, can be written as

$$Y_i = \beta'_0 + \beta'_1\xi'_{1i} + \beta'_2\xi'_{2i} + \dots + \beta'_k\xi'_{ki} + \varepsilon_i, \quad i = 1,2,\dots,n \qquad (10.3)$$

where

$$\sum_{j=1}^{n} \xi'_{1j} = \sum_{j=1}^{n} \xi'_{2j} = \dots = \sum_{j=1}^{n} \xi'_{kj} = 0$$

and

$$\Sigma\xi_r'\xi_s' = \sum_i \xi_{ri}'\xi_{si}' = 0 \text{ for } r \neq s.$$

The normal equations resulting from a least squares differentiation of the orthogonal model are

$$\begin{array}{ccccccccccc}
n\beta_0' & + & 0 & + & 0 & + & \cdots & + & 0 & = & \sum_i Y_i \\
0 & + & \hat{\beta}_1'\sum_i(\xi_{1i}')^2 & + & 0 & + & \cdots & + & 0 & = & \sum_i \xi_{1i}'Y_i \\
0 & + & 0 & + & \hat{\beta}_2'\sum_i(\xi_{2i}')^2 & + & \cdots & + & 0 & = & \sum_i \xi_{2i}'Y_i \\
\vdots & & \vdots & & & & \ddots & & & & \vdots \\
0 & + & 0 & + & 0 & + & \cdots & + & \hat{\beta}_k'\sum_i(\xi_{ki}')^2 & = & \sum_i \xi_{ki}'Y_i
\end{array} \tag{10.4}$$

The normal equations (10.4) are just Eq. (9.56) with all higher order terms and those involving cross-product terms equal to zero. The solution to the system (10.4) is

$$\begin{aligned}
\hat{\beta}_0' &= \sum_{i=1}^{n} Y_i/n = \overline{Y}, \\
\hat{\beta}_1' &= \sum_{i=1}^{n} \xi_{1i}'Y_i / \sum_{i=1}^{n} (\xi_{1i}')^2, \\
&\vdots \\
\hat{\beta}_k' &= \sum_{i=1}^{n} \xi_{ki}'Y_i / \sum_{i=1}^{n} (\xi_{ki}')^2 .
\end{aligned} \tag{10.5}$$

When the values of the independent variable X in the original polynomial model in Eq. (10.1) are $X = 1,2,\ldots,n$, then the values of the orthogonal polynomials will depend only on n and can thus be determined and tabulated for various values of n. Values of ξ_{ki}' for $n = 3$ to $n = 75$ and $k = 1$ to 5 are contained in Ref. [3]. Tables extended up to $n = 104$ are contained in Anderson and Houseman [1].

If the values of the independent variable are $X_1 = 1$, $X_2 = 2,\ldots,$ $X_n = n$, then the first three orthogonal polynomials are, for $i = 1,2,\ldots,n$,

$$\xi'_{1i} = \lambda_1(X_i-\overline{X}) = \lambda_1\xi_{1i} ,$$

$$\xi'_{2i} = \lambda_2\left((X_i-\overline{X})^2 - \frac{n^2-1}{12}\right) = \lambda_2\xi_{2i} , \tag{10.6}$$

$$\xi'_{3i} = \lambda_3\left((X_i-\overline{X})^3 - (X_i-\overline{X})\frac{3n^2-7}{20}\right) = \lambda_3\xi_{3i} ,$$

where the λ_i are constants, depending on n, chosen so that the values of the ξ'_{ji} are integers reduced to their lowest terms. The quantities $\xi_{1i},\xi_{2i},\xi_{3i}$, on the right-hand side are also orthogonal: however, they contain fractional values. It will be more convenient, computationally, to use the $\xi'_{1i},\xi'_{2i},\ldots,\xi'_{ki}$.

A recursive formula that yields the orthogonal polynomials is

$$\xi'_{r+1} = \lambda_{r+1}\ \xi_{r+1}, \tag{10.7}$$

where

$$\xi_{r+1} = \xi_1\xi_r - \frac{r^2(n^2-r^2)}{4(4r^2-1)}\xi_{r-1} ,$$

and $\xi'_0 = \xi_0 = 1$. Thus, for example, $\xi_1 = (X-\overline{X})$ from which we obtain

$$\xi_2 = \xi_{1+1} = \xi_1\xi_1 - \frac{1^2(n^2-1^2)}{4(4-1)}\xi_0$$

$$= (X-\overline{X})^2 - \frac{n^2-1}{12} .$$

Hence

$$\xi'_2 = \lambda_2\left((X-\overline{X})^2 - \frac{n^2-1}{12}\right). \tag{10.8}$$

In similar fashion

$$\xi'_3 = \lambda_3\left((X - \overline{X})^3 - (X - \overline{X})\frac{3n^2-7}{20}\right) , \tag{10.9}$$

$$\xi_4' = \lambda_4 \left[(X - \bar{X})^4 - (X - \bar{X})^2 \left(\frac{3n^2 - 13}{14} \right) + \frac{3(n^2-1)(n^2-9)}{560} \right], \quad (10.10)$$

$$\xi_5' = \lambda_5 \left[(X - \bar{X})^5 - (X - \bar{X})^3 \left(\frac{5n^2 - 35}{18} \right) + (X - \bar{X}) \frac{15n^4 - 230n^2 + 407}{1008} \right], \quad (10.11)$$

where $\bar{X}$ is the mean value of $X_i = i$ for $i = 1, 2, \ldots, n$.

A partial listing of orthogonal polynomials, extracted from Fisher and Yates [3], is contained in Table 10.1. For values of n = 3 through n = 8, the table gives all the values of each orthogonal polynomial. For $n \geq 9$, the table only gives the last or bottom half of the ξ_i' values. If ξ_i' is such that i is odd, then the remainder of the ξ_i' values are the negatives of the bottom half. If ξ_i' is such that i is even, then the ξ_i' are symmetric. Thus the values of ξ_1' for n = 10 are -9, -7, -5, -3, -1, 1, 3, 5, 7, and 9. The values of ξ_2' for n = 10 are 6, 2, -1, -3, -4, -4, -3, -1, 2, and 6. The first value at the bottom of each column of ξ' values is the sum of squares of the corresponding ξ' values. Thus if n = 5, then for ξ_1' we have

$\sum_{i=1}^{5} (\xi_{1i}')^2 = (-2)^2 + (-1)^2 + (0)^2 + (1)^2 + (2)^2 = 10.$ The second value at the bottom of each column is the corresponding λ value. For example, for n = 14 and ξ_4' we have $\lambda_4 = 7/12$.

To use the method of orthogonal polynomials, first the $\hat{\beta}_i'$ are computed by means of Eqs. (10.5) to yield the result

$$\hat{Y} = \hat{\beta}_0' + \hat{\beta}_1'\xi_1' + \ldots + \hat{\beta}_k'\xi_k' . \quad (10.12)$$

The ξ_i' are obtained from Eq. (10.7) and then substituted into (10.12). Like powers of X can then be grouped to yield

$$\hat{Y} = \hat{\beta}_0 + \hat{\beta}_1 X + \hat{\beta}_2 X^2 + \ldots + \hat{\beta}_k X^k . \quad (10.13)$$

TABLE 10.1

ξ'_i Values[a]

	n=3		n=4			n=5				n=6					n=7				
	ξ'_1	ξ'_2	ξ'_1	ξ'_2	ξ'_3	ξ'_1	ξ'_2	ξ'_3	ξ'_4	ξ'_1	ξ'_2	ξ'_3	ξ'_4	ξ'_5	ξ'_1	ξ'_2	ξ'_3	ξ'_4	ξ'_5
															-3	+5	-1	+3	-1
										-5	+5	-5	+1	-1	-2	0	+1	-7	+4
						-2	+2	-1	+1	-3	-1	+7	-3	+5	-1	-3	+1	+1	-5
			-3	+1	-1	-1	-1	+2	-4	-1	-4	+4	+2	-10	0	-4	0	+6	0
	-1	+1	-1	-1	+3	0	-2	0	+6	+1	-4	-4	+2	+10	+1	-3	-1	+1	+5
	0	-2	+1	-1	-3	+1	-1	-2	-4	+3	-1	-7	-3	-5	+2	0	-1	-7	-4
	+1	+1	+3	+1	+1	+2	+2	+1	+1	+5	+5	+5	+1	+1	+3	+5	+1	+3	+1
$\sum_i (\xi'_i)^2$	2	6	20	4	20	10	14	10	70	70	84	180	28	252	28	84	6	154	84
λ	1	3	2	1	10/3	1	1	5/6	35/12	2	3/2	5/3	7/12	21/10	1	1	1/6	7/12	7/20

TABLE 10.1 (Continued)

	n=8					n=9					n=10					n=11				
	ξ_1'	ξ_2'	ξ_3'	ξ_4'	ξ_5'	ξ_1'	ξ_2'	ξ_3'	ξ_4'	ξ_5'	ξ_1'	ξ_2'	ξ_3'	ξ_4'	ξ_5'	ξ_1'	ξ_2'	ξ_3'	ξ_4'	ξ_5'
	-7	+7	-7	+7	-7															
	-5	+1	+5	-13	+23															
	-3	-3	+7	-3	-17															
	-1	-5	+3	+9	-15	0	-20	0	+18	0	+1	-4	-12	+18	+6	0	-10	0	+6	0
	+1	-5	-3	+9	+15	+1	-17	-9	+9	+9	+3	-3	-31	+3	+11	+1	-9	-14	+4	+4
	+3	-3	-7	-3	+17	+2	-8	-13	-11	+4	+5	-1	-35	-17	+1	+2	-6	-23	-1	+4
	+5	+1	-5	-13	-23	+3	+7	-7	-21	-11	+7	+2	-14	-22	-14	+3	-1	-22	-6	-1
	+7	+7	+7	+7	+7	+4	+28	+14	+14	+4	+9	+6	+42	+18	+6	+4	+6	-6	-6	-6
																+5	+15	+30	+6	+3
$\sum_i (\xi_i')^2$	168	168	264	616	2184	60	2772	990	2002	468	330	132	8580	2860	780	110	858	4290	286	156
λ	2	1	2/3	7/12	7/10	1	3	5/6	7/12	3/20	2	1/2	5/3	5/12	1/10	1	1	5/6	1/12	1/40

	n=12					n=13					n=14				
	ξ'_1	ξ'_2	ξ'_3	ξ'_4	ξ'_5	ξ'_1	ξ'_2	ξ'_3	ξ'_4	ξ'_5	ξ'_1	ξ'_2	ξ'_3	ξ'_4	ξ'_5
						0	-14	0	+84	0	+1	-8	-24	+108	+60
	+1	-35	-7	+28	+20	+1	-13	-4	+64	+20	+3	-7	-67	+63	+145
	+3	-29	-19	+12	+44	+2	-10	-7	+11	+26	+5	-5	-95	-13	+139
	+5	-17	-25	-13	+29	+3	-5	-8	-54	+11	+7	-2	-98	-92	+28
	+7	+1	-21	-33	-21	+4	+2	-6	-96	-18	+9	+2	-66	-132	-132
	+9	+25	-3	-27	-57	+5	+11	0	-66	-33	+11	+7	+11	-77	-187
	+11	+55	+33	+33	+33	+6	+22	+11	+99	+22	+13	+13	+143	+143	+143
$\sum_i (\xi'_i)^2$	572	12012	5148	8008	15912	182	2002	572	68068	6188	910	728	97240	136136	235144
λ	2	3	2/3	7/24	3/20	1	1	1/6	7/12	7/120	2	1/2	5/3	7/12	7/30

(Continued)

TABLE 10.1 (Continued)

	n=15					n=16					n=17				
	ξ'_1	ξ'_2	ξ'_3	ξ'_4	ξ'_5	ξ'_1	ξ'_2	ξ'_3	ξ'_4	ξ'_5	ξ'_1	ξ'_2	ξ'_3	ξ'_4	ξ'_5
	0	-56	0	+756	0	+1	-21	-63	+189	+45	0	-24	0	+36	0
	+1	-53	-27	+621	+675	+3	-19	-179	+129	+115	+1	-23	-7	+31	+55
	+2	-44	-49	+251	+1000	+5	-15	-265	+23	+131	+2	-20	-13	+17	+88
	+3	-29	-61	-249	+751	+7	-9	-301	-101	+77	+3	-15	-17	-3	+83
	+4	-8	-58	-704	-44	+9	-1	-267	-201	-33	+4	-8	-18	-24	+36
	+5	+19	-35	-869	-979	+11	+9	-143	-221	-143	+5	+1	-15	-39	-39
	+6	+52	+13	-429	-1144	+13	+21	+91	-91	-143	+6	+12	-7	-39	-104
	+7	+91	+91	+1001	+1001	+15	+35	+455	+273	+143	+7	+25	+7	-13	-91
											+8	+40	+28	+52	+104
$\Sigma(\xi'_i)^2$	280	37128	39780	6466460	10581480	1360	5712	1007760	470288	201552	408	7752	3876	16796	100776
λ	1	3	5/6	35/12	21/20	2	1	10/3	7/12	1/10	1	1	1/6	1/12	1/20

This table is reproduced for n = 3 to n = 20 from Table XXIII of Fisher & Yates: Statistical Tables for Biological, Agricultural and Medical Research published by Longman Group Ltd. London (previously published by Oliver and Boyd Ltd, Edinburgh) and by permission of the authors and publishers.

10.2 FITTING A QUADRATIC BY ORTHOGONAL POLYNOMIALS

The quadratic form to be minimized is

$$\sum_i e_i^2 = \sum_i (Y_i - \hat{\beta}'_0 - \hat{\beta}'_1\xi'_1 - \hat{\beta}'_2\xi'_2)^2 \tag{10.14}$$

which, when differentiated partially with respect to each β'_i in turn and set equal to zero, gives the system of normal equations,

$$\begin{aligned}
n\beta'_0 &+ 0 &+ 0 &= \sum_i Y_i \\
0 &+ \hat{\beta}'_1\sum_i(\xi'_{1i})^2 &+ 0 &= \sum_i \xi'_{1i}Y_i \\
0 &+ 0 &+ \hat{\beta}'_2\sum_i(\xi'_{2i})^2 &= \sum_i \xi'_{2i}Y_i ,
\end{aligned}$$

due to orthogonality of the ξ'_i. Thus

$$\hat{\beta}'_0 = \overline{Y} ,$$

$$\hat{\beta}'_1 = \sum_i \xi'_{1i}Y_i / \sum_i(\xi'_{1i})^2 ,$$

$$\hat{\beta}'_2 = \sum_i \xi'_{2i}Y_i / \sum_i(\xi'_{2i})^2 .$$

A very strong feature of the orthogonal polynomial method is that the $\hat{\beta}'_i$ are computed independently of each other. In fitting a quadratic, for example, the coefficient $\hat{\beta}'_1$ is the same as it is in fitting a simple linear model. If orthogonal polynomials are not used, then the coefficients all have to be recomputed every time the degree of the fitted polynomial is changed. The orthogonal polynomial technique thus supplies a very expedient method of determining what degree polynomial best fits the data.

The error sum of squares in fitting a quadratic can be written as

$$\begin{aligned}
SS_E &= \sum_i (Y_i - \hat{Y}_i)^2 \\
&= \sum_i (Y_i - \overline{Y})^2 - (\hat{\beta}'_1)^2\sum_i(\xi'_{1i})^2 - (\hat{\beta}'_2)^2\sum_i(\xi'_{2i})^2
\end{aligned}$$

according to Eq. (9.57). The reduction in sum of squares due to β_j' is $(\hat{\beta}_j')^2 \sum_i (\xi_{ji}')^2$, j = 1,2. Furthermore, the reduction due to β_2' is independent of the reduction due to β_1' and one can thus test to see if the added reduction in using a quadratic over a linear model is significant. These ideas are discussed in the next section for the general k^{th} degree model.

10.3 TESTS OF SIGNIFICANCE

In the general orthogonal model in Eq. (10.2) the residual sum of squares is, according to Section 9.4,

$$\begin{aligned} SS_E &= \sum_i (Y_i - \hat{Y}_i)^2 \\ &= \sum_i (Y_i - \overline{Y})^2 - (\hat{\beta}_1')^2 \sum_i (\xi_{1i}')^2 - (\hat{\beta}_2')^2 \sum_i (\xi_{2i}')^2 - \cdots - (\hat{\beta}_k')^2 \sum_i (\xi_{ki}')^2 \\ &= SS_{TC} - (\hat{\beta}_1')^2 S_1 - (\hat{\beta}_2')^2 S_2 - \cdots - (\hat{\beta}_k')^2 S_k \quad , \end{aligned}$$

where $(\hat{\beta}_j')^2 S_j$ is the reduction in sums of squares due to fitting β_j' or due to fitting the jth power of the independent variable X. Furthermore, due to the orthogonality of ξ_j's the reductions are independent and each of the random variables

$$F = (\hat{\beta}_j')^2 S_j / \hat{\sigma}^2, \quad j = 1,2,\ldots,k \quad , \tag{10.15}$$

where $\hat{\sigma}^2 = SS_{Res.}/(n-k-1) = \sum_i (Y_i - \hat{Y}_i)^2/(n-k-1)$ has an F-distribution with 1 and (n-k-1) degrees of freedom if $\beta_j' = 0$. Thus the statistic (10.15) can be used to test H_0: $\beta_j' = 0$, the rejection region being $F > F_{1,n-k-1,1-\alpha}$.

The method of orthogonal polynomials is not useful if the problem is merely to fit a linear model. However, if the problem is to fit first a linear, then a quadratic, and so on, until a "best" fit is found then orthogonal polynomials are very helpful. The procedure would be to first fit a linear model using the statistic $F = (\hat{\beta}_1')^2 S_1 / [\sum_i (Y_i - \hat{Y}_i)^2/(n-2)]$ to test H_0: $\beta_1' = 0$. If H_0 is rejected, then compute $\hat{\beta}_2'$ and test H_0: $\beta_2' = 0$ by using

$F = (\hat{\beta}_2')^2 S_2 / [\sum_i (Y_i - \hat{Y}_i)^2/(n-3)]$. If H_0: $\beta_2' = 0$ is rejected, then fit a cubic, etc. A general rule is to continue until two consecutive coefficients, say $\hat{\beta}_q'$ and $\hat{\beta}_{q+1}'$, are not significantly different from 0, in which case the best fitting polynomial would be of degree q-1. The big advantage in using this procedure is that each time the degree is increased by 1 only one additional coefficient needs to be computed, that is, going from degree 4 to degree 5 requires only the computation of $\hat{\beta}_5' = \sum_i \xi_{5i}' Y_i / \sum_i (\xi_{5i}')^2$. It should be noted that each time a power of X is added, that is, each time an additional β_j' is fitted, the degrees of freedom of the residual sum of squares are decreased by 1. Thus the degrees of freedom of the denominator of the F-statistic change as the degree of the fitted polynomial changes.

If m Y values are associated with each of n equally spaced X's, then a lack of fit sum of squares, such as was discussed in Section 9.2.5, can be computed and used to test lack of fit at each step in the procedure that uses orthogonal polynomials. The procedure is continued until the lack of fit mean square is not significantly larger than the residual mean square. This procedure can be used in fitting any function to a set of data. The function is first expanded in a Taylor series. The lowest degree polynomial that adequately fits the data can then be determined.

Example 10.1 Use the method of orthogonal polynomials to fit a third-order polynomial in X to the Y data of Example 9.18.

The model is $Y = \beta_0 + \beta_1 X + \beta_2 X^2 + \beta_3 X^3 + \varepsilon$ which is replaced by the orthogonal model

$$Y = \beta_0' + \beta_1'\xi_1' + \beta_2'\xi_2' + \beta_3'\xi_3' + \varepsilon$$

where the β_i' and ξ_i' are as previously defined. To find the values of the orthogonal polynomials, Table 10.1 is used with n = 13. The values of the new independent variables ξ_i', ξ_2', ξ_3' and other computations are summarized in the following table:

X	Y	ξ_1'	ξ_2'	ξ_3'	$Y\xi_1'$	$Y\xi_2'$	$Y\xi_3'$
1	17	-6	22	-11	-102	374	-187
2	21	-5	11	0	-105	231	0
3	22	-4	2	6	-88	44	132
4	27	-3	-5	8	-81	-135	216
5	36	-2	-10	7	-72	-360	252
6	49	-1	-13	4	-49	-637	196
7	56	0	-14	0	0	-784	0
8	64	1	-13	-4	64	-832	-256
9	80	2	-10	-7	160	-800	-560
10	86	3	-5	-8	258	-430	-688
11	88	4	2	-6	352	176	-528
12	92	5	11	0	460	1012	0
13	94	6	22	11	564	2068	1034

	Y	ξ'_1	ξ'_2	ξ'_3	$Y\xi'_1$	$Y\xi'_2$	$Y\xi'_3$
Σ	732	0	0	0	1361	-73	-389
λ		1	1	1/6			
$\Sigma(\xi'_i)^2$		182	2002	572			
$\sum_i y_i^2$	51,712						

The $\hat{\beta}'_i$ are calculated as follows:

$$\hat{\beta}'_0 = \frac{\Sigma y}{n} = \frac{732}{13} = 56.31 \quad ,$$

$$\hat{\beta}'_1 = \frac{\Sigma y\xi'_1}{\Sigma(\xi'_1)^2} = \frac{1361}{182} = 7.47802 \quad ,$$

$$\hat{\beta}'_2 = \frac{\Sigma y\xi'_2}{\Sigma(\xi'_2)^2} = \frac{-73}{2002} = -0.0365 \quad ,$$

$$\hat{\beta}'_3 = \frac{\Sigma y\xi'_3}{\Sigma(\xi'_3)^2} = \frac{-389}{572} = -0.6801 \quad .$$

The sums of squares are then calculated as shown below and are used to prepare the analysis of variance table.

$$SS_T = \Sigma y^2 = 51,712 \quad ,$$

$$SS_{\beta'_0} = \hat{\beta}'_0\Sigma y = (732)^2/13 = 41,217.231 \quad ,$$

$$SS_{\beta'_1} = \hat{\beta}'_1\Sigma\xi'_1 y = (1361)^2/182 = 10,177.588 \quad ,$$

$$SS_{\beta'_2} = \hat{\beta}'_2\Sigma y\xi'_2 = (73)^2/2002 = 2.6618 \quad ,$$

$$SS_{\beta'_3} = \hat{\beta}'_3\Sigma y\xi'_3 = (389)^2/572 = 264.5472 \quad ,$$

$$SS_{Res.} = SS_T - SS_{\beta'_0} - SS_{\beta'_1} - SS_{\beta'_2} - SS_{\beta'_3} = 49.972 \quad .$$

The AOV table below summarizes the sums of squares needed to test the coefficients in the orthogonal model and the corresponding residual sums of squares.

Analysis of Variance

Source	d.f.	SS	MS
Total	13	51,712	
Constant	1	41,217.231	41,217.231
residual	12	10,494.769	874.561
Linear	1	10,177.588	10,177.588
residual	11	317.181	28.835
Quadratic	1	2.662	2.662
residual	10	314.519	31.452
Cubic	1	264.547	264.547
residual	9	49.972	5.552

The mean squares for each term are compared to the corresponding residual mean squares by an F-test as shown below.

F-tests

Regression term	$F_{(computed)}$	$F_{(tabular)}$	Comment
Constant	f = 47.1	$F_{1,12,0.95} = 4.75$	Signif.
Linear	f = 353.0	$F_{1,11,0.95} = 4.84$	Signif.
Quadratic	f = 0.085	$F_{1,10,0.95} = 4.96$	Signif.
Cubic	f = 47.6	$F_{1,9,0.95} = 5.12$	Signif.

As shown above, the quadratic term is not significant. Polynomials of higher degree than the third could be tested until two consecutive $\hat{\beta}_i'$ were insignificant but we will not do that here.

If ζ_1', ζ_2', and ζ_3' are evaluated in terms of X according to Eqs. (10.6) with $\lambda_1 = 1$, $\lambda_2 = 1$, $\lambda_3 = 1/6$, one obtains

$$\xi_1' = X-7 \quad ,$$

$$\xi_2' = X^2 - 14X + 35 \quad ,$$

$$\xi_3' = (1/6)X^3 - (7/2)X^2 + (61/3)X - 28 \quad .$$

Substituting these three polynomials in

$$Y = 56.31 + 7.47802\xi_1' - 0.0365\xi_2' - 0.6801\xi_3'$$

one obtains

$$Y = 21.7277 - 5.8458X + 2.34449X^2 - 0.1134X^3$$

the prediction equation in terms of the original independent variable.

Example 10.2 The relation between absorption peak height Y of the $-C \equiv N$ group in infrared spectroscopy and acetonitrile concentration X in a $CH_3CN\text{-}NaBr\text{-}H_2O$ system is $Y = \beta_0 + \beta_1 X + \beta_2 X^2$. For the data below, determine the value of the estimators of β_0, β_1, and β_2 using the method of orthogonal polynomials. Construct an AOV table and test the combined significance of the regression coefficients $\hat{\beta}_1$ and $\hat{\beta}_2$ using a one-sided, 5% F-test to compare the mean square due to regression to the mean square due to the deviations about regression.

X, $-C \equiv N$ concentration:	0	1	2	3	4	5	6
Y, peak height:	2	1	3	5	9	13	20

Replacing the model above by the orthogonal model, we have

$$Y = \beta_0' + \beta_1'\xi_1' + \beta_2'\xi_2' + \varepsilon$$

where

$$\hat{\beta}_0' = \sum_i y_i/n, \quad \hat{\beta}_1' = \frac{\sum_i y_i\xi_{1i}'}{\sum_i (\xi_{1i}')^2} \quad \text{and} \quad \hat{\beta}_2' = \frac{\sum_i y_i\xi_{2i}'}{\sum_i (\xi_{2i}')^2} \quad .$$

The Y_i data and the required sums and products are tabulated as follows:

X	Y	ξ'_1	ξ'_2	$Y\xi'_1$	$Y\xi'_2$
0	2	-3	5	-6	10
1	1	-2	0	-2	0
2	3	-1	-3	-3	-9
3	5	0	-4	0	-20
4	9	1	-3	9	-27
5	13	2	0	26	0
6	20	3	5	60	100
Σ	53	0	0	84	54
λ		1	1		
$\sum_i (\xi'_{ji})^2$		28	84		
$\sum_i y_i^2$	689				

$$\hat{\beta}'_0 = \frac{\sum_i y_i}{n} = \frac{53}{7} = 7.571 \quad ,$$

$$\hat{\beta}'_1 = \frac{\sum_i y_i \xi'_{1i}}{\sum_i (\xi'_{1i})^2} = \frac{84}{28} = 3.000 \quad ,$$

$$\hat{\beta}'_2 = \frac{\sum_i y_i \xi'_{2i}}{\sum_i (\xi'_{2i})^2} = \frac{54}{84} = 0.643 \quad ,$$

$$SS_T = \sum_i y_i^2 = 689.0 \quad ,$$

$$SS_{\beta'_0} = \hat{\beta}'_0 \sum_i y_i = 401.286 \quad ,$$

$$SS_{\beta'_1} = \hat{\beta}'_1 \sum_i y_i \xi'_{1i} = 252.0 \quad ,$$

$$SS_{\beta'_2} = \hat{\beta}'_2 \sum_i y_i \xi'_{2i} = 34.714 \quad .$$

The analysis of variance table is prepared as before:

Analysis of Variance

Source	d.f.	SS	MS
Total	7	689.0	
Constant	1	401.286	401.286
residual	6	287.714	47.95
Linear	1	252.0	252.0
residual	5	35.714	7.143
Quadratic	1	34.714	34.714
residual	4	1.0	0.25

The resulting mean squares for fitting each term are handled by the F-test as before:

F-Test (one-sided, 5%)

Regression term	$F_{(computed)}$	$F_{(tabular)}$	Comment
Constant	8.369	$F_{1,6,0.95} = 5.99$	Signif.
Linear	35.279	$F_{1,5,0.95} = 6.61$	Signif.
Quadratic	138.856	$F_{1,4,0.95} = 7.71$	Signif.

Next, the ξ_1' and ξ_2' are calculated and the orthogonal model converted to the original form:

$$\xi_1' = \lambda_1(X - \overline{X}) = X - 3 \quad ,$$

$$\xi_2' = \lambda_2\left((X - \overline{X})^2 - \frac{n^2-1}{12}\right) = X^2 - 6X + 5 \quad .$$

Using ξ_1' and ξ_2' above we obtain

$$\hat{Y} = \hat{\beta}_0' + \hat{\beta}_1'\xi_1' + \hat{\beta}_2'\xi_2'$$

$$= 7.571 + 3.0(X-3) + 0.643(X^2 - 6X + 5)$$

$$= 1.786 - 0.858X + 0.643X^2 ,$$

which is the prediction equation in terms of the original variable X.

PROBLEMS

10.1 For the data of Problem 9.2 obtain the required regression equation by orthogonal polynomials. Show how to test the validity of β_0' and β_1'. Show what comparisons must be made and the corresponding null hypotheses and acceptance regions. Would you expect the values of C_{P_0} and a to be different from those obtained by linear regression? Why?

10.2 The total amount of impurities present in the overhead product from an acetic acid purification tower, Y, has been estimated to be (among other things) a function of the amount of subcooling of the reflux, X. For the data below, use the method of orthogonal polynomials to find the relation between these variables. Test each term for significance at the 5% level.

Y	$\bar{Y}$	X
2, 4, 4	3.33	4
7, 8	7.5	8
10, 10	10	12
14, 15, 13	14	16
18, 18, 16	17.33	20

10.3 In a process for making quick-cured superphosphate fertilizer, phosphoric and sulfuric acids, water, and crushed phosphate rock are fed into a rod mill on a continuous basis. The production rate Y in lb_m/min is a function of the rotation X, in revolutions/hour as determined by the following data:

X	Y
10	5.6
20	10.8
30	43.0
40	71.4
50	83.2

Use the method of orthogonal polynomials to determine the exact form of the relationship between production rate and rotation speed. (Hint: start with a cubic equation.) Test the significance of each term at the 1% and the 5% levels.

10.4 The total number of theoretical trays needed in the primary distillation section in the Keyes process for purifying ethanol by extractive distillation has been studied in terms of the internal reflux ratio above the feed. For the data below, use the method of orthogonal polynomials to find the correct functional relationship. Test each term for its significance.

L/V	Number of trays
1/1	8.4
2/1	7.1
3/1	6.0
4/1	5.4
5/1	4.9

10.5 For the coded data below, fit an appropriate polynomial in R to the Z data by any technique. Does each term in your relationship contribute significantly to the reduction in the error sum of sum of squares at the 0.05 level?

Z:	4.2	0.9	0.0	1.1	4.0
R:	-2	-1	0	1	2

10.6 - 10.12 Work Problems 9.1, 9.3, 9.7, 9.8, 9.18, 9.23, and 9.28 by the method of orthogonal polynomials. Test each term for significance at the level prescribed by your instructor.

10.13 - 10.18 For the data of Problem 9.20 between 10 and 90% recovery, find the equations of each of the ASTM distillation curves so represented by the method of orthogonal polynomials.

REFERENCES

1 Anderson, R. L., and Houseman, E. E., Tables of Orthogonal Polynomial Values Extended to N = 104. Research Bulletin 297, Agricultural Experiment Station, Iowa State University, Ames, Iowa. (Apr. 1942, reprinted Mar. 1963.)

2 Carmer, S. G., and Seif, R. D., Calculation of orthogonal coefficients when treatments are unequally replicated and/or unequally spaced, Agron. J., 55, 387-389 (1963).

3 Fisher, R. A., and Yates, F., Statistical Tables for Biological, Agricultural and Medical Research, Oliver and Boyd, Edinburgh, (1938).

4 Grandage, A., Orthogonal coefficients for unequal intervals. Biometrics, 14, 287-289 (1958).

11
EXPERIMENTAL DESIGN

11.1 INTRODUCTION

It is important at this point to consider the manner in which the experimental data were collected as this greatly influences the choice of the proper technique for data analysis. Before going any further it is well to point out that the person performing the data analysis should be fully aware of several things:

(1) What was to have been found out?

(2) What is considered a significant answer?

(3) How are the data to be collected and what are the factors which influence the responses?

If an experiment has been properly designed or planned, the data will have been collected in the most efficient form for the problem being considered. Experimental design is the sequence of steps initially taken to insure that the data will be obtained in such a way that its analysis will lead immediately to valid statistical inferences. Before a design can be chosen the following questions must be answered:

(1) How is the effect to be measured?

(2) What factors influence the effect?

(3) How many of the factors will be considered at one time?

(4) How many replications (repetitions) of the experiment will be required?

(5) What type of data analysis is required (regression, AOV, etc.)?

(6) What level of difference in effects is considered significant?

The purpose of statistically designing an experiment is to collect the maximum amount of relevant information with a minimum expenditure of time and resources. It is important to remember also that the design of the experiment should be as simple as possible and consistent with the requirements of the problem. Three principles to be considered are replication, randomization, and control.

Replication is merely a complete repetition of the basic experiment. It provides an estimate of the magnitude of the experimental error. It also makes tests of significance of effects possible.

Randomization is the means used to eliminate any bias in the experimental units and/or treatment combinations. If the data are random it is safe to assume that they are independently distributed. Errors associated with experimental units which are adjacent in time or space will tend to be correlated, thus violating the assumption of independence. Randomization helps to make this correlation as small as possible so that the analyses can be carried out as though the assumption of independence were true.

An experimental unit is a unit to which a single treatment combination is applied in a single replication of the experiment. The term treatment (or treatment combinations) means the experimental conditions which are imposed on an experimental unit in a particular experiment. For example, in an experiment which studies a chemical process, the purpose may be to determine how temperature affects the response obtained from the process. Temperature would be a treatment. Another term, used interchangeably with treatment, is the term factor. This is especially true in the case of a factor which is quantitative, such as temperature. In such a case, the different values of the quantitative factor are referred to as levels of the factor. The factors are often called independent variables. In our example the different temperatures used in the experiment would be levels of the factor (or treatment) temperature. The one-way classification model studied in Chapter 8 can be referred to as a one-factor model. Similarly, the two-way classification model can be referred to as a two-factor experiment. Another example of a treatment is the amount of catalyst used in a chemical conversion process. A treatment combination could be the combination of a certain catalyst at a particular temperature. A treatment combination may consist of a single value of a single variable or it may represent a combination of factors.

Experimental error, which has previously been mentioned, measures the failure of two identical experiments to yield identical results. The experimental error is composed of many things: measurement errors, observation errors, variation in experimental units, errors associated with the particular experiment, and all factors not being studied which could possibly have an effect on the outcomes of the experiment.

Occasionally two or more effects in an experiment are <u>confounded</u>. That is, it is impossible to separate their individual effects when the statistical analysis of the data is performed. It might appear that confounding would be rigorously avoided but this is not always true. In complex experiments where several levels of many effects are to be used, confounding often produces a significant decrease in the size of the experiment. Of course, the analysis of the data is more difficult and the effectiveness of the experiment may be seriously impaired if any data are missing.

<u>Control</u> refers to the way in which the experimental units in a particular design are balanced, blocked, and grouped. <u>Balancing</u> means the assignment of the treatment combinations to the experimental units in such a way that a balanced or symmetric configuration is obtained. For example, in a two-factor experiment involving a certain mechanical process in textile production, the two

factors may be line tension and winding speed. If the same number of responses are obtained for each tension-speed combination then we say the design is balanced. Otherwise it is unbalanced or we simply say that there is missing data. Blocking is the assignment of the experimental units to blocks in such a manner that the units within any particular block are as homogeneous as possible. Grouping refers to the placement of homogeneous experimental units into different groups to which separate treatments may be assigned.

Example 11.1 Consider the problem of testing the effect of electrolyte (HBF_4) concentration on the available power of $Mn_3(PO_4)_2$ - Pb dry charge batteries. Suppose 20 batteries are available. In this case the experimental units are the batteries. Suppose there are four concentrations of HBF_4 available: 100, 90, 80, and 70. The treatment (or factor) is % HBF_4 and the levels are 100, 90, 80, and 70. The 20 batteries can be grouped so that there are 5 batteries corresponding to each concentration of HBF_4. (The batteries for each concentration are chosen at random.)

Example 11.2 Suppose an experiment is conducted to study the effect of fuel rate, burner angle, and steam-fuel ratio on the roof temperature of a steel furnace. In this case we have three factors (or treatments): (1) fuel rate, (2) burner angle, and (3) steam-fuel ratio. If two levels of each factor are used then we have what is commonly called a 3-factor factorial with two levels of each factor or a 2^3 factorial. In this case we must consider the added effect of the various combinations of two or three factors. These combination effects are called interactions.

In any experiment the number of replicates depends on the variance measured on a per experimental unit basis, the probability of committing both Type I and Type II errors, and the desired magnitude of the difference between two treatment means.

Consider the hypothesis H_0: $\mu_1 - \mu_2 = 0$. In this case, the statistic T_0 defined by

$$T_0 = \frac{\bar{X}_1 - \bar{X}_2 - 0}{\sqrt{2\sigma^2/r}} = \frac{\bar{X}_1 - \bar{X}_2 - \delta}{(2\sigma^2/r)^{1/2}} + \frac{\delta}{(2\sigma^2/r)^{1/2}}$$

has a normal distribution with mean 0 and variance 1. We will write this more compactly by saying "T_0 has an N(0,1) distribution" or "T_0 is N(0,1)." The alternative is H_A: $\mu_1 - \mu_2 = \delta$. Under this alternative, we have

$$T_A = \frac{\bar{X}_1 - \bar{X}_2 - \delta}{(2\sigma^2/r)^{1/2}} \tag{11.1}$$

which is N(0,1). The distribution of $T_0 = (\bar{X}_1 - \bar{X}_2)/\sqrt{2\sigma^2/r}$ is shown in Fig. 11.1 for both hypotheses in the case when $\delta > 0$.

Suppose that the level of significance is fixed at α. Note that, as $t_{1-\alpha}$ increases, Type I error decreases, but Type II error in-

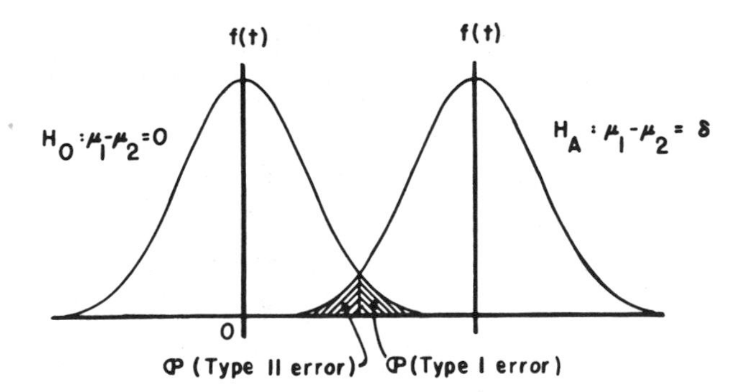

Fig. 11.1 Error types for testing differences between treatment means.

creases. In order for Type I error = α, we must have $(T_0 \geq t_{1-\alpha} | \delta=0) = \alpha$. If we also specify β = Type II error, then we must have

$$(T_0 < t_{1-\alpha} | \delta) = \left(T_A < t_{1-\alpha} - \frac{\delta}{(2\sigma^2/r)^{1/2}} \middle| \delta\right) = (T_A < t_\beta | \delta) = \beta \quad ,$$

according to Eq. (11.1). This implies

$$\frac{\delta}{(2\sigma^2/r)^{1/2}} = t_{1-\alpha} - t_\beta \quad . \tag{11.2}$$

As we usually do not know σ^2, we use S^2 as an estimator for it and determine the number of replicates required, r, by

$$r = \frac{2S^2}{\delta^2} (t_{1-\alpha} - t_\beta)^2 \quad . \tag{11.3}$$

The use of Eq. (11.3) to obtain the number of replications r required for a given probability P of obtaining a significant result requires a trial and error procedure as the degrees of freedom for the t-values $t_{1-\alpha}$ and t_β both depend on r. For those cases where the true difference δ and the true standard error per unit $\sigma/\sqrt{r}$ are known, the number of replicates required for 1- and 2-tailed tests may be readily found in Table 11.1 which has been taken from Stillito [4].

For the case where the number of replicates to be used is known, Eq. (11.3) may be rearranged and used to find the size of the true difference δ which can be detected provided that the true standard error per unit, σ, or its corresponding estimate, s, is known. The true difference δ which can be detected with probability β for a given size α is given by

$$\delta = \sigma\sqrt{2/r}\,(t_{1-\alpha} - t_\beta) \quad . \tag{11.4}$$

TABLE 11.1

Replications Required for a Given Probability of Obtaining a Significant Result for One-Class Data[a]

Upper figure: Test of significance at 5% level, probability 80%
Middle figure: Test of significance at 5% level, probability 90%
Lower figure: Test of significance at 1% level, probability 95%

One-tailed tests

True difference (δ) as percent of the mean	True standard error per unit (σ) as percent of the mean														
	2	3	4	5	6	7	8	9	10	11	12	14	16	18	20
5	3	6	9	13	19	25	33	41	50						
	4	7	12	18	26	35	45								
	7	13	22	33	47										
10	2	2	3	4	6	7	9	11	13	16	19	25	33	41	50
	2	3	4	5	8	9	12	15	18	22	26	35	45		
	3	4	7	9	13	17	22	27	33	40	47				
15	2	2	2	3	3	4	5	6	7	8	9	12	15	19	23
	2	2	3	3	4	5	6	7	9	10	12	16	21	26	31
	2	3	4	5	7	8	10	13	15	18	22	29	37	47	

(Continued)

TABLE 11.1 (continued)

True difference (δ) as percent of the mean	2	3	4	5	6	7	8	9	10	11	12	14	16	18	20
20	2	2	2	2	2	3	3	4	4	5	6	7	9	11	13
	2	2	2	2	3	3	4	5	5	6	7	9	12	15	18
	2	3	3	4	4	5	7	8	9	11	13	17	22	27	33
25	2	2	2	2	2	2	3	3	3	4	4	5	6	7	9
	2	2	2	2	2	3	3	3	4	5	5	7	8	10	12
	2	2	3	3	3	4	5	6	7	8	9	11	14	18	22
30	2	2	2	2	2	2	2	2	3	3	3	4	5	6	7
	2	2	2	2	2	2	3	3	3	4	4	5	6	7	9
	2	2	2	3	3	3	4	4	5	6	7	8	10	13	15
Two-tailed tests															
5	4	7	11	17	24	32	41								
	5	9	15	22	31	42									
	7	14	24	38											

10	2	3	4	5	7	9	11	14	17	20	24	32	41		
	2	3	5	7	9	12	15	18	22	27	31	42			
	3	5	7	11	14	19	24	30	37	45					
15	2	2	3	3	4	5	6	7	8	10	11	15	19	24	29
	2	2	3	4	5	6	7	9	11	13	15	19	25	31	39
	3	3	4	6	7	9	12	14	17	21	24	33	42		
20	2	2	2	3	3	3	4	5	5	6	7	9	11	14	17
	2	2	2	3	3	4	5	6	7	8	9	12	15	18	22
	2	3	3	4	5	6	7	9	11	12	14	19	24	30	37
25	2	2	2	2	2	3	3	3	4	4	5	6	7	9	11
	2	2	2	2	3	3	4	4	5	5	6	8	10	12	15
	2	2	3	3	4	5	5	6	7	9	10	13	16	20	24
30	2	2	2	2	2	2	3	3	3	4	4	5	6	7	8
	2	2	2	2	2	3	3	3	4	4	5	6	7	8	11
	2	2	3	3	3	4	4	5	6	6	7	9	12	14	17

Notes. In constructing the table, it was assumed that the number of degrees of freedom for error is 3(r - 1); this would apply in a randomized blocks experiment with 4 treatments.

[a]This table is reproduced from Stillitto, G. P., Research, 1948, 1, 520-525 by permission of Butterworth & Co. (Publishers) Limited ©.

The probability P of obtaining a significant result can be obtained from the formula

$$P = 1 - (1/2)p_{\beta} \quad , \qquad (11.5)$$

where p_{β} is the probability corresponding to t_{β} in the t-table. The quantities t_{β}, $t_{1-\alpha}$, δ, σ, and r are the same as used in Eqs. (11.3) and (11.4).

Example 11.3 In an experiment utilizing paired observations on colorimetric determination of Cr^{+7} at two different wavelengths, the estimate of the true standard error per unit was 8% of the overall mean. Ten replicate sets of data were obtained. The degrees of freedom available for estimating σ^2 was 24. What is the probability of obtaining a significant result when the true difference in means is 14% of the overall mean? Let the level of significance be $\alpha = 0.05$.

The estimate of the true standard error of the difference between the means, $\bar{d} = \bar{x}_1 - \bar{x}_2$, of the results obtained at different wavelengths is

$$(2s^2/r)^{1/2} = \sqrt{2(8)^2/10} = 3.578 \quad .$$

To test the significance of $\bar{d} = \bar{x}_1 - \bar{x}_2$, we use a two-tailed t-test with 24 degrees of freedom, for which t = 2.064. Therefore, if $\bar{d}$ is to be significant,

$$\bar{d} = \bar{x}_1 - \bar{x}_2 \geq \sqrt{\frac{2}{r}}\, st_{1-\alpha} = 3.758(2.064) = 7.385 \quad .$$

From Eq. (11.2), using the fact that we have a two-tailed test, we have

$$t_\beta = \frac{-\delta}{(2s^2/r)^{1/2}} + t_{1-\alpha/2}$$

$$= \frac{-14}{[2(8)^2/10]^{1/2}} + 2.064$$

$$= \frac{-14}{3.578} + 2.064$$

$$= -1.851 \quad .$$

From the t-table, Table IV, we find that the probability of T being less than -1.851 or greater than +1.851 is 0.08, that is, the probability of not finding a significant difference is 0.08. Thus, the probability of finding a significant difference is 1 - 0.08 = 0.92, which is also the power of the test. Consequently the chances are quite high that a significant difference between the means of the two types of determinations would be observed in this experiment.

Example 11.4 In Example 11.3 suppose $\alpha = 0.05$. How many replicates are required in order to have 4 out of 5 chances of obtaining a significant result if we wish to detect a true difference which is 15% of the mean.

We have $\alpha = 0.05$, σ = 8% of the mean, $\delta = 15$, and the probability of obtaining a significant result is 80%. Thus, since we have a two-tailed test, from Table 11.1 we see that $r = 6$ replicates are needed. The answer could also have been obtained by using Eq. (11.3).

In some experiments the experimental units fall into two categories, e.g., defective and nondefective. One might then be interested in comparing the percentages of units that fall into one of the classes under two different treatments. Table 11.2 gives the number of replications required in such experiments where the units fall into one of two classes.

Example 11.5 Two different methods of manufacturing compression rings for 1/4-in. copper tubing are being used. Method I produces

TABLE 11.2

Replications Required for a Given Probability of Obtaining a Significant Result for Two-Class Data[a]

Upper figure: Test of significance at 5% level, probability 80%
Middle figure: Test of significance at 5% level, probability 90%
Lower figure: Test of significance at 1% level, probability 95%

One-tailed tests

P_1 = smaller % success	$\delta = P_2 - P_1$ = larger *minus* smaller percentage of success													
	5	10	15	20	25	30	35	40	45	50	55	60	65	70
5	330	105	55	35	25	20	16	13	11	9	8	7	6	6
	460	145	76	48	34	26	21	17	15	13	11	9	8	7
	850	270	140	89	63	47	37	30	25	21	19	17	14	13
10	540	155	76	47	32	23	19	15	13	11	9	8	7	6
	740	210	105	64	44	33	25	21	17	14	12	11	9	8
	1370	390	195	120	81	60	46	37	30	25	21	19	16	14

15	710	200	94	56	38	27	21	17	14	12	10	8	7	6
	990	270	130	77	52	38	29	22	19	16	13	10	10	8
	1820	500	240	145	96	69	52	41	33	27	22	20	17	14
20	860	230	110	63	42	30	22	18	15	12	10	8	7	6
	1190	320	150	88	58	41	31	24	20	16	14	11	10	8
	2190	590	280	160	105	76	57	44	35	28	23	20	17	14
25	980	260	120	69	45	32	24	19	15	12	10	8	7	--
	1360	360	165	96	63	44	33	25	21	16	14	11	9	--
	2510	660	300	175	115	81	60	46	36	29	23	20	16	--
30	1080	280	130	73	47	33	24	19	15	12	10	8	--	--
	1500	390	175	100	65	46	33	25	21	16	13	11	--	--
	2760	720	330	185	120	84	61	47	36	28	22	19	--	--
35	1160	300	135	75	48	33	24	19	15	12	9	--	--	--
	1600	410	185	105	67	46	33	25	20	16	12	--	--	--
	2960	750	340	190	125	85	61	46	35	27	21	--	--	--
40	1210	310	135	76	48	33	24	18	14	11	--	--	--	--
	1670	420	190	105	67	46	33	24	19	14	--	--	--	--
	3080	780	350	195	125	84	60	44	33	25	--	--	--	--

(Continued)

TABLE 11.2 (Continued)

P_1 = smaller % success	5	10	15	20	25	30	35	40	45	50	55	60	65	70
45	1230	310	135	75	47	32	22	17	13	--	--	--	--	--
	1710	430	190	105	65	44	31	22	17	--	--	--	--	--
	3140	790	350	190	120	81	57	41	30	--	--	--	--	--
50	1230	310	135	73	45	30	21	15	--	--	--	--	--	--
	1710	420	185	100	63	41	29	21	--	--	--	--	--	--
	3140	780	340	185	115	76	52	37	--	--	--	--	--	--
Two-tailed tests														
5	420	130	69	44	31	24	20	16	14	12	10	9	9	7
	570	175	93	59	42	32	25	21	18	15	13	11	10	9
	960	300	155	100	71	54	42	34	28	24	21	19	16	14

10	680	195	96	59	41	30	23	19	16	13	11	10	9	7
	910	260	130	79	54	40	31	24	21	18	15	13	11	10
	1550	440	220	135	92	68	52	41	34	28	23	21	18	15
15	910	250	120	71	48	34	26	21	17	14	12	10	9	8
	1220	330	160	95	64	46	35	27	22	19	16	13	11	10
	2060	560	270	160	110	78	59	47	37	31	25	21	19	16
20	1090	290	135	80	53	38	28	22	18	15	13	10	9	7
	1460	390	185	105	71	51	38	29	23	20	16	14	11	10
	2470	660	310	180	120	86	64	50	40	32	26	21	19	15
25	1250	330	150	88	57	40	30	23	19	15	13	10	9	--
	1680	440	200	115	77	54	40	31	24	20	16	13	11	--
	2840	740	340	200	130	92	68	52	41	32	26	21	18	--
30	1380	360	160	93	60	42	31	23	19	15	12	10	--	--
	1840	480	220	125	80	56	41	31	24	20	16	13	--	--
	3120	810	370	210	135	95	69	53	41	32	25	21	--	--
35	1470	380	170	96	61	42	31	23	18	14	11	--	--	--
	1970	500	225	130	82	57	41	31	23	19	15	--	--	--
	3340	850	380	215	140	96	69	52	40	31	23	--	--	--

(Continued)

TABLE 11.2 (Continued)

P_1 = smaller success	% 5	10	15	20	25	30	35	40	45	50	55	60	65	70
40	1530	390	175	97	61	42	30	22	17	13	--	--	--	--
	2050	520	230	130	82	56	40	29	22	18	--	--	--	--
	3480	880	390	220	140	95	68	50	37	28	--	--	--	--
45	1560	390	175	96	60	40	28	21	16	--	--	--	--	--
	2100	520	230	130	80	54	38	27	21	--	--	--	--	--
	3550	890	390	215	135	92	64	47	34	--	--	--	--	--
50	1560	390	170	93	57	38	26	19	--	--	--	--	--	--
	2100	520	225	125	77	51	35	24	--	--	--	--	--	--
	3550	880	380	210	130	86	59	41	--	--	--	--	--	--

[a]This table has been reproduced from Table 2.2, pp. 24-25, in Experimental Design, 2nd ed., by W. G. Cochrane and G. M. Cox. Copyright 1957 by John Wiley & Sons, Inc. Reprinted by permission of John Wiley and Sons, Inc.

about 20% defectives. If method II produces as low as 5% the manufacturer wishes to be fairly certain of obtaining a significant result in comparing the two methods. Suppose the test is to be two-tailed at the 5% level and the manufacturer is satisfied with a probability of 0.80 of obtaining a significant result. How many rings are needed to carry out the test?

We have

P_1 = smaller percentage of defectives = 5%
δ = size of difference to be detected
= 20% - 5%
= 15% .

From Table 11.2, for two-tailed tests we read the upper figure corresponding to $P_1 = 5$ and $\delta = 15$ to obtain $r = 69$. The upper figure corresponds to $\alpha = 0.05$ and the given probability of 0.80. The size of each sample is 69, so that a total of 138 rings need to be obtained.

Various factors affect the size of an experimental unit. These are practical considerations which are used to determine the optimum amount of material to be studied or used in each of the optimum number of replicates, the nature of the experimental material, the nature of the treatments, and the cost of each of the experimental units.

To sum up, the requirements for a good experiment are that the treatment comparisons should:

(1) be free from systematic error;
(2) be made sufficiently precise, which depends on
  (a) the intrinsic variability of the experimental material,
  (b) the intrinsic variability of the measuring devices and techniques,
  (c) the number of experimental units available, and
  (d) the experimental design;
(3) have a wide range of validity;

(4) not be unnecessarily complicated, consistent with the desired results; and

(5) be easy to manipulate so as to obtain the standard error between two treatment combinations.

11.2 SOURCES OF ERROR

There are two important sources of experimental error. These are the variability of the material itself and poor experimental technique. The accuracy of the experiment can be increased by increasing the size of the experiment, by refining the experimental technique, and by handling the material so as to reduce the variability of the different effects. At this point, two terms often confused should be defined: accuracy is the approach to the true value of a datum point: precision is the repeatability of the results of a particular experiment. It should be noted here that when the degrees of freedom decrease, the limits of error for a true difference between treatment means are increased and the probability of getting a significant result is decreased accordingly.

Before starting into the discussion of experimental design it is well to reiterate the two basic assumptions used in making an analysis of variance. These are that the treatment and the environmental effects are additive and that the experimental errors are normally and independently distributed with mean 0 and variance σ^2. This is usually written as NID$(0,\sigma^2)$. This is saying that the observations comprise a random sample.

11.3 COMPLETELY RANDOMIZED DESIGNS

In completely randomized designs, the treatments are allocated entirely by chance. The design is completely flexible. Any number of treatments or replications may be used. You may vary the replications from treatment to treatment and all available experimental material can be utilized. Several advantages of this method are that the statistical analysis of the results is simple even for the case of missing data. The relative loss of information due to missing data is less for the completely random design than for any other design. All variation among the experimental units goes into the experimental error.

The completely randomized design should be used when the experimental material is homogeneous or in anticipation of missing data points or in small experiments when an increase in accuracy from other designs does not outweigh the loss in experimental error degrees of freedom.

11.3.1 Analysis of Variance

The analysis of variance in the completely randomized design is based on the model

$$Y_{ij} = \mu + \tau_j + \varepsilon_{ij}; \quad i = 1,2,\ldots,n; \quad j = 1,2,\ldots,t \; ; \qquad (11.6)$$

where the assumptions inherent in the model are linearity, normality, additivity, independence, and homogeneity of variances among the treatments. The model equation (11.6) is actually the one-way AOV model discussed in Chapter 8. There are a total of nt experimental units, and n units are assigned to each of t treatments. There are two models with which we are concerned: Model I requires $\sum_{j=1}^{t} \tau_j = 0$. This model is concerned only with the t treatments present in the experiment. Model II requires the τ_j to be random variables which are NID$(0,\sigma_\tau^2)$. In this case we are concerned with the population of treatments of which only t (which are sampled at random) are present in the experiment.

The hypothesis of interest in model I, fixed effects, is H_0: $\tau_1 = \tau_2 = \ldots = \tau_t = 0$, that is, that of no treatment effect. In model II, we are still interested in the hypothesis of no treatment effect, however, the τ_j are random variables with mean 0 and variance σ_τ^2. In this case the hypothesis of no treatment effect is H_0: $\sigma_\tau^2 = 0$.

The analysis of variance for completely randomized designs with equal numbers of observations per treatment, n, is shown in Table 11.3. For the estimation of σ_τ^2 which is of interest in the case of model II, we use the estimator S_τ^2 given by

$$S^2 = (MS_{Tr} - MS_E)/n \quad .$$

This follows from the expected mean squares in Table 11.3, $E(MS_{Tr}) = \sigma^2 + n\sigma_\tau^2$ and $E(MS_E) = \sigma^2$, which imply $E[MS_{Tr} - MS_E)/n] = \sigma_\tau^2$.

TABLE 11.3

Analysis of Variance for Completely Randomized Design with Equal Numbers of Observations per Treatment

Source	d.f.	SS	EMS
Mean	1	$(\sum_{ij}\sum Y_{ij})^2/tn = SS_M$	---
Among treatments	t-1	$\dfrac{\sum_{j=1}^{t}\left(\sum_{i=1}^{n} Y_{ij}\right)^2}{n} - SS_M = SS_{Tr}$	Model I: $\sigma^2 + n\dfrac{\sum_j^t \tau_j^2}{t-1}$ Model II: $\sigma^2 + n\sigma_\tau^2$
Within treatments	t(n-1)	$\sum_{ij}\sum Y_{ij}^2 - SS_M - SS_{Tr} = SS_E$	σ^2
Total	tn	$\sum_{ij}\sum Y_{ij}^2 = SS_T$	

To estimate the variance of the mean corresponding to the jth treatment, in the case of model I, we use

$$\text{VAR}(j^{th}\text{ mean}) = \hat{V}(\overline{Y}_{\cdot j}) = S^2_{\overline{Y}} = \frac{MS_E}{n} . \tag{11.7}$$

This and other ideas were presented in Chapter 8 in discussing the one-way AOV model.

A point estimate of the true mean effect of the jth treatment, $\hat{\mu}_j = \mu + \tau_j$, is

$$\hat{\mu}_j = \mu + \tau_j = \overline{y}_j \quad . \tag{11.8}$$

Example 11.6 A young summer engineer was assigned the task of testing the effect of electrolyte (HBF_4) concentration on the available power of $Mn_3(PO_4)_2$ - Pb dry charge batteries. He elected to perform the required experimental work in a completely random design. The following data were obtained from small scale tests:

Results, amps min ↓	Treatment, % HBF_4 →			
	40	30	20	10
	62	61	58	56
	64	60	60	54
	64	61	59	54
	62	61	58	55
	63	62	58	55

(a) Prepare a complete AOV table for model I and model II interpretations and describe their differences.

(b) Are there any significant differences between treatments?

(c) What is the variance of the difference between the second and fourth treatment means?

The treatment sums and means are as follows:

$y_{\cdot j}$:	315	305	293	274	for decreasing % HBF_4
$\overline{y}_{\cdot j}$:	63.0	61.0	58.6	54.8	for decreasing % HBF_4 .

Also,

$$y_{\bullet\bullet} = \sum_{ij}\sum y_{ij} = 1187; \quad \bar{y} = \text{overall mean} = 59.35 .$$

The sums of squares obtained from Table 11.3 are

$$SS_T = \sum_{ij}\sum y_{ij}^2 = 70,647.$$

$$SS_M = (\sum_{ij}\sum y_{ij})^2/nt = (1187)^2/20 = 70,448.45$$

$$SS_{Tr} = \sum_j (y_{\bullet j}^2)/n - SS_M ,$$

$$= \frac{(315)^2 + (305)^2 + (293)^2 + (274)^2}{5} - 70,448.45$$

$$= 186.55$$

$$SS_E = SS_T - SS_M - SS_{Tr}$$

$$= 70,647. - 70,448.45 - 186.55 = 12.00$$

(a) AOV Table:

Source	d.f.	SS	MS	EMS	Model
Mean	1	70,448.45	70,448.45	---	---
Treatments	3	186.55	62.18	$\sigma^2+\frac{5}{3}\sum_{j=1}^{4}\tau_j^2$	I
				$\sigma^2+5\sigma_\tau^2$	II
Error	16	12.0	0.75	σ^2	---
Total	20	70,647.			

Model I assumes 4 fixed treatments applied to fixed experimental units in a random manner. Model II assumes assignment of each of 4 randomly selected treatments to 5 randomly selected experimental units.

(b) H_0: $\tau_j = 0$ (or H_0: $\sigma_\tau^2 = 0$).

$$F_{3,16,0.99} = 5.29 \quad .$$

$$f = \frac{MS_{Tr}}{MS_E} = \frac{62.18}{0.75} = 82.41 \quad .$$

As $f > 5.29$, we reject the null hypothesis and conclude that there is very likely a difference in the available power related to HBF_4 concentration in the electrolyte.

(c) To calculate the variance of the difference between two treatment means, $2\sigma^2/r$, where r = number of replicates (in this case, r = 5), we first estimate σ^2 by $S^2 = MS_E$. Therefore, an estimate of the variance of the difference between the second and fourth treatment means is $2(0.75)/5 = 0.20$.

The analysis of variance in a completely randomized design with unequal numbers of observations per treatment is given in Table 11. where n_0 is defined as

$$n_0 = \frac{\sum_{j=1}^{t} n_j - \sum_{j=1}^{t} n_j^2 / \sum_{j=1}^{t} n_j}{t - 1} \quad . \tag{11.9}$$

The statistic $F = MS_{Tr}/MS_E$, which has an F-distribution with $(t-1)$ and $\Sigma(n_j-1)$ degrees of freedom, is used to test H_0: $\tau_1 = \tau_2 = \cdots = \tau_t = 0$ and H_0: $\sigma_\tau^2 = 0$, depending on which model is assumed. The use of this statistic was discussed in Chapter 8 together with the use of the AOV technique as a way of partitioning the total variation into several components, one corresponding to

TABLE 11.4

Analysis of Variance for Completely Randomized Design with Unequal Numbers of Observations per Treatment

Source	d.f.	SS	EMS
Mean	1	$(\sum_{ij}\sum Y_{ij})^2/\sum_j n_j = SS_M$	
Among treatments	t-1	$SS_{Tr} = \sum_j[(\sum_i Y_{ij})^2/n_j]-SS_M$	Model I: $\sigma^2+\frac{\sum_{j=1}^{t} n_j\tau_j^2}{t-1}$ Model II: $\sigma^2+n_0\sigma_\tau^2$
Experimental error	$\sum_j(n_j-1)$	$\sum_{ij}\sum Y_{ij}^2 - SS_M - SS_{Tr} = SS_E$	σ^2
Total	$\sum_j n_j$	$\sum_{ij}\sum Y_{ij}^2 = SS_T$	

each source of variation. Furthermore, it was seen in Chapter 8 that the expected mean squares can be used as a guide to the appropriate F-ratio to use in testing particular hypotheses. In this chapter we will not dwell on finding expected mean squares but will rather present numerous useful models and summarize the necessary techniques pertinent to each model in terms of an AOV table. Illustrative examples will also be presented.

11.3.2 Subsampling in a Completely Randomized Design

For the case of subsampling in a completely randomized design we use the model

$$Y_{ijk} = \mu + \tau_j + \varepsilon_{ij} + \eta_{ijk}; \quad i = 1,2,\ldots,n;$$
$$j = 1,2,\ldots,t; \quad k = 1,2,\ldots,m \quad , \qquad (11.10)$$

where we assume that the ε_{ij}, the experimental errors, are NID$(0,\sigma^2)$ and that the η_{ijk}, the sampling errors, are NID$(0,\sigma_\eta^2)$. ε_{ij} is the true effect of the ith unit subjected to the jth treatment and η_{ijk} is the effect of the kth sample taken from the ith unit which is subjected to the jth treatment. The analysis of variance for subsampling m times in a completely randomized design with n observations is as shown in Table 11.5. The experimental error variance σ^2 is estimated by

$$s^2 = \frac{\text{experimental error MS - sampling error MS}}{m} = \frac{MS_{EE} - MS_{SE}}{m}. \quad (11.11)$$

TABLE 11.5

Analysis of Variance for Subsampling in a Completely Randomized Design (Equal Subclass Numbers)

Source	d.f.	SS	EMS
Mean	1	$(\sum_{ijk}\sum\sum Y_{ijk})^2/tnm = SS_M$	---
Treatments	t-1	$\sum_j(\sum_{ik}\sum Y_{ijk})^2/mn - SS_M = SS_{Tr}$	$\begin{cases} EMS_I \\ EMS_{II} \end{cases}$
Experimental error	t(n-1)	$\sum_{ij}\sum(\sum_k Y_{ijk})^2/m - SS_M - SS_{Tr} = SS_{EE}$	$\sigma_\eta^2 + m\sigma^2$
Sampling error	tn(m-1)	$\sum_{ijk}\sum\sum Y_{ijk}^2 - SS_M - SS_{Tr} - SS_{EE} = SS_{SE}$	σ_η^2
Total	tnm	$\sum_{ijk}\sum\sum Y_{ijk}^2 = SS_T$	

The variance of a treatment mean is given by

$$\mathrm{VAR}(\overline{Y}_{.j.}) = S^2_{\overline{Y}.j.} = MS_{EE}/mn \quad . \tag{11.12}$$

Note that in Table 11.5 two expected mean squares, EMS_I and EMS_{II}, are designated for the treatments. These correspond to model I and model II as previously discussed. The mean squares for treatments for models I and II are

$$E(MS_I) = \sigma^2_\eta + m\sigma^2 + nm \sum_{j=1}^{t} \tau_j^2/(t-1)$$

and

$$E(MS_{II}) = \sigma^2_\eta + m\sigma^2 + nm\sigma^2_\tau \quad .$$

It should also be noted that the sum of squares for the sampling error is a pooled sum of squares among samples on the same experimental unit. The F-ratios for testing H_0: $\tau_j = 0$, $i = 1,2,\ldots,t$ and H_0: $\sigma^2_\tau = 0$ can be easily obtained by considering the expected mean squares.

Note that in a subsampling situation, n samples are taken and then each is divided into m subsamples. The subsamples do not constitute independent observations but merely provide an estimate of the sampling error as shown in Table 11.5. Repeated observations (temperature, flow rate, residence time, attendance in large lecture classes, etc.) by their very nature cannot be subdivided. As discussed later, such items lead to estimates of interactions between the independent variables. Consider the following example.

Example 11.7 Refractive indices of samples of the liquid on 3 trays in the methanol-water distillation column were measured in duplicate. The samples were taken simultaneously and divided prior to analysis. Is there a difference in the liquid composition (as indicated by the refractive index) on the trays or is any apparent difference due to sampling error?

Sample	Tray 1	Tray 2	Tray 3
1	1.3280	1.3375	1.3390
	1.3283	1.3375	1.3391
2	1.3300	1.3380	1.3389
	1.3329	1.3385	1.3391
3	1.3290	1.3380	1.3388
	1.3293	1.3379	1.3389

The SAS program (SAS User's Guide: Statistics, 1982 Edition) for this situation is below.

```
(put initial JCL cards/statements here)
DATA;
INPUT TRAY 1 SAMPLE 3 INDEX 5-10;
CARDS;
(data here according to INPUT format)
PROC PRINT;
PROC GLM;
CLASSES TRAY SAMPLE;
MODEL INDEX = TRAY SAMPLE(TRAY)/SS3;
TITLE 'SUBSAMPLING IN ONE-WAY AOV';
(put final JCL cards/statements here)
```

The results are below.

```
                              SAS

                  SUBSAMPLING IN ONE-WAY AOV

              GENERAL LINEAR MODELS PROCEDURE

                  CLASS LEVEL INFORMATION

                  CLASS      LEVELS     VALUES

                  TRAY          3       1 2 3

                  SAMPLE        3       1 2 3

          NUMBER OF OBSERVATIONS IN DATA SET = 18
```

SAS

SUBSAMPLING IN ONE-WAY AOV

GENERAL LINEAR MODELS PROCEDURE

DEPENDENT VARIABLE: SUB

SOURCE	DF	SUM OF SQUARES	MEAN SQUARE	F VALUE
MODEL	8	0.00032877	0.00004110	83.02
ERROR	9	0.00000446	0.00000050	PR > F
CORRECTED TOTAL	17	0.00033323		0.0001

SOURCE	DF	TYPE III SS	F VALUE	PR > F
TRAY	2	0.00031670	319.54	0.0001
SAMPLE(TRAY)	6	0.00001207	4.06	0.0299

From the f-values, we conclude that for this Model I situation (trays are a fixed, not a random variable), experimental error (SAMPLE(TRAY)) is first compared to sampling error (ERROR) by

$$f = MS_{EE}/MS_{SE} = 4.059 \cong 4.06$$

which indicates that the experimental error is significant (H_0: $\sigma^2 = 0$ is rejected) compared to sampling error. The treatment (TRAY) effects are tested by calculating

$$f = MS_{Tr}/MS_{EE} = 319.54$$

and comparing it to $F_{2,9,0.95} = 4.26$ we reject H_0: $\Sigma\tau_j^2 = 0$. The f = 319.54 in the SAS output was obtained by using the mean square for sampling error. The 82.3 version of the SAS base product does not compare error terms. You must pick the correct one and then compute the f-value for treatments.

11.3.3 Interaction

Two-way analysis of variance (and higher classifications) lead to the presence of interactions. If, for example, an additive A is added to a lube oil stock to improve its resistance to oxidation and another additive, B, is added to inhibit corrosion by the stock under load or stress, it is entirely possible that the performance of the lube oil in a standard ball-and-socket wear test will be different from that expected if only one additive were present. In other words, the presence of one additive may adversely or helpfully affect the action of the other additive in modifying the properties of the lube oil. These are termed antagonistic and synergistic effects, respectively. It is important to consider the presence of such interactions in any treatment of multiply classified data. To do this, the two-way analysis of variance table is set up as shown in Tables 11.6 and 11.7.

11.3.4 Model for Two-Way Analysis of Variance

In some cases, it is highly desirable to use factorial treatment combinations in completely randomized designs. Consider the case where T treatments are actually combinations of a levels of factor A and b levels of factor B. Then the statistical model is given as

$$Y_{ijk} = \mu + \alpha_i + \beta_j + (\alpha\beta)_{ij} + \varepsilon_{ijk} , \qquad (11.13)$$

$$i = 1,2,\ldots,a; \quad j = 1,2,\ldots,b; \quad k = 1,2,\ldots,n ,$$

where μ is the true mean effect, α_i is the true effect of the ith level of factor A, β_j is the true effect of the jth level of factor B, $(\alpha\beta)_{ij}$ is the true effect of the interaction of the ith level of factor A with the jth level of factor B, and ε_{ijk} is the true effect of the kth experimental unit subjected to the $(ij)^{th}$ treatment combination. This is actually the two-way AOV model of Chapter 8 with n observations per $\alpha\beta$ combination. The usual assumption is made that the ε_{ijk} are NID$(0,\sigma^2)$.

There are four possible sets of assumptions which can be made concerning the true treatment effects. The first set of assumptions is that we are concerned with only the fixed effects, α and β, of factors A and B. That is, we are concerned only with the a levels of factor A and b levels of factor B. This is our model I (fixed effects). These assumptions can be summarized by

$$\sum_i \alpha_i = \sum_j \beta_j = \sum_i (\alpha\beta)_{ij} = \sum_j (\alpha\beta)_{ij} = 0 \quad . \tag{11.14}$$

The analysis of variance for α and β fixed is given in Table 11.6. Compare Table 11.6 with Table 8.3.

TABLE 11.6

Analysis of Variance for Factorial Treatment Combinations in Completely Randomized Design, Model I

Source	d.f.	SS	EMS
Mean	1	$(\sum_{ijk}\sum\sum Y_{ijk})^2/abn = SS_M$	---
A	a-1	$\sum_i(\sum_{jk}\sum Y_{ijk})^2/bn - SS_M = SS_A$	$\sigma^2+nb\sum_i\alpha_i^2/(a-1)$
B	b-1	$\sum_j(\sum_{ik}\sum Y_{ijk})^2/an - SS_M = SS_B$	$\sigma^2+na\sum_j\beta_j^2/(b-1)$
AB	(a-1)(b-1)	$\sum_{ij}\sum(\sum_k Y_{ijk})^2/n-SS_M-SS_A-SS_B = SS_{AB}$	$\sigma^2+\frac{n\sum_{ij}\sum(\alpha\beta)_{ij}^2}{(a-1)(b-1)}$
Error	ab(n-1)	$\sum_{ijk}\sum\sum Y_{ijk}^2-SS_A-SS_B-SS_{AB}-SS_M = SS_E$	σ^2
Total	abn	$\sum_{ijk}\sum\sum Y_{ijk}^2 = SS_T$	

The second set of assumptions to be considered is that where α and β are random effects. This is model II. Here we are concerned with the population of levels of factor A of which only a random sample are present in the experiment and the population of levels of factor B again of which only a random sample are present in the experiment. These assumptions are summarized as follows:

$$\alpha_i \text{ are NID}(0,\sigma_\alpha^2) \quad , \tag{11.15a}$$

$$\beta_j \text{ are NID}(0,\sigma_\beta^2) \quad , \tag{11.15b}$$

$$(\alpha\beta)_{ij} \text{ are NID}(0,\sigma_{\alpha\beta}^2) \quad . \tag{11.15c}$$

The corresponding analysis of variance for this model is given in Table 11.7.

TABLE 11.7

Analysis of Variance for Factorial Treatment Combinations in a Completely Randomized Design, Model II

Source	d.f.	SS	EMS
Mean	1	$(\sum_{ijk}\sum\sum Y_{ijk})^2/abn = SS_M$	
A	a-1	$\sum_i(\sum_{jk}\sum Y_{ijk})^2/bn-SS_M = SS_A$	$\sigma^2+n\sigma_{\alpha\beta}^2+nb\sigma_\alpha^2$
B	b-1	$\sum_j(\sum_{ik}\sum Y_{ijk})^2/an-SS_M = SS_B$	$\sigma^2+n\sigma_{\alpha\beta}^2+na\sigma_\beta^2$
AB	(a-1)(b-1)	$\sum_{ij}\sum(\sum_k Y_{ijk})^2/n-SS_M-SS_A-SS_B = SS_{AB}$	$\sigma^2+n\sigma_{\alpha\beta}^2$
Error	ab(n-1)	$\sum_{ijk}\sum\sum Y_{ijk}^2-SS_A-SS_B-SS_{AB}-SS_M = SS_E$	σ^2
Total	abn	$\sum_{ijk}\sum\sum Y_{ijk}^2 = SS_T$	

The other two sets of assumptions yield what are called mixed models. In the third model (III) α is considered fixed and β is random, and in the fourth model (IV) α is considered random and β fixed. Assumptions underlying these two models are the following:

Model III:

$$\sum_i \alpha_i = \sum_i (\alpha\beta)_{ij} = 0, \quad \beta_j \text{ are NID}(0,\sigma_\beta^2); \tag{11.16a}$$

Model IV:

$$\sum_j \beta_j = \sum_j (\alpha\beta)_{ij} = 0, \tag{11.16b}$$

Examination of the expected mean squares in Tables 11.6 and 11.7 will indicate the proper F-test for hypotheses concerning the various treatments and treatment combinations. These are summarized in Table 11.8. For an example of a two-factor model see Example 8.4. In that

TABLE 11.8

F-Ratios for Hypothesis Testing in Completely Randomized Design with Factorial Treatment Combinations

Source	Model I	Model II	Model III	Model IV
Effects	a,b:fixed	a,b:random	a fixed b random	a random b fixed
Mean	---	---	---	---
A	MS_A/MS_E	MS_A/MS_{AB}	MS_A/MS_{AB}	MS_A/MS_E
B	MS_B/MS_E	MS_B/MS_{AB}	MS_B/MS_E	MS_B/MS_{AB}
AB	MS_{AB}/MS_E	MS_{AB}/MS_E	MS_{AB}/MS_E	MS_{AB}/MS_E
E	---	---	---	---

case n=1 and there was no estimate of σ^2. However, if the assumption of no interaction is used the $\sigma^2_{\alpha\beta}$ can be taken to be σ^2.

A few comments are in order regarding testing the significance of main effects in models which contain interactions. If H_0: $(\alpha\beta)_{ij} = 0$, i = 1,2,...a; j = 1,2,...b; is rejected, then H_0: $\alpha_i = 0$, i = 1,2,...a can be tested but the results usually are of no interest. When interactions are present, the best treatment combinations are of prime interest, rather than the best levels of A or B. If there is a significant interaction present, the acceptance of H_0: $\alpha_i = 0$ should be interpreted as meaning that there are no differences in the various levels of A when averaged over the levels of B. Similarly, if H_0: $\beta_j = 0$, j = 1,2,,,,b, is accepted, the interpretation is that there is no difference in the levels of B when they are averaged over the levels of A.

The use of the SAS System (SAS User's Guide: Statistics, 1982 Edition) to handle any two-way AOV situation is illustrated below. When multiple observations are available for each treatment combination (or "cell"), you can get the interaction term provided that those extra observations came from replications of the basic experiment and not from subsamples. To get the interaction term, add var1*var2 to the MODEL statement as shown in this example.

Example 11.8 The effects of 5-bromodeoxyuridine (A) and cytochlasin B (B) on normal development of sucrase specific activity in the embryonic chicken were studied. Thirty fertilized eggs were selected for study. Ten of the eggs remained uninjected (N) during incubation. Ten eggs were injected with 5-bromodioxyuridine prior to the 10th day of incubation and 10 were injected with cytochlasin B prior to the 10th day of incubation. The embryos were removed at embryonic age 13 days to 17 days and duodenal loop tissue was analyzed for specific invertase activity. Data representing g glucose released/mg protein/hr were obtained. What can be said concerning the age effect, the treatment effect, or any interaction?

Treatment	Embryonic age, days				
	13	14	15	16	17
N	56.3	60.5	64.3	86.1	97.5
	55.4	61.2	63.9	86.0	98.0
A	51.2	54.4	62.3	78.7	95.5
	50.8	55.6	61.7	78.3	96.5
B	55.4	61.5	62.0	72.8	81.9
	54.8	60.5	63.0	71.2	80.1

The program is below.

```
(put initial JCL cards/statements here)
DATA;
INPUT AGE 1-2 TRT $ 4 SPACT 6-9;
CARDS;
(data here according to INPUT format)
PROC PRINT;
PROC ANOVA;
CLASSES AGE TRT;
MODEL SPACT = AGE TRT AGE * TRT;
TITLE 'EFFECT OF AGE & TREATMENT ON SUCRASE SPECIFIC ACTIVITY';
(put final JCL cards/statements here)
```

The resulting analysis of variance shows by comparing the calculated f-values for this Model I situation with the tabular values that all the hypotheses expressed in Eq. (11.14) should be rejected.

Note that an interaction term (AGE*TRT) could be obtained because repeated observations for the same treatment combinations are available. The SAS System does not know the difference between Model I and Model II hypotheses so the appropriate f-values must be calculated from the mean square provided in the output.

SAS

ANALYSIS OF VARIANCE PROCEDURE

DEPENDENT VARIABLE: SPACT

SOURCE	DF	SUM OF SQUARES	MEAN SQUARE	F VALUE
MODEL	14	6440.83466667	460.05961905	1061.68
ERROR	15	6.50000000	0.43333333	PR > F
CORRECTED TOTAL	29	6447.33466667		0.0001

SOURCE	DF	ANOVA SS	F VALUE	PR > F
TRT	2	226.16266667	260.96	0.0001
AGE	4	5825.07466667	3360.62	0.0001
TRT*AGE	8	389.59733333	112.38	0.0001

For the situation when an unequal number or no observation exists for a cell, the PROC ANOVA command must be replaced by PROC GLM. The MODEL statement must also be modified by adding

/ SS1 SS2 SS3 SS4; after the interaction term.

Type I functions (SS1 output) add each source of variation sequentially to the model in the order listed in the MODEL statement. If the observations are balanced, Type I results are the same as those from PROC ANOVA. Type II functions (SS2 output) are adjusted sums of squares. For example, for effect U which may be a main effect or an interaction, SS2 is adjusted for the effect of V (another main effect) if and only if V does <u>not</u> contain U. As an example, main effects A and B are not adjusted for A*B because A*B contains both A and B. A <u>is</u> adjusted for the presence of B because B does not contain A. Similarly, B is adjusted for A and A*B is adjusted for A and B. These sums of squares are appropriate for the case where there are no interactions.

Type III functions (SS3 output) are weighted mean squares analyses and are used when it is necessary to compare main effects in the presence of suspected significant interactions. Each effect is adjusted for all others. Main effects A and B will be adjusted for A*B. Type IV functions (SS4 output) are used for missing data (empty cells). For unequal observations in the cells, specify SS3 as a Type III function is needed for the AOV. If data are entirely missing for a cell, specify SS4 for the Type IV function. For empty cells, omit the entire data line. Otherwise, the SAS System inserts 0.0 in the cell for the missing value.

Example 11.9 Results of a laboratory analysis for the specific rate constant for the saponification of ethyl acetate by NaOH at $0^{\circ}C$ are given below. One of the runs was performed with an electric stirrer, while the others were hand stirred. The results were analyzed by the differential and integral methods. Does the electric stirrer make a significant difference in the results? Does the method of analysis of data make a significant difference?

	With stirrer		Without stirrer	
Integral	0.02058	0.02121	0.01849	0.01816
method	0.02214	0.02073	0.01951	0.01884
Differential	0.01995	0.02003	0.01725	0.01752
method	0.01968	0.01982	0.01696	0.01726

The use of the SAS program for two-way analysis of variance with interaction gave the results below.

SAS

EFFECT OF ANALYSIS TECHNIQUE AND STIRRING METHOD ON RATE CONSTANT

ANALYSIS OF VARIANCE PROCEDURE

DEPENDENT VARIABLE: RATE

SOURCE	DF	SUM OF SQUARES	MEAN SQUARE	F VALUE
MODEL	3	0.00003452	0.00001151	71.03
ERROR	12	0.00000194	0.00000016	PR > F
CORRECTED TOTAL	15	0.00003646		0.0001

SOURCE	DF	ANOVA SS	F VALUE	PR > F
METHOD	1	0.00000643	39.67	0.0001
STIR	1	0.00002809	173.39	0.0001
METHOD*STIR	1	0.00000000	0.02	0.8936

We interpret the F-tests as follows for this Model I situation. As $F_{1,12,0.95} = 4.75$, we reject the first two hypotheses of Eq. (11.14) and accept the third. Thus we say that the rate constants are dependent on stirring and analysis method but that there is no apparent interaction between the two variables.

11.3.5 Model for the Three-way Analysis of Variance

In some instances, the data obtained are the result of three factors and their interactions. The general model for this situation is

$$Y_{ijk\ell} = \mu + \alpha_i + \beta_j + \gamma_k + (\alpha\beta)_{ij} + (\alpha\gamma)_{ik} + (\beta\gamma)_{jk} + (\alpha\beta\gamma)_{ijk} + \varepsilon_{ijk\ell}, \tag{11.17}$$

$i = 1,2,\ldots,a,\quad j = 1,2,\ldots,b,\quad k = 1,2,\ldots,c,\quad \ell = 1,2,\ldots,n$

TABLE 11.9

Three-Way Analysis of Variance

Source	d.f.	SS	MS	EMS
Between group means	a-1	$SS_G = \sum_i (y_{i...}^2/bcn) - (y_{....}^2/N)$	$SS_G/(a-1)$	$\sigma^2 + bc \sum_{i=1}^{a} \alpha_i^2/(a-1)$
Between block means	b-1	$SS_B = \sum_j (y_{.j..}^2/acn) - (y_{....}^2/N)$	$SS_B/(b-1)$	$\sigma^2 + ac \sum_{j=1}^{b} \beta_j^2/(b-1)$
Between treatment means	c-1	$SS_T = \sum_k (y_{..k.}^2/abn) - (y_{....}^2/N)$	$SS_T/(c-1)$	$\sigma^2 + ab \sum_{k=1}^{c} \gamma_k^2/(c-1)$
INTERACTIONS				
G×B	(a-1)(b-1)	$SS_{GB} = \sum\sum_{ij} (y_{ij..}^2/cn) - \sum_i (y_{i...}^2/bcn) - \sum_j (y_{.j..}^2/acn) + (y_{....}^2/N)$	$SS_{GB}/(a-1)(b-1)$	$\sigma^2 + c \sum_{i=1}^{a} \sum_{j=1}^{b} (\alpha\beta)_{ij}^2/(a-1)(b-1)$
G×T	(a-1)(c-1)	$SS_{GT} = \sum\sum_{ik} (y_{i.k.}^2/bn) - \sum_i (y_{i...}^2/bcn) - \sum_k (y_{..k.}^2/abn) + (y_{....}^2/N)$	$SS_{GT}/(a-1)(c-1)$	$\sigma^2 + b \sum_{i=1}^{a} \sum_{k=1}^{c} (\alpha\gamma)_{ik}^2/(a-1)(c-1)$

B×T	(b-1)(c-1)	$SS_{BT} = \sum_{jk}\sum (y^2_{\cdot jk\cdot}/an) - \sum_{j} (y^2_{\cdot j\cdot\cdot}/acn) - \sum_{k} (y^2_{\cdot\cdot k\cdot}/abn) + (y^2_{\cdot\cdot\cdot\cdot}/N)$	$SS_{BT}/(b-1)(c-1)$	$\sigma^2 + a \sum_{j=1}^{b} \sum_{k=1}^{c} (\beta\gamma)^2_{jk}/(b-1)(c-1)$
G×B×T	(a-1) × (b-1) × (c-1)	$SS_{GBT} = \sum_{ijk}\sum\sum (y^2_{ijk\cdot}/n) - \sum_{ij}\sum (y^2_{ij\cdot\cdot}/cn) - \sum_{ik}\sum (y^2_{i\cdot k\cdot}/bn) - \sum_{jk}\sum (y^2_{\cdot jk\cdot}/an) + \sum_{i} (y^2_{i\cdot\cdot\cdot}/bcn) + \sum_{j} (y^2_{\cdot j\cdot\cdot}/acn) + \sum_{k} (y^2_{\cdot\cdot k\cdot}/abn) - (y^2_{\cdot\cdot\cdot\cdot}/N)$	$\frac{SS_{GBT}}{(a-1)(b-1)(c-1)}$	$\sigma^2 + \frac{\sum_{i=1}^{a} \sum_{j=1}^{b} \sum_{k=1}^{c} (\alpha\beta\gamma)^2_{ijk}}{(a-1)(b-1)(c-1)}$
Residual	N-abc	$SS_R = \sum_{ijk\ell}\sum\sum\sum y^2_{ijk\ell} - \sum_{ijk}\sum\sum (y^2_{ijk\cdot}/n)$	$SS_R/(N-abc)$	σ^2
Total Corrected	N-1	$SS_{TC} = \sum_{ijk\ell}\sum\sum\sum y^2_{ijk\ell} - (y^2_{\cdot\cdot\cdot\cdot}/N)$	$SS_{TC}/(N-1)$	

where the three main effects α, β, and γ can be referred to as groups, blocks, and treatments. The random effects $\varepsilon_{ijk\ell}$ are assumed to be independently and normally distributed. The other terms denote the corresponding interactions among the three factors.

The assumptions for the three-way model involving the main effects and interactions are

$$\sum_i \alpha_i = \sum_j \beta_j = \sum_k \gamma_k = 0 \quad , \qquad (11.18a)$$

$$\sum_i (\alpha\beta)_{ij} = \sum_j (\alpha\beta)_{ij} = 0 \quad \text{[similarly for } (\alpha\gamma) \text{ and } (\beta\gamma)], \qquad (11.18b)$$

$$\sum_i (\alpha\beta\gamma)_{ijk} = \sum_j (\alpha\beta\gamma)_{ijk} = \sum_k (\alpha\beta\gamma)_{ijk} = 0 \quad . \qquad (11.18c)$$

The presentation of the data in tabular form is similar to the previous situations and will only be illustrated in the example to follow. The three-way AOV is summarized in Table 11.9. In this case the total number of observations is $N = abcn$ and the computational expressions for the sums of squares are given directly in the AOV table.

An example to illustrate these computations will now be given.

Example 11.10 The table below gives the results of an experiment on the amounts of niacin found in peas after three treatments (R_1, R_2, R_3) on blanched and processed peas of different sieve sizes (A, B, C).

Blanched Peas

	A			B			C		
	R_1	R_2	R_3	R_1	R_2	R_3	R_1	R_2	R_3
	65	90	44	59	70	83	88	80	123
	87	94	92	63	65	95	60	81	95
	48	86	88	81	78	88	96	105	100
	28	70	75	80	85	99	87	130	131
	20	78	80	76	74	81	68	122	121
	22	65	70	85	73	98	75	130	115
	24	75	73	64	61	95	96	121	127
	47	98	88	96	71	95	98	125	99
	28	95	77	91	63	90	84	172	101
	42	66	72	65	54	76	82	133	111
Sums	411	817	759	760	694	900	834	1199	1123

Processed Peas

	A			B			C		
	R_1	R_2	R_3	R_1	R_2	R_3	R_1	R_2	R_3
	62	106	126	150	138	150	146	52	100
	113	107	193	112	120	112	172	97	133
	171	79	122	136	135	126	138	112	125
	135	122	115	120	126	123	124	116	124
	123	125	126	118	120	125	113	121	115
	132	96	110	134	132	110	121	99	122
	120	111	98	125	135	125	125	120	112
	117	116	115	114	124	120	116	121	116
	153	124	112	112	137	110	165	122	99
	132	126	109	120	125	125	137	134	105
Sums	1258	1112	1226	1241	1292	1226	1357	1094	1151

From the table we get a = 2, b = 3, c = 3, n = 10, N = 180;

$$\Sigma\Sigma\Sigma\Sigma y_{ijk\ell}^2 = (65)^2 + \ldots + (105)^2 = 2{,}054{,}828;\ y^2_{....} = 340{,}550{,}116.$$

Thus SS_{TC} is found to be 162,883. The sums from the previous table are now placed in a table as follows:

	Blanched Peas		
	A	B	C
R_1	411	760	834
R_2	817	694	1199
R_3	759	900	1123
Sums	1987	2354	3156

	Processed Peas		
	A	B	C
R_1	1258	1241	1357
R_2	1112	1292	1094
R_3	1226	1226	1151
Sums	3596	3759	3602

The following three two-way tables can now be constructed by forming the appropriate sums from the above table.

	Blanched	Processed	Sums
A	1987	3596	5583
B	2354	3759	6113
C	3156	3602	6758
Sums	7497	10,957	

	Blanched	Processed
R_1	2005	3856
R_2	2710	3498
R_3	2782	3603

	A	B	C	Sums
R_1	1669	2001	2191	5861
R_2	1929	1986	2293	6208
R_3	1985	2126	2274	6385

The necessary statistics can now be calculated:

$$\Sigma y^2_{i\cdot\cdot\cdot} = (7497)^2 + (10{,}957)^2 = 176{,}260{,}858.$$

$$\Sigma y^2_{\cdot j\cdot\cdot} = (5583)^2 + (6113)^2 + (6758)^2 = 114{,}209{,}222.$$

$$\Sigma y^2_{\cdot\cdot k\cdot} = (5861)^2 + (6208)^2 + (6385)^2 = 113{,}658{,}810.$$

$$\Sigma\Sigma y^2_{ij\cdot\cdot} = (1987)^2 + \ldots + (3602)^2 = 59{,}485{,}522.$$

$$\Sigma\Sigma y^2_{i\cdot k\cdot} = (2005)^2 + \ldots + (3603)^2 = 59{,}189{,}998.$$

$$\Sigma\Sigma y^2_{\cdot jk\cdot} = (1669)^2 + \ldots + (2274)^2 = 38{,}144{,}306.$$

$$\Sigma\Sigma\Sigma y^2_{ijk\cdot} = (411)^2 + \ldots + (1151)^2 = 20{,}073{,}704.$$

Dividing by the appropriate degrees of freedom and combining the results according to Table 11.9 we get the following:

$$SS_{TC} = 2{,}054{,}828 - 1{,}891{,}945 = 162{,}883.$$

$$SS_G = 1{,}958{,}454 - 1{,}891{,}945 = 66{,}509.$$

$$SS_B = 1{,}903{,}487 - 1{,}891{,}945 = 11{,}542.$$

$$SS_T = 1{,}894{,}314 - 1{,}891{,}945 = 2{,}369.$$

$$SS_{GB} = 1{,}982{,}851 - 1{,}958{,}454 - 1{,}903{,}487 + 1{,}891{,}945 = 12{,}855.$$

$$SS_{GT} = 1{,}973{,}000 - 1{,}958{,}454 - 1{,}894{,}314 + 1{,}891{,}945 = 12{,}177.$$

$$SS_{BT} = 1{,}907{,}215 - 1{,}903{,}487 - 1{,}894{,}314 + 1{,}891{,}945 = 1{,}359.$$

$$SS_{GBT} = 2{,}007{,}370 - 1{,}982{,}851 - 1{,}973{,}000 - 1{,}907{,}215 + 1{,}958{,}454 + 1{,}903{,}487 + 1{,}894{,}314 - 1{,}891{,}945 = 8{,}614.$$

$$SS_R = 2{,}054{,}828 - 2{,}007{,}370 = 47{,}458.$$

Summarizing these results in an AOV table we have:

Source	d.f.	SS	MS
Between group means	1	66,509	66,509.00
Between block means	2	11,542	5,771.00
Between treatments means	2	2,369	1,184.50

Interactions			
G × B	2	12,855	6,427.50
G × T	2	12,177	6,088.50
B × T	4	1,359	339.75
G × B × T	4	8,614	2,153.5
Residual	162	47,458	292.95
Total Corrected	179	162,883	909.96

It is clear from the EMS column of Table 8.5 which quotients need to be formed to test hypotheses about various parameters in the model. For example, to test the hypothesis H_0: $(\alpha\beta\gamma)_{ijk} = 0$ for all i, j, and k we use the statistic $F = MS_{GBT}/MS_R$ since this statistic has an F-distribution if and only if H_0 is true. The rejection region is $f > F_{\nu_1,\nu_2,1-\alpha}$ where $\nu_1 = (a-1)(b-1)(c-1)$ and $\nu_2 = N - abc$.

In our example $MS_{GBT} = 2,153.5$, $MS_R = 292.95$ and $f = 2,153.5/292.95 = 7.351$. Now $F_{4,162,0.95} = 2.37$ and we reject H_0 since $7.351 > 2.37$.

11.3.6 Four-way Analysis of Variance

We now consider an example of a 4-factor model, each factor at two levels.

Example 11.11 A completely randomized experiment was conducted to study the effect on the roof temperature Y of a steel furnace, of the following four factors, each at two levels:

F: fuel rate (α)
O: oxygen in waste gas (β)
B: burner angle (γ)
S: steam-fuel ratio (δ)

The model for this experiment is supposed to be adequately represented by

$$Y_{ijk\ell} = \mu + \alpha_i + \beta_j + \gamma_k + \delta_\ell + (\alpha\beta)_{ij} + (\alpha\gamma)_{ik} + (\alpha\delta)_{i\ell}$$

$$+ (\beta\gamma)_{jk} + (\beta\delta)_{j\ell} + (\gamma\delta)_{k\ell} + \varepsilon_{ijk\ell} \ .$$

The key feature of this assumed model is of course that all population interactions of order higher than 1 are assumed to equal zero, which means that the corresponding single-degree-of-freedom mean squares are assumed to estimate only error.

The observed values of roof temperature were as follows where subscript 1 refers to the low-level and 2 to the high level of the variable involved:

$y_{1111} = 1138$	$y_{1211} = 1065$	$y_{2111} = 1152$	$y_{2211} = 1125$
$y_{1112} = 1206$	$y_{1212} = 1103$	$y_{2112} = 1309$	$y_{2212} = 1373$
$y_{1121} = 1082$	$y_{1221} = 1046$	$y_{2121} = 1091$	$y_{2221} = 1129$
$y_{1122} = 1198$	$y_{1222} = 1168$	$y_{2122} = 1359$	$y_{2222} = 1301$.

On the basis of these observations, prepare a complete AOV table. The sums of squares and expected mean squares for selected main effects and interactions are shown below as a guide.

$$SS_F = \frac{\sum_i y_{i\cdots}^2}{8} - \frac{y_{\cdots\cdot}^2}{16} \ ,$$

$$EMS_F = \sigma^2 + 8\sum_i \alpha_i^2 \ ,$$

$$SS_{FO} = \frac{\sum\sum_{ij} y_{ij\cdot\cdot}^2}{4} - \frac{\sum_i y_{i\cdots}^2}{8} - \frac{\sum_j y_{\cdot j\cdot\cdot}^2}{8} + \frac{y_{\cdots\cdot}^2}{16} \ ,$$

$$EMS_{FO} = \sigma^2 + 4\sum_{ij}\sum(\alpha\beta)^2_{ij} ,$$

$$SS_{FOB} = \frac{\sum\sum\sum_{ijk} y^2_{ijk.}}{2} - \frac{\sum\sum_{ij} y^2_{ij..}}{4} - \frac{\sum\sum_{ik} y^2_{i.k.}}{4} - \frac{\sum\sum_{jk} y^2_{.jk.}}{4}$$

$$+ \frac{\sum_i y^2_{i...}}{8} + \frac{\sum_j y^2_{.j..}}{8} + \frac{\sum_k y^2_{..k.}}{8} - \frac{y^2_{....}}{16} ,$$

$$EMS_{FOB} = \sigma^2 ,$$

$$SS_{FOBS} = \sum\sum\sum\sum_{ijk\ell} y^2_{ijk\ell} - \frac{\sum\sum\sum_{ijk} y^2_{ijk.}}{2} - \frac{\sum\sum\sum_{ik\ell} y^2_{i.k\ell}}{2}$$

$$- \frac{\sum\sum\sum_{jk\ell} y^2_{.jk\ell}}{2} - \frac{\sum\sum\sum_{ij\ell} y^2_{ij.\ell}}{2} + \frac{\sum\sum_{ij} y^2_{ij..}}{4} + \frac{\sum\sum_{ik} y^2_{i.k.}}{4}$$

$$+ \frac{\sum\sum_{i\ell} y^2_{i..\ell}}{4} + \frac{\sum\sum_{k\ell} y^2_{..k\ell}}{4} + \frac{\sum\sum_{j\ell} y^2_{.j.\ell}}{4} - \frac{\sum_i y^2_{i...}}{8}$$

$$- \frac{\sum_j y^2_{.j..}}{8} - \frac{\sum_k y^2_{..k.}}{8} - \frac{\sum_\ell y^2_{...\ell}}{8} + \frac{y^2_{....}}{16} ,$$

$$EMS_{FOBS} = \sigma^2 .$$

In addition,

1. Test the plausibility of the assumption regarding error by an F-test pitting the four degrees of freedom for second-order interactions against the single degree of freedom for the third-order interaction. A first-order interaction contains 2 factors, a second-order interaction contains 3 factors, etc.

2. If things turn out satisfactorily in (1), pool the five degrees of freedom for error as planned and test all ten main effects and first-order interactions against this pooled error.

AOV Table

Source	d.f.	SS(=MS as d.f.=1)	EMS
F	1	43,368.06	$\sigma^2 + 8\sum_i \alpha_i^2$ *
O	1	3,164.06	$\sigma^2 + 8\sum_j \beta_j^2$
B	1	588.06	$\sigma^2 + 8\sum_k \gamma_k^2$
S	1	88,357.06	$\sigma^2 + 8\sum_\ell \delta_\ell^2$ *
FO	1	4,192.56	$\sigma^2 + 4\sum\sum_{ij} (\alpha\beta)_{ij}^2$
FB	1	232.56	$\sigma^2 + 4\sum\sum_{ik} (\alpha\gamma)_{ik}^2$
FS	1	15,687.56	$\sigma^2 + 4\sum\sum_{i\ell} (\alpha\delta)_{i\ell}^2$ *
OB	1	175.56	$\sigma^2 + 4\sum\sum_{jk} (\beta\gamma)_{jk}^2$
OS	1	52.56	$\sigma^2 + 4\sum\sum_{j\ell} (\beta\delta)_{j\ell}^2$
BS	1	1,743.06	$\sigma^2 + 4\sum\sum_{k\ell} (\gamma\delta)_{k\ell}^2$
FOB	1	1,743.06	σ^2
FOS	1	22.56	σ^2
FBS	1	588.06	σ^2
OBS	1	1,425.06	σ^2
FOBS	1	3,108.06	σ^2
	15	164,448.44	

* = Significant at 0.05 level

We have

1. $$\frac{MS_{FOB} + MS_{FOS} + MS_{FBS} + MS_{OBS}}{4} = \frac{3778.74}{4} = 944.685 = S_1$$

$$\frac{S_1}{MS_{FOBS}} = \frac{944.685}{3108.06} = 0.30394 \text{ which is less than } F_{4,1,0.95} = 225$$

so we accept H_0: interactions of order greater than 1 estimate error only.

2. $$\frac{MS_{FOB} + MS_{FOS} + MS_{FBS} + MS_{OBS} + MS_{FOBS}}{5} = \frac{6886.8}{5} = 1377.3 = S_2$$

$F_{1,5,0.95} = 6.61$

$\frac{MS_F}{S_2} = \frac{43368.06}{1377.3} = 31.48$ so we reject H_0: fuel rate is not significant.

$\frac{MS_0}{S_2} = \frac{3164.06}{1377.3} = 2.297$ so we accept H_0: oxygen content in waste gas is not significant.

$\frac{MS_B}{S_2} = \frac{558.06}{1377.3} = 0.427$ so we accept H_0: burner angle is not significant.

$\frac{MS_S}{S_2} = \frac{88356.56}{1377.3} = 64.151$ so we reject H_0: steam/fuel ratio is not significant.

$\frac{MS_{F0}}{S_2} = \frac{4192.56}{1377.3} = 3.044$ so we accept H_0: interaction between fuel rate and oxygen content is not significant.

$\frac{MS_{FB}}{S_2} = \frac{232.56}{1377.3} = 0.169$ so we accept H_0: interaction between fuel rate and burner angle is not significant.

$\frac{MS_{FS}}{S_2} = \frac{15,687.56}{1,377.3} = 11.390$ so we reject H_0: interaction between fuel rate and steam ratio is not significant.

$\frac{MS_{OB}}{S_2} = \frac{175.56}{1377.3} = 0.127$ so we accept H_0: interaction between oxygen content and burner angle is not significant.

$\frac{MS_{OS}}{S_2} = \frac{52.56}{1377.3} = 0.038$ so we accept H_0: interaction between oxygen content and steam/fuel ratio is not significant.

$\frac{MS_{BS}}{S_2} = \frac{1743.06}{1377.3} = 1.266$ so we accept H_0: interaction between burner angle and steam/fuel ratio is not significant.

11.3.7 Nested Designs

In experimental designs such as the two-way and three-way AOV designs, the factors are "crossed" with each other, i.e., there are observations present for each treatment combination. For example, in the two-way AOV design in (11.3.4), every level of factor A occurs with every level of factor B. There are situations where the factors are not crossed but factors may be "nested" inside one another.

In comparing 4 methods for doing a chemical analysis, suppose that a batch of a certain substance (chemical) is split into 60 parts with 15 parts allocated to each method. For each method, 3 technicians will perform 5 replicate analyses on the chemical. The total number of observations obtained will be 4(3)(5) = 60. If there are 3 different technicians for each method (that is, a total of 15 technicians) then the technician effect is nested within method. If the same 3 technicians were used across methods, then the method and technician factors would be crossed.

As another example, consider an experiment intended to measure a certain flow characteristic Y at a certain location in a whirl pool tank, as a function of I nozzle apertures. In this case there might be J runs for each nozzle aperture. That is, the run effect is nested within aperture.

The model for 2 factors A and B where factor B is nested within factor A is

$$Y_{ijk} = \mu + \alpha_i + \beta_{ij} + \varepsilon_{ijk} \quad ;$$

$$i = 1,2,..,a; \quad j = 1,2,...,b; \quad k = 1,2,...,n \quad ; \qquad (11.19)$$

where μ is the main effect, α_i is the true effect of the ith level of A, β_{ij} is the true effect of the jth level of factor B nested within the ith level of factor A, and ε_{ijk} is the random error corresponding to the kth unit for the jth level of B within the ith level of A. We assume ε_{ijk} are NID(0, σ^2). We further assume the design is balanced.

The two-factor nested design is sometimes referred to as a two-fold nested design. The AOV sums of squares and associated degrees of freedom are the same as those in the AOV table for subsampling in a completely randomized design in Table 11.5. Replacing t with a, n with b, and m with n yields the AOV table for the two-fold nested design. One main difference is that factors A and B may both be fixed in the two-fold nested design whereas only factor A may have that option in the completely randomized subsampling design. The expected mean squares for the two-fold nested design when both A and B are fixed are:

$$E(MS_A) = \sigma^2 + nb \sum_{i=1}^{a} \alpha_i^2/(I-1) \quad ,$$

$$E(MA_{B(A)}) = \sigma^2 + n \sum_{i=1}^{a} \sum_{j=1}^{b} \beta_{ij}^2/(I(J-1)), \text{ and}$$

$$E(MS_E) = \sigma^2 \quad .$$

The F-ratio for testing the appropriate hypotheses in the two-fold nested design are:

A and B fixed: $F = MS_A/MS_E$

A and B random: $F = MS_A/MS_{B(A)}$ and

A fixed, B random: $F = MS_A/MS_{B(A)}$

The case of A random and B fixed is rarely used and is thus not considered here.

If there are 3 factors (A, B, and C), then C could be nested within B, and the combination of C and B could be nested within A. This would be a three-fold nested design. Designs with more factors could be defined accordingly. Nested designs are also known as hierarchical designs. Furthermore, there are designs which contain factors, some of which are crossed, and some of which are nested. For example, in a three-factor experiment A and B could be crossed and that combination nested within C. Such designs will not be covered in this book.

We now consider an example for a two-fold nested design.

Example 11.12 Consider the following AOV table, computed on the basis of an experiment intended to measure a certain flow characteristic Y at a certain location in a whirlpool tank, as a function of I nozzle apertures:

Source	d.f.	MS	EMS
Among nozzle apertures	5	12,489	$\sigma_\varepsilon^2 + 3\sigma_\beta^2 + 30\sigma_\alpha^2$
Among runs, within nozzle apertures	54	3,339	$\sigma_\varepsilon^2 + 3\sigma_\beta^2$
Among measurements, within runs	120	627	σ_ε^2

1. Write out the model equations that probably were assumed for this experiment.
2. How many nozzle settings were used? Runs per setting? Measurements per run?
3. State, as precisely as you can, what μ, α_i, β_{ij}, and ε_{ijk} signify for this experiment.
4. Perform the F-test suggested by the AOV table of the null hypothesis that nozzle aperture has no effect on the flow characteristic Y.
5. Perform the F-test of the null hypothesis that σ_β^2 is three times as large as σ_ε^2 (use a two-sided F-test).
6. Give a CI for $\sigma_\beta^2/\sigma_\varepsilon^2$.
7. Estimate $V[\bar{Y}_{3..}]$.
8. Assuming that $Y_{3..} = 193.7$, find a CI for $\mu_3 = \mu + \alpha_3$.
9. Assume that the estimates of σ_ε^2 and σ_β^2 are exact. Assume that the overhead cost of a run is \$2.00, and that the cost of a measurement is 50¢. Find the most efficient of all experimental plans costing as much as the given plan.

Solution

1. $Y_{ijk} = \mu + \alpha_i + \beta_{ij} + \varepsilon_{ijk}$.
2. I = 6 = nozzle settings,

   J = 10 = number of runs per setting,

   K = 3 = measurements per run.
3. μ = overall population mean of entire experiment,

   α_i = contribution to Y_{ijk} due to treatment (nozzle settings),

   β_{ij} = contribution to Y_{ijk} due to run to run variations within any one treatment,

   ε_{ijk} = contribution to Y_{ijk} due to measurement variation in a run at any treatment level.

4. $\frac{MS_{n.a.}}{MS_{run}} = \frac{12489}{3339} = f = 3.7403.$

$H_0: \sigma_\alpha^2 = 0$.

$f < F_{5,54,0.95}$ is the test criterion and as 3.7403 is not $< F_{5,54,0.95} = 2.384$, we reject H_0.

5. $H_0: \sigma_\beta^2/\sigma_\varepsilon^2 = 3$ or: $H_0: \sigma_\beta^2 = 3\sigma_\varepsilon^2$.

$$\frac{MS_{runs}/(\sigma_\varepsilon^2 + 3\sigma_\beta^2)}{MS_{meas.}/\sigma_\varepsilon^2} = \frac{MS_{runs}/10}{MS_{meas.}} = \frac{\chi^2_{runs}/10}{\chi^2_{meas.}} = f \quad .$$

$$f = \frac{3339/10}{627} = \frac{333.9}{627} = 0.5325 \quad .$$

$F_{54,120,0.05} < f < F_{54,120,0.95}$ is the test criterion.

$\frac{1}{F_{120,54,0.95}} < f < F_{54,120,0.95}$ is found from tabular F-values as $\frac{1}{1.488} < 0.5325 < 1.464$ which reduces to the partial inequality $0.672 \not< 0.5325 < 1.464$, so we reject H_0 and conclude that there is a significant variation due to replications (runs).

6. $H_0: \sigma_\beta^2/\sigma_\varepsilon^2 = \ell.$

Accept if

$$F_{54,120,0.05} < \frac{MS_{runs}/(1+3\ell)}{MS_{meas.}} < F_{54,120,0.95}$$

$$0.672 < \frac{5.32536}{1+3\ell} < 1.464 \quad .$$

For 90% CI limits, $\frac{5.32536}{1+3\ell} = 1.464$ so $\ell_L = 0.879$

and $\frac{5.32536}{1+3\ell} = 0.672$ so $\ell_U = 2.308$ to yield

$0.879 < \ell < 2.308$ as the 90% CI for ℓ.

7. Estimate $V[\overline{Y}_{3}..] = \frac{MS_{runs}}{JK} = \frac{3339}{30} = 111.3$.

8. $E[\overline{Y}_{3}..] = \mu_3 = \mu + \alpha_3$.

$$\mu_3 : \overline{Y}_3 \pm t_{54,0.05}\sqrt{\frac{MS_{runs}}{JK}} ,$$

$$\mu_3 : 193.7 \pm t_{54,0.05}\sqrt{\frac{3339}{30}} ,$$

$$\mu_3 : 193.7 \pm t_{54,0.05} \sqrt{111.3} ,$$

$$\mu_3 : 193.7 \pm 2.0043(10.55) ,$$

$$\mu_3 : 193.7 \pm 21.14 ,$$

$172.6 < \mu_3 < 214.8$ is a 95% CI for μ_3.

9. Cost/nozzle opening = \$2/run (10 runs) + $\frac{\$0.5}{\text{meas.}}\left(\frac{3\text{ meas.}}{\text{run}}\right)$ (10 runs) = \$35.

Must minimize $\frac{\sigma_\beta^2}{J} + \frac{\sigma_\varepsilon^2}{JK} = Var[\overline{Y}_{3..}] = 111.3$ subject to $2J + JK/2 = 35$.

$$\sigma_\varepsilon^2 = 627 .$$

$$\sigma_\beta^2 = \frac{3339-627}{3} = \frac{2712}{3} = 904 .$$

$$\frac{904}{J} + \frac{627}{JK} = 111.3 \text{ or } 904K + 627 = 111.3JK .$$

Also: $2J + \frac{JK}{2} = 35$ or $K = \frac{70 - 4J}{J}$,

so $\frac{904(70-4J)}{J} + 627 = 111.3\, J \left(\frac{70 - 4J}{J}\right)$

and: $904(70 - 4J) + 627J = 111.3(70J - 4J^2)$

or: $445.2J^2 - 10780J + 63280 = 0$.

$\frac{dJ}{dK} = 0 = 890.4J - 10780$, so $J = 10780/890.4 = 12.107$

and $K = 1.782$.

As fractional runs and measurements are impossible, use $J = 12$, $K = 2$ for best results costing no more than $35.

11.4 RANDOMIZED COMPLETE BLOCK DESIGN

In the case of the randomized complete block design, the experimental units are divided into groups, each of which is a single trial or replication. Here the major variations are kept between blocks so as to keep the experimental error in each group small. Because of this grouping or blocking, as it is more often called, we get more accuracy than with the completely randomized design. In the randomized complete block design any number of treatments or replications may be used. The statistical analysis is easy even for missing data. In the case where the error sum of squares for some treatments is larger than for other treatments we can still get an unbiased estimate for testing any specific combination of treatment means. This is a decided advantage. It should be noted here that the replications provide unbiased comparisons of differences between replications. The method of using the randomized complete block design is to group the experimental units into blocks and then randomly assign the treatments within each block.

The basic assumption for a randomized complete block design with one observation per experimental unit is that the observations may be represented by the model given as

$$Y_{ij} = \mu + \tau_i + \beta_j + \varepsilon_{ij},$$
$$i = 1,2,\ldots,t; \quad j = 1,2,\ldots,b, \tag{11.20}$$

where μ is the true mean effect, β_j is the true effect of the jth block, and τ_i is the true effect of the ith treatment. The additional assumption made is that the ε_{ij} are NID$(0,\sigma^2)$.

11.4.1 Analysis of Variance, RCB

The analysis of variance for a randomized complete block design with one observation per experimental unit is given in Table 11.10. As in the case of the completely randomized design, either model I or model II may be assumed with respect to the β_j and τ_i. The hypothesis H_0: $\tau_i = 0$ is of primary importance and may be tested by computing

$$f = \frac{\text{MS for treatments}}{MS_E} \quad . \tag{11.21}$$

TABLE 11.10

Analysis of Variance of Randomized Complete Block Design with One Observation per Experimental Unit

Source	d.f.	SS	EMS
Mean	1	$(\sum_{ij}\sum Y_{ij})^2/bt = SS_M$	---
Blocks	b-1	$\sum_j B_j^2/t - SS_M = SS_B$	$\sigma^2 + t\sum_j \beta_j^2/(b-1)$
Treatments	t-1	$\sum_i T_i^2/b - SS_M = SS_{Tr}$	$\sigma^2 + b\sum_i \tau_i^2/(t-1)$
Error	(b-1)(t-1)	$\sum_{ij}\sum Y_{ij}^2 - SS_M - SS_B - SS_{Tr} = SS_E$	σ^2
Total	bt	$\sum_{ij}\sum Y_{ij}^2$	

In Table 11.10 the B_j are the individual block sums ($Y_{.j}$) and the T_i are the individual treatment sums ($Y_{i.}$).

If the value of f so computed by Eq. (11.21) exceeds $F_{\nu_1,\nu_2,1-\alpha}$ then the hypothesis will be rejected and the conclusion is that there are significant differences among the treatments. The hypothesis H_0: $\beta_j = 0$, although of minor importance, may be tested in a similar manner using F = (MS for blocks)/MS_E. However, due to the manner in which the experiment is set up, the hypothesis H_0: $\beta_j = 0$ should not be tested except as a check on the blocking of the experiment. The whole purpose of a RCB design is to reduce experimental error and get a more efficient test of H_0: $\tau_i = 0$, $i = 1,2,\ldots,t$. It is wise to check the efficacy of blocking by an F-test on the blocks. If that test gives an insignificant f-value, the blocking was almost certainly faulty. In that case, the entire experiment should be repeated with more careful attention to assignment of the treatment to the experimental units.

Example 11.13 In an effort to determine the effect of temperature on degradation of a particular polymer proposed for use in manned spacecraft, 5 temperature levels were used in a randomized complete block design. The temperature effect was measured as ppm (parts per million) total hydrocarbons produced for samples of consistent mass. The tests were conducted using 5 g sample/liter in each test chamber to simulate the weight to volume ratio expected in the actual mission. Five replicates were made at each temperature. The means of the data were as follows:

t, ^{o}C	Total ppm hydrocarbons
0	20
10	140
20	260
30	300
40	280

Compute the AOV table, and if temperature is significant, quantify the temperature-hydrocarbon emission results by the method of orthogonal polynomials. The calculations below were made using all the data, not the means.

AOV Table

Source	d.f.	SS	MS	EMS
Blocks	4	10,000	2,500	$\sigma^2 + \frac{5}{4}\sum_j \beta_j^2$
Temperature	4	11,200	2,800	$\sigma^2 + \frac{5}{4}\sum_i \tau_i^2$
Linear	1	9,248	9,248	
Quadratic	1	1,851	1,851	
Deviation	2	101	50.5	
Error	16	24,000	1,500	σ^2
	24			

From Table 10.1, we have the linear and quadratic orthogonal polynomial coefficients for 5 treatments as

$$\xi_1' = -2, -1, 0, 1, 2 \quad ,$$

$$\xi_2' = 2, -1, -2, -1, 2 \quad ,$$

which were used to partition the treatment sum of squares as

$$SS_1 = \frac{(\sum_j \xi_1' T_j)^2}{n\Sigma(\xi_1')^2} \quad ,$$

$$SS_2 = \frac{(\sum_j \xi_2' T_j)^2}{n\Sigma(\xi_2')^2} \quad ,$$

and

$$SS_{Dev} = SS_{Tr} - SS_1 - SS_2 \quad .$$

For this example where n = 5 observations per treatment, SS_1 = 9,248; SS_2 = 1,851; and SS_{Dev} = 101.

The hypothesis $H_{0_1}: \tau_i = 0$ can be tested by comparing $F_{4,16,0.95}$ with $F = MS_{Tr}/MS_E$. We have

$$F_{4,16,0.95} = 3.01 \quad ,$$

$$f = \frac{MS_{Tr}}{MS_E} = \frac{2800}{1500} = 1.866 \quad .$$

Since $f < F_{4,16,0.95}$, we accept H_{0_1}, that the treatments (temperatures) are not significant.

H_{0_2}: linear regression term is insignificant. We have

$$F_{1,16,0.95} = 4.49 \quad ,$$

$$f = \frac{MS_1}{MS_E} = \frac{9248}{1500} = 6.17 \quad .$$

As f = 6.17 > 4.49, we reject H_{0_2} and conclude that the linear regression term is significant.

H_{0_3}: quadratic regression term is insignificant. We have

$$F_{1,16,0.95} = 4.49 \quad ,$$

$$f = \frac{MS_2}{MS_E} = \frac{1851}{1500} = 1.234 \quad .$$

As f = 1.234 < 4.49, we must accept H_{0_3} that the quadratic regression term is insignificant. It should be noted here that as the temperature effect is insignificant at the 5% significance level, partitioning of the treatment sum of squares by orthogonal polynomials (or by any other method) is futile. Too many students unfortunately ignore the qualitative result (insignificant temperature effect) in favor of a supposedly valid semi-quantitative effect.

11.4.2 Missing Data, RCB

There are two common cases of missing data in the randomized complete block design: (1) a complete block is missing and (2) the treatment is completely missing. In the first case, if at least two blocks still remain, the analysis of variance proceeds as usual with one less number of blocks than originally thought to be available. If at least two treatments still remain, we again proceed in the regular manner.

A more common situation is the one in which one observation is missing. A correction for a single missing observation is made by assigning a value for the missing observation which minimizes the experimental error sum of squares when the standard analysis of variance is performed. For this technique an expression for the experimental sum of squares is obtained and differentiated with respect to the missing observation M, equated to 0, and solved for M. The resulting estimate of the missing observation is

$$M = \frac{tT + bB - S}{(t-1)(b-1)} \tag{11.22}$$

where

t = number of treatments,

b = number of blocks,

T = ΣY with same treatment as M,

B = ΣY in the same block as M,

S = sum of all actual observations.

This value M is then entered into the table of data and the analysis proceeds as before with one important exception.

The degrees of freedom associated with the experimental error and the total degrees of freedom must be decreased by one to account for the simulated datum point. After this is done, the treatment mean square calculated from the data with the estimated observation inserted is greater than the expected value of the experimental error mean square. Therefore, any test of hypothesis which does not correct for this fact will be biased. The correction for bias, Z, is made by decreasing the treatment sum of squares by an amount

$$Z = \frac{[B-(t-1)M]^2}{t(t-1)} \tag{11.23}$$

to give a new corrected treatment sum of squares which is then used in the analysis of variance.

Example 11.14 For the RCB data below in which blocks are different Δt's and the treatments are different feeding arrangements to an evaporator system, test the hypothesis that the true treatment means are equal at the 5% level.

	Blocks			
Treatments	B_1	B_2	B_3	B_4
T_1	3.6	3.8	3.0	4.0
T_2	4.1	3.7	3.9	4.2
T_3	3.0	3.3	3.4	3.8
T_4	---	4.5	4.0	4.8

A value for the missing datum point of steam economy may be estimated from Eq. (11.25) (Section 11.4.4).

We have

$$M = \frac{tT + bB - S}{(t-1)(n-1)}$$

$$= \frac{4 \times 13.3 + 4 \times 10.7 - 57.1}{3 \times 3} = 4.32 \quad ,$$

$$SS_T = \sum_{ij}\sum y_{ij}^2 = 239.5924 \quad ,$$

$$\sum_{ij}\sum y_{ij} = 61.42 \quad ,$$

$$SS_M = (\sum_{ij}\sum y_{ij})^2/bt = (61.42)^2/16 = 235.7760 \quad ,$$

$$SS_B = \sum_j B_j^2/t - SS_M = 946.4204/4 - 235.7760 = 0.82908 \quad ,$$

$$SS_{Tr} = \sum_i T_i^2/b - SS_M = 952.8844/4 - 235.7760 = 2.44508$$

Due to the calculation of the missing datum point, the above value of SS_{Tr} must be corrected by subtracting the following correction term Z from it [see Eq. (11.23):

$$Z = \frac{[B-(t-1)M]^2}{t(t-1)}$$

$$= \frac{(10.7-3(4.3))^2}{3(4)} = \frac{(-2.2)^2}{12} = \frac{4.84}{12}$$

$$= 0.4033 \quad .$$

The new value of SS_{Tr} is then SS'_{Tr} which is

$$SS'_{Tr} = SS_{Tr} - Z = 2.44508 - 0.4033 = 2.04178$$

Also

$$SS_E = SS_T - SS_M - SS_B - SS'_{Tr}$$

$$= 239.5924 - 235.7760 - 0.8325 - 2.04178 = 0.942095$$

Analysis of Variance

Source	d.f.	SS	MS	f
Mean	1	235.7760	---	---
Blocks	3	0.82908	0.27636	2.347
Treatments	3	2.04178	0.68059	5.782
Error	8	0.942095	0.11776	---
Total	15	239.5924		

To test the H_0 that there is no difference in treatment means, $f = \frac{MS'_{Tr}}{MS_E}$ will be compared with

$$F_{3,8,0.95} = 4.07$$

Since $f_{treatments} > F_{3,8,0.95}$, the null hypothesis that the treatment means are equal is rejected at the 95% level and it may be said that there is a significant difference in steam economy due to different feeding arrangements.

Since $f_{blocks} < F_{3,8,0.95}$, the blocks do not have a significant effect on steam economy. This is not as it should be since the blocks (Δt's) should constitute the major source of variation in a RCB design. It is known that Δt changes affect steam economy significantly for a given feeding arrangement. The experiment might have been more informative with feeding arrangement as the major source of variation (blocks) since it is known that feeding arrangement affects the steam economy. Although no interaction tests were possible with this design, a certain feeding arrangement -Δt combination might have been very significant in this experiment and could have affected the AOV accordingly.

11.4.3 Paired Observations, RCB

We can consider the method of paired observations, discussed in Section 7.7, as a randomized block with two blocks. In this case the test for equality of means H_0: $\mu_1 = \mu_2$ is obtained by computing the statistic

$$T = (\overline{X}_1 - \overline{X}_2)/S_{\overline{X}_1 - \overline{X}_2} , \tag{11.24}$$

where the acceptance region is given by

$$t_{n-1,\alpha/2} < T < t_{n-1,1-\alpha/2} . \tag{11.25}$$

It can be shown that the F-statistic, using t = 2, in the RCB design is the square of the T-statistic given in Eq. (11.24).

11.4.4 Subsampling in a Randomized Complete Block Design

When subsampling is used in a randomized complete block design, the corresponding statistical model is

$$Y_{ijk} = \mu + \tau_i + \beta_j + \varepsilon_{ij} + \eta_{ijk},$$

$$i = 1,2,\ldots,t, \quad j = 1,2,\ldots,b, \quad k = 1,2,\ldots,n, \tag{11.26}$$

and the corresponding analysis of variance is presented in Table 11.11. In the calculation for Table 11.11, B_j refers to the total of all observations in the jth block and T_i is the total of all observations subjected to the ith treatment. The among-cells sum of squares used in calculating the sampling error sum of squares is given as

$$SS_{AC} = \sum_{i=1}^{t} \sum_{j=1}^{b} Y_{ij.}^2/n - Y_{...}^2/btn . \tag{11.27}$$

Treatment effects can be tested as before by using an F-ratio, which is the mean square for treatments divided by the experimental error mean square, and comparing it with the tabular value of F, for the corresponding degrees of freedom and for a given significance level.

Example 11.15 In an experiment designed as a randomized complete block, the effect of % Co on the tensile strength of steel was

TABLE 11.11

Analysis of Variance for Randomized Complete Block Design with Subsampling in the Experimental Units (Model I)

Source	d.f.	SS	EMS
Mean	1	$(\sum_{ijk}\sum\sum Y_{ijk})^2/btn = SS_M$	---
Blocks	b-1	$\sum_j B_j^2/tn - SS_M = SS_B$	$\sigma_\eta^2 + n\sigma^2 + tn\sum_j^b \beta_j^2/(b-1)$
Treatments	t-1	$\sum_i T_i^2/bn - SS_M = SS_{Tr}$	$\sigma_\eta^2 + n\sigma^2 + bn\sum_i^t \tau_i^2/(t-1)$
Experimental error	(b-1)(t-1)	$SS_{AC} - SS_B - SS_{Tr} = SS_{EE}$	$\sigma_\eta^2 + n\sigma^2$
Sampling error	bt(n-1)	$\sum_{ijk}\sum\sum Y_{ijk}^2 - SS_M - SS_{AC} = SS_{SE}$	σ_η^2
Total	btn	$\sum_{ijk}\sum\sum Y_{ijk}^2 = SS_T$	

evaluated. Four levels of % Co were used. Three different crucibles were used in the alloying process. Each crucible was sampled once and the samples divided before analysis to give the following data which have been presented in thousands of psi for convenience in making the calculations.

		Crucible		
		1	2	3
% Co	1	49,50	44,45	53,56
	2	60,62	53,56	64,65
	3	64,67	63,65	74,78
	4	71,75	65,67	76,80

It is believed that a reasonable model for this experiment is

$$Y_{ijk} = \mu + \tau_i + \beta_j + \varepsilon_{ij} + \eta_{ijk}$$
$$i = 1,2,3,4;\ j = 1,2,3;\ k = 1,2$$

where i, j, and k correspond to % Co, crucible, and replication, respectively. The model then gives the observed tensile strengths Y_{ijk} in terms of the overall mean μ; the block or crucible effect β_j, the treatment or % Co effect τ_i; and includes terms for experimental error, ε_{ij}, and subsampling error, η_{ijk}.

For these data,

(a) Compute the values for the AOV table.

(b) Test the homogeneity of measurement error by testing the 1 and 2% Co data vs the rest of the data.

(c) Test the homogeneity of the (experimental plus measurement) error by testing the effects of the linear and quadratic terms vs the cubic term.

(d) Test the significance of the linear, quadratic, and cubic components of the treatment sum of squares using the (experimental and measurement) error sum of squares.

Using the model for subsampling in randomized complete blocks (<u>SAS for Linear Models</u>), the following program is prepared.

```
(put initial JCL cards/statements here)
DATA ALLOY;
INPUT CRUC 1 CO 3 TENSTR 5-6;
CARDS;
(put data here according to INPUT format)
PROC PRINT;
PROC GLM;
CLASSES CO CRUC CO(CRUC);
MODEL TENSTR = CO CRUC CO(CRUC);
TITLE 'EFFECT OF CRUCIBLE AND COBALT CONCENTRATION ON TENSILE
      STRENGTH';
(put final JCL cards/statements here)
```

SAS

EFFECT OF CRUCIBLE AND COBALT CONCENTRATION ON TENSILE STRENGTH

GENERAL LINEAR MODELS PROCEDURE

DEPENDENT VARIABLE: TENSTR

SOURCE	DF	SUM OF SQUARES	MEAN SQUARE	F VALUE
MODEL	11	2366.83333333	215.16666667	57.38
ERROR	12	45.00000000	3.75000000	PR > F
CORRECTED TOTAL	23	2411.83333333		0.0001

SOURCE	DF	TYPE I SS	F VALUE	PR > F
CRUC	2	485.33333333	64.71	0.0001
CO	3	1847.50000000	164.22	0.0001
CO(CRUC)	6	34.00000000	1.51	0.2552

The output consists of the AOV table above. The CO(CRUC) component in the CLASSES and MODEL statements results in the experimental error term. The ERROR term in the output is the sampling error term. We first compare experimental error to sampling error by

$$f = MS_{EE}/MS_{SE} = (34.0/6)/(45.0/12) = 1.511$$

which is not significant so H_0: $\sigma^2 = 0$ is accepted. When the treatments (% Co) are tested vs. sampling error as the major error source,

$$f_{Tr} = MS_{Tr}/MS_{SE} = (1847.5/3)/(45/12) = 164.22$$

which is significant so the hypothesis of no treatment effects (H_0: $\Sigma\tau_i^2 = 0$) is rejected. The same result is obtained if the two error terms are pooled to give

$$MS_{PE} = (SS_{EE} + SS_{SE})/(df_{EE} + df_{SE}) = 4.3888$$

with $f_{Tr} = 140.32$ when using the pooled error term to increase precision.

An alternate solution is presented below in which the analysis of variance is computed by means of orthogonal polynomials to obtain a simultaneous estimate of the significance of each proposed term in the regression model. The analysis of variance is prepared as shown in Table 11.12. The necessary calculations to prepare that table are below.

From Table 10.1, we use the orthogonal polynomial coefficients for a third-degree polynomial and (realizing that the reduction in treatment sum of squares due to the fitting of a third-degree polynomial may be partitioned into linear, quadratic, and cubic terms) we define a set of orthogonal contrasts as

$$C_j^L = -3y_{1j.} - y_{2j.} + y_{3j.} + 3y_{4j.} \quad ,$$

$$C_j^Q = y_{1j.} - y_{2j.} - y_{3j.} + y_{4j.} \quad ,$$

$$C_j^C = -y_{1j.} + 3y_{2j.} - 3y_{3j.} + y_{4j.} \quad ,$$

$$C_{\bullet}^L = \sum_j C_j^L = -3y_{1..} - y_{2..} + y_{3..} + 3y_{4..} \quad ,$$

$$C_{\bullet}^Q = \sum_j C_j^Q = y_{1..} - y_{2..} - y_{3..} + y_{4..} \quad ,$$

$$C_{\bullet}^C = \sum_j C_j^C = -y_{1..} + 3y_{2..} - 3y_{3..} + y_{4..} \quad .$$

Two contrasts, $L_1 = \sum_{j=1}^{n} a_j y_j$ and $L_2 = \sum_{j=1}^{n} b_j y_j$ are said to be orthogonal if $\sum_{i=1}^{n} a_i b_i = 0$. For example, in the above, C_j^L and C_j^Q are orthogonal since

$$\sum_{j=1}^{4} a_j b_j = (-3)(1) + (-1)(-1) + (1)(-1) + (3)(1) = 0 \quad .$$

Remember that the dot subscript means that the term corresponding to that subscript has already been summed.

(a) The AOV table is computed as follows:

$$C_1^L = -3(99) - 122 + 131 + 3(146) = 150$$

$$C_2^L = -3(89) - 109 + 128 + 3(132) = 148$$

$$C_3^L = -3(109) - 129 + 152 + 3(156) = 164$$

$$C_1^Q = 99 - 122 - 131 + 146 = -8$$

$$C_2^Q = 89 - 109 - 128 + 132 = -16$$

$$C_3^Q = 109 - 129 - 152 + 156 = -16$$

$$C_1^C = -99 + 3(122) - 3(131) + 146 = 20$$

$$C_2^C = -89 + 3(109) - 3(128) + 132 = -14$$

$$C_3^C = -109 + 3(129) - 3(152) + 156 = -22$$

$$C_\bullet^L = 462, \quad (C_\bullet^L)^2/120 = 1778.7$$

$$C_\bullet^Q = -40, \quad (C_\bullet^Q)^2/24 = 66.67$$

$$C_\bullet^C = -16, \quad (C_\bullet^C)^2/120 = 2.13$$

$$\sum_j (C_j^L)^2 = 71{,}300$$

$$\sum_j (C_j^Q)^2 = 576$$

$$\sum_j (C_j^C) = 1080$$

$$\sum_j (C_j^L)^2/40 - (C_\bullet^L)^2/120 = 1782.5 - 1778.7 = 3.80$$

$$\sum_j (C_j^Q)^2/8 - (C_\bullet^Q)^2/24 = 72.00 - 66.67 = 5.33$$

$$\sum_j (C_j^C)^2/40 - (C_\bullet^C)^2/120 = 27.00 - 2.13 = 24.87$$

$$\frac{\sum_j y_{\bullet j \bullet}^2}{8} = \frac{(498)^2 + (458)^2 + (546)^2}{8} = 94{,}485.5$$

$$\frac{y_{\bullet\bullet\bullet}^2}{24} = \frac{(1502)^2}{24} = 94{,}000.2$$

$$\frac{\sum_i y_{i\bullet\bullet}^2}{6} = \frac{(297)^2 + (360)^2 + (411)^2 + (434)^2}{6} = 95{,}847.7$$

$$\frac{1}{2}\left(\sum_{i=1}^{2}\sum_{j=1}^{3} y_{ij\bullet}^2\right) = (73{,}009)1/2 = 36{,}504.5$$

$$\frac{1}{2}\left(\sum_{i=3}^{4}\sum_{j=1}^{3} y_{ij\bullet}^2\right) = (119{,}725)1/2 = 59{,}862.5$$

$$\frac{1}{2}\left(\sum_{i=1}^{4}\sum_{j=1}^{3} y_{ij\bullet}^2\right) = (192{,}734)1/2 = 96{,}367$$

$$\sum_{i=1}^{2}\sum_{j=1}^{3}\sum_{k=1}^{2} y_{ijk}^2 = 36{,}517$$

$$\sum_{i=3}^{4}\sum_{j=1}^{3}\sum_{k=1}^{2} y_{ijk}^2 = 59{,}895$$

$$\sum_{i=1}^{4}\sum_{j=1}^{3}\sum_{k=1}^{2} y_{ijk}^2 = 96{,}412 \quad .$$

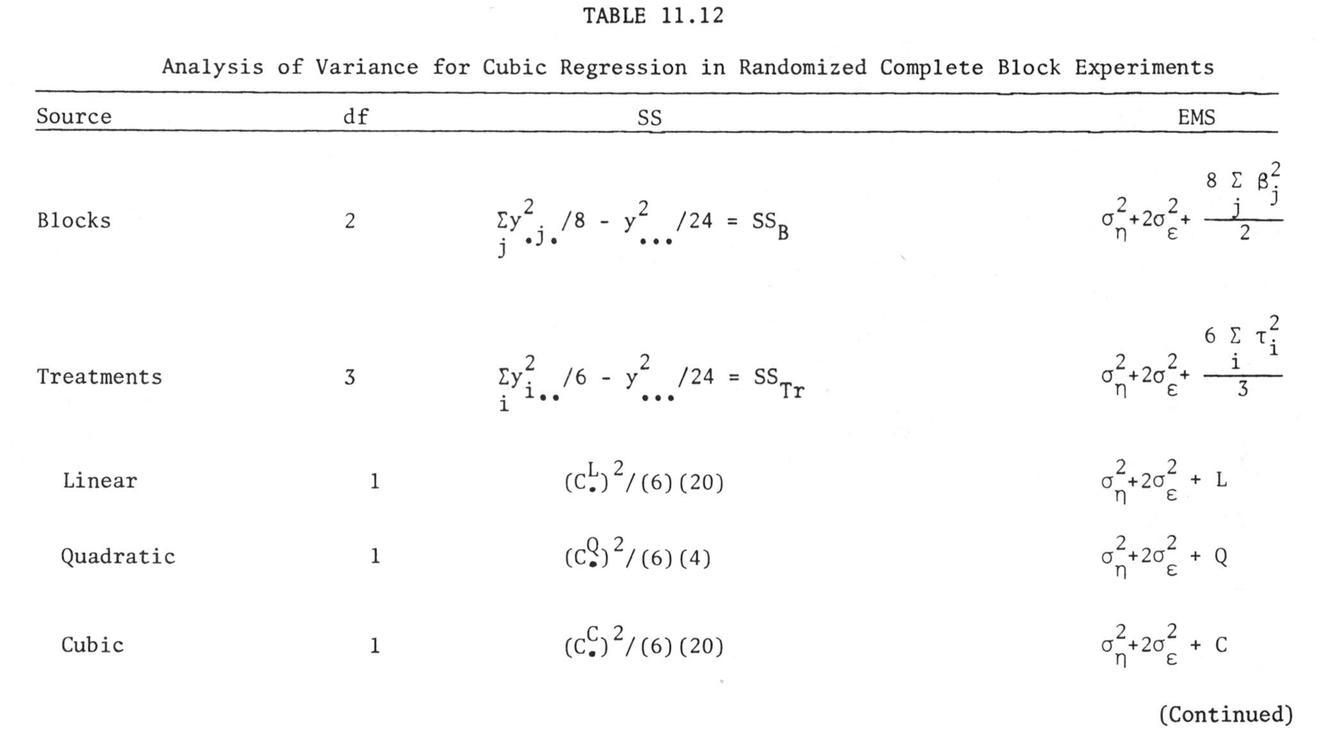

TABLE 11.12

Analysis of Variance for Cubic Regression in Randomized Complete Block Experiments

Source	df	SS	EMS
Blocks	2	$\sum_j y_{\cdot j \cdot}^2/8 - y_{\cdots}^2/24 = SS_B$	$\sigma_\eta^2 + 2\sigma_\varepsilon^2 + \frac{8 \sum_j \beta_j^2}{2}$
Treatments	3	$\sum_i y_{i \cdot\cdot}^2/6 - y_{\cdots}^2/24 = SS_{Tr}$	$\sigma_\eta^2 + 2\sigma_\varepsilon^2 + \frac{6 \sum_i \tau_i^2}{3}$
Linear	1	$(C_\cdot^L)^2/(6)(20)$	$\sigma_\eta^2 + 2\sigma_\varepsilon^2 + L$
Quadratic	1	$(C_\cdot^Q)^2/(6)(4)$	$\sigma_\eta^2 + 2\sigma_\varepsilon^2 + Q$
Cubic	1	$(C_\cdot^C)^2/(6)(20)$	$\sigma_\eta^2 + 2\sigma_\varepsilon^2 + C$

(Continued)

TABLE 11.12 (Continued)

Source	df	SS	EMS
(Experimental + measurement) Error	6	$\sum_{ij}\sum y^2_{ij\cdot}/2 - \sum_j y^2_{\cdot j\cdot}/8 - \sum_i y^2_{i\cdot\cdot}/6 + y^2_{\cdot\cdot\cdot}/24$	$\sigma^2_\eta+2\sigma^2_\varepsilon$
Linear	2	$\sum_j (C^L_j)^2/(2)(20) - (C^L_\cdot)^2/(6)(20)$	$\sigma^2_\eta+2\sigma^2_\varepsilon$
Quadratic	2	$\sum_j (C^Q_j)^2/(2)(4) - (C^Q_\cdot)^2/(6)(4)$	$\sigma^2_\eta+2\sigma^2_\varepsilon$
Cubic	2	$\sum_j (C^C_j)^2/(2)(20) - (C^C_\cdot)^2/(6)(20)$	$\sigma^2_\eta+2\sigma^2_\varepsilon$
Measurement error	12	$\sum_{ijk}\sum\sum y^2_{ijk} - \sum_{ij}\sum y^2_{ij\cdot}/2$	σ^2_η
1% & 2%	6	$\sum_{i=1}^{2}\sum_j\sum_k y^2_{ijk} - \sum_{i=1}^{2}\sum_j y^2_{ij\cdot}/2$	σ^2_η
3% & 4%	6	$\sum_{i=3}^{4}\sum_j\sum_k y^2_{ijk} - \sum_{i=3}^{4}\sum_j y^2_{ij\cdot}/2$	σ^2_η
Total corrected	23	$\sum_{ijk}\sum\sum y^2_{ijk} - y^2_{\cdot\cdot\cdot}/24$	

AOV Table

Source	df	SS	MS
Blocks	2	485.3	242.65
Treatments	3	1847.5	615.83
Linear	1	1778.7	1778.7
Quadratic	1	66.67	66.67
Cubic	1	2.13	2.13
(Experimental + Measurement) Error	6	34.0	5.67
Linear	2	3.8	1.9
Quadratic	2	5.33	2.667
Cubic	2	24.87	12.435
Measurement error	12	45.0	3.75
1% & 2%	6	12.5	2.083
3% & 4%	6	32.5	5.417
Total corrected	23	2411.8	

(b) $\dfrac{MS_{ME_{3,4}}}{MS_{ME_{1,2}}}$ has an F distribution with 6 and 6 df.

$$\frac{5.417}{2.083} = 2.6005 = \frac{MS_{ME_{3,4}}}{MS_{ME_{1,2}}}$$

$$F_{6,6,0.95} = 4.28 \quad .$$

As $\dfrac{MS_{ME_{3,4}}}{MS_{ME_{1,2}}} = 2.6005 < F_{6,6,0.95} = 4.28$, we "accept" H_0: measurement error is homogeneous.

(c) Pool the linear and quadratic (experimental + measurement) error terms as MS_{E_1} and test vs. the corresponding cubic term as MS_{E_2}.

$$MS_{E_1} = \frac{(df_L)(MS_{E_L}) + (df_Q)(MS_{E_Q})}{df_L + df_Q} = \frac{3.8 + 5.33}{4} = 2.283 \quad .$$

$$MS_{E_2} = MS_{E_C} = 12.435 \quad .$$

$$\frac{MS_{E_1}}{MS_{E_2}} = \frac{2.283}{12.435} = 0.1836 \quad .$$

$\dfrac{MS_{E_1}}{MS_{E_2}}$ has an F distribution with 4 and 2 d.f.

Now $F_{4,2,0.05} = 0.144$, $F_{4,2,0.95} = 19.25$, and as $0.1441 < 0.1836 < 19.25$ we accept H_0: (exp. error + measurement error) is homogeneous.

(d) $\dfrac{MS_{T_L}}{MS_E}$ has an F distribution with 1 and 6 d.f.

$\dfrac{MS_{T_L}}{MS_E} = \dfrac{1778.7}{5.67} = 313.704$ so we reject H_0: linear term is insignificant.

$\dfrac{MS_{T_Q}}{MS_E}$ has an F distribution with 1 and 6 d.f.

$\dfrac{66.67}{5.67} = 11.7584$ so we reject H_0: quadratic term is insignificant.

$$\frac{MS_{T_C}}{MS_E} \text{ has an F distribution with 1 and 6 d.f.}$$

$$\frac{2.13}{5.67} = 0.3757 \quad .$$

$F_{1,6,0.95} = 5.99$ so we accept H_0: cubic term is insignificant.

We can say that the cubic component adds little and the relationship is at least quadratic.

Example 11.16 In a randomized complete block experiment intended to determine the effect of percent cobalt on steel tensile strength, 3 different percentages of cobalt were annealed in each of 4 ovens. The 12 resulting specimens were then divided into 3 samples each, and the resulting 36 samples tested for strength. The 36 strength values thus obtained were as follows (values are in 1000's psi):

		Ovens			
		1	2	3	4
% CO	1	50,47,55	43,42,51	49,53,57	45,49,55
	2	64,59,55	48,55,62	66,63,65	67,63,60
	3	60,66,69	59,60,70	67,83,75	63,72,70

Convert the data to standard form by averaging triplets. Give the resulting model and prepare the complete AOV table, based on triplet means. Test H_0: $\tau_i = 0$.

(a) Model: $Y_{ij} = \mu + \tau_i + \beta_j + \varepsilon_{ij}$

Treatments %CO	Triplet Means Ovens 1	2	3	4
1	50.667	45.333	53.000	49.667
2	59.333	55.000	64.667	63.333
3	65.000	63.000	75.000	68.333

b = ovens (blocks) = 4

t = percents (treatments) = 3

$$\frac{\sum_i y_{i.}^2}{b} = \frac{39{,}468.5769 + 58{,}725.2829 + 73{,}621.5969}{4} \quad ,$$

$$\frac{\sum_i y_{i.}^2}{b} = \frac{171{,}815.5}{4} = 42{,}953.9 \quad ,$$

$$\frac{y_{..}^2}{bt} = \frac{(712.333)^2}{(4)\ (3)} = 42{,}284.9 \quad ,$$

$$\frac{\sum_j y_{.j}^2}{t} = \frac{30{,}624 + 26{,}677.669 + 37{,}120.573 + 32{,}881.657}{3}$$

$$\frac{\sum_j y_{.j}^2}{t} = 42{,}435. \quad ,$$

$$SS_B = 42{,}435. - 42{,}284.9 = 150.1 = \sum_j y_{.j}^2/t - y_{..}^2/bt \quad ,$$

$$SS_{Tr} = 42{,}953.9 - 42{,}284.9 = 669.0 = \sum_i y_{i.}^2/b - y_{..}^2/bt \quad ,$$

$$\sum_{ij}\sum y_{ij}^2 = 43{,}124.730 \quad ,$$

$$SS_E = \sum_{ij}\sum y_{ij}^2 - \frac{\sum_j y_{.j}^2}{t} - \frac{\sum_i y_{i.}^2}{b} + \frac{y_{..}^2}{bt} = 20.7 \quad \text{(see Table 11.10)} \quad ,$$

$$SS_{TC} = \sum y_{ij}^2 - \frac{y_{..}^2}{bt} = 43{,}124.7 - 42{,}284.9 = 839.8 \quad .$$

AOV Table

Source of variation	d.f.	SS	MS	EMS
Treatments	2	669	334.5	$\sigma^2+b\sum_i\tau_i^2/(t-1)$
Blocks	3	150.1	75.05	$\sigma^2+t\sum_j\beta_j^2/(b-1)$
Error	6	20.7	3.45	σ^2
Total corrected	11	839.8		

H_0: $\tau_i = 0$; test by $\frac{MS_{Tr}}{MS_E}$ which has an F-distribution with 2 and 6 d.f.

As $f = \frac{MS_{Tr}}{MS_E} = \frac{334.5}{3.45} = 96.96$ and $F_{2,6,0.95} = 5.14$, we reject H_0.

It is often advantageous when using randomized complete block designs to consider factorial treatment combinations. Two important cases are the two-factor case for which the statistical model is given by

$$Y_{ijk} = \mu + \rho_i + \alpha_j + \beta_k + (\alpha\beta)_{jk} + \varepsilon_{ijk} ,$$

$$i = 1,2,\ldots,r, \quad j = 1,2,\ldots,a, \quad k = 1,2,\ldots,b \quad . \qquad (11.28)$$

and the three-factor case for which the statistical model is

$$Y_{ijk\ell} = \mu + \rho_i + \alpha_j + \beta_k + (\alpha\beta)_{jk} + \gamma_\ell + (\alpha\gamma)_{j\ell} + (\beta\gamma)_{k\ell} + (\alpha\beta\gamma)_{jk\ell} + \varepsilon_{ijk\ell} ,$$

$$i = 1,2,\ldots,r, \quad j = 1,2,\ldots,a, \quad k = 1,2,\ldots,b, \quad \ell = 1,2,\ldots,c \quad . \qquad (11.29)$$

In these equations, μ is the true mean effect, ρ_i is the true effect of the ith replicate, and the various terms involving α_j, β_k, γ_l, $(\alpha\beta)_{jk}$, etc., are the true effects of the several factors and their interactions. The ε's are then the true effects of the experimental units. The analysis of variance proceeds in a manner analogous to that used for factorial treatment combinations in the completely randomized design. The interested student is referred to Ostle and Mensing [3] for the analysis of multifactor factorials in a randomized complete block design with subsampling and several determinations per subsampling unit.

11.5 LATIN SQUARE DESIGNS

A design that is particularly useful in industrial experimentation is the Latin square design. We have seen how randomized blocks can be used to reduce experimental error by eliminating a source of variation in which there is no interest, that is, if a block effect exists then the corresponding test procedure used to test treatment effects is more sensitive to actual treatment differences. Also, the probability of a type II error is decreased. If there are two sources of variation to control, a Latin square provides a very good method of analyzing treatment differences. The Latin square can thus be viewed as a method of assessing treatment effects when a double type of blocking is imposed on the experimental units. In this sense we can say that the Latin square is an extension of the randomized complete block design.

Example 11.17 Consider an experiment involving the study of tire life of four brands of tires. There are 16 tires, 4 of each brand, available. Furthermore, there are 4 cars available. The treatment in this case is tire brand. There are two other factors that effect tire life, (1) car type and (2) position of the tire on the car. Label the cars I, II, III, and IV and tire position as RF,

LF, RR, and LR. The 4 cars can be set up as rows and the 4 positions as columns and the tires can then be assigned at random subject to the condition that each brand occur once in each row and each column. This example will be analyzed further in Example 11.18.

In a Latin square the two types of block are referred to as rows and columns. The differences among rows and among columns represent two sources of variation which the design is meant to remove. The experimental error per unit usually increases as the size of the square increases. This leads us to the conclusion that in a small Latin square such as a 3 × 3 or a 4 × 4, a substantial reduction in experimental error must be obtained over that obtainable with the completely randomized or the randomized complete block design to counteract for the concomitant loss of degrees of freedom for error in the Latin square design. Latin squares are often used when different treatments are applied in sequence to the same experimental units. In this case the rows become successive periods or applications and the columns are the experimental units.

To form a Latin square design, assign one variable as rows, the second as columns, the third as treatments. Then randomly assign values of the row and column variables as shown below.

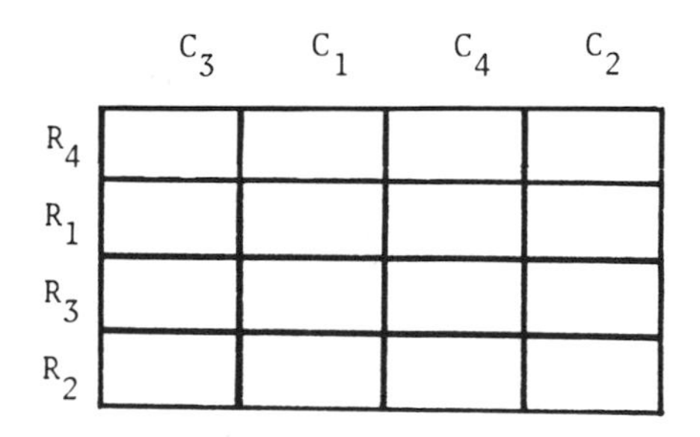

Each small square above represents one experimental unit to which will be assigned (in this case of a 4 × 4 Latin square) one of the four treatments. For the sake of simplicity, start at the upper left corner and assign the treatments in any order across the first row or down the first column. Then permute the treatments across (or down) the rest of the square. Shown below are four simple 4×4 Latin squares. Designation of the levels of the row and column variables have been omitted for clarity.

A	B	C	D
B	C	D	A
C	D	A	B
D	A	B	C

Column Permutation

A	B	C	D
D	A	B	C
C	D	A	B
B	C	D	A

Row Permutation

A	B	C	D
D	C	B	A
C	D	A	B
B	A	D	C

Helical Permutation

A	B	C	D
B	A	D	C
C	D	A	B
D	C	B	A

Alternating Forward Reverse Permutation

Diagonal and other permutations are possible. It is not necessary to start with any particular order of treatments for the first row or column. The important feature to observe is that each row and each column contains each treatment once and only once.

11.5.1 Analysis of Variance for the Latin Square

The model for a general $m \times m$ Latin square with one observation per experimental unit is represented by

$$Y_{ij(k)} = \mu + \rho_i + \gamma_j + \tau_k + \varepsilon_{ij(k)} , \qquad (11.30)$$

$$i = 1,2,\ldots,m, \; j = 1,2,\ldots,m, \; k = 1,2,\ldots,m,$$

where

$$\sum_i \rho_i = \sum_j \gamma_j = \sum_k \tau_k = 0 \text{ and the } \varepsilon_{ij(k)} \text{ are NID}(0,\sigma^2).$$

The variables ρ_i, γ_j, and τ_k are the true effects associated with the ith row, j^{th} column, and k^{th} treatment. The reason for the parentheses ( ) around the k in Eq. (11.30) is that there are actually m^2 observations and not m^3 observations such as are present in the usual 3-factor design with one replication. The analysis of variance for an $m \times m$ Latin square with one observation per experimental unit corresponding to the model given in Eq. (11.30) is shown in Table 11.13 where R_i, C_j, and T_k represent the indicated row, column, and treatment totals.

It seems as though the Latin square gives us a very powerful technique for assessing treatment differences in the presence of two other sources of variation. This is true: however, a price must be paid and in this case the price is the assumption of no interactions in the model in Eq. (11.30). This assumption must be considered by the experimenter, and if in doubt it may be necessary to check it out by use of one of the previously considered models. Note that the Latin square calls for the same number of rows, columns, and treatments.

Example 11.18 Four brands (A, B, C, D) of tires were evaluated by a Latin square design in the following manner:

Sixteen tires, 4 of each brand, were put on 4 cars (I, II, III, IV), and in such a way that

(1) Each car had a tire from every brand.
(2) Each brand was tried in a RF, LF, RR, and LR position.
(3) The exact disposition of the 16 tires was determined by the random selection of a 4×4 Latin square, which turned out to be

	RF	LF	RR	LR
I	A	B	C	D
II	B	D	A	C
III	C	A	D	B
IV	D	C	B	A

TABLE 11.13

Analysis of Variance for an m x m Latin Square with One Observation per Experimental Unit

Source	d.f.	SS	EMS
Mean	1	$(\sum_{ij}\sum Y_{ij(k)})^2/m^2 = SS_M$ or $(\sum_{ik}\sum Y_{i(j)k})^2/m^2 = SS_M$	
Rows	m-1	$\sum_i R_i^2/m - SS_M = SS_R$	$\sigma^2 + \frac{m}{m-1}\sum_i \rho_i^2$
Columns	m-1	$\sum_j C_j^2/m - SS_M = SS_C$	$\sigma^2 + \frac{m}{m-1}\sum_j \gamma_j^2$
Treatments	m-1	$\sum_k T_k^2/m - SS_M = SS_{Tr}$	$\sigma^2 + \frac{m}{m-1}\sum_k \tau_k^2$
Experimental error	(m-1)(m-2)	$\sum_{ijk}\sum\sum Y^2_{ij(k)} - SS_M - SS_R - SS_C - SS_{Tr}$	σ^2
Total	m^2	$\sum_{ijk}\sum\sum Y^2_{ij(k)} = SS_T$	

The cars were driven 5,000 miles, and tire wear, as measured by a standard procedure, turned out to be

31	33	47	54
36	53	42	54
51	43	62	49
81	78	72	84

1. What car had the hot-rod driver? Does the presence of the hot-rod driver invalidate the experiment?
2. Test for the equality of tire effects, the equality of position effects, and the equality of car effects.

	Position								
Car	RF		LF		RR		LR		Row Σ
I	A	31	B	33	C	47	D	54	165
II	B	36	D	53	A	42	C	54	185
III	C	51	A	43	D	62	B	49	205
IV	D	81	C	78	B	72	A	84	315
Col Σ		199		207		223		241	870

We have

$$\frac{y^2_{\cdot\cdot(\cdot)}}{16} = \frac{(870)^2}{16} = 47,306.25 \quad ,$$

$$\frac{\sum_j y^2_{\cdot j(\cdot)}}{4} = \frac{190,260}{4} = 47,565 \quad ,$$

$$\frac{\sum_i y^2_{i\cdot(\cdot)}}{4} = \frac{202,700.}{4} = 50,675. \quad ,$$

$$\frac{\sum_k y^2_{\cdot\cdot(k)}}{4} = \frac{(200)^2+(190)^2+(230)^2+(250)^2}{4} \quad ,$$

$$= 47,875 \quad .$$

Also,

$$\text{SS for cars:}\quad \frac{\sum_i y^2_{i\cdot(\cdot)}}{4} - \frac{y^2_{\cdot\cdot(\cdot)}}{16} = 3368.75 \quad ,$$

$$\text{SS for positions:}\quad \frac{\sum_j y^2_{\cdot j(\cdot)}}{4} - \frac{y^2_{\cdot\cdot(\cdot)}}{16} = 258.75 \quad ,$$

$$\text{SS for tires:}\quad \frac{\sum_k y^2_{\cdot\cdot(k)}}{4} - \frac{y^2_{\cdot\cdot(\cdot)}}{16} = 568.75 \quad ,$$

$$SS_{TC} = 51{,}540. - 47{,}306.25 = \sum_{ij}\sum y^2_{ij(k)} - \frac{y^2_{\cdot\cdot(\cdot)}}{16} = 4233.75 \quad ,$$

$$SS_E = SS_{TC} - (SS_{cars} + SS_{pos} + SS_{tires}) = 37.50 \quad .$$

1. Car IV had the hot-rod driver as there was much more tire wear on that car than on any other. His presence does not invalidate the experiment as we may consider him to be the "upper limit of wear" while the driver of Car I may be the "lower limit of wear," Cars II and III being somewhere in between. Anyway, the design tests each tire in each position by each driver and the AOV eliminates bias of tire position and car effects. The AOV table is given below.

Source of variation	df	SS	MS
Blocks (cars)	3	3368.75	1122.917
Treatments (tires)	3	568.75	189.583
Positions	3	258.75	86.25
Exp. error	6	37.50	6.25
Total corrected	15	4233.75	

2. The hypotheses involving various effects and the corresponding decisions follow.

(a) H_{0_1}: tire effects are equal

$$F = \frac{\text{Tire MS}}{MS_E}\ ;\ F_{3,6,0.95} = 4.76\ .$$

$$f = \frac{189.583}{6.25} = 30.33, \text{ so we reject } H_{0_1}\ .$$

(b) H_{0_2}: position effects are equal

$$F = \frac{\text{Pos.MS}}{MS_E}\ ;\ F_{3,6,0.95} = 4.76\ .$$

$$f = \frac{86.25}{6.25} = 13.80, \text{ so we reject } H_{0_2}\ .$$

(c) H_{0_3}: car effects are equal

$$F = \frac{\text{Car MS}}{MS_E}\ ;\ F_{3,6,0.95} = 4.76\ .$$

$$f = \frac{1122.917}{6.25} = 179.67, \text{ so we reject } H_{0_3}.$$

11.5.2 Missing Data, LS

If a single observation is missing in an experiment conducted according to an m × m Latin square, its value may be estimated by using

$$M = \frac{m(R+C+T) - 2S}{(m-1)(m-2)} \tag{11.31}$$

where

$R = \Sigma Y$ in same row as M,

$C = \Sigma Y$ in same column as M,

$T = \Sigma Y$ with same treatment as M,

S = sum of all actual observations,

m = number of rows = number of columns = number of treatments.

As before, the degrees of freedom associated with the experimental error must be decreased by one. The treatment sum of squares must be corrected for bias by subtracting a correction

$$Z = \frac{[S - R - C - (m-1)T]^2}{(m-1)^2(m-2)} \tag{11.32}$$

from the calculated or original treatment sum of squares before any tests of hypotheses may be made.

Example 11.19 A 4 × 4 Latin square design was used in a series of tests to determine the effect of water pressure, air flow rate, and the number of nozzles in operation on the overall efficiency of the scrubber-impinger for the control of particulate effluents from cotton lint cleaners. The result of these tests are tabulated in Table 11.14. The statistical evaluation of the data is shown in Table 11.15.

The analysis of variance is summarized in Table 11.16. Each variable was then tested to determine whether it had a significant effect on efficiency at a significance level of $\alpha = 0.05$. The tabular F-value which corresponds to this significance level is 4.76. As the calculated F's were all less than 4.76, the null hypothesis of no significant effect on efficiency was accepted for each variable over the range studied.

In this example, none of the major variables has any significant effect on performance of the air pollution control device. The most plausible explanation is that either the effects of the propor

TABLE 11.14

Results for Testing the Effect of Water Pressure, Air Flow, and Nozzle Operation on Efficiency

Air flow, cfm	Nozzle operation	Efficiency
6400	N_A	95
6400	N_C	85
6400	N_D	67
6400	N_B	94
5120	N_A	90
5120	N_C	--[a]
5120	N_D	83
5120	N_B	96
2560	N_A	81
2560	N_C	80
2560	N_D	83
2560	N_B	95
4160	N_A	95
4160	N_C	88
4160	N_D	88
4160	N_B	88

[a]Power failure during first part of test.

TABLE 11.15

Latin Square for the Effect of Water Pressure, Air Flow, and Nozzle Operation on Efficiency

Nozzle operation	Air flow rate, cfm				
	5120	2560	4160	6400	$\sum_j \eta_j$
N_A	$\eta = 90$	$\eta = 81$	$\eta = 95$	$\eta = 95$	361
N_C	$\eta = 90^a$	$\eta = 80$	$\eta = 88$	$\eta = 85$	343
N_D	$\eta = 83$	$\eta = 83$	$\eta = 88$	$\eta = 67$	321
N_B	$\eta = 96$	$\eta = 95$	$\eta = 88$	$\eta = 94$	373
$\sum_i \eta_i$	359	339	359	341	1398

[a]Calculated efficiency, η = collection efficiency.

TABLE 11.16

Analysis of Variance for Scrubber Tests

Source	d.f.	SS	MS	F
Mean	1	122,150.25	--	--
Nozzle operation	3	204.5	68.167	0.729
Air flow rate	3	90.75	30.25	0.323
Water pressure on nozzle taps	3	106.75 (98.75)[b]	35.583 (32.917)[b]	(0.352)[b]
Error	5[a]	467.75	93.55	--
Total	15[a]	122,823.		

[a]One d.f. subtracted for missing datum point

[b]Adjusted to account for bias from missing datum point

variables were not examined or that the ranges used in the experiment for these variables were too small.

11.6 GRAECO-LATIN SQUARE

The concept of a Latin square design is easily extended to that of a Graeco-Latin square. In this type of design equal numbers of treatments are tested on equal numbers of experimental units at different times or in different locations. The use of the Graeco-Latin square results in a tremendous saving in time and money in that the total number of observations is greatly reduced. However, this design, as in the case of the Latin square, should be used with caution as no interactions are tolerated.

A Graeco-Latin can be obtained by superimposing one Latin square using Latin letters on another Latin square involving Greek letters in such a way that each Greek-Latin letter pair occurs once and only once. The construction of Graeco-Latin squares has been studied and it has been shown that $n \times n$ Graeco-Latin squares can be constructed provided n is not equal to 2 or 6.

Graeco-Latin squares for the cases n = 3,4, and 5 are given below:

Aα	Bγ	Cβ
Bβ	Cα	Aγ
Cγ	Aβ	Bα

A α	B γ	C δ	D β
B β	A δ	D γ	C α
C γ	D α	A β	B δ
D δ	C β	B α	A γ

A α	B γ	C ε	D β	E δ
B β	C δ	D α	E γ	A ε
C γ	D ε	E β	A δ	B α
D δ	E α	A γ	B ε	C β
E ε	A β	B δ	C α	D γ

In each case rows represent one variable, columns a second variable, Greek letters a third variable, and Latin letters represent treatments (a fourth variable). Thus a Graeco-Latin square can be viewed as a design subject to three types of blocking. An $n \times n$ Graeco-Latin square requires n^2 observations as compared with n^4 observations which are required by a four-factor design where all factor level combinations are present. The AOV table for a Graeco-Latin square is similar to that of a Latin square with one added variable and will not be given here. It should be observed, though, that each of the 4 variables (factors) has n-1 degrees of freedom associated with it and consequently the experimental error sum of squares has degrees of freedom $= n^2 - 1 - 4(n-1) = (n-1)(n-3)$.

The model for a Graeco-Latin square is

$$Y_{ij(k\ell)} = \mu + \rho_i + \gamma_j + \beta_k + \tau_\ell + \varepsilon_{ij(k\ell)}, \qquad (11.33)$$

$$i,j,k,\ell = 1,2,\ldots,n \quad ,$$

such that

$$\Sigma\rho_i = \Sigma\gamma_j = \Sigma\beta_k = \Sigma\tau_\ell = 0 \quad .$$

Although i,j,k, and ℓ range from 1 to n, there are only n^2 observations present.

11.7 FACTORIAL EXPERIMENTS

We have considered a wide variety of experimental designs, both in Chapter 8 and in the present chapter. The type of design which has come to be called a factorial design has found great popularity in industrial investigations. We have already stated that the terms independent variable, factor, and treatment are often used interchangeably although in the language of experimental design the term factor is preferred. Very often it is desired to study various factors in a single experiment. In studying a chemical process it may be desired to study the factors pressure, temperature, and type of catalyst. Suppose further that one uses 3 different pressures,

4 temperatures, and 2 catalysts. This yields a 3-factor experiment with 3 levels of the first factor, 4 levels of the second, and 2 levels of the third.

There are various reasons for using a factorial experiment. First, it allows us to study the interactions (or independence) of the various factors. For example, is there an added effect in using pressure in combination with several temperatures? Second is the gain in time and effort since all observations may be used to study the effects of all factors. For example, in the above problem, if one conducted three 1-factor experiments then the observations in each experiment would yield information about the corresponding factor only and more experimental units would be needed to obtain the same accuracy as using the 3-factor experiment. Lastly, the results have wider application since each factor has been studied with varying levels of the other factors.

Since factorial experiments are used quite often we now present some notation and topics pertaining to such experiments.

11.7.1 Main Effects

Consider a 2-factor experiment with 2 levels of each factor. The two factors are denoted by a and b and the two levels of a and b are denoted by a_0, a_1 and b_0, b_1, respectively. The quantities a_0b_0, a_0b_1, a_1b_0, and a_1b_1 denote the average response from the corresponding level combination as well as the particular combination. The experiment can be displayed as follows:

	a_0	a_1
b_0	a_0b_0	a_1b_0
b_1	a_0b_1	a_1b_1

The effect of factor a at level b_0 of factor b is given by $a_1b_0 - a_0b_0$. Similarly the effect of factor a at level b_1 of factor b is given by $a_1b_1 - a_0b_1$. The <u>main effect</u> of factor a is then defined by

$$[(a_1b_0 - a_0b_0) + (a_1b_1 - a_0b_1)]/2 = (a_1 - a_0)(b_1 + b_0)/2.$$

In the same fashion the main effect of b is given by $(a_1 + a_0)$ $(b_1 - b_0)/2$.

The effects of factor a at b_0 and b_1 will not be the same unless a and b are independent of each other; that is, unless a and b do not interact. A measure of interaction between a and b is given by

$$[(a_1b_1 - a_1b_0) - (a_0b_1 - a_0b_0)]/2 = [(a_1b_1 - a_0b_1) - (a_1b_0 - a_0b_0)]/2.$$

If there is no interaction then the two terms within each set of brackets should be the same and the interaction would then be zero.

The effect of one factor at a given level of a second factor is sometimes called a <u>simple effect</u>. Thus $a_1b_0 - a_0b_0$ is a simple effect. A main effect of factor a is then the average simple effect, averaged over all levels of the other factor. There is no interaction between a and b if all the simple effects corresponding to a are equal to the main effect. These ideas can be extended to higher order factorial experiments.

The two-way and three-way AOV models discussed in Chapter 8 correspond to factorial experiments. The model $Y_{ijk} = \mu + \alpha_i + \beta_j + (\alpha\beta)_{ij} + \varepsilon_{ijk}$ considered in Section 11.3.3 is a 2-factor factorial with n replications. There are a levels of factor a and b levels of factor b. Table 11.6 gives the appropriate AOV table.

11.7.2 Confounding

We shall discuss briefly an idea which is very useful, especially in factorial type experiments. This is the idea of confounding. In working with randomized block experiments where each block is treated as a factor, it may not be possible to have a complete replicate of the experiment within one block. Confounding is used to reduce the size of blocks by sacrificing accuracy on the higher order interactions. It reduces the experimental error by the use of more homogeneous blocks than those of complete replicates. The disadvantages of confounding are that it reduces the replications on the confounded treatment comparisons and increases complexity of the calculations. If a term is confounded, it will be impossible to separate its effects from those of the effects of the blocks. The error term in a confounded factorial is made up of interactions between the treatments and the incomplete blocks. If confounding has been effective, the error for a third-order interaction (being composed of comparisons among blocks) will be greater than the error for the rest of the experiments. The main effect of a factor will not be confounded; that is, main effects will be kept free of block effects if every block contains each level of the factor the same number of times.

With only one replication we cannot obtain an estimate of error from the interaction of treatments and blocks. If, however, certain higher order interactions are negligible, their mean squares in the analysis of variance will behave exactly like components of the error mean square. Therefore they can be used to provide an estimate of error. If some of these interactions happen to be quite large, the error mean square that is used as the experimental error mean square will overestimate the true error variance and the fact that the interactions are large might not be discovered. Consider the case of the factorial design involving two levels of three different factors as shown below:

Factorial effect	(1)	a	b	c	ab	ac	bc	abc
M (mean)	+	+	+	+	+	+	+	+
A	-	+	-	-	+	+	-	+
B	-	-	+	-	+	-	+	+
C	-	-	-	+	-	+	+	+
AB	+	-	-	+	+	-	-	+
AC	+	-	+	-	-	+	-	+
BC	+	+	-	-	-	-	+	+
ABC	-	+	+	+	-	-	-	+

a, b, c, etc., are the second-level effects of corresponding factors A, B, C, etc. (1) is used as the first level of all factors. The first level of any factor is indicated by a lack of the corresponding lower case letter. In other words, bc in a 2^3 experiment means the first level of a and the second levels of b and c.

The simple effect of A at the second level of B is ab - b and the simple effect of A at the first level of B is a - (1). In like manner the main effect of C is found from

$$4 \times \text{Main effect of C} = c + ac + bc + abc - (1) - a - b - ab. \tag{11.34}$$

In a 2^n factorial experiment the interaction effect, say (BCDE), is expressed by

$$(BCDE) = \frac{1}{2^{n-1}} [(a+1)(b-1)(c-1)(d-1)(e-1)(f+1)]. \tag{11.35}$$

In other words, write out the product $\prod_i (i-1)$ for each letter i of the interaction and multiply by $\prod_i (i+1)$ for each letter i not in the interaction. In this case, as before, the factorial main effects are A, B, C, and so on. The contribution of any effect to the sum of squares is the sum of squares due to that effect divided by $r2^n$, where r is the number of replications and n is the order of the factorial.

Example 11.20 One of your colleagues is interested in the effect of flow rate and temperature on the retention time for the gas chromatographic analysis of 1-nitropropane. Four flow rates (60, 90, 120, 150 ml H_2/min) and five temperatures (40, 50, 60, 70, 80°C) are to be investigated. Five columns are available, all having the same packing. To change from one flow rate to another requires about 2 hr, while a temperature change can be accomplished in about 15 min. Consequently a reason exists to do the five temperature treatments for each flow rate together.

(a) Give a description of how you would physically carry out the experiment. Five replications are needed.
(b) Give the model for the experiment, the consequent AOV, and a discussion.

This is a factorial (4 × 5) experiment with a total of 20 treatments. There are several ways in which the physical layout of those treatments can be approached, all of which are concerned with assigning the flow rates, temperatures, or replications to the five columns. One of the first decisions to be made is, "Do we want the same degree of sensitivity in estimating flow rate effects, and vice versa?" Because of the difficulty in changing flow rates, it is suggested that we may have to reduce the sensitivity of the test for flow rate effects and to compensate for this loss in precision by increasing the sensitivity of the test for temperature effects. Having decided to assign all five temperatures as subplots to a given main plot flow rate, we now have to decide whether to assign flow rate or replication to the five columns on the basis of sensitivity desired and information available from past experience. Let us assume that the five columns are five separate individual columns; in which case there is reason to suspect they may not be exactly alike and may not give the same results.

If we decide to assign flow rate to the columns, we can rerandomize these rates at the beginning of each replication; or, we can maintain the same flow rate in a given column for all five replications. In the latter case, flow rate is completely confounded with column and it is not possible to calculate a valid error for testing flow rate effects. The only advantage of this is the experiment could be run in a shorter time since no changes in flow rate would have to be made during the course of the experiment. In the first case, where flow rates are rerandomized at the beginning of each replication, the identity of the replications is not obvious, except with respect to time (which we will assume has no effect on the results obtained). Because we have no reason for pairing replications, the AOV for this design would be as follows, where flow rates are considered main plots and temperatures are the subplots assigned to main plots in a split fashion:

Source	d.f.
Flow rates	3
Reps/flow rates	16
Temperatures	4
T × FR	12
Pooled error	64

One would expect the above design to be less sensitive in testing flow rate effects because the error term has been inflated by an unidentified portion due to replication. An alternative design that is suggested for use pulls out this unidentified replication effect in such a way as to make replication correspond to a physical feature in the experiment. This is done by considering the columns as the replication and assigning flow rates and temperatures within a single column. The model for this design is

$$Y_{ijk} = \mu + FR_i + R_j + \varepsilon_{ij} + T_k + (FR \times T)_{ik} + \delta_{ijk} \tag{11.36}$$

$$i = 1,2,3,\ldots,n, \quad j = 1,2,\ldots,m, \quad k = 1,2,\ldots,h,$$

for which the abbreviated analysis of variance is

Source	d.f.
Reps	4
Flow rates	3
Error a (ε_{ij})	12
Temperature	4
T × FR	12
Error b (δ_{ijk})	64

11.8 OTHER DESIGNS

11.8.1 Split-Plot Designs

In a two-factor experiment, involving factors A and B, it may be desirable to get more precise information on factor B and its interaction AB with A than on factor A. For example, in a randomized complete block design suppose one has 4 levels of factor (treatment) A and 4 blocks. This is a one-factor experiment replicated over 4 blocks. Suppose there is a second factor B that is present in the experiment and that it is of primary importance. Moreover, its interaction with factor A is also important. Suppose further that factor B has 2 levels and that each level of B occurs with each level of A within each block. The configuration might look as follows:

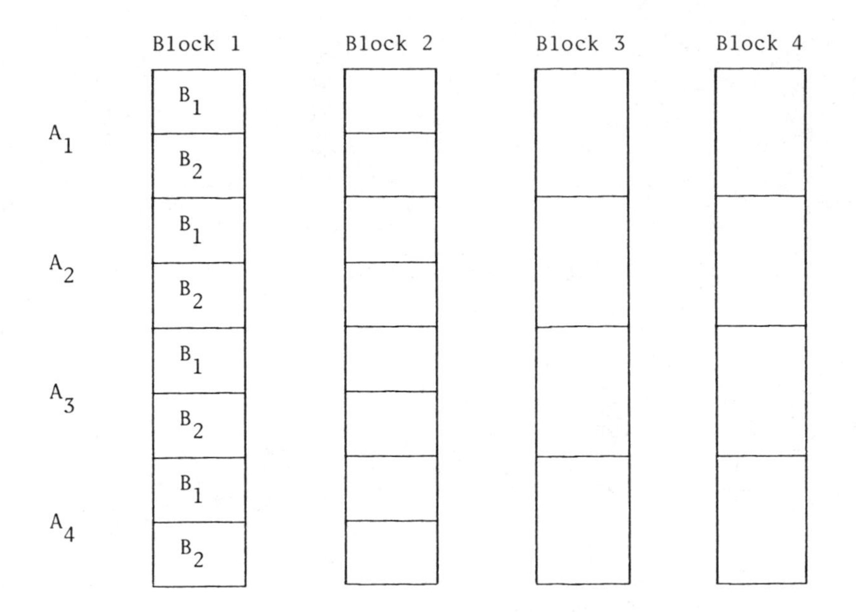

Such a design is called a split-plot experiment in randomized blocks. One can think of the main plots as being those that accommodate the levels of A and subplots those that accommodate the levels of B. Thus the main plots are "split" to accommodate the levels of the actually important factor B.

Example 11.21 In a plastics curing operation, temperature can be readily controlled at 2 levels and pressure at 4 levels. In order to speed up the experimental work needed to evaluate the effects of these two variables on surface finish of the resulting products from the molding process, four molds can be used. In this experiment, molds are the blocks, the whole plots are temperature level, and the subplots are pressure levels. This is done because the primary variable affecting surface finish and strength of the finished parts is pressure. The use of the split-plot design will give us some information about the temperature pressure interaction.

The model for a split plot in randomized blocks is

$$Y_{ijk} = \mu + \alpha_i + \beta_j + \varepsilon_{ij} + \tau_k + (\alpha\tau)_{ik} + \delta_{ijk} , \qquad (11.37)$$

$$i = 1,2,\ldots,a; \quad j = 1,2,\ldots,r; \quad k = 1,2,\ldots,b,$$

$$\varepsilon_{ij} \text{ are NID}(0,\sigma_\alpha^2); \quad \delta_{ijk} \text{ are NID}(0,\sigma_\tau^2) ,$$

where β stands for blocks or replicates, α stands for main plot treatment, and τ stands for subplot treatment.

In split-plot designs certain main effects are confounded. This is in contrast to factorial experiments where interaction effects are confounded. If the subplots are regarded as the experimental units, then the subplot treatments (levels of factor B) are applied to blocks of two units (i.e., the main or whole plot). Therefore, the differences among these blocks (main plots) are confounded with the differences among the levels of the individual treatments. The number of degrees of freedom for experimental error mean square is smaller for whole plot comparisons than for subplot comparisons. The average experimental error over all treatments is the same for randomized complete blocks as for split plots. Therefore no net gain in precision results from the use of a split-plot design. But we do have an increase in the precision of estimation of the effects of B and AB coming from a corresponding loss of precision of A. Therefore, a split-plot design should be used if B and AB are of greater interest than A or if the A effect cannot be measured on small amounts of materials. The analysis of variance for a split-plot design from a randomized complete block with r replications is shown in Table 11.17.

The sums of squares indicated in Table 11.17 are straightforward to compute and are given by

$$SS_R = \sum_{j=1}^{r} y_{\cdot j \cdot}^2/ab - y_{\cdots}^2/rab \quad ,$$

$$SS_A = \sum_{i=1}^{a} y_{i \cdot\cdot}^2/rb - y_{\cdots}^2/rab \quad ,$$

$$SS_{WPE} = \sum_{i=1}^{a} \sum_{j=1}^{r} y_{ij\cdot}^2/b - y_{\cdots}^2/rab - SS_R - SS_A \quad ,$$

$$SS_B = \sum_{k=1}^{b} y_{\cdot\cdot k}^2/ra - y_{\cdots}^2/rab \quad ,$$

$$SS_{AB} = \sum_{j=1}^{r} \sum_{k=1}^{b} y_{\cdot jk}^2/a - y_{\cdots}^2/rab - SS_A - SS_B \quad ,$$

$$SS_{SPE} = SS_{TC} - SS_R - SS_A - SS_{WPE} - SS_B - SS_{AB} \quad .$$

TABLE 11.17

Analysis of Variance for Split Plots in Randomized Complete Block Experiment of r Replications

Source	d.f.	SS	MS
Whole Plots			
Replicates (blocks)	r-1	SS_R	MS_R
A	a-1	SS_A	MS_A
Whole plot error	(r-1)(a-1)	SS_{WPE}	MS_{WPE}
Subplots			
B	(b-1)	SS_B	MS_B
AB	(a-1)(b-1)	SS_{AB}	MS_{AB}
Subplot error	a(r-1)(b-1)	SS_{SPE}	MS_{SPE}
Total (corrected)	rab-1	SS_{TC}	

If the α effects and the τ effects are fixed, then the F-ratios needed for testing H_0: all $\alpha_i = 0$, H_0: all $\tau_k = 0$, and H_0: all

$(\alpha\tau)_{ik} = 0$ are: $F = MS_A/MS_{WPE}$, $F = MS_B/MS_{SPE}$, and $F = MS_{AB}/MS_{SPE}$, respectively. Each F-statistic has the degrees of freedom corresponding to the mean squares used.

It is worth noting that the concept of "splitting" plots may be carried on for several stages, obtaining split-split-plot designs, and so on. Furthermore, split plots can be used in relation to other designs. For example, Table 11.18 gives a breakdown of the degrees of freedom of a split plot in a Latin square experiment with r replications. The corresponding sums of squares are omitted, but are easily obtained.

TABLE 11.18

Analysis of Variance for Split Plots in Latin Square Experiments with r Replications

Source	d.f.
Whole Plots	
Rows	a-1
Columns	a-1
A	a-1
Whole plot error	(a-1)(a-2)
Subplots	
B	b-1
AB	(a-1)(b-1)
Subplot error	a(a-1)(b-1)
Total	$a^2(b-1)$

11.8.2 Incomplete Block Designs

In using a factorial design it may happen that the number of factors is so great that the size of the experiment becomes excessively large. Furthermore, the precision of the experiment, in terms of experimental error, may get out of hand. One way to get around the problem is to use fractional replication or to use what are called incomplete block designs. Our aim here is not to present a comprehensive treatment of such designs. Instead we will present a brief discussion of incomplete block designs and conclude with an example of a balanced incomplete block design.

Incomplete block designs are those designs arranged in groups smaller than a whole replication in order to eliminate heterogeneity to a greater extent than is possible, say, with a randomized complete block or with a Latin square design. The term incomplete block means that not all treatment combinations are present in every block. Incomplete block designs may be balanced or unbalanced. If they are balanced, every pair of treatments occurs once and only once in the same block. Thus, all treatment pairs are compared with approximately the same precision. If the blocks cannot be balanced in separate replications, we have balanced incomplete blocks. If every treatment pair occurs once in the same row and once in the same column, the design is called a lattice square. For details as to the use and calculation methods pertaining to incomplete block designs, lattice square, and other less frequently used experimental designs, the interested students are advised to consult the text by Cochran and Cox [2], Chapters 9 through 13.

A balanced incomplete block can be viewed as an arrangement of treatments and blocks such that

i. every block contains k experimental units,
ii. the number of treatments, t, exceeds the number of plots, k, in a block,
iii. every treatment appears in r blocks out of the possible b blocks,
iv. every pair of treatments appears in the same number of blocks (λ denotes this number).

Example 11.22 Because of the disappointing results obtained as described in Example 11.19, a second set of tests of that system were carried out in an attempt to determine the effect of water pressure and air flow rate on the scrubber-impinger's operating efficiency at higher inlet particulate concentrations. Table 11.19 tabulates the individual results from these tests. This series of tests was performed as a balanced incomplete block as shown in Table 11.20. In the incomplete block layout of Table 11.20 we see that there are $b = 3$ blocks corresponding to 6400, 5120, and 4160 and $t = 3$ treatments labeled as 1, 2, 3. Furthermore, $k = 2$, $r = 2$, and $\lambda = 1$. The entries in Table 11.20 are the averages over three replications.

As indicated in Table 11.21, the errors were pooled before testing the adjusted water pressure and air rate effects. Since the calculated F's were all less than the corresponding tabular values, the null hypotheses of no significant effects coming from air flow rates and water pressure changes on collection efficiency were accepted for each variable over the range studied.

TABLE 11.19

Results for the Analysis of the Effect of Air Flow Rate and Water Pressure on Efficiency at High Inlet Particulate Concentration Levels

Air flow, cfm (block)	Water pressure level (Treatment)	Efficiency, $\%\eta^a$
4160	2	53
6400	2	83
5120	3	96
4160	1	87
6400	3	83
5120	3	86
4160	2	97
6400	3	91

[a] η = collection efficiency.

TABLE 11.19 (Continued)

Air flow, cfm (block)	Water pressure level (Treatment)	Efficiency, $\%\eta^{a}$
5120	3	92
4160	1	92
6400	2	97
5120	1	95
4160	2	90
6400	3	98
5120	1	92
4160	1	93
6400	2	98
5120	1	92

TABLE 11.20

Balanced Incomplete Block Design for the Effect of Air Flow and Water Pressure on the Nozzle Taps on Efficiency[a]

Levels of water pressure on each nozzle tap	Air flow rate, cfm: 6400	5120	4160	$\sum_j \eta_j$
3	$\eta = 91$	$\eta = 91$	X	182
2	$\eta = 93$	X	$\eta = 80$	173
1	X	$\eta = 93$	$\eta = 91$	184
$\sum_i \eta_i$	184	184	171	539

[a]Efficiencies shown are averages of 3 observations at each set of conditions.

TABLE 11.21

Analysis of Variance for the Second Series Of Air Pollution Control Tests

Source of variation	d.f.	SS	MS	F
Mean	1	144,901.38	---	---
Air rate (crude)	---	174.12	---	---
Water pressure (adjusted)	2	131.44	65.72	0.535
Water pressure (crude)	---	107.45	---	---
Air rate (adjusted)	2	198.11	99.055	0.806
Replication error	2	42.02	21.01	0.147
Experimental error	10	1,432.05	143.21	---
Pooled error	12	1,474.07	122.84	---
Total	18	146,705.		

11.8.3 Box-Wilson Composite Rotatable Design

An efficient design type for use in multifactor investigations is the Box-Wilson composite rotatable design. This design requires relatively few experiments and yields much information in a form convenient for reduction to a mathematical model of the process.

Example 11.23 Consider for example a batch reaction between two partially miscible liquids. Independent variables which would be expected to have major effect on conversion to the desired product, product quality, etc., are temperature (T), time (θ), and agitator speed (S). In order to assess the variable effects, five levels of each are selected for investigation.

In the classical experimental design a standard set of conditions would be selected (T_3, θ_3, S_3) and each variable would be varied while holding the other two constant at the standard value. The experimental conditions would be:

(T_1, θ_3, S_3) (T_2, θ_3, S_3) (T_4, θ_3, S_3) (T_5, θ_3, S_3)

(T_3, θ_1, S_3) (T_3, θ_2, S_3) (T_3, θ_4, S_3) (T_3, θ_5, S_3)

(T_3, θ_3, S_1) (T_3, θ_3, S_2) (T_3, θ_3, S_4) (T_3, θ_3, S_5)

(T_3, θ_3, S_3)

The results of these experiments would then be correlated to give the effect of each of the independent variables (temperature, time, and agitator speed) on each of the dependent variables which the experimenter chose to monitor. However, these correlations are not necessarily valid for other levels of the variables held constant. For instance, is it logical that a given change in time at level 5 of temperature will have the same effect on conversion as that same change in time at level 1 of temperature? Certainly doubt would exist as to the magnitude of the interactions and as to the reproducibility of the results, about which no information has been obtained.

Statistical designs for the above investigation might be of the factorial, fractional factorial, or Box-Wilson types. Each will provide information on variable interactions and data precision but to varying degrees. In a full factorial design, all combinations of the variables are investigated, yielding maximum information about interactions and good estimates of data precision or reproducibility. However, a large number of experiments is required. In the case of three variables, each at five levels, $(5)^3$ or 125 experiments are required. It is possible to greatly reduce the number of experiments without sacrificing much in the way of information by the use of fractional factorial or Box-Wilson designs.

The design with specific spacings of the variable levels depending on the range of each and the number of variables is shown schematically in Fig. 11.2, whose axes are T, θ, S with variation centered at T_3, θ_3, S_3. It will be noted that interaction effects may be ob-

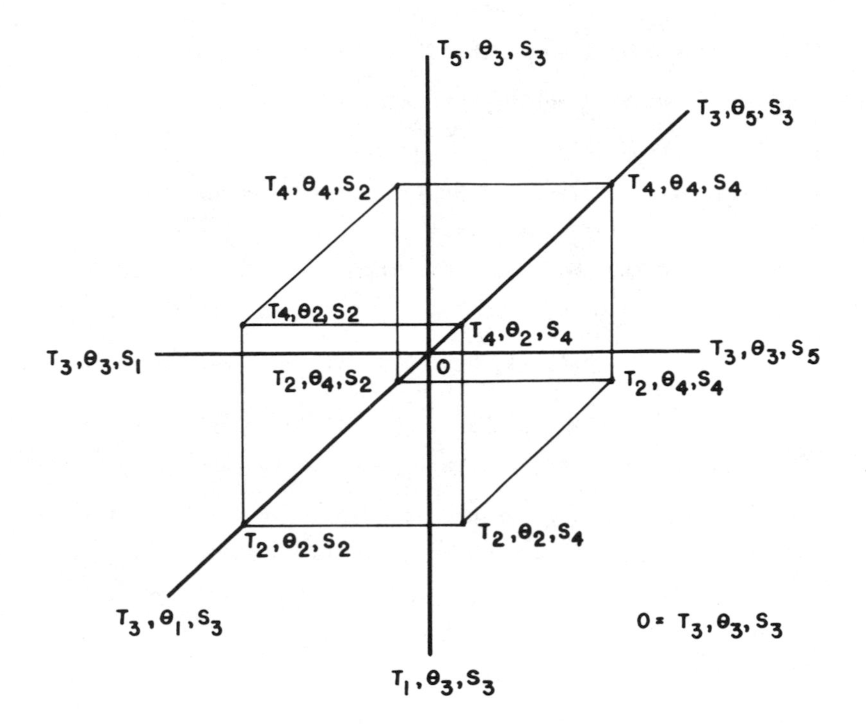

Fig. 11.2 Box-Wilson composite rotatable design.

tained. For instance, an estimate of the temperature-time interaction may be obtained by averaging the values on the right and left planes of the cube, i.e., averaging out the S effect. Thus we get the following values:

$(T_4, \theta_4)\bar{S}_X$ $\qquad$ $(T_4, \theta_2)\bar{S}_X$

$(T_2, \theta_4)\bar{S}_X$ $\qquad$ $(T_2, \theta_2)\bar{S}_X$

The top line difference gives us $(\theta_4 - \theta_2)$ at $T = T_4$ and the bottom line difference is $(\theta_4 - \theta_2)$ at $T = T_2$. If these differences differ significantly, there is a significant $T\theta$ interaction, i.e., the effect of time variation at the T_4 level of temperature is not the same as that of time variation at the T_2 level of temperature. Estimates of other interactions may be made similarly.

Main effects are obtained by comparing results at the various levels as was done in the classical design. Note however that each 2 and 4 level value for the main effect is the average of a set of four observations and thus more reliable than a single observation. For instance the mean value of T at $T = T_4$ is the mean of the values at (T_4, θ_2, S_2), (T_4, θ_4, S_2), (T_4, θ_4, S_4), (T_4, θ_2, S_4) and that at $T = T_2$ is the mean of (T_2, θ_2, S_2), (T_2, θ_4, S_2), (T_2, θ_4, S_4), and (T_2, θ_2, S_4). While θ and S do vary, they do so in a manner such that their effects cancel. A characteristic of statistically designed experiments is that each observation is used over and over again in establishing the variable effects. Therein lies its efficiency and superiority over classical designs.

Thus in comparing this statistical design with the classical ones we observe that for the two additional observations (15 vs 13) in the statistical design we get better estimates of the main effects, estimates of the two variable interactions, and an estimate of the observation error. As the number of variables increases the advantages become greater.

It must be remembered, however, that increased efficiency is not free. The number of degrees of freedom for experimental error is lower than for classical designs. More important, the loss of a single datum point may and indeed often will not only prejudice the results but may necessitate redoing the entire experiment. In selecting any design, these potential disadvantages must be weighed against the potential gains.

This example and No. 11.20 constitute examples of but two of the rapidly growing numbers of statistical experimental designs. Many of these were developed in answer to a particular applied problem.

They are, however, universally adaptable provided that the same type of conditions exist. They, as most of the newer designs, are much more efficient than the so-called classical methods (completely random, Latin square, etc.). It is hoped that the serious student will not stop here, but will, having by now attained some background in the methods for designing effective and efficient experiments, go to the literature for specialized designs when the need arises.

11.9 DESIGN EFFICIENCY

One of the greatest problems a practicing engineer has is that of selecting the single design or designs for a particular application which are most efficient and at the same time most effective for his particular needs. As mentioned earlier, effectiveness refers to the ability of a design to answer the questions that you have regarding a process, operation, effect, and so on. Efficiency refers to how speedily they are answered with regard to resource utilization and with what accuracy and precision. Most of all, efficiency describes how well the existing data are utilized.

Let us define the variance of a treatment mean in general as the mean square for experimental error divided by the number of observations per treatment. We can, by comparing the variance of a treatment mean for any two designs, compare them with regard to their relative efficiency. Consider, for example, the randomized complete block design where we have t observations per treatment and b blocks. For these conditions the variance of a treatment mean can be calculated as V_1. Now let us change the layout of the entire RCB experiment. Instead of having t observations per treatment, let us have w observations per treatment and let us have d blocks instead of b blocks. Under these new conditions the variance of a treatment mean is V_2. The relative efficiency of the original design as compared to the revised design is equal to $100V_2/V_1$. In order to compare the efficiency of the randomized complete block design to the efficiency of a completely randomized design we must, at least on paper, manipulate the data involved so that it appears that the treatments were assigned at random to the experimental units and that there were no blocks involved at all. We would then estimate the experimental error mean square under these conditions of complete randomization and from it calculate the relative efficiency of the randomized complete block design with respect to the simulated completely randomized design. This is $100V_{CR}/V_{RCB}$. Comparisons of this type can be made for any two designs.

You will find that, as the designs become more complex, the effectiveness of the design increases rapidly. We are able to find out more things from the results of the experiment. However, while this is going on the relative efficiencies are changing in a manner that reflects the increase in accuracy of the estimate of experimental error and its corresponding decrease in sensitivity. With this goes a decreasing probability of obtaining a significant calculated F-value. A convenient calculational formula ([2], p. 112) to enable you to estimate the mean square for experimental error for a simulated completely random design from known randomized complete block data is

$$V_{CR} = \frac{(b-1)MS_B + b(t-1)MS_E}{bt-1} \tag{11.38}$$

The estimation ([2], p. 127) of the mean square for experimental error for a simulated randomized complete block design from known Latin square data is found from

$$V_{RCB} = \frac{(m-1)MS_R + (m^2+1)MS_E}{m^2-1} \tag{11.39}$$

Of course, the calculation of relative efficiencies after you have performed the experiment is foolish: It profits you absolutely nothing! What one should realize is that some designs are more effective and more efficient than others in certain applications. Learn them well and calculation of efficiency will be a problem of minor concern.

11.10 ANALYSIS OF COVARIANCE

In discussing analysis of variance and experimental designs, we have seen how, by incorporating an additional factor or by the use of blocks, effects which are of no interest in an experiment can be eliminated. This results in a smaller denominator in the F-ratio thus resulting in a test which is more sensitive to the effects which are of interest. That in turn increases the power of the test (i.e., the probability of rejecting a false hypothesis). The analysis of covariance is another technique which may be used to accomplish the same objective.

As an example, suppose one is interested in comparing several teaching methods. A number of students are randomly assigned to each of the teaching methods. The dependent variable is Y, the final score obtained by the students on a final examination. The effect of student IQ on the final score can be eliminated by using IQ as a covariate, that is, by adjusting the final score for IQ.

As a second example, suppose one is studying the carbon content Y (in grams) in 4 types of steel. There are 5 batches, not all of the same weight, for each type of steel. The determinations of carbon content are to be made by one analyst. The weight of each batch, X, can be used as a covariate and differences in carbon content can be studied by an analysis of covariance. (Note that in this case, however, one also could do a one-way AOV on weight percent. See Problem 8.8.)

Consider a one-way AOV where the dependent (response) variable is Y and the independent (concomitant) variable is X. The usual AOV model on the Y's is

$$Y_{ij} = \mu + \alpha_i + \varepsilon_{ij}, \quad i = 1,2,\ldots,t; \quad j = 1,2,\ldots,n_i \quad .$$

The analysis of covariance model in a completely randomized design is

$$Y_{ij} = \mu + \alpha_i + \beta(X_{ij} - \overline{X}) + \varepsilon_{ij} \quad ,$$
$$i = 1,2,\ldots,t; \quad j = 1,2,\ldots,n_i \quad . \tag{11.40}$$

The model in (11.40) can be written as $Y_{ij} = \nu + \alpha_i + \beta X_{ij} + \varepsilon_{ij}$, with $\nu = \mu - \beta\overline{X}$. The form in (11.40) is more customary and is the one discussed here. It is assumed that the ε_{ij} are NID(0, σ^2).

From (11.40) it is seen that $E(Y_{ij}) = \mu_{ij} = \mu + \alpha_i + \beta(X_{ij} - \overline{X})$. Thus the means of the Y's for each fixed i lie on a straight line. The slope of the line is constant for each of the t groups; there may be t different lines due to different values for the α_i but they are parallel.

The usual assumptions as used for the completely randomized design are made. An additional and important assumption is that the values of X are not affected by the treatments. This may not always hold in practice so the analyst should be aware of this assumption and take care in interpreting the results of a covariance analysis.

The results of a covariance analysis in the one-way AOV are summarized in the following table. The notation in Table 11.22 is as follows:

$$n = \sum_{i=1}^{t} n_i \quad ,$$

$$T_{xx} = \sum_{i=1}^{t} \sum_{j=1}^{n_i} (\overline{X}_{i.} - \overline{X}_{..})^2 = \sum_{i=1}^{t} \frac{X_{i.}^2}{n_i} - \frac{X_{..}^2}{n} \, ,$$

$$T_{yy} = \sum_{i=1}^{t} \frac{Y_{i.}^2}{n_i} - \frac{Y_{..}^2}{n} \, , \qquad T_{xy} = \sum_{i=1}^{t} \frac{X_{i.}Y_{i.}}{n_i} - \frac{X_{..}Y_{..}}{n} \, ,$$

$$E_{xx} = \sum_{i=1}^{t} \sum_{j=1}^{n_i} (X_{ij} - \overline{X}_{i.})^2 = \sum_{i=1}^{t} \sum_{j=1}^{n_i} X_{ij}^2 - \sum_{i=1}^{t} \frac{X_{i.}^2}{n_i} \, ,$$

TABLE 11.22

Results of Covariance Analysis in the One-Way AOV

Source	d.f.	SS_{x^2}	SS_{xy}	SS_{y^2}	SS'_y	d.f.	MS'_y	F
Treatment	t-1	T_{xx}	T_{xy}	T_{yy}	SS'_{yTr}	t-1	MS'_{yTr}	$\frac{MS'_{yTr}}{MS'_{yE}}$
Error	n-t	E_{xx}	E_{xy}	E_{yy}	SS'_{yE}	n-t-1	MS'_{yE}	
Total	n-1	$S_{xx} = T_{xx} + E_{xx}$	$S_{xy} = T_{xy} + E_{xy}$	$S_{yy} = T_{yy} + E_{yy}$	SS'_{yT}			

$$E_{yy} = \sum_{i=1}^{t} \sum_{j=1}^{n_i} Y_{ij}^2 - \sum_{i=1}^{t} \frac{Y_{i.}^2}{n_i}, \quad E_{xy} = \sum_{i=1}^{t} \sum_{j=1}^{n_i} X_{ij} Y_{ij} - \sum_{i=1}^{t} \frac{X_{i.} Y_{i.}}{n_i},$$

$$SS'_{yT} = S_{yy} - S_{xy}^2/S_{xx},$$

$$SS'_{yE} = E_{yy} - E_{xy}^2/E_{xx},$$

$$SS'_{yTr} = SS'_{yT} - SS'_{yE} = T_{yy} - S_{xy}^2/S_{xx} + E_{xy}^2/E_{xx}.$$

The F ratio to use in testing H_0: $\alpha_1 = \alpha_2 = \ldots = \alpha_t = 0$, after adjusting for the X values is

$$F = \frac{SS'_{yTr}/(t-1)}{SS'_{yE}/(n-t-1)} = \frac{MS'_{yTr}}{MS'_{yE}}.$$

The hypothesis H_0: $\alpha_1 = \alpha_2 = \ldots = \alpha_t = 0$ is rejected at the significance level α if $f \geq F_{t-1,n-t-1,1-\alpha}$. This is a test of the hypothesis that there are no differences among the effects of the t treatments after adjusting for the concomitant variable (covariate) X.

The estimate of β is

$$\hat{\beta} = E_{xy}/E_{xx}$$
$$= \sum_{i=1}^{t} \sum_{j=1}^{n_i} (X_{ij} - \overline{X}_{i.})(Y_{ij} - \overline{Y}_{i.}) / \sum_{i=1}^{t} \sum_{j=1}^{n_i} (X_{ij} - \overline{X}_{i.})^2$$

When doing a covariance analysis, it is assumed that $\beta \neq 0$. If $\beta = 0$, then it is not necessary to analyze the data by covariance. Often, the experimenter will want to test the null hypothesis H_0: $\beta = 0$. This may be done by computing

$$F = \frac{E_{xy}^2/E_{xx}}{SS'_{yE}/(n-t-1)} = \frac{E_{xy}^2/E_{xx}}{MS'_{yE}}$$

and rejecting H_0: $\beta = 0$ if $f \geq F_{1,n-t-1,1-\alpha}$.

In addition to the F-test on treatment differences adjusted for X, one is often interested in the adjusted treatment means and their standard errors. The adjusted treatment means are

$$\text{adj. } \overline{Y}_{i.} = \overline{Y}_{i.} - \hat{\beta}(\overline{X}_{i.} - \overline{X}_{..}), \quad i = 1,2,\ldots,t \quad ,$$

and the estimated variances are,

$$\hat{V}(\text{adj. } \overline{Y}_{i.}) = MS'_{yE}\left[\frac{1}{n_i} + \frac{(\overline{X}_{i.} - \overline{X}_{..})^2}{E_{xx}}\right], \quad i = 1,2,\ldots,t \quad .$$

The estimate of the variance of the difference of two adjusted means is

$$\hat{V}(\text{adj. } \overline{Y}_{i.} - \text{adj. } \overline{Y}_{j.}) = MS'_{yE}\left[\frac{1}{n_i} + \frac{1}{n_j} + \frac{(\overline{X}_{i.} - \overline{X}_{j.})^2}{E_{xx}}\right] .$$

As has been seen, analysis of covariance combines analysis of variance with regression analysis. Analysis of covariance may be used with many other designs besides the one-way AOV, however, we will not elaborate any further. We now present an example of analysis of covariance in a one-way AOV.

Example 11.24 Three lab groups varied the tube-side water rate X in a laboratory-scale, two-pass, double-pipe heat exchanger in order to determine the local inside heat-transfer film coefficient Y as a function of water rate. The 3 groups conducting this experiment did not use the same steam rates (STEAM) for this experiment. In order to evaluate the effect of different steam rates, we perform an analysis of covariance for these data using water rate as the covariate. The SAS program (SAS User's Guide: Statistics, 1982 Edition) is below followed by the program output.

```
(put initial JCL cards/statements here)
DATA DP;
INPUT STEAM  X  Y;
CARDS;

    0.5     749    241
    0.5     749    259
    0.5    1498    335
    0.5    1498    317
    0.5    1498    344
    0.5    2246    380
    0.5    2246    369
    0.5    2246    370
    0.5    2995    453
    0.5    2995    433
    0.5    3744    497
    2.0    1470    593
    2.0    4410    550
    2.0    5880    670
    2.0    7350    700
    2.0    8090    753
    2.0    8820    766
    2.0    1470    321
    2.0    2940    442
    2.0    4410    594
    2.0    5880    647
    2.0    5880    662
    5.0    1470    305
    5.0    4410    540
    5.0    5880    690
    5.0    7350    710
    5.0    8090    771
    5.0    8820    794
    5.0    1470    300
    5.0    2940    412
    5.0    4410    641
    5.0    5880    719
    5.0    5880    722

PROC GLM;
CLASS STEAM;
MODEL Y = STEAM X/SOLUTION;
(put final JCL cards/statements here)
```

GENERAL LINEAR MODELS PROCEDURE

DEPENDENT VARIABLE: Y

SOURCE	DF	SUM OF SQUARES	MEAN SQUARE	F VALUE
MODEL	3	898732.53734567	299577.51244856	87.99
ERROR	29	98733.52326040	3404.60425036	PR > F
CORRECTED TOTAL	32	997466.06060606		0.0001

SOURCE	DF	TYPE III SS	F VALUE	PR > F
STEAM	2	15326.16040168	2.25	0.1234
X	1	471760.65855779	138.57	0.0001

PARAMETER		ESTIMATE	T FOR H0: PARAMETER=0	PR > |T|	STD ERROR OF ESTIMATE
INTERCEPT		241.50623113 B	11.83	0.0001	20.41649812
STEAM	2	60.14348519 B	2.04	0.0503	29.44223788
	5	51.59803065 B	1.75	0.0903	29.44223788
	0.5	0.00000000 B	.	.	.
X		0.05971472	11.77	0.0001	0.00507287

Note: The X'X matrix has been defined singular and a generalized inverse has been employed to solve the normal equations. The above estimates represent only one of many possible solutions to the normal equations. Estimates followed by the letter B are biased and do not estimate the parameter but are "the best linear unbiased estimates" for some linear combination of parameters (or are zero). The expected value of the biased estimators may be obtained from the general form of estimable functions. For the biased estimators, the STD ERR is that of the biased estimator and the T value tests H_0: E(Biased Estimator) = 0. Estimates not followed by the letter B are "the best linear unbiased estimates" for the parameter.

To test for differences among levels of the steam rate adjusted for the covariate (water rate) we use the Type III SS to obtain

$$f = MS'_{Tr}/MS'_{yE} = \frac{15,326.16040/2}{98,733.52326/29} = 2.25$$

As $f < F_{2,29,0.95} \cong 3.33$, we accept the hypothesis of no treatment (steam rate) effect as adjusted for the covariate (water rate). To test the null hypothesis of no effect of water rate on the heat-transfer coefficient, we can use either the Type III SS to calculate

$$f = \frac{E_{xy}^2/E_{xx}}{MS'_{yE}} = \frac{471,760.65856/1}{98,733.52326/29} \simeq 138.57 \text{ vs. } F_{1,29,0.95} \cong 4.19$$

or

$$T = 11.77 \text{ vs. } t_{29,0.975} = 2.045 \quad .$$

By either test, the null hypothesis that the water rate does not affect the heat transfer coefficient (H_0: $\beta = 0$) is rejected as it should be. Regardless of flow regime (laminar, transition, or turbulent), the local inside heat-transfer coefficient is known to be affected by the inside fluid (water in this case) rate. This can be shown by the Seider-Tate, Hausen, and Dittus-Boelter equations, respectively.

PROBLEMS

11.1 An experiment is planned which will measure the effect of applying various amounts of a chemical which will partially close the stomata on plant leaves on the transpiration rate. As the research engineer responsible for the pilot plant producing this new material, you are working closely with the plant scientist who is running the field tests of your material. The usual transpiration rate for sorghum is 5 g/dm^2 hr for the wind velocity, degree of cloudiness, relative humidity, and dry bulb temperature on the South Plains of Texas. This rate may fluctuate by 0.5 g/dm^2 hr under normal daytime conditions. How many replicates will be required to detect, 9 times out of 10, a difference of 1.5 g/dm^2 hr in transpiration rate? The significance level you and the plant physiologist have agreed to use is 5%.

11.2 Another experiment related to water conservation is concerned with the application of naturally occurring lipids to the soil between irrigated plant rows to cut down on surface evaporation. The effect of the control agent can be determined as a function of water content 6 in. below the surface. The average moisture content of an untreated soil is 25% $\pm$ 1.4%. If a difference in water content of 2% moisture is to be detected 4 out of 5 times, how many replications will be needed? The significance level is 5%.

11.3 Using the data of Problem 7.29, how many observations are needed for a 5% test with a 80% probability of finding a significant result?

11.4 Using the data of Problem 7.7, if $\alpha = 0.01$ and a 95% probability of finding a significant result is desired, how many observations are needed? For $\alpha = 0.05$? What do you conclude regarding this experiment?

11.5 The results of a laboratory determination for the purity of feed, distillate and bottoms, with feed entering on tray 3 in a 6 sieve-tray, Scott fractional distillation column, are given below. Three samples were taken and analyzed in triplicate determinations for the feed, distillate and bottoms. Does subsampling make a significant difference in the results?

Methanol Purity, %

Feed	Distillate	Bottoms
6.25	64.8	2.06
6.24	64.7	2.03
6.24	64.7	1.98
6.51	64.5	2.55
6.48	64.5	2.62
6.60	64.3	2.49
6.21	64.8	2.91
6.13	64.9	2.97
6.17	64.8	2.94

11.6 The Visbreaker[R] Unit was fed 3 kinds of heavy oil: 30% hydro-treated, 50% hydro-treated and 70% hydro-treated. Samples of the flash tower bottoms were taken each day for 4 days. Distillation at 10 mm Hg was used to analyze the initial boiling point, °F, of the bottoms product. The distillation was carried out four times for each sample.

Sample	30%	50%	70%
1	425	457	510
	431	462	507
	436	460	500
	433	455	505
2	431	460	500
	423	456	510
	427	463	495
	429	465	498
3	428	482	505
	437	476	511
	436	480	506
	431	475	513
4	433	470	513
	435	476	505
	425	467	507
	430	465	510

Prepare the analysis of variance for these data and interpret the results.

11.7 In the gas absorption experiment in our laboratory, one-liter samples of the effluent (cleaned) gas are taken for ammonia analysis at equilibrium for each set of operating conditions. Four 40 ml aliquots are taken from each one-liter sample bag. Each aliquot is contacted with 10 ml double distilled, CO_2-free water. The water is analyzed for ammonia by specific ion electrode. For the past semester, the results below, in ppm ammonia, were obtained for an absorber water flow rate of 8 GPH and a total gas flow rate of 4 CFM (70°F, 14.7 psia). Does inlet ammonia concentration affect absorber efficiency?

	% Ammonia Entering Gas Phase						
	4	8	12	16	20	$y_{i..}$	$y_{i..}^2$
	12	21	37	48	60		
	13	24	37	42	62	725	525,625
	18	26	31	50	57		
	17	20	33	46	71		
	14	25	36	43	66		
	15	27	34	49	64	728	529,984
	13	22	31	51	59		
	10	24	39	41	65		
	11	19	36	49	63		
	14	23	37	52	61	734	538,756
	12	21	34	48	58		
	19	28	32	47	70		
$y_{.j.}$	168	280	417	566	756		
$y_{.j.}^2$	28,224	78,400	173,889	320,356	571,536		

$\sum_j y_{.j.}^2 = 1,172,405$ $\qquad$ $\sum_i y_{i..}^2 = 1,594,365$

$y_{...} = 2187$ $\qquad$ $SS_M = 79,716.15$

$\sum_i \sum_j y_{ij.}^2 = 390,959$ $\qquad$ $SS_T = 98,307$

11.8 For one physical chemistry experiment, the teaching assistant made up a stock buffer solution. He took three samples once a week thereafter to check the pH. Two determinations were made on each weekly sample. Did the solution pH vary with time?

Sample	Week 1	Week 2	Week 3
1	5.2	4.9	5.0
	5.3	4.8	5.0
2	4.9	5.0	5.3
	5.2	5.0	5.0
3	4.8	5.2	5.1
	5.1	5.2	5.2

11.9 Invertase (β-D-fructofuranoside fructohydrolase) activity has been studied in the developing embryonic chicken. Duodenal loop tissue obtained from embryonic chicks ranging from 13 days to hatching was homogenated. Two sub-samples from each homogenated tissue sample were then assayed for specific invertase activity. It has been proposed that invertase activity neither increases nor decreases from day 13 to 15. Does the following invertase activity data, mg glucose released/mg protein/hr, support this opinion?

Sample	Age, Days 13	14	15
1	56.8, 55.4	60.5, 60.0	62.1, 62.0
2	40.3, 41.7	59.8, 59.2	63.2, 63.0
3	54.3, 55.5	57.6, 56.9	59.8, 60.2
4	55.2, 55.7	59.3, 58.7	63.4, 63.1
5	59.2, 58.9	62.0, 61.8	67.3, 67.7

11.10 In another study of invertase activity in the developing embryonic chicken, duodenal loop samples were obtained from silver laced Winedot-Eastern Ray (A) and silver laced Winedot-AVA Conda (B) embryonic chicks from age 13 to age 17 days. The duodenal tissue was homogenized and assayed for specific invertase activity. Does age influence invertase activity?

Is there any significant variation in invertase activity between the 2 varieties of chicken?

Variety	Age, Days				
	13	14	15	16	17
A	56.3	60.5	64.3	86.1	97.5
B	53.4	61.2	64.5	87.0	94.2

11.11 Results of a laboratory determination for the cotton dust weight (μg) present on air sampling filters at a yarn mill are given below using 5 vertical elutriators during shift A and shift B. The data was collected at 2 different sites in the mill: the warehouse and the bale opening area. Is there a significant (α = 0.05) difference in the results between shifts? Does the site location make a significant difference?

Site	Shift A	Shift B
Warehouse	610	350
	830	130
	630	380
	660	460
	490	400
Bale Opening	430	170
	380	250
	690	270
	500	360
	330	370

11.12 The sampling program described in Problem 11.11 was continued for the other two shifts with the results in μg shown below. It should be noted that the samples from all 4 shifts were obtained over a six-hour period and that the elutriator flow rates ranged from 7.43 to 7.59 liters/min. as specified by 29CFR1910.1043.

Site	Shift C	Shift B
Warehouse	515	635
	635	355
	485	855
	465	655
	405	685
Bale Opening	845	870
	765	815
	605	495
	610	530
	670	460

Prepare the analysis of variance table for these new data and interpret the results.

11.13 Using all the data of Problems 11.11 and 11.12, prepare the analysis of variance table and determine, at the $\alpha = 0.05$ level, the significance of site, shift, and any possible interaction between the two.

11.14 Ten different paints are to be evaluated for use in a chemical plant. In order to determine the effectiveness of each paint as an atmospheric corrosion inhibitor, 40 pieces of 18WF60 structural framework are selected on the plant site for experimental purposes. Ten test sections are located near the methanol oxidizers, 10 near the coal-fired powerhouse, 10 near the HNO_3 chamber process, and the last 10 near the synthetic NH_3 compressors. The following data were obtained when the paints were assigned at random within each of the 10 locations: (Data represent a function of discoloration, weight loss, and cracking after ten months).

		Paint No.									
		1	2	3	4	5	6	7	8	9	10
Location No.	1	2	5	8	6	1	3	8	6	4	4
	2	3	4	7	5	2	5	8	12	5	4
	3	3	5	10	5	1	7	7	2	6	2
	4	5	5	9	2	2	8	8	5	3	3

(a) What does the above design accomplish?

(b) Write the AOV showing expected mean squares for this experiment.

(c) What is the standard error of a treatment mean for these data?

(d) Are the paints equally effective?

(e) Based on your results in (d), what would you do next?

11.15 A new engineer would like to investigate the effect of four temperatures on the yield of a certain product. Since he would like his experiment to be complete, he used four different pressures as blocks in an RCB design. In the time available, he could make 16 replications so he assigned all four temperatures to each block. Give the model for this experiment and define all terms in it. What modifications, if any, would you make in the design of this experiment? If you recommend any changes, give the resulting model and tell specifically how each change is beneficial.

11.16 A plant scale experiment is being designed to compare the effects of a number of changes in reaction conditions (temperature, pressure, catalyst concentration, regeneration frequency, and throughput) on the yield of ethylene glycol via the du Pont process. Because of the various process lags involved, considerable quantities of raw materials will be used. Give two general design approaches to this problem and discuss them as you would with the production superintendent.

11.17 A randomized block experiment was conducted to investigate the effects of the concentration of detergent (D) and sodium carbonate (S) on the working and suspending power of a solution. The data below refer to performance on working tests, with the higher figure indicating better performance. Each figure is the total of three observations. The levels were equally spaced.

	D_1	D_2	D_3	Sum
S_1	437	673	925	2035
S_2	711	1082	1157	2950
S_3	814	1146	1123	3083
	1962	2901	3205	8068

Compute the analysis of variance and test the significance of the D and S terms using the appropriate error terms. Total sum of squares is 2,875,729.

11.18 Given the following analysis of variance from a randomized complete block experiment,

(a) What is the variance of a treatment mean?

(b) What is the variance of the difference between any two treatment means?

Source	d.f.	MS
Blocks	5	81
Treatments	6	190
Error	30	40

11.19 A certain pilot plant operation involves the saponification of a particular fatty alcohol. This reaction can be carried out in either of 2 different reactors. As the engineer in charge of this project, you feel that the raw material from one supplier is contaminated. Before you left for vacation, you told your technicians to run 6 batches from each of the 2 possible suppliers through the process under normal conditions. This will give you 6 replications on each of the 2 treatments (fatty alcohol sources) for the t-test you have in mind. Your summer technicians decided to speed things up and run 3 batches through each reactor. Now that you're back from your trip, how are you going to analyze the data which your technicians have obtained? Include in your answer the following: model, hypotheses to be tested and the necessary F-ratios, and the type of effects (fixed, random, mixed) being evaluated.

11.20 A test of paint primers used for corrosion inhibition gave the following results in a RCB experiment:

Source	d.f.	MS
Blocks	9	241.59
Primers	4	269.67
Error	36	97.41

Are differences in primers significant at the 5% level?

11.21 Data were obtained in a RCB experiment in which the treatments were feeding arrangement in a triple-effect evaporator system. The effect of feed direction [forward, backward, mixed (2,3,1)], mixed (2,1,3,) were measured in terms of steam economy, lb evaporated/lb steam used. The results were as follows:

		Treatments			
		T_1	T_2	T_3	T_4
Blocks	B_1	3.6	4.1	3.0	--
	B_2	3.8	3.7	3.3	4.5
	B_3	3.0	3.9	3.4	4.0
	B_4	4.0	4.2	3.8	4.8

For this experiment, the total Δt allowed was used to group the data into blocks. Prepare the analysis of variance table for these data and test the hypothesis that the true treatment means are equal at the 5% significance level.

11.22 In an experiment to determine the effect of acid pickling in the galvanizing process, ten samples were used. Four samples were dipped in the pickling solution once; three samples were dipped twice; two samples were dipped three times; and the other sample was dipped four times. The number of dippings was randomly assigned to the samples. The data below were obtained where the quality index is a measure of adherence of the zinc in the next step in the process.

Sample (arbitrary numbering)	Number of dippings	Quality index
1	1	2
2	1	2
3	1	3
4	1	5
5	2	5
6	2	6
7	2	7
8	3	9
9	3	11
10	4	10

The uncorrected sum of squares for the quality index is 454. $SS_M = 360$.

(a) Set up the analysis of variance for this experiment.

(b) Compute the relation between number of dippings and quality index and test the regression coefficients.

11.23 Discussions with the design engineer revealed that half the peel strength values presented in Problem 8.26 (those marked with an asterisk) were obtained when using samples of one thickness. The other values (unmarked) were obtained using samples of a different thickness. You are required to include this new variable (sample thickness) in your analysis of variance.

(a) Give the appropriate model for this three-way classification.
(b) Prepare the complete analysis of variance table.
(c) State and test the null hypotheses involved.

11.24 As part of your senior project, you wish to study the effects of temperature, pressure, and molar feed rate ratios (liquid to gas) on the efficiency of a formaldehyde absorber. Three pressures, two temperatures, and three feed rate ratios are to be evaluated.

Six small bench scale absorbers are available and suitable for operation at any of the desired pressures. Temperature can be controlled at either level on each. Evaluation of any particular feed rate ratio will require one afternoon allowing for start-up and shut-down. In this semester you estimate having an average of three afternoons per week for ten weeks available. Then you must stop all experimental results, correlate and analyze your data, and write up the report.

Design an experiment to obtain the maximum amount of information in the available time. Discuss the reasons for your choice of designs. Describe how you would set up the experiment and analyze the data.

11.25 Given the following analysis of variance for fixed treatment levels:

Source	d.f.	MS
Among treatments	9	400
Among experimental units within treatments	70	80
Among samples within experimental units	320	5

(a) How many samples were used for each experimental unit?

(b) What are the expected mean squares for these sources of variations?

(c) Test the hypothesis of equality of treatment effects at $\alpha = 0.05$.

(d) What is your estimate of the difference between any two treatment means?

11.26 You designed an experiment to measure the wearing quality of sweaters as a function of the amount of drawing and crimping used in the process where the tow was originally made. This experiment started out as a Latin square with amount of crimp as columns and the degree of drawing as rows. The treatments applied are surface finishes on the finished yarn. The original design is shown below. When the experiment was over, you realized from the lab notebook that your technician had applied treatment A in the R4C2 position and treatment C in the R4C4 position. This is just the reverse of what you had intended. To further complicate matters, you expect a much stronger gradient among rows than among columns. Indicate the various possible methods of data analysis. Discuss the relative merits of each.

		Crimp			
		C_1	C_2	C_3	C_4
	R_1	A	B	C	D
Draw	R_2	D	A	B	C
	R_3	C	D	A	B
	R_4	B	C	D	A

11.27 In an industrial experiment investigating the effects of four different processes (I, II, III, IV), you believe you can eliminate any technician and/or day bias by using a Latin square design. Weekdays are rows and technicians are used as columns. Your design is laid out as below. You showed this design to your supervisor. He thought that a day-technician interaction exists. Does your design take this interaction into consideration? If so, how? For the given design, show the model used defining all terms. Also show the source and corresponding degrees of freedom for all error sources in an AOV table.

		Technician			
		1	2	3	4
	M	I	II	III	IV
Days	Tu	IV	I	II	III
	Th	III	IV	I	II
	F	II	III	IV	I

11.28 Give at least four methods of increasing the precision of an experiment. Explain briefly the principle on which each is based.

(e) Estimate the efficiency of this design relative to one in which 10 experimental units per treatment and 3 samples per experimental unit are used.

11.29 In a nested sampling arrangement for the determination of percent solids from the viscosity of a polyester spinning solution, duplicate determinations were made on two samples from each of three Marco mixers. A partial analysis of variance is below for the data obtained.

Source	Sum of squares
Between mixers	25.9
Between samples from mixers	0.42
Between determinations per sample	0.66

(a) Complete the analysis of variance showing the degrees of freedom, mean square, and expected mean square for each treatment.
(b) Compute the variance of the overall mean on a per determination basis.
(c) If the cost of the experiment were the same, what would have been the effect of taking four samples from each mixer and making only one determination of solids content per sample?

11.30 An experiment was conducted to determine the relative effectiveness of three different solvents in a vapor de-greaser. Since cleaning efficiency is a function not only of the solvent used but also of the depth of immersion of the parts in the vapor and the amount of stripped grease already in the vat, the experiment was conducted as a 3 × 3 × 3 factorial with 2 observations per factorial combination. The resulting sums of squares from the analysis of de-greasing efficiency are given:

Source	Sum of squares
Solvent (S)	400
Immersion (I)	640
Stripped grease conc (C)	1660
S × I	620
S × C	720
I × C	420
S × I × C	1160
Error	2700
Total	8320

(a) Complete the analysis of variance table.

(b) Discuss the factors involved (S, I, C) as to whether they are fixed, random, or either and present the appropriate models and expected mean squares.

(c) This experiment could have been performed as a Latin square with stripped grease concentration as rows, immersion depth as columns, and solvent type as treatments. Discuss the advantages and disadvantages of such a design relative to the one actually used.

11.31 The following experiment was conducted in the University Hospital at Iowa City. The problem was to investigate the influence of an anaesthetic on the consumption of oxygen by various brain tissues. Eight dogs were used in the experiment with a two-stage operation being performed, samples of the three brain tissues being removed before and after administration of the anaesthetic. Each tissue sample was used to make up two 1-g subsamples and these were placed in separate flasks for the measurement of O_2 consumption. Readings of O_2 consumptions were taken every 30 minutes over a 4.5-hr period giving, in all, eight readings per subsample. Thus we have an experiment including 8 dogs, 3 brain tissues, 2 treatments (with and without anaesthetic), 2 samples per treatment, 8 observations in time, and a total of 768 observations. Write out your version of the analysis of variance for this experiment showing the appropriate degrees of freedom; consider dogs as replicates. Indicate the proper F-tests for brain tissues, treatments, and the interaction of the two factors.

11.32 In an experiment to measure the variability in breaking strengths of tire cords made at two different plants, Akutowicz and Truax [1] obtained the data given. In this experiment, bobbins of tire cord were randomly selected in two groups of 8 from regular

Coded raw data[a] for Problem 11.32

		0 yd		500 yd		1000 yd		1500 yd		2000 yd		2500 yd	
Adjacent breaks		1	2	1	2	1	2	1	2	1	2	1	2
Bobbin Plant I	1	-1	-5	-2	-8	-2	3	-3	-4	0	-1	-12	4
	2	1	10	1	2	2	2	10	-4	-4	3	4	8
	3	2	-3	5	-5	1	-1	-6	1	2	5	7	5
	4	6	10	1	5	0	5	-2	-2	1	1	5	9
	5	-1	-8	5	-10	1	-5	1	-4	5	-5	3	6
	6	-1	-10	-9	-8	-2	2	0	-3	-8	-1	-2	-4
	7	-9	-2	5	-2	7	-2	-2	-2	-1	2	10	5
	8	0	2	-5	-2	5	3	10	-1	4	1	7	-1
Bobbin Plant II	1	10	8	-5	6	2	13	7	15	17	14	18	11
	2	9	12	6	15	15	12	18	16	13	10	9	11
	3	0	8	12	6	2	0	5	4	18	8	6	8
	4	5	9	2	16	15	5	21	18	15	11	18	15
	5	-1	-1	11	19	12	10	1	20	13	9	4	6
	6	7	16	15	11	12	12	8	12	22	11	12	21
	7	-5	1	-2	10	12	15	2	13	10	10	7	5
	8	10	9	10	15	9	16	12	11	13	20	11	15

[a]The numbers are 1/10 lb_F deflections from 21.5 lb_F.

production runs at two different plants. The breaking strength was recorded for adjacent pairs of cord at 500-yd intervals over the length of each bobbin to obtain these data.

(a) Write the model involved and the corresponding expected mean squares.
(b) Perform the complete analysis of variance to determine the effects of bobbins, spacing, position, and plant on the quality of tire cord as measured by breaking strength. Interpret the results.

11.33 As a result of an economic analysis of finishing methods for automotive transmission gears, Yokota [5] showed that the grinding technique was less expensive than the older honing method. Qualities of the finished products were equal. In an experiment to determine the optimum operating conditions for grinding, grinding worm movement, depth of cut, feed rate, and stock for finishing were varied over two levels each. The results of this experiment are shown below as they affected the noise abatement rating. Higher ratings (coded given) indicate lower transmission gear noise levels.

	Factor				Coded rating	
Exp. No.	Grinding worm movement	Depth of cut (mm./rev)	Feed rate (mm./rev)	Stock for finishing (mm.)	x_1	x_2
1	Fixed	0.020	1.8	0.020	6	8
2	Fixed	0.020	2.2	0.040	7	11
3	Fixed	0.040	1.8	0.020	8	8
4	Fixed	0.040	2.2	0.040	9	13
5	Shifted	0.020	1.8	0.040	11	13
6	Shifted	0.020	2.2	0.020	9	13
7	Shifted	0.040	1.8	0.040	15	19
8	Shifted	0.040	2.2	0.020	14	16

Which factor was of greatest significance in noise reduction? Present the complete analysis of variance table in support of your answer.

11.34 You would like to investigate the effect of the amount and quality of ammonium chloride on the yield of an organic chemical from a batch of base material. Interest is in the comparison between the use of finely ground and coarse ammonium chloride, the effect of a 10% increase in the charge of ammonium chloride to the batch, and the effect of two apparently identical units of the plant in which the actual conversion of base to the organic chemical product takes place.

Each batch of the base material was sufficient to make only two batches of the product. Facilities are available to do a fair job of mixing two batches of the base material in order to obtain a fairly uniform base material which will make four batches of the product. From chemical considerations there is unlikely to be any interaction between amount or quality of ammonium chloride and unit of plant, i.e., if five grade ammonium chloride produced a given increase in yield in one unit it should also, apart from the experimental error, produce an equivalent increase in yield in the other unit. The three-factor interaction between quality of ammonium chloride, amount of ammonium chloride, and plant unit is even less likely to be appreciable.

Give the design you would use including the physical layout, the model, and AOV. Four replications are needed.

11.35 One of your associates is working on the adsorption of Cl_2 in nuclear submarine atmospheres by various materials--activated charcoal, MnO_2, etc. A total of six adsorbents are to be examined. Facilities are available for performing tests on up to eight adsorbents simultaneously. The research program allows time for testing 48 materials. Although four tests per material will give the required precision, he decides to use eight tests each in an attempt to improve the accuracy and precision of his results. Lay out the possible designs this young engineer can use. Give and explain the model for each. Discuss each with regard to efficiency, ease of data handling, and reliability of results.

11.36 Describe the use of a Graeco-Latin square for the tires-car-driver setup in Example 11.16. Give the model, the AOV, and make the appropriate statistical tests for the major variables at the 5% level.

11.37 Apply a Graeco-Latin square to the experiment in Example 11.16 by adding a car and driver and cycling in the spare tire in regular rotation. Write out the model and show how to compute the analysis of variance table.

11.38 In a pilot-scale gas absorber (6 ft long, 3 in ID, 1/2-in ceramic berl saddles), the number of transfer units NTU was investigated for varying ammonia concentrations in air (NH3). Various combinations of the ratio of the molar liquid to gas phase rates L/G were used. Does ammonia concentration affect performance as measured by the number of transfer units? Does the phase rate ratio affect the results? Use analysis of covariance for this problem.

NH3	L/G	NTU
15	5.8	0.91
15	14.6	0.77
15	20.4	0.77
15	29.2	0.76
5	22.3	3.60
5	16.8	4.70
5	13.4	6.60
5	8.9	5.00
10	6.8	1.38
10	13.6	1.40
10	20.4	1.52
10	34.0	1.65

11.39 In an experiment to measure the film coefficient HF for steam condensing on the outside surface of a single horizontal tube, various steam rates (MS) were used. The covariate is the cooling water flow rate V. Do the steam and/or the water rates affect HF?

MS	V	HF
0.80	1.00	785
0.80	0.90	1010
0.80	0.80	1121
0.80	0.70	1293
0.80	0.60	1505
0.80	0.80	1149
0.80	0.50	1853

MS	V	HF
0.80	0.40	2310
0.80	0.80	1211
0.80	0.30	3120
0.80	0.20	4562
0.60	0.80	792
0.60	1.00	946
0.60	0.60	731
0.60	0.40	684
0.60	0.40	693
0.30	0.95	847
0.30	0.75	800
0.30	0.63	738
0.30	0.45	715
0.30	0.20	429
0.45	1.00	4049
0.45	0.78	4000
0.45	0.59	4883
0.45	0.39	5053
0.45	0.18	5054
0.70	0.20	4530
0.70	0.30	3670
0.70	0.50	3510
0.70	0.70	2900
0.70	0.80	3050
0.70	1.00	2690
0.70	1.10	2510

REFERENCES

1. Akutowicz, F., and Truax, H. M.; Establishing Control of Tire Cord Testing Laboratories, Ind. Qual. Control, 13, no. 2, 4-5 (1956). Copyright American Society for Quality Control, Inc. Reprinted by permission.

2. Cochran, W. G., and Cox, G. M.; Experimental Designs, 2nd ed., pp. 20-27, John Wiley & Sons, Inc., New York (1957).

3. Ostle, B. and Mensing, R. W.; Statistics in Research, 3rd ed., pp. 395-408, Iowa State University Press, Ames (1975).

4. Stillitto, G. F.; Research, 1, 520-525 (1948). Copyright Butterworth & Co. (Publishers) Limited. Reproduced with permission.

5. Yokota, Y.; Quality Control of Gear Grinding, Ind. Qual. Control, 20, 18-19 (1964). Copyright American Society for Quality Control, Inc. Data reprinted by permission.

Appendix A
MATRIX ALGEBRA

A.1 DEFINITIONS

A matrix is an array of numbers or quantities indicated by $\underline{A}$ with entries a_{ij} as given by

$$\underline{A} = \begin{pmatrix} a_{11} & a_{12} & a_{13} \\ a_{21} & a_{22} & a_{23} \\ a_{31} & a_{32} & a_{33} \end{pmatrix} .$$

The underline indicates a matrix just as an overbar or arrow is used to indicate a vector. The subscripts on the entries indicate the row and column position of the entry.

A matrix can be partitioned as rows,

$$\underline{A} = \begin{pmatrix} A_1 \\ A_2 \\ A_3 \end{pmatrix} \quad \text{where} \quad \begin{aligned} A_1 &= (a_{11} \quad a_{12} \quad a_{13}) \\ A_2 &= (a_{21} \quad a_{22} \quad a_{23}) \\ A_3 &= (a_{31} \quad a_{32} \quad a_{33}) \end{aligned}$$

or as columns,

$\underline{A} = (A_1, A_2, A_3)$ where

$$A_1 = \begin{pmatrix} a_{11} \\ a_{21} \\ a_{31} \end{pmatrix}, \quad A_2 = \begin{pmatrix} a_{12} \\ a_{22} \\ a_{32} \end{pmatrix}, \quad A_3 = \begin{pmatrix} a_{13} \\ a_{23} \\ a_{33} \end{pmatrix} .$$

A.2 ELEMENTARY OPERATIONS

The elementary operations of addition and subtraction are illustrated below for $\underline{A}$ and $\underline{B}$:

$$\underline{A} = \begin{pmatrix} a_{11} & a_{12} & a_{13} \\ a_{21} & a_{22} & a_{23} \\ a_{31} & a_{32} & a_{33} \end{pmatrix} \quad \text{and} \quad \underline{B} = \begin{pmatrix} b_{11} & b_{12} & b_{13} \\ b_{21} & b_{22} & b_{23} \\ b_{31} & b_{32} & b_{33} \end{pmatrix} .$$

Addition and subtraction:

$$\underline{C} = \underline{A} + \underline{B} = \{a_{ij} + b_{ij}\} = \{c_{ij}\} ,$$

$$\underline{C} = \underline{A} - \underline{B} = \{a_{ij} - b_{ij}\} = \{c_{ij}\} .$$

In order for two matrices to be conformable for multiplication, the column order of the first and the row order of the second must be identical. For example, the square (no. of rows = no. of columns) matrix above is a 3 × 3 matrix. It can only be multiplied on the right by another 3 × 3 matrix, by a 3 × N matrix (rectangular matrix with 3 rows), or by a scalar quantity. Briefly stated,

$$\underset{3\times3}{\underline{P}} = \underset{3\times3}{\underline{A}} \; \underset{3\times3}{\underline{B}} = \begin{pmatrix} P_{11} & P_{12} & P_{13} \\ P_{21} & P_{22} & P_{23} \\ P_{31} & P_{32} & P_{33} \end{pmatrix} ,$$

where

$$P_{11} = a_{11}b_{11} + a_{12}b_{21} + a_{13}b_{31} \quad ,$$

$$P_{12} = a_{11}b_{12} + a_{12}b_{22} + a_{13}b_{32} \quad ,$$

$$P_{13} = a_{11}b_{13} + a_{12}b_{23} + a_{13}b_{33} \quad , \text{ etc.}$$

In general

$$P_{ij} = a_{i1}b_{1j} + \ldots + a_{im}b_{mj}$$

or

$$P_{ij} = \sum_{k=1}^{m} a_{ik}b_{kj} \quad ,$$

When these elementary operations are combined with matrix multiplication, we have the following laws of matrix algebra:

1st Distributive Law: $\mathbf{A}(\mathbf{B} + \mathbf{C}) = \mathbf{AB} + \mathbf{AC}$

2nd Distributive Law: $(\mathbf{A} + \mathbf{B})\mathbf{C} = \mathbf{AC} + \mathbf{BC}$

Associative Law: $\mathbf{A}(\mathbf{BC}) = (\mathbf{AB})\mathbf{C}$.

From these three laws you should realize that the order in which matrix operations are carried out is quite important.

A.3 CONJUGATE OF A MATRIX

The conjugate of a matrix is denoted by $\overline{\mathbf{A}}$. To get a conjugate of a matrix, replace all the complex numbers in $\mathbf{A}$ by their complex conjugates to get $\overline{\mathbf{A}}$. Some rules for conjugates are

$$\overline{\overline{\mathbf{A}}} = \mathbf{A} \quad ,$$

$$\overline{(\mathbf{A} + \mathbf{B})} = \overline{\mathbf{A}} + \overline{\mathbf{B}}$$

$$\overline{\mathbf{AB}} = \overline{\mathbf{A}} \cdot \overline{\mathbf{B}} \quad .$$

A.4 TRANSPOSE OF A MATRIX

The transpose of a matrix $\underline{A}_{mxn}$ is $\underline{A}'_{nxm}$. It is obtained by merely interchanging rows and columns. For example, the 1st row of $\underline{A}$ becomes the 1st column of $\underline{A}'$, etc. Some rules which apply to transpose matrices are

$$(\underline{A}')' = \underline{A} \quad ,$$
$$(k\underline{A})' = k\underline{A}' \quad ,$$
$$(\underline{A} + \underline{B})' = \underline{A}' + \underline{B}' \quad ,$$
$$(\underline{AB})' = \underline{B}' \cdot \underline{A}' \quad ,$$

If a matrix is symmetric, $\underline{A} = \underline{A}'$.

A.5 DETERMINANTS

The determinant of a 2 × 2 matrix $\underline{A}$, denoted by $|\underline{A}|$, is given by $|\underline{A}| = a_{11}a_{22} - a_{21}a_{12}$. For a 3 × 3 matrix, the determinant is given by

$$|\underline{A}| = a_{11}a_{22}a_{33} + a_{12}a_{23}a_{31} + a_{21}a_{32}a_{13} - a_{13}a_{22}a_{31}$$
$$- a_{11}a_{32}a_{23} - a_{33}a_{21}a_{12}.$$

The minor M_{ij} of an element a_{ij} of the matrix $\underline{A}$ is defined to be the matrix consisting of all the elements remaining after deleting the i-th row and j-th column. For example, for the matrix $\underline{A}$ on page 482, the minor of a_{22} is given by

$$M_{22} = \begin{matrix} a_{11} & a_{13} \\ a_{31} & a_{33} \end{matrix}$$

The cofactor of an element a_{ij} is defined by $\alpha_{ij} = (-1)^{i+j}|M_{ij}|$. The cofactor matrix for $\underline{A}$ is the matrix obtained by replacing each element a_{ij} by its cofactor α_{ij}.

For determinants of order more than 3, add a constant k times some row (or column) to another row (or column). Continue to do this until you have a row or column with all factors zero except for one 1. Then the value of the original determinant = cofactor of the 1_{ij} above. Repeat as necessary to get a second or third order determinant which can be easily evaluated. Consider the example below:

$$\begin{vmatrix} 2 & 3 & -2 & 4 \\ 3 & -2 & 1 & 2 \\ 3 & 2 & -3 & 4 \\ -2 & 4 & 0 & 5 \end{vmatrix} .$$

To evaluate this determinant, add twice the second row to the first row to get

$$\begin{vmatrix} 8 & -1 & 0 & 8 \\ 3 & -2 & 1 & 2 \\ 3 & 2 & 3 & 4 \\ -2 & 4 & 0 & 5 \end{vmatrix} .$$

Then add three times the second row to the third row to get

$$\begin{vmatrix} 8 & -1 & 0 & 8 \\ 3 & -2 & 1 & 2 \\ 12 & -4 & 0 & 10 \\ -2 & 4 & 0 & 5 \end{vmatrix} ,$$

which reduces to

$$(-1)^{2+3} \begin{vmatrix} 8 & -1 & 8 \\ 12 & -4 & 10 \\ -2 & 4 & 5 \end{vmatrix} = -80.$$

The value of the determinant of a matrix product is the product of the determinants. Notationally we have

$$|\underline{A}\ \underline{B}| = |\underline{A}| \cdot |\underline{B}| .$$

A.6 ADJOINT OF A MATRIX

The adjoint of a matrix, denoted adj $\underline{A}$, is the transpose of the cofactor matrix. For example, if a matrix $\underline{A}$ is given as

$$\underline{A} = \begin{bmatrix} 1 & 2 & 3 \\ 1 & 3 & 4 \\ 1 & 4 & 3 \end{bmatrix},$$

the corresponding cofactor matrix is

$$\begin{pmatrix} \begin{vmatrix} 3 & 4 \\ 4 & 3 \end{vmatrix} & -\begin{vmatrix} 1 & 4 \\ 1 & 3 \end{vmatrix} & \begin{vmatrix} 1 & 1 \\ 3 & 4 \end{vmatrix} \\ -\begin{vmatrix} 2 & 3 \\ 4 & 3 \end{vmatrix} & \begin{vmatrix} 1 & 3 \\ 1 & 3 \end{vmatrix} & -\begin{vmatrix} 1 & 2 \\ 1 & 4 \end{vmatrix} \\ \begin{vmatrix} 2 & 3 \\ 3 & 4 \end{vmatrix} & -\begin{vmatrix} 1 & 3 \\ 1 & 4 \end{vmatrix} & \begin{vmatrix} 1 & 2 \\ 1 & 3 \end{vmatrix} \end{pmatrix} = \begin{pmatrix} -7 & 1 & 1 \\ 6 & 0 & -2 \\ -1 & -1 & 1 \end{pmatrix},$$

and the adjoint of $\underline{A}$ is given as

$$\text{adj } \underline{A} = \begin{pmatrix} -7 & 6 & -1 \\ 1 & 0 & -1 \\ 1 & -2 & 1 \end{pmatrix}.$$

A.7 MATRIX INVERSION

The inverse of a square matrix $\underline{A}$ when it exists is written as $\underline{A}^{-1}$, and is the unique matrix such that $\underline{A}\,\underline{A}^{-1} = \underline{A}^{-1}\underline{A} = \underline{I}$ where $\underline{I}$ is the square matrix having ones in the ii positions and zeroes elsewhere. The inverse matrix $\underline{A}^{-1}$ can be computed from

$$\underline{A}^{-1} = \frac{\text{adj } \underline{A}}{|\underline{A}|}.$$

The inverse of a matrix product can be obtained from

$$(\underline{AB})^{-1} = \underline{B}^{-1} \cdot \underline{A}^{-1}.$$

A.8 MATRICES FOR SOLUTIONS OF SIMULTANEOUS EQUATIONS

Cramer's Rule may be applied for the solution of nonhomogeneous linear equations by replacing the column for any x_i by the right-hand side when the equations are arranged in standard form, for example,

$$2x_1 + x_2 + 5x_3 + x_4 = 5 \quad ,$$

$$x_1 + x_2 - 3x_3 - 4x_4 = -1 \quad ,$$

$$3x_1 + 6x_2 - 2x_3 + x_4 = 8 \quad ,$$

$$2x_1 + 2x_2 + 2x_3 - 3x_4 = 2 \quad .$$

To solve for x_i, then, use $x_i = \dfrac{|\underline{A}_i|}{|\underline{A}|}$.

As an example, $|\underline{A}_1|$ is

$$\begin{vmatrix} 5 & 1 & 5 & 1 \\ -1 & 1 & -3 & -4 \\ 8 & 6 & -2 & 1 \\ 2 & 2 & 2 & -3 \end{vmatrix}$$

and $|\underline{A}|$ is

$$\begin{vmatrix} 2 & 1 & 5 & 1 \\ 1 & 1 & -3 & -4 \\ 3 & 6 & -2 & 4 \\ 2 & 2 & 2 & -3 \end{vmatrix} \quad , \qquad \text{so } x_1 = \frac{|\underline{A}_1|}{|\underline{A}\ |} \quad .$$

For systems of homogeneous equations, application of Cramer's Rule gives only the trivial solution $x_1 = x_2 = x_3 = \ldots x_n = 0$ for systems of rank n.

A.9 PARTIAL DERIVATIVE OF A MATRIX

If $\underline{A}$ is an n × n matrix and if each a_{ij} is a function of x, then the partial derivative of the matrix, $\frac{\partial \underline{A}}{\partial x}$, is the matrix whose ij^{th} entry is $\frac{\partial a_{ij}(x)}{\partial x}$. As an example,

$$\underline{A} = \begin{pmatrix} x^2 & x+1 & 3 \\ 1 & 2x-1 & x^3 \\ 0 & x & -2 \end{pmatrix}, \quad \frac{\partial \underline{A}}{\partial x} = \begin{pmatrix} 2x & 1 & 0 \\ 0 & 2 & 3x^2 \\ 0 & 1 & 0 \end{pmatrix}.$$

Appendix B
INTRODUCTION TO SAS

B.1 DEFINITIONS

1. Data Value: A single measurement, i.e., boiling point, resistivity, tensile strength.
2. Observation: A set of data values for the same item, process, individual, e.g. temperature, ohms, psi.
3. Variable: A set of data values for the same measurement, i.e. the average boiling points of all distillate samples collected during a shift, the resistance rating of heaters of a given type, the tensile strengths of all specimens from a mill run of steel strip stock.
4. Variable Name: A name chosen for each variable, containing from one to eight characters. If alphameric, the variable name must begin with a letter. Some examples are AVGTEMP, RESIST, TENSILE.
5. Data Set: A collection of observations, such as the temperature at 10% intervals along the boiling point curves for distillate samples, the actual resistances of each heater with a specific rating in a shipment from a single supplier, the physical properties of rolled steel strip.

B.2 GENERAL INFORMATION

Jobs within the SAS System are made up of statements and procedures which provide information or ask the SAS System to perform some activity. The first word of a SAS statement tells the SAS System what activity you want to perform while the rest of the statement (if required) gives the SAS System more information about how you want the activity performed. Every SAS statement ends with a semicolon.

B.3 CREATING A SAS DATA SET

To get your data into a SAS data set, you need:

a DATA statement
an INPUT statement
either CARDS or INFILE statement

The DATA statement begins with the word DATA and then gives the name you chose for the data set, i.e.

```
DATA PHYSPROP;
```

If you leave out the data set name, the SAS System will create one for you. This may create a problem, especially if more than one activity is performed in a SAS job. In that event, the SAS System will use the most recently named data set for a subsequent procedure. You may really want the activity performed on the original or some other data set instead. Always label data sets and specify the one the SAS System is to use. Note that data set names are limited to eight characters.

The INPUT statement is used to describe your data to the SAS System. Therefore you need to know how your data are arranged. Begin with the word INPUT followed by the variable name for the first variable to be read. If the variable contains non-numeric characters, follow the variable name with a dollar sign. Then give the column(s) in which the data values occur. Repeat for additional variables, i.e.

```
INPUT ASTM $ 10-14 TENSILE 16-18 BREAK 20-22 ROCK 24-27;
```

Variables that you do not wish to read in for a particular job may be skipped, i.e.

```
INPUT ASTM $ 10-14 BREAK 20-22;
```

Note that to avoid reading in all variables, you must know exactly in which columns your data are located.

If all variables are to be read in and the values are separated by at least one blank space, column designations may be omitted. However, any missing data values must be indicated by a period. Otherwise the data entries are advanced into the wrong columns. If the breaking strength was not available for a specimen and no period were inserted, the Rockwell hardness would be read in for the breaking strength. If the data set consists of data values for only one observation, you may want to put more than one data value on a card. To tell the SAS System to read them all, instead of the first one only, use:

```
INPUT name @@;
```

If the data is part of your job, the CARDS statement follows the DATA and INPUT statements in your SAS program. Simply use:

```
CARDS;
```

This tells the SAS System that the data comes next. Our mainframe operating system interprets data as part of the program as though they are cards. Your system may be different so check with the computer center personnel for exact instructions. If the data values are on disk, you must tell both the computer's operating system using JCL and the SAS System where to find the data. The data set name in this example is PHYSPROP and must be used by the INFILE statement, thus:

```
INFILE PHYSPROP;
```

which goes before the input statement. Entering data in a SAS job, then, could be illustrated by the following 3 examples:

```
(initial JCL cards/statements here)
DATA PHYSPROP;
INPUT ASTM $ 10-14 TENSILE 16-18 BREAK 20-22 ROCK 24-27;
CARDS
bbbbbbbbbD1022b121b175b62.4
bbbbbbbbbA0147b072b094b12.1
```

or

```
(JCL cards/statements here)
DATA PHYSPROP;
INFILE PHYSPROP;
INPUT ASTM $ 10-14 TENSILE 16-18;
```

or

```
(JCL cards/statements here)
DATA TENSTR;
INPUT TENSILE @@;
CARDS;
047   062   072   091   058   107
101   121   091   116   082   086
```

(note that in this example, all the data values belong to the same variable TENSILE which is the only variable in the data set TENSTR).

B.4 HOW TO USE SAS PROCEDURES

SAS procedures are used to analyze and process a SAS data set which you have created. The SAS statement is started with the word PROC to tell the SAS System that you want a procedure executed. You may string together several procedures that use the same data set. After PROC, you must give the name of the procedure you want used such as PRINT to print the data set or MEANS to calculate means. Unless you follow this by DATA=data set name, the SAS System will use the most recently created data set.

The SAS System will perform the indicated procedure on all variables unless otherwise specified. To indicate which variables are to be processed you must follow the PROC statement with a variable statement VAR which lists the variables to use. For example:

```
PROC PRINT DATA=PHYSPROP;
VAR ASTM TENSILE;
PROC MEANS DATA=PHYSPROP;
VAR TENSILE BREAK ROCK;
```

will print the values for ASTM and TENSILE in PHYSPROP and will calculate means, standard deviation, etc. for TENSILE, BREAK, and ROCK.

B.5 HISTOGRAMS AND FREQUENCY PLOTS

The procedure PROC FREQ can be used to obtain frequency tables. To get frequency tables for all variables in the most recently created data set, use:

```
PROC FREQ;
```

If you want to use a previous data set and specify only some variables, specify the data set name and use a TABLES statement:

```
PROC FREQ DATA=PHYSPROP;
TABLES TENSILE;
```

The SAS System will go to data set PHYSPROP and make frequency tables for the variable TENSILE.

To create a bar chart histogram using frequency, use:

```
PROC CHART DATA=data set name;
VBAR variable name/TYPE=FREQ;
```

SAS will create a vertical bar chart for the designated variable. It will create its own scales. If you want to specify the scale for the abscissa, you can indicate the range to be used and the distance between classes using the midpoints option as in this example:

```
VBAR TENSILE/TYPE=FREQ MIDPOINTS=35 TO 155 BY 10;
```

This statement tells the SAS System to make a frequency bar chart for the variable TENSILE using an abscissa scale of 35, 45, ...,145, 155. To get an estimate of the range and size of the interval to use, you can either let the SAS System do a plot and then guess at modifications or you can use PROC FREQ results to get an idea of the distribution of your data.

To obtain the data you need to plot frequency polygons, you must use the frequency table which you obtained from PROC FREQ. Prepare a program to execute PROC FREQ and hold the output by using the command appropriate for your system. After the program has been executed, call a copy of the output into your active file with an appropriate command. Locate the frequency table in your active file and delete all other lines. You may also have to delete any page header information if the table runs over to a new page. Now your active file should contain only the numbers in the frequency table.

Renumber the remaining lines. List the last line of the frequency table and you will note the presence of a large empty space before the first column. Count these spaces and eliminate them as needed. Now relist the last line followed by a column number display command so you can find out exactly where your data are stored. Write down the columns where each variable is located so you can tell the SAS System where to find the data for PROC PLOT. Save this entire data set (table). Now clear your active file and start writing your new SAS program. When you have reached the place in the SAS program where the data are to be inserted, call in the frequency table and then proceed to finish the SAS program. Save the program on disk or tape, edit it as necessary, and then run it. Note that the cumulative relative frequency (CRF) will be plotted on the ordinate vs. the named variable (VAR) as the abscissa. You must already have named the variable in your SAS program.

Appendix C
TABLES OF STATISTICAL FUNCTIONS

TABLE I

Binomial Cumulative Distribution

This table gives values of the cumulative binomial distribution function,

$$P(X \leq x) = \sum_{k=0}^{x} \binom{n}{k} p^k (1-p)^{n-k} .$$

$P(X \leq x)$ is the probability of x or less successes in n independent binomial trials with probability p of success on a single trial.

The table gives values of the distribution function for $x = 0,1,\ldots,n$; $n = 2,3,\ldots,25$; $p = 0.1$ through 0.9.

All probabilities entered as 0.000 in the table are actually larger than 0 but less than 0.0005. All probabilities entered as 1.000, except for those when $x = n$, are actually larger than 0.9995 but less than 1.

BINOMIAL CUMULATIVE DISTRIBUTION

		p								
n	x	.1	.2	.3	.4	.5	.6	.7	.8	.9
2	0	0.810	0.640	0.490	0.360	0.250	0.160	0.090	0.040	0.010
2	1	0.990	0.960	0.910	0.840	0.750	0.640	0.510	0.360	0.190
2	2	1.000	1.000	1.000	1.000	1.000	1.000	1.000	1.000	1.000
3	0	0.729	0.512	0.343	0.216	0.125	0.064	0.027	0.008	0.001
3	1	0.972	0.896	0.784	0.648	0.500	0.352	0.216	0.104	0.028
3	2	0.999	0.992	0.973	0.936	0.875	0.784	0.657	0.488	0.271
3	3	1.000	1.000	1.000	1.000	1.000	1.000	1.000	1.000	1.000
4	0	0.656	0.410	0.240	0.130	0.063	0.026	0.008	0.002	0.000
4	1	0.948	0.819	0.652	0.475	0.313	0.179	0.084	0.027	0.004
4	2	0.996	0.973	0.916	0.821	0.687	0.525	0.348	0.181	0.052
4	3	1.000	0.998	0.992	0.974	0.937	0.870	0.760	0.590	0.344
4	4	1.000	1.000	1.000	1.000	1.000	1.000	1.000	1.000	1.000
5	0	0.590	0.328	0.168	0.078	0.031	0.010	0.002	0.000	0.000
5	1	0.919	0.737	0.528	0.337	0.188	0.087	0.031	0.007	0.000
5	2	0.991	0.942	0.837	0.683	0.500	0.317	0.163	0.058	0.009
5	3	1.000	0.993	0.969	0.913	0.812	0.663	0.472	0.263	0.081
5	4	1.000	1.000	0.998	0.990	0.969	0.922	0.832	0.672	0.410
5	5	1.000	1.000	1.000	1.000	1.000	1.000	1.000	1.000	1.000
6	0	0.531	0.262	0.118	0.047	0.016	0.004	0.001	0.000	0.000
6	1	0.886	0.655	0.420	0.233	0.109	0.041	0.011	0.002	0.000
6	2	0.984	0.901	0.744	0.544	0.344	0.179	0.070	0.017	0.001
6	3	0.999	0.983	0.930	0.821	0.656	0.456	0.256	0.099	0.016
6	4	1.000	0.998	0.989	0.959	0.891	0.767	0.580	0.345	0.114
6	5	1.000	1.000	0.999	0.996	0.984	0.953	0.882	0.738	0.469
6	6	1.000	1.000	1.000	1.000	1.000	1.000	1.000	1.000	1.000
7	0	0.478	0.210	0.082	0.028	0.008	0.002	0.000	0.000	0.000
7	1	0.850	0.577	0.329	0.159	0.062	0.019	0.004	0.000	0.000
7	2	0.974	0.852	0.647	0.420	0.227	0.096	0.029	0.005	0.000
7	3	0.997	0.967	0.874	0.710	0.500	0.290	0.126	0.033	0.003
7	4	1.000	0.995	0.971	0.904	0.773	0.580	0.353	0.148	0.026
7	5	1.000	1.000	0.996	0.981	0.938	0.841	0.671	0.423	0.150
7	6	1.000	1.000	1.000	0.998	0.992	0.972	0.918	0.790	0.522
7	7	1.000	1.000	1.000	1.000	1.000	1.000	1.000	1.000	1.000
8	0	0.430	0.168	0.058	0.017	0.004	0.001	0.000	0.000	0.000
8	1	0.813	0.503	0.255	0.106	0.035	0.009	0.001	0.000	0.000
8	2	0.962	0.797	0.552	0.315	0.145	0.050	0.011	0.001	0.000
8	3	0.995	0.944	0.806	0.594	0.363	0.174	0.058	0.010	0.000
8	4	1.000	0.990	0.942	0.826	0.637	0.406	0.194	0.056	0.005
8	5	1.000	0.999	0.989	0.950	0.855	0.685	0.448	0.203	0.038
8	6	1.000	1.000	0.999	0.991	0.965	0.894	0.745	0.497	0.187
8	7	1.000	1.000	1.000	0.999	0.996	0.983	0.942	0.832	0.570
8	8	1.000	1.000	1.000	1.000	1.000	1.000	1.000	1.000	1.000

BINOMIAL CUMULATIVE DISTRIBUTION

		p								
n	x	.1	.2	.3	.4	.5	.6	.7	.8	.9
9	0	0.387	0.134	0.040	0.010	0.002	0.000	0.000	0.000	0.000
9	1	0.775	0.436	0.196	0.071	0.020	0.004	0.000	0.000	0.000
9	2	0.947	0.738	0.463	0.232	0.090	0.025	0.004	0.000	0.000
9	3	0.992	0.914	0.730	0.483	0.254	0.099	0.025	0.003	0.000
9	4	0.999	0.980	0.901	0.733	0.500	0.267	0.099	0.020	0.001
9	5	1.000	0.997	0.975	0.901	0.746	0.517	0.270	0.086	0.008
9	6	1.000	1.000	0.996	0.975	0.910	0.768	0.537	0.262	0.053
9	7	1.000	1.000	1.000	0.996	0.980	0.929	0.804	0.564	0.225
9	8	1.000	1.000	1.000	1.000	0.998	0.990	0.960	0.866	0.613
9	9	1.000	1.000	1.000	1.000	1.000	1.000	1.000	1.000	1.000
10	0	0.349	0.107	0.028	0.006	0.001	0.000	0.000	0.000	0.000
10	1	0.736	0.376	0.149	0.046	0.011	0.002	0.000	0.000	0.000
10	2	0.930	0.678	0.383	0.167	0.055	0.012	0.002	0.000	0.000
10	3	0.987	0.879	0.650	0.382	0.172	0.055	0.011	0.001	0.000
10	4	0.998	0.967	0.850	0.633	0.377	0.166	0.047	0.006	0.000
10	5	1.000	0.994	0.953	0.834	0.623	0.367	0.150	0.033	0.002
10	6	1.000	0.999	0.989	0.945	0.828	0.618	0.350	0.121	0.013
10	7	1.000	1.000	0.998	0.988	0.945	0.833	0.617	0.322	0.070
10	8	1.000	1.000	1.000	0.998	0.989	0.954	0.851	0.624	0.264
10	9	1.000	1.000	1.000	1.000	0.999	0.994	0.972	0.893	0.651
10	10	1.000	1.000	1.000	1.000	1.000	1.000	1.000	1.000	1.000
11	0	0.314	0.086	0.020	0.004	0.000	0.000	0.000	0.000	0.000
11	1	0.697	0.322	0.113	0.030	0.006	0.001	0.000	0.000	0.000
11	2	0.910	0.617	0.313	0.119	0.033	0.006	0.001	0.000	0.000
11	3	0.981	0.839	0.570	0.296	0.113	0.029	0.004	0.000	0.000
11	4	0.997	0.950	0.790	0.533	0.274	0.099	0.022	0.002	0.000
11	5	1.000	0.988	0.922	0.753	0.500	0.247	0.078	0.012	0.000
11	6	1.000	0.998	0.978	0.901	0.726	0.467	0.210	0.050	0.003
11	7	1.000	1.000	0.996	0.971	0.887	0.704	0.430	0.161	0.019
11	8	1.000	1.000	0.999	0.994	0.967	0.881	0.687	0.383	0.090
11	9	1.000	1.000	1.000	0.999	0.994	0.970	0.887	0.678	0.303
11	10	1.000	1.000	1.000	1.000	1.000	0.996	0.980	0.914	0.686
11	11	1.000	1.000	1.000	1.000	1.000	1.000	1.000	1.000	1.000
12	0	0.282	0.069	0.014	0.002	0.000	0.000	0.000	0.000	0.000
12	1	0.659	0.275	0.085	0.020	0.003	0.000	0.000	0.000	0.000
12	2	0.889	0.558	0.253	0.083	0.019	0.003	0.000	0.000	0.000
12	3	0.974	0.795	0.493	0.225	0.073	0.015	0.002	0.000	0.000
12	4	0.996	0.927	0.724	0.438	0.194	0.057	0.009	0.001	0.000
12	5	0.999	0.981	0.882	0.665	0.387	0.158	0.039	0.004	0.000
12	6	1.000	0.996	0.961	0.842	0.613	0.335	0.118	0.019	0.001
12	7	1.000	0.999	0.991	0.943	0.806	0.562	0.276	0.073	0.004
12	8	1.000	1.000	0.998	0.985	0.927	0.775	0.507	0.205	0.026
12	9	1.000	1.000	1.000	0.997	0.981	0.917	0.747	0.442	0.111
12	10	1.000	1.000	1.000	1.000	0.997	0.980	0.915	0.725	0.341

BINOMIAL CUMULATIVE DISTRIBUTION

		p								
n	x	.1	.2	.3	.4	.5	.6	.7	.8	.9
12	11	1.000	1.000	1.000	1.000	1.000	0.998	0.986	0.931	0.718
12	12	1.000	1.000	1.000	1.000	1.000	1.000	1.000	1.000	1.000
13	0	0.254	0.055	0.010	0.001	0.000	0.000	0.000	0.000	0.000
13	1	0.621	0.234	0.064	0.013	0.002	0.000	0.000	0.000	0.000
13	2	0.866	0.502	0.202	0.058	0.011	0.001	0.000	0.000	0.000
13	3	0.966	0.747	0.421	0.169	0.046	0.008	0.001	0.000	0.000
13	4	0.994	0.901	0.654	0.353	0.133	0.032	0.004	0.000	0.000
13	5	0.999	0.970	0.835	0.574	0.291	0.098	0.018	0.001	0.000
13	6	1.000	0.993	0.938	0.771	0.500	0.229	0.062	0.007	0.000
13	7	1.000	0.999	0.982	0.902	0.709	0.426	0.165	0.030	0.001
13	8	1.000	1.000	0.996	0.968	0.867	0.647	0.346	0.099	0.006
13	9	1.000	1.000	0.999	0.992	0.954	0.831	0.579	0.253	0.034
13	10	1.000	1.000	1.000	0.999	0.989	0.942	0.798	0.498	0.134
13	11	1.000	1.000	1.000	1.000	0.998	0.987	0.936	0.766	0.379
13	12	1.000	1.000	1.000	1.000	1.000	0.999	0.990	0.945	0.746
13	13	1.000	1.000	1.000	1.000	1.000	1.000	1.000	1.000	1.000
14	0	0.229	0.044	0.007	0.001	0.000	0.000	0.000	0.000	0.000
14	1	0.585	0.198	0.047	0.008	0.001	0.000	0.000	0.000	0.000
14	2	0.842	0.448	0.161	0.040	0.006	0.001	0.000	0.000	0.000
14	3	0.956	0.698	0.355	0.124	0.029	0.004	0.000	0.000	0.000
14	4	0.991	0.870	0.584	0.279	0.090	0.018	0.002	0.000	0.000
14	5	0.999	0.956	0.781	0.486	0.212	0.058	0.008	0.000	0.000
14	6	1.000	0.988	0.907	0.692	0.395	0.150	0.031	0.002	0.000
14	7	1.000	0.998	0.969	0.850	0.605	0.308	0.093	0.012	0.000
14	8	1.000	1.000	0.992	0.942	0.788	0.514	0.219	0.044	0.001
14	9	1.000	1.000	0.998	0.982	0.910	0.721	0.416	0.130	0.009
14	10	1.000	1.000	1.000	0.996	0.971	0.876	0.645	0.302	0.044
14	11	1.000	1.000	1.000	0.999	0.994	0.960	0.839	0.552	0.158
14	12	1.000	1.000	1.000	1.000	0.999	0.992	0.953	0.802	0.415
14	13	1.000	1.000	1.000	1.000	1.000	0.999	0.993	0.956	0.771
14	14	1.000	1.000	1.000	1.000	1.000	1.000	1.000	1.000	1.000
15	0	0.206	0.035	0.005	0.000	0.000	0.000	0.000	0.000	0.000
15	1	0.549	0.167	0.035	0.005	0.000	0.000	0.000	0.000	0.000
15	2	0.816	0.398	0.127	0.027	0.004	0.000	0.000	0.000	0.000
15	3	0.944	0.648	0.297	0.091	0.018	0.002	0.000	0.000	0.000
15	4	0.987	0.836	0.515	0.217	0.059	0.009	0.001	0.000	0.000
15	5	0.998	0.939	0.722	0.403	0.151	0.034	0.004	0.000	0.000
15	6	1.000	0.982	0.869	0.610	0.304	0.095	0.015	0.001	0.000
15	7	1.000	0.996	0.950	0.787	0.500	0.213	0.050	0.004	0.000
15	8	1.000	0.999	0.985	0.905	0.696	0.390	0.131	0.018	0.000
15	9	1.000	1.000	0.996	0.966	0.849	0.597	0.278	0.061	0.002
15	10	1.000	1.000	0.999	0.991	0.941	0.783	0.485	0.164	0.013
15	11	1.000	1.000	1.000	0.998	0.982	0.909	0.703	0.352	0.056
15	12	1.000	1.000	1.000	1.000	0.996	0.973	0.873	0.602	0.184
15	13	1.000	1.000	1.000	1.000	1.000	0.995	0.965	0.833	0.451

BINOMIAL CUMULATIVE DISTRIBUTION

n	x	p .1	.2	.3	.4	.5	.6	.7	.8	.9
15	14	1.000	1.000	1.000	1.000	1.000	1.000	0.995	0.965	0.794
15	15	1.000	1.000	1.000	1.000	1.000	1.000	1.000	1.000	1.000
16	0	0.185	0.028	0.003	0.000	0.000	0.000	0.000	0.000	0.000
16	1	0.515	0.141	0.026	0.003	0.000	0.000	0.000	0.000	0.000
16	2	0.789	0.352	0.099	0.018	0.002	0.000	0.000	0.000	0.000
16	3	0.932	0.598	0.246	0.065	0.011	0.001	0.000	0.000	0.000
16	4	0.983	0.798	0.450	0.167	0.038	0.005	0.000	0.000	0.000
16	5	0.997	0.918	0.660	0.329	0.105	0.019	0.002	0.000	0.000
16	6	0.999	0.973	0.825	0.527	0.227	0.058	0.007	0.000	0.000
16	7	1.000	0.993	0.926	0.716	0.402	0.142	0.026	0.001	0.000
16	8	1.000	0.999	0.974	0.858	0.598	0.284	0.074	0.007	0.000
16	9	1.000	1.000	0.993	0.942	0.773	0.473	0.175	0.027	0.001
16	10	1.000	1.000	0.998	0.981	0.895	0.671	0.340	0.082	0.003
16	11	1.000	1.000	1.000	0.995	0.962	0.833	0.550	0.202	0.017
16	12	1.000	1.000	1.000	0.999	0.989	0.935	0.754	0.402	0.068
16	13	1.000	1.000	1.000	1.000	0.998	0.982	0.901	0.648	0.211
16	14	1.000	1.000	1.000	1.000	1.000	0.997	0.974	0.859	0.485
16	15	1.000	1.000	1.000	1.000	1.000	1.000	0.997	0.972	0.815
16	16	1.000	1.000	1.000	1.000	1.000	1.000	1.000	1.000	1.000
17	0	0.167	0.023	0.002	0.000	0.000	0.000	0.000	0.000	0.000
17	1	0.482	0.118	0.019	0.002	0.000	0.000	0.000	0.000	0.000
17	2	0.762	0.310	0.077	0.012	0.001	0.000	0.000	0.000	0.000
17	3	0.917	0.549	0.202	0.046	0.006	0.000	0.000	0.000	0.000
17	4	0.978	0.758	0.389	0.126	0.025	0.003	0.000	0.000	0.000
17	5	0.995	0.894	0.597	0.264	0.072	0.011	0.001	0.000	0.000
17	6	0.999	0.962	0.775	0.448	0.166	0.035	0.003	0.000	0.000
17	7	1.000	0.989	0.895	0.641	0.315	0.092	0.013	0.000	0.000
17	8	1.000	0.997	0.960	0.801	0.500	0.199	0.040	0.003	0.000
17	9	1.000	1.000	0.987	0.908	0.685	0.359	0.105	0.011	0.000
17	10	1.000	1.000	0.997	0.965	0.834	0.552	0.225	0.038	0.001
17	11	1.000	1.000	0.999	0.989	0.928	0.736	0.403	0.106	0.005
17	12	1.000	1.000	1.000	0.997	0.975	0.874	0.611	0.242	0.022
17	13	1.000	1.000	1.000	1.000	0.994	0.954	0.798	0.451	0.083
17	14	1.000	1.000	1.000	1.000	0.999	0.988	0.923	0.690	0.238
17	15	1.000	1.000	1.000	1.000	1.000	0.998	0.981	0.882	0.518
17	16	1.000	1.000	1.000	1.000	1.000	1.000	0.998	0.977	0.833
17	17	1.000	1.000	1.000	1.000	1.000	1.000	1.000	1.000	1.000
18	0	0.150	0.018	0.002	0.000	0.000	0.000	0.000	0.000	0.000
18	1	0.450	0.099	0.014	0.001	0.000	0.000	0.000	0.000	0.000
18	2	0.734	0.271	0.060	0.008	0.001	0.000	0.000	0.000	0.000
18	3	0.902	0.501	0.165	0.033	0.004	0.000	0.000	0.000	0.000
18	4	0.972	0.716	0.333	0.094	0.015	0.001	0.000	0.000	0.000
18	5	0.994	0.867	0.534	0.209	0.048	0.006	0.000	0.000	0.000
18	6	0.999	0.949	0.722	0.374	0.119	0.020	0.001	0.000	0.000
18	7	1.000	0.984	0.859	0.563	0.240	0.058	0.006	0.000	0.000

BINOMIAL CUMULATIVE DISTRIBUTION

		p								
n	x	.1	.2	.3	.4	.5	.6	.7	.8	.9
18	8	1.000	0.996	0.940	0.737	0.407	0.135	0.021	0.001	0.000
18	9	1.000	0.999	0.979	0.865	0.593	0.263	0.060	0.004	0.000
18	10	1.000	1.000	0.994	0.942	0.760	0.437	0.141	0.016	0.000
18	11	1.000	1.000	0.999	0.980	0.881	0.626	0.278	0.051	0.001
18	12	1.000	1.000	1.000	0.994	0.952	0.791	0.466	0.133	0.006
18	13	1.000	1.000	1.000	0.999	0.985	0.906	0.667	0.284	0.028
18	14	1.000	1.000	1.000	1.000	0.996	0.967	0.835	0.499	0.098
18	15	1.000	1.000	1.000	1.000	0.999	0.992	0.940	0.729	0.266
18	16	1.000	1.000	1.000	1.000	1.000	0.999	0.986	0.901	0.550
18	17	1.000	1.000	1.000	1.000	1.000	1.000	0.998	0.982	0.850
18	18	1.000	1.000	1.000	1.000	1.000	1.000	1.000	1.000	1.000
19	0	0.135	0.014	0.001	0.000	0.000	0.000	0.000	0.000	0.000
19	1	0.420	0.083	0.010	0.001	0.000	0.000	0.000	0.000	0.000
19	2	0.705	0.237	0.046	0.005	0.000	0.000	0.000	0.000	0.000
19	3	0.885	0.455	0.133	0.023	0.002	0.000	0.000	0.000	0.000
19	4	0.965	0.673	0.282	0.070	0.010	0.001	0.000	0.000	0.000
19	5	0.991	0.837	0.474	0.163	0.032	0.003	0.000	0.000	0.000
19	6	0.998	0.932	0.666	0.308	0.084	0.012	0.001	0.000	0.000
19	7	1.000	0.977	0.818	0.488	0.180	0.035	0.003	0.000	0.000
19	8	1.000	0.993	0.916	0.667	0.324	0.088	0.011	0.000	0.000
19	9	1.000	0.998	0.967	0.814	0.500	0.186	0.033	0.002	0.000
19	10	1.000	1.000	0.989	0.912	0.676	0.333	0.084	0.007	0.000
19	11	1.000	1.000	0.997	0.965	0.820	0.512	0.182	0.023	0.000
19	12	1.000	1.000	0.999	0.988	0.916	0.692	0.334	0.068	0.002
19	13	1.000	1.000	1.000	0.997	0.968	0.837	0.526	0.163	0.009
19	14	1.000	1.000	1.000	0.999	0.990	0.930	0.718	0.327	0.035
19	15	1.000	1.000	1.000	1.000	0.998	0.977	0.867	0.545	0.115
19	16	1.000	1.000	1.000	1.000	1.000	0.995	0.954	0.763	0.295
19	17	1.000	1.000	1.000	1.000	1.000	0.999	0.990	0.917	0.580
19	18	1.000	1.000	1.000	1.000	1.000	1.000	0.999	0.986	0.865
19	19	1.000	1.000	1.000	1.000	1.000	1.000	1.000	1.000	1.000
20	0	0.122	0.012	0.001	0.000	0.000	0.000	0.000	0.000	0.000
20	1	0.392	0.069	0.008	0.001	0.000	0.000	0.000	0.000	0.000
20	2	0.677	0.206	0.035	0.004	0.000	0.000	0.000	0.000	0.000
20	3	0.867	0.411	0.107	0.016	0.001	0.000	0.000	0.000	0.000
20	4	0.957	0.630	0.238	0.051	0.006	0.000	0.000	0.000	0.000
20	5	0.989	0.804	0.416	0.126	0.021	0.002	0.000	0.000	0.000
20	6	0.998	0.913	0.608	0.250	0.058	0.006	0.000	0.000	0.000
20	7	1.000	0.968	0.772	0.416	0.132	0.021	0.001	0.000	0.000
20	8	1.000	0.990	0.887	0.596	0.252	0.057	0.005	0.000	0.000
20	9	1.000	0.997	0.952	0.755	0.412	0.128	0.017	0.001	0.000
20	10	1.000	0.999	0.983	0.872	0.588	0.245	0.048	0.003	0.000
20	11	1.000	1.000	0.995	0.943	0.748	0.404	0.113	0.010	0.000
20	12	1.000	1.000	0.999	0.979	0.868	0.584	0.228	0.032	0.000
20	13	1.000	1.000	1.000	0.994	0.942	0.750	0.392	0.087	0.002
20	14	1.000	1.000	1.000	0.998	0.979	0.874	0.584	0.196	0.011

BINOMIAL CUMULATIVE DISTRIBUTION

n	x	p .1	.2	.3	.4	.5	.6	.7	.8	.9
20	15	1.000	1.000	1.000	1.000	0.994	0.949	0.762	0.370	0.043
20	16	1.000	1.000	1.000	1.000	0.999	0.984	0.893	0.589	0.133
20	17	1.000	1.000	1.000	1.000	1.000	0.996	0.965	0.794	0.323
20	18	1.000	1.000	1.000	1.000	1.000	0.999	0.992	0.931	0.608
20	19	1.000	1.000	1.000	1.000	1.000	1.000	0.999	0.988	0.878
20	20	1.000	1.000	1.000	1.000	1.000	1.000	1.000	1.000	1.000
21	0	0.109	0.009	0.001	0.000	0.000	0.000	0.000	0.000	0.000
21	1	0.365	0.058	0.006	0.000	0.000	0.000	0.000	0.000	0.000
21	2	0.648	0.179	0.027	0.002	0.000	0.000	0.000	0.000	0.000
21	3	0.848	0.370	0.086	0.011	0.001	0.000	0.000	0.000	0.000
21	4	0.948	0.586	0.198	0.037	0.004	0.000	0.000	0.000	0.000
21	5	0.986	0.769	0.363	0.096	0.013	0.001	0.000	0.000	0.000
21	6	0.997	0.891	0.551	0.200	0.039	0.004	0.000	0.000	0.000
21	7	0.999	0.957	0.723	0.350	0.095	0.012	0.001	0.000	0.000
21	8	1.000	0.986	0.852	0.524	0.192	0.035	0.002	0.000	0.000
21	9	1.000	0.996	0.932	0.691	0.332	0.085	0.009	0.000	0.000
21	10	1.000	0.999	0.974	0.826	0.500	0.174	0.026	0.001	0.000
21	11	1.000	1.000	0.991	0.915	0.668	0.309	0.068	0.004	0.000
21	12	1.000	1.000	0.998	0.965	0.808	0.476	0.148	0.014	0.000
21	13	1.000	1.000	0.999	0.988	0.905	0.650	0.277	0.043	0.001
21	14	1.000	1.000	1.000	0.996	0.961	0.800	0.449	0.109	0.003
21	15	1.000	1.000	1.000	0.999	0.987	0.904	0.637	0.231	0.014
21	16	1.000	1.000	1.000	1.000	0.996	0.963	0.802	0.414	0.052
21	17	1.000	1.000	1.000	1.000	0.999	0.989	0.914	0.630	0.152
21	18	1.000	1.000	1.000	1.000	1.000	0.998	0.973	0.821	0.352
21	19	1.000	1.000	1.000	1.000	1.000	1.000	0.994	0.942	0.635
21	20	1.000	1.000	1.000	1.000	1.000	1.000	0.999	0.991	0.891
21	21	1.000	1.000	1.000	1.000	1.000	1.000	1.000	1.000	1.000
22	0	0.098	0.007	0.000	0.000	0.000	0.000	0.000	0.000	0.000
22	1	0.339	0.048	0.004	0.000	0.000	0.000	0.000	0.000	0.000
22	2	0.620	0.154	0.021	0.002	0.000	0.000	0.000	0.000	0.000
22	3	0.828	0.332	0.068	0.008	0.000	0.000	0.000	0.000	0.000
22	4	0.938	0.543	0.165	0.027	0.002	0.000	0.000	0.000	0.000
22	5	0.982	0.733	0.313	0.072	0.008	0.000	0.000	0.000	0.000
22	6	0.996	0.867	0.494	0.158	0.026	0.002	0.000	0.000	0.000
22	7	0.999	0.944	0.671	0.290	0.067	0.007	0.000	0.000	0.000
22	8	1.000	0.980	0.814	0.454	0.143	0.021	0.001	0.000	0.000
22	9	1.000	0.994	0.908	0.624	0.262	0.055	0.004	0.000	0.000
22	10	1.000	0.998	0.961	0.772	0.416	0.121	0.014	0.000	0.000
22	11	1.000	1.000	0.986	0.879	0.584	0.228	0.039	0.002	0.000
22	12	1.000	1.000	0.996	0.945	0.738	0.376	0.092	0.006	0.000
22	13	1.000	1.000	0.999	0.979	0.857	0.546	0.186	0.020	0.000
22	14	1.000	1.000	1.000	0.993	0.933	0.710	0.329	0.056	0.001
22	15	1.000	1.000	1.000	0.998	0.974	0.842	0.506	0.133	0.004
22	16	1.000	1.000	1.000	1.000	0.992	0.928	0.687	0.267	0.018
22	17	1.000	1.000	1.000	1.000	0.998	0.973	0.835	0.457	0.062

BINOMIAL CUMULATIVE DISTRIBUTION

		p								
n	x	.1	.2	.3	.4	.5	.6	.7	.8	.9
22	18	1.000	1.000	1.000	1.000	1.000	0.992	0.932	0.668	0.172
22	19	1.000	1.000	1.000	1.000	1.000	0.998	0.979	0.846	0.380
22	20	1.000	1.000	1.000	1.000	1.000	1.000	0.996	0.952	0.661
22	21	1.000	1.000	1.000	1.000	1.000	1.000	1.000	0.993	0.902
22	22	1.000	1.000	1.000	1.000	1.000	1.000	1.000	1.000	1.000
23	0	0.089	0.006	0.000	0.000	0.000	0.000	0.000	0.000	0.000
23	1	0.315	0.040	0.003	0.000	0.000	0.000	0.000	0.000	0.000
23	2	0.592	0.133	0.016	0.001	0.000	0.000	0.000	0.000	0.000
23	3	0.807	0.297	0.054	0.005	0.000	0.000	0.000	0.000	0.000
23	4	0.927	0.501	0.136	0.019	0.001	0.000	0.000	0.000	0.000
23	5	0.977	0.695	0.269	0.054	0.005	0.000	0.000	0.000	0.000
23	6	0.994	0.840	0.440	0.124	0.017	0.001	0.000	0.000	0.000
23	7	0.999	0.928	0.618	0.237	0.047	0.004	0.000	0.000	0.000
23	8	1.000	0.973	0.771	0.388	0.105	0.013	0.001	0.000	0.000
23	9	1.000	0.991	0.880	0.556	0.202	0.035	0.002	0.000	0.000
23	10	1.000	0.997	0.945	0.713	0.339	0.081	0.007	0.000	0.000
23	11	1.000	0.999	0.979	0.836	0.500	0.164	0.021	0.001	0.000
23	12	1.000	1.000	0.993	0.919	0.661	0.287	0.055	0.003	0.000
23	13	1.000	1.000	0.998	0.965	0.798	0.444	0.120	0.009	0.000
23	14	1.000	1.000	0.999	0.987	0.895	0.612	0.229	0.027	0.000
23	15	1.000	1.000	1.000	0.996	0.953	0.763	0.382	0.072	0.001
23	16	1.000	1.000	1.000	0.999	0.983	0.876	0.560	0.160	0.006
23	17	1.000	1.000	1.000	1.000	0.995	0.946	0.731	0.305	0.023
23	18	1.000	1.000	1.000	1.000	0.999	0.981	0.864	0.499	0.073
23	19	1.000	1.000	1.000	1.000	1.000	0.995	0.946	0.703	0.193
23	20	1.000	1.000	1.000	1.000	1.000	0.999	0.984	0.867	0.408
23	21	1.000	1.000	1.000	1.000	1.000	1.000	0.997	0.960	0.685
23	22	1.000	1.000	1.000	1.000	1.000	1.000	1.000	0.994	0.911
23	23	1.000	1.000	1.000	1.000	1.000	1.000	1.000	1.000	1.000
24	0	0.080	0.005	0.000	0.000	0.000	0.000	0.000	0.000	0.000
24	1	0.292	0.033	0.002	0.000	0.000	0.000	0.000	0.000	0.000
24	2	0.564	0.115	0.012	0.001	0.000	0.000	0.000	0.000	0.000
24	3	0.786	0.264	0.042	0.004	0.000	0.000	0.000	0.000	0.000
24	4	0.915	0.460	0.111	0.013	0.001	0.000	0.000	0.000	0.000
24	5	0.972	0.656	0.229	0.040	0.003	0.000	0.000	0.000	0.000
24	6	0.993	0.811	0.389	0.096	0.011	0.001	0.000	0.000	0.000
24	7	0.998	0.911	0.565	0.192	0.032	0.002	0.000	0.000	0.000
24	8	1.000	0.964	0.725	0.328	0.076	0.008	0.000	0.000	0.000
24	9	1.000	0.987	0.847	0.489	0.154	0.022	0.001	0.000	0.000
24	10	1.000	0.996	0.926	0.650	0.271	0.053	0.004	0.000	0.000
24	11	1.000	0.999	0.969	0.787	0.419	0.114	0.012	0.000	0.000
24	12	1.000	1.000	0.988	0.886	0.581	0.213	0.031	0.001	0.000
24	13	1.000	1.000	0.996	0.947	0.729	0.350	0.074	0.004	0.000
24	14	1.000	1.000	0.999	0.978	0.846	0.511	0.153	0.013	0.000
24	15	1.000	1.000	1.000	0.992	0.924	0.672	0.275	0.036	0.000
24	16	1.000	1.000	1.000	0.998	0.968	0.808	0.435	0.089	0.002

BINOMIAL CUMULATIVE DISTRIBUTION

n	x	p .1	.2	.3	.4	.5	.6	.7	.8	.9
24	17	1.000	1.000	1.000	0.999	0.989	0.904	0.611	0.189	0.007
24	18	1.000	1.000	1.000	1.000	0.997	0.960	0.771	0.344	0.028
24	19	1.000	1.000	1.000	1.000	0.999	0.987	0.889	0.540	0.085
24	20	1.000	1.000	1.000	1.000	1.000	0.996	0.958	0.736	0.214
24	21	1.000	1.000	1.000	1.000	1.000	0.999	0.988	0.885	0.436
24	22	1.000	1.000	1.000	1.000	1.000	1.000	0.998	0.967	0.708
24	23	1.000	1.000	1.000	1.000	1.000	1.000	1.000	0.995	0.920
24	24	1.000	1.000	1.000	1.000	1.000	1.000	1.000	1.000	1.000
25	0	0.072	0.004	0.000	0.000	0.000	0.000	0.000	0.000	0.000
25	1	0.271	0.027	0.002	0.000	0.000	0.000	0.000	0.000	0.000
25	2	0.537	0.098	0.009	0.000	0.000	0.000	0.000	0.000	0.000
25	3	0.764	0.234	0.033	0.002	0.000	0.000	0.000	0.000	0.000
25	4	0.902	0.421	0.090	0.009	0.000	0.000	0.000	0.000	0.000
25	5	0.967	0.617	0.193	0.029	0.002	0.000	0.000	0.000	0.000
25	6	0.991	0.780	0.341	0.074	0.007	0.000	0.000	0.000	0.000
25	7	0.998	0.891	0.512	0.154	0.022	0.001	0.000	0.000	0.000
25	8	1.000	0.953	0.677	0.274	0.054	0.004	0.000	0.000	0.000
25	9	1.000	0.983	0.811	0.425	0.115	0.013	0.000	0.000	0.000
25	10	1.000	0.994	0.902	0.586	0.212	0.034	0.002	0.000	0.000
25	11	1.000	0.998	0.956	0.732	0.345	0.078	0.006	0.000	0.000
25	12	1.000	1.000	0.983	0.846	0.500	0.154	0.017	0.000	0.000
25	13	1.000	1.000	0.994	0.922	0.655	0.268	0.044	0.002	0.000
25	14	1.000	1.000	0.998	0.966	0.788	0.414	0.098	0.006	0.000
25	15	1.000	1.000	1.000	0.987	0.885	0.575	0.189	0.017	0.000
25	16	1.000	1.000	1.000	0.996	0.946	0.726	0.323	0.047	0.000
25	17	1.000	1.000	1.000	0.999	0.978	0.846	0.488	0.109	0.002
25	18	1.000	1.000	1.000	1.000	0.993	0.926	0.659	0.220	0.009
25	19	1.000	1.000	1.000	1.000	0.998	0.971	0.807	0.383	0.033
25	20	1.000	1.000	1.000	1.000	1.000	0.991	0.910	0.579	0.098
25	21	1.000	1.000	1.000	1.000	1.000	0.998	0.967	0.766	0.236
25	22	1.000	1.000	1.000	1.000	1.000	1.000	0.991	0.902	0.463
25	23	1.000	1.000	1.000	1.000	1.000	1.000	0.998	0.973	0.729
25	24	1.000	1.000	1.000	1.000	1.000	1.000	1.000	0.996	0.928
25	25	1.000	1.000	1.000	1.000	1.000	1.000	1.000	1.000	1.000

TABLE II

Poisson Cumulative Distribution

This table gives values of F(x) where

$$F(x) = P(X \leq x) = \sum_{k=0}^{x} e^{-\lambda}\lambda^k/k!$$

All entries of 0.000 in the table are actually larger than 0.0 but less than 0.0005. All entries of 1.000 are actually larger than 0.9995 but less than 1.

POISSON CUMULATIVE DISTRIBUTION

λ \ x	0	1	2	3	4	5	6	7
0.01	0.990	1.000	1.000	1.000	1.000	1.000	1.000	1.000
0.02	0.980	1.000	1.000	1.000	1.000	1.000	1.000	1.000
0.03	0.970	1.000	1.000	1.000	1.000	1.000	1.000	1.000
0.04	0.961	0.999	1.000	1.000	1.000	1.000	1.000	1.000
0.05	0.951	0.999	1.000	1.000	1.000	1.000	1.000	1.000
0.06	0.942	0.998	1.000	1.000	1.000	1.000	1.000	1.000
0.07	0.932	0.998	1.000	1.000	1.000	1.000	1.000	1.000
0.08	0.923	0.997	1.000	1.000	1.000	1.000	1.000	1.000
0.09	0.914	0.996	1.000	1.000	1.000	1.000	1.000	1.000
0.10	0.905	0.995	1.000	1.000	1.000	1.000	1.000	1.000
0.15	0.861	0.990	0.999	1.000	1.000	1.000	1.000	1.000
0.20	0.819	0.982	0.999	1.000	1.000	1.000	1.000	1.000
0.25	0.779	0.974	0.998	1.000	1.000	1.000	1.000	1.000
0.30	0.741	0.963	0.996	1.000	1.000	1.000	1.000	1.000
0.35	0.705	0.951	0.994	1.000	1.000	1.000	1.000	1.000
0.40	0.670	0.938	0.992	0.999	1.000	1.000	1.000	1.000
0.45	0.638	0.925	0.989	0.999	1.000	1.000	1.000	1.000
0.50	0.607	0.910	0.986	0.998	1.000	1.000	1.000	1.000
0.55	0.577	0.894	0.982	0.998	1.000	1.000	1.000	1.000
0.60	0.549	0.878	0.977	0.997	1.000	1.000	1.000	1.000
0.65	0.522	0.861	0.972	0.996	0.999	1.000	1.000	1.000
0.70	0.497	0.844	0.966	0.994	0.999	1.000	1.000	1.000
0.75	0.472	0.827	0.959	0.993	0.999	1.000	1.000	1.000
0.80	0.449	0.809	0.953	0.991	0.999	1.000	1.000	1.000
0.85	0.427	0.791	0.945	0.989	0.998	1.000	1.000	1.000
0.90	0.407	0.772	0.937	0.987	0.998	1.000	1.000	1.000
0.95	0.387	0.754	0.929	0.984	0.997	1.000	1.000	1.000
1.00	0.368	0.736	0.920	0.981	0.996	0.999	1.000	1.000
1.10	0.333	0.699	0.900	0.974	0.995	0.999	1.000	1.000
1.20	0.301	0.663	0.879	0.966	0.992	0.998	1.000	1.000
1.30	0.273	0.627	0.857	0.957	0.989	0.998	1.000	1.000
1.40	0.247	0.592	0.833	0.946	0.986	0.997	0.999	1.000
1.50	0.223	0.558	0.809	0.934	0.981	0.996	0.999	1.000
1.60	0.202	0.525	0.783	0.921	0.976	0.994	0.999	1.000
1.70	0.183	0.493	0.757	0.907	0.970	0.992	0.998	1.000
1.80	0.165	0.463	0.731	0.891	0.964	0.990	0.997	0.999
1.90	0.150	0.434	0.704	0.875	0.956	0.987	0.997	0.999
2.00	0.135	0.406	0.677	0.857	0.947	0.983	0.995	0.999
2.10	0.122	0.380	0.650	0.839	0.938	0.980	0.994	0.999
2.20	0.111	0.355	0.623	0.819	0.928	0.975	0.993	0.998
2.30	0.100	0.331	0.596	0.799	0.916	0.970	0.991	0.997
2.40	0.091	0.308	0.570	0.779	0.904	0.964	0.988	0.997
2.50	0.082	0.287	0.544	0.758	0.891	0.958	0.986	0.996
2.60	0.074	0.267	0.518	0.736	0.877	0.951	0.983	0.995
2.70	0.067	0.249	0.494	0.714	0.863	0.943	0.979	0.993
2.80	0.061	0.231	0.469	0.692	0.848	0.935	0.976	0.992
2.90	0.055	0.215	0.446	0.670	0.832	0.926	0.971	0.990
3.00	0.050	0.199	0.423	0.647	0.815	0.916	0.966	0.988

POISSON CUMULATIVE DISTRIBUTION

λ \ x	8	9	10	11	12	13	14	15
2.20	1.000	1.000	1.000	1.000	1.000	1.000	1.000	1.000
2.30	0.999	1.000	1.000	1.000	1.000	1.000	1.000	1.000
2.40	0.999	1.000	1.000	1.000	1.000	1.000	1.000	1.000
2.50	0.999	1.000	1.000	1.000	1.000	1.000	1.000	1.000
2.60	0.999	1.000	1.000	1.000	1.000	1.000	1.000	1.000
2.70	0.998	0.999	1.000	1.000	1.000	1.000	1.000	1.000
2.80	0.998	0.999	1.000	1.000	1.000	1.000	1.000	1.000
2.90	0.997	0.999	1.000	1.000	1.000	1.000	1.000	1.000
3.00	0.996	0.999	1.000	1.000	1.000	1.000	1.000	1.000

λ \ x	0	1	2	3	4	5	6	7
3.20	0.041	0.171	0.380	0.603	0.781	0.895	0.955	0.983
3.40	0.033	0.147	0.340	0.558	0.744	0.871	0.942	0.977
3.60	0.027	0.126	0.303	0.515	0.706	0.844	0.927	0.969
3.80	0.022	0.107	0.269	0.473	0.668	0.816	0.909	0.960
4.00	0.018	0.092	0.238	0.433	0.629	0.785	0.889	0.949
4.20	0.015	0.078	0.210	0.395	0.590	0.753	0.867	0.936
4.40	0.012	0.066	0.185	0.359	0.551	0.720	0.844	0.921
4.60	0.010	0.056	0.163	0.326	0.513	0.686	0.818	0.905
4.80	0.008	0.048	0.143	0.294	0.476	0.651	0.791	0.887
5.00	0.007	0.040	0.125	0.265	0.440	0.616	0.762	0.867
5.20	0.006	0.034	0.109	0.238	0.406	0.581	0.732	0.845
5.40	0.005	0.029	0.095	0.213	0.373	0.546	0.702	0.822
5.60	0.004	0.024	0.082	0.191	0.342	0.512	0.670	0.797
5.80	0.003	0.021	0.072	0.170	0.313	0.478	0.638	0.771
6.00	0.002	0.017	0.062	0.151	0.285	0.446	0.606	0.744
6.20	0.002	0.015	0.054	0.134	0.259	0.414	0.574	0.716
6.40	0.002	0.012	0.046	0.119	0.235	0.384	0.542	0.687
6.60	0.001	0.010	0.040	0.105	0.213	0.355	0.511	0.658
6.80	0.001	0.009	0.034	0.093	0.192	0.327	0.480	0.628
7.00	0.001	0.007	0.030	0.082	0.173	0.301	0.450	0.599
7.20	0.001	0.006	0.025	0.072	0.156	0.276	0.420	0.569
7.40	0.001	0.005	0.022	0.063	0.140	0.253	0.392	0.539
7.60	0.001	0.004	0.019	0.055	0.125	0.231	0.365	0.510
7.80	0.000	0.004	0.016	0.040	0.112	0.210	0.330	0.481
8.00	0.000	0.003	0.014	0.042	0.100	0.191	0.313	0.453
8.50	0.000	0.002	0.009	0.030	0.074	0.150	0.256	0.386
9.00	0.000	0.001	0.006	0.021	0.055	0.116	0.207	0.324
9.50	0.000	0.001	0.004	0.015	0.040	0.089	0.165	0.269
10.00	0.000	0.000	0.003	0.010	0.029	0.067	0.130	0.220
10.50	0.000	0.000	0.002	0.007	0.021	0.050	0.102	0.179
11.00	0.000	0.000	0.001	0.005	0.015	0.038	0.079	0.143
11.50	0.000	0.000	0.001	0.003	0.011	0.028	0.060	0.114
12.00	0.000	0.000	0.001	0.002	0.008	0.020	0.046	0.090
12.50	0.000	0.000	0.000	0.002	0.005	0.015	0.035	0.070
13.00	0.000	0.000	0.000	0.001	0.004	0.011	0.026	0.054

POISSON CUMULATIVE DISTRIBUTION

λ \ x	8	9	10	11	12	13	14	15
3.20	0.994	0.998	1.000	1.000	1.000	1.000	1.000	1.000
3.40	0.992	0.997	0.999	1.000	1.000	1.000	1.000	1.000
3.60	0.988	0.996	0.999	1.000	1.000	1.000	1.000	1.000
3.80	0.984	0.994	0.998	0.999	1.000	1.000	1.000	1.000
4.00	0.979	0.992	0.997	0.999	1.000	1.000	1.000	1.000
4.20	0.972	0.989	0.996	0.999	1.000	1.000	1.000	1.000
4.40	0.964	0.985	0.994	0.998	0.999	1.000	1.000	1.000
4.60	0.955	0.980	0.992	0.997	0.999	1.000	1.000	1.000
4.80	0.944	0.975	0.990	0.996	0.999	1.000	1.000	1.000
5.00	0.932	0.968	0.986	0.995	0.998	0.999	1.000	1.000
5.20	0.918	0.960	0.982	0.993	0.997	0.999	1.000	1.000
5.40	0.903	0.951	0.977	0.990	0.996	0.999	1.000	1.000
5.60	0.886	0.941	0.972	0.988	0.995	0.998	0.999	1.000
5.80	0.867	0.929	0.965	0.984	0.993	0.997	0.999	1.000
6.00	0.847	0.916	0.957	0.980	0.991	0.996	0.999	0.999
6.20	0.826	0.902	0.949	0.975	0.989	0.995	0.998	0.999
6.40	0.803	0.886	0.939	0.969	0.986	0.994	0.997	0.999
6.60	0.780	0.869	0.927	0.963	0.982	0.992	0.997	0.999
6.80	0.755	0.850	0.915	0.955	0.978	0.990	0.996	0.998
7.00	0.729	0.830	0.901	0.947	0.973	0.987	0.994	0.998
7.20	0.703	0.810	0.887	0.937	0.967	0.984	0.993	0.997
7.40	0.676	0.788	0.871	0.926	0.961	0.980	0.991	0.996
7.60	0.648	0.765	0.854	0.915	0.954	0.976	0.989	0.995
7.80	0.620	0.741	0.835	0.902	0.945	0.971	0.986	0.993
8.00	0.593	0.717	0.816	0.888	0.936	0.966	0.983	0.992
8.50	0.523	0.653	0.763	0.849	0.909	0.949	0.973	0.986
9.00	0.456	0.587	0.706	0.803	0.876	0.926	0.959	0.978
9.50	0.392	0.522	0.645	0.752	0.836	0.898	0.940	0.967
10.00	0.333	0.458	0.583	0.697	0.792	0.864	0.917	0.951
10.50	0.279	0.397	0.521	0.639	0.742	0.825	0.888	0.932
11.00	0.232	0.341	0.460	0.579	0.689	0.781	0.854	0.907
11.50	0.191	0.289	0.402	0.520	0.633	0.733	0.815	0.878
12.00	0.155	0.242	0.347	0.462	0.576	0.682	0.772	0.844
12.50	0.125	0.201	0.297	0.406	0.519	0.628	0.725	0.806
13.00	0.100	0.166	0.252	0.353	0.463	0.573	0.675	0.764

λ \ x	16	17	18	19	20	21	22	23
6.00	1.000	1.000	1.000	1.000	1.000	1.000	1.000	1.000
6.20	1.000	1.000	1.000	1.000	1.000	1.000	1.000	1.000
6.40	1.000	1.000	1.000	1.000	1.000	1.000	1.000	1.000
6.60	0.999	1.000	1.000	1.000	1.000	1.000	1.000	1.000
6.80	0.999	1.000	1.000	1.000	1.000	1.000	1.000	1.000
7.00	0.999	1.000	1.000	1.000	1.000	1.000	1.000	1.000
7.20	0.999	0.999	1.000	1.000	1.000	1.000	1.000	1.000
7.40	0.998	0.999	1.000	1.000	1.000	1.000	1.000	1.000
7.60	0.998	0.999	1.000	1.000	1.000	1.000	1.000	1.000
7.80	0.997	0.999	1.000	1.000	1.000	1.000	1.000	1.000
8.00	0.996	0.998	0.999	1.000	1.000	1.000	1.000	1.000

POISSON CUMULATIVE DISTRIBUTION

λ \ x	16	17	18	19	20	21	22	23
8.50	0.993	0.997	0.999	0.999	1.000	1.000	1.000	1.000
9.00	0.989	0.995	0.998	0.999	1.000	1.000	1.000	1.000
9.50	0.982	0.991	0.996	0.998	0.999	1.000	1.000	1.000
10.00	0.973	0.986	0.993	0.997	0.998	0.999	1.000	1.000
10.50	0.960	0.978	0.988	0.994	0.997	0.999	0.999	1.000
11.00	0.944	0.968	0.982	0.991	0.995	0.998	0.999	1.000
11.50	0.924	0.954	0.974	0.986	0.992	0.996	0.998	0.999
12.00	0.899	0.937	0.963	0.979	0.988	0.994	0.997	0.999
12.50	0.869	0.916	0.948	0.969	0.983	0.991	0.995	0.998
13.00	0.835	0.890	0.930	0.957	0.975	0.986	0.992	0.996

λ \ x	24	25	26	27	28	29	30	31
12.00	0.999	1.000	1.000	1.000	1.000	1.000	1.000	1.000
12.50	0.999	0.999	1.000	1.000	1.000	1.000	1.000	1.000
13.00	0.998	0.999	1.000	1.000	1.000	1.000	1.000	1.000

λ \ x	2	3	4	5	6	7	8	9
13.50	0.000	0.001	0.003	0.008	0.019	0.041	0.079	0.135
14.00	0.000	0.000	0.002	0.006	0.014	0.032	0.062	0.109
14.50	0.000	0.000	0.001	0.004	0.010	0.024	0.048	0.088
15.00	0.000	0.000	0.001	0.003	0.008	0.018	0.037	0.070
16.00	0.000	0.000	0.000	0.001	0.004	0.010	0.022	0.043
17.00	0.000	0.000	0.000	0.001	0.002	0.005	0.013	0.026
18.00	0.000	0.000	0.000	0.000	0.001	0.003	0.007	0.015
19.00	0.000	0.000	0.000	0.000	0.001	0.002	0.004	0.009
20.00	0.000	0.000	0.000	0.000	0.000	0.001	0.002	0.005
21.00	0.000	0.000	0.000	0.000	0.000	0.000	0.001	0.003
22.00	0.000	0.000	0.000	0.000	0.000	0.000	0.001	0.002
23.00	0.000	0.000	0.000	0.000	0.000	0.000	0.000	0.001

λ \ x	10	11	12	13	14	15	16	17
13.50	0.211	0.304	0.409	0.518	0.623	0.718	0.798	0.861
14.00	0.176	0.260	0.358	0.464	0.570	0.669	0.756	0.827
14.50	0.145	0.220	0.311	0.413	0.518	0.619	0.711	0.790
15.00	0.118	0.185	0.268	0.363	0.466	0.568	0.664	0.749
16.00	0.077	0.127	0.193	0.275	0.368	0.467	0.566	0.659
17.00	0.049	0.085	0.135	0.201	0.281	0.371	0.468	0.564
18.00	0.030	0.055	0.092	0.143	0.208	0.287	0.375	0.469
19.00	0.018	0.035	0.061	0.098	0.150	0.215	0.292	0.378
20.00	0.011	0.021	0.039	0.066	0.105	0.157	0.221	0.297
21.00	0.006	0.013	0.025	0.043	0.072	0.111	0.163	0.227
22.00	0.004	0.008	0.015	0.028	0.048	0.077	0.117	0.169
23.00	0.002	0.004	0.009	0.017	0.031	0.052	0.082	0.123
24.00	0.001	0.003	0.005	0.011	0.020	0.034	0.056	0.087
25.00	0.001	0.001	0.003	0.006	0.012	0.022	0.038	0.060

POISSON CUMULATIVE DISTRIBUTION

λ \ x	18	19	20	21	22	23	24	25
13.50	0.908	0.942	0.965	0.980	0.989	0.994	0.997	0.998
14.00	0.883	0.923	0.952	0.971	0.983	0.991	0.995	0.997
14.50	0.853	0.901	0.936	0.960	0.976	0.986	0.992	0.996
15.00	0.819	0.875	0.917	0.947	0.967	0.981	0.989	0.994
16.00	0.742	0.812	0.868	0.911	0.942	0.963	0.978	0.987
17.00	0.655	0.736	0.805	0.861	0.905	0.937	0.959	0.975
18.00	0.562	0.651	0.731	0.799	0.855	0.899	0.932	0.955
19.00	0.469	0.561	0.647	0.725	0.793	0.849	0.893	0.927
20.00	0.381	0.470	0.559	0.644	0.721	0.787	0.843	0.888
21.00	0.302	0.384	0.471	0.558	0.640	0.716	0.782	0.838
22.00	0.232	0.306	0.387	0.472	0.556	0.637	0.712	0.777
23.00	0.175	0.238	0.310	0.389	0.472	0.555	0.635	0.708
24.00	0.128	0.180	0.243	0.314	0.392	0.473	0.554	0.632
25.00	0.092	0.134	0.185	0.247	0.318	0.394	0.473	0.553

λ \ x	26	27	28	29	30	31	32	33
13.50	0.999	1.000	1.000	1.000	1.000	1.000	1.000	1.000
14.00	0.999	0.999	1.000	1.000	1.000	1.000	1.000	1.000
14.50	0.998	0.999	0.999	1.000	1.000	1.000	1.000	1.000
15.00	0.997	0.998	0.999	1.000	1.000	1.000	1.000	1.000
16.00	0.993	0.996	0.998	0.999	0.999	1.000	1.000	1.000
17.00	0.985	0.991	0.995	0.997	0.999	0.999	1.000	1.000
18.00	0.972	0.983	0.990	0.994	0.997	0.998	0.999	0.999
19.00	0.951	0.969	0.980	0.988	0.993	0.996	0.998	0.999
20.00	0.922	0.948	0.966	0.978	0.987	0.992	0.995	0.997
21.00	0.883	0.917	0.944	0.963	0.976	0.985	0.991	0.994
22.00	0.832	0.877	0.913	0.940	0.959	0.973	0.983	0.989
23.00	0.772	0.827	0.873	0.908	0.936	0.956	0.971	0.981
24.00	0.704	0.768	0.823	0.868	0.904	0.932	0.953	0.969
25.00	0.629	0.700	0.763	0.818	0.863	0.900	0.929	0.950

λ \ x	34	35	36	37	38	39	40	41
18.00	1.000	1.000	1.000	1.000	1.000	1.000	1.000	1.000
19.00	0.999	1.000	1.000	1.000	1.000	1.000	1.000	1.000
20.00	0.998	0.999	1.000	1.000	1.000	1.000	1.000	1.000
21.00	0.997	0.998	0.999	0.999	1.000	1.000	1.000	1.000
22.00	0.994	0.996	0.998	0.999	0.999	1.000	1.000	1.000
23.00	0.988	0.993	0.996	0.997	0.999	0.999	1.000	1.000
24.00	0.979	0.987	0.992	0.995	0.997	0.998	0.999	0.999
25.00	0.966	0.978	0.985	0.991	0.994	0.997	0.998	0.999

λ \ x	42	43	44	45	46	47	48	49
24.00	1.000	1.000	1.000	1.000	1.000	1.000	1.000	1.000
25.00	0.999	1.000	1.000	1.000	1.000	1.000	1.000	1.000

TABLE III

Standard Normal Cumulative Distribution

This table gives values of the standard normal cumulative distribution function

$$F(z) = P(Z \leq z) = \int_{-\infty}^{z} \frac{1}{\sqrt{2\pi}} e^{-t^2/2} dt$$

for values of z = -4.00(0.01)4.00.

STANDARD NORMAL CUMULATIVE DISTRIBUTION

z	F(z)	z	F(z)	z	F(z)
-4.00	0.00003	-3.60	0.00016	-3.20	0.00069
-3.99	0.00003	-3.59	0.00017	-3.19	0.00071
-3.98	0.00003	-3.58	0.00017	-3.18	0.00074
-3.97	0.00004	-3.57	0.00018	-3.17	0.00076
-3.96	0.00004	-3.56	0.00019	-3.16	0.00079
-3.95	0.00004	-3.55	0.00019	-3.15	0.00082
-3.94	0.00004	-3.54	0.00020	-3.14	0.00084
-3.93	0.00004	-3.53	0.00021	-3.13	0.00087
-3.92	0.00004	-3.52	0.00022	-3.12	0.00090
-3.91	0.00005	-3.51	0.00022	-3.11	0.00094
-3.90	0.00005	-3.50	0.00023	-3.10	0.00097
-3.89	0.00005	-3.49	0.00024	-3.09	0.00100
-3.88	0.00005	-3.48	0.00025	-3.08	0.00104
-3.87	0.00005	-3.47	0.00026	-3.07	0.00107
-3.86	0.00006	-3.46	0.00027	-3.06	0.00111
-3.85	0.00006	-3.45	0.00028	-3.05	0.00114
-3.84	0.00006	-3.44	0.00029	-3.04	0.00118
-3.83	0.00006	-3.43	0.00030	-3.03	0.00122
-3.82	0.00007	-3.42	0.00031	-3.02	0.00126
-3.81	0.00007	-3.41	0.00032	-3.01	0.00131
-3.80	0.00007	-3.40	0.00034	-3.00	0.00135
-3.79	0.00008	-3.39	0.00035	-2.99	0.00139
-3.78	0.00008	-3.38	0.00036	-2.98	0.00144
-3.77	0.00008	-3.37	0.00038	-2.97	0.00149
-3.76	0.00008	-3.36	0.00039	-2.96	0.00154
-3.75	0.00009	-3.35	0.00040	-2.95	0.00159
-3.74	0.00009	-3.34	0.00042	-2.94	0.00164
-3.73	0.00010	-3.33	0.00043	-2.93	0.00169
-3.72	0.00010	-3.32	0.00045	-2.92	0.00175
-3.71	0.00010	-3.31	0.00047	-2.91	0.00181
-3.70	0.00011	-3.30	0.00048	-2.90	0.00187
-3.69	0.00011	-3.29	0.00050	-2.89	0.00193
-3.68	0.00012	-3.28	0.00052	-2.88	0.00199
-3.67	0.00012	-3.27	0.00054	-2.87	0.00205
-3.66	0.00013	-3.26	0.00056	-2.86	0.00212
-3.65	0.00013	-3.25	0.00058	-2.85	0.00219
-3.64	0.00014	-3.24	0.00060	-2.84	0.00226
-3.63	0.00014	-3.23	0.00062	-2.83	0.00233
-3.62	0.00015	-3.22	0.00064	-2.82	0.00240
-3.61	0.00015	-3.21	0.00066	-2.81	0.00248

STANDARD NORMAL CUMULATIVE DISTRIBUTION

z	F(z)	z	F(z)	z	F(z)
-2.80	0.00256	-2.40	0.00820	-2.00	0.02275
-2.79	0.00264	-2.39	0.00842	-1.99	0.02330
-2.78	0.00272	-2.38	0.00866	-1.98	0.02385
-2.77	0.00280	-2.37	0.00889	-1.97	0.02442
-2.76	0.00289	-2.36	0.00914	-1.96	0.02500
-2.75	0.00298	-2.35	0.00939	-1.95	0.02559
-2.74	0.00307	-2.34	0.00964	-1.94	0.02619
-2.73	0.00317	-2.33	0.00990	-1.93	0.02680
-2.72	0.00326	-2.32	0.01017	-1.92	0.02743
-2.71	0.00336	-2.31	0.01044	-1.91	0.02807
-2.70	0.00347	-2.30	0.01072	-1.90	0.02872
-2.69	0.00357	-2.29	0.01101	-1.89	0.02938
-2.68	0.00368	-2.28	0.01130	-1.88	0.03005
-2.67	0.00379	-2.27	0.01160	-1.87	0.03074
-2.66	0.00391	-2.26	0.01191	-1.86	0.03144
-2.65	0.00402	-2.25	0.01222	-1.85	0.03216
-2.64	0.00415	-2.24	0.01255	-1.84	0.03288
-2.63	0.00427	-2.23	0.01287	-1.83	0.03362
-2.62	0.00440	-2.22	0.01321	-1.82	0.03438
-2.61	0.00453	-2.21	0.01355	-1.81	0.03515
-2.60	0.00466	-2.20	0.01390	-1.80	0.03593
-2.59	0.00480	-2.19	0.01426	-1.79	0.03673
-2.58	0.00494	-2.18	0.01463	-1.78	0.03754
-2.57	0.00508	-2.17	0.01500	-1.77	0.03836
-2.56	0.00523	-2.16	0.01539	-1.76	0.03920
-2.55	0.00539	-2.15	0.01578	-1.75	0.04006
-2.54	0.00554	-2.14	0.01618	-1.74	0.04093
-2.53	0.00570	-2.13	0.01659	-1.73	0.04182
-2.52	0.00587	-2.12	0.01700	-1.72	0.04272
-2.51	0.00604	-2.11	0.01743	-1.71	0.04363
-2.50	0.00621	-2.10	0.01786	-1.70	0.04457
-2.49	0.00639	-2.09	0.01831	-1.69	0.04551
-2.48	0.00657	-2.08	0.01876	-1.68	0.04648
-2.47	0.00676	-2.07	0.01923	-1.67	0.04746
-2.46	0.00695	-2.06	0.01970	-1.66	0.04846
-2.45	0.00714	-2.05	0.02018	-1.65	0.04947
-2.44	0.00734	-2.04	0.02068	-1.64	0.05050
-2.43	0.00755	-2.03	0.02118	-1.63	0.05155
-2.42	0.00776	-2.02	0.02169	-1.62	0.05262
-2.41	0.00798	-2.01	0.02222	-1.61	0.05370

STANDARD NORMAL CUMULATIVE DISTRIBUTION

z	F(z)	z	F(z)	z	F(z)
-1.60	0.05480	-1.20	0.11507	-0.80	0.21186
-1.59	0.05592	-1.19	0.11702	-0.79	0.21476
-1.58	0.05705	-1.18	0.11900	-0.78	0.21770
-1.57	0.05821	-1.17	0.12100	-0.77	0.22065
-1.56	0.05938	-1.16	0.12302	-0.76	0.22363
-1.55	0.06057	-1.15	0.12507	-0.75	0.22663
-1.54	0.06178	-1.14	0.12714	-0.74	0.22965
-1.53	0.06301	-1.13	0.12924	-0.73	0.23270
-1.52	0.06426	-1.12	0.13136	-0.72	0.23576
-1.51	0.06552	-1.11	0.13350	-0.71	0.23885
-1.50	0.06681	-1.10	0.13567	-0.70	0.24196
-1.49	0.06811	-1.09	0.13786	-0.69	0.24510
-1.48	0.06944	-1.08	0.14007	-0.68	0.24825
-1.47	0.07078	-1.07	0.14231	-0.67	0.25143
-1.46	0.07215	-1.06	0.14457	-0.66	0.25463
-1.45	0.07353	-1.05	0.14686	-0.65	0.25785
-1.44	0.07493	-1.04	0.14917	-0.64	0.26109
-1.43	0.07636	-1.03	0.15150	-0.63	0.26435
-1.42	0.07780	-1.02	0.15386	-0.62	0.26763
-1.41	0.07927	-1.01	0.15625	-0.61	0.27093
-1.40	0.08076	-1.00	0.15866	-0.60	0.27425
-1.39	0.08226	-0.99	0.16109	-0.59	0.27760
-1.38	0.08379	-0.98	0.16354	-0.58	0.28096
-1.37	0.08534	-0.97	0.16602	-0.57	0.28434
-1.36	0.08691	-0.96	0.16853	-0.56	0.28774
-1.35	0.08851	-0.95	0.17106	-0.55	0.29116
-1.34	0.09012	-0.94	0.17361	-0.54	0.29460
-1.33	0.09176	-0.93	0.17619	-0.53	0.29806
-1.32	0.09342	-0.92	0.17879	-0.52	0.30153
-1.31	0.09510	-0.91	0.18141	-0.51	0.30503
-1.30	0.09680	-0.90	0.18406	-0.50	0.30854
-1.29	0.09853	-0.89	0.18673	-0.49	0.31207
-1.28	0.10027	-0.88	0.18943	-0.48	0.31561
-1.27	0.10204	-0.87	0.19215	-0.47	0.31918
-1.26	0.10383	-0.86	0.19489	-0.46	0.32276
-1.25	0.10565	-0.85	0.19766	-0.45	0.32636
-1.24	0.10749	-0.84	0.20045	-0.44	0.32997
-1.23	0.10935	-0.83	0.20327	-0.43	0.33360
-1.22	0.11123	-0.82	0.20611	-0.42	0.33724
-1.21	0.11314	-0.81	0.20897	-0.41	0.34090

STANDARD NORMAL CUMULATIVE DISTRIBUTION

z	F(z)	z	F(z)	z	F(z)
-0.40	0.34458	0.0	0.50000	0.40	0.65542
-0.39	0.34827	0.01	0.50399	0.41	0.65910
-0.38	0.35197	0.02	0.50798	0.42	0.66276
-0.37	0.35569	0.03	0.51197	0.43	0.66640
-0.36	0.35942	0.04	0.51595	0.44	0.67003
-0.35	0.36317	0.05	0.51994	0.45	0.67364
-0.34	0.36693	0.06	0.52392	0.46	0.67724
-0.33	0.37070	0.07	0.52790	0.47	0.68082
-0.32	0.37448	0.08	0.53188	0.48	0.68439
-0.31	0.37828	0.09	0.53586	0.49	0.68793
-0.30	0.38209	0.10	0.53983	0.50	0.69146
-0.29	0.38591	0.11	0.54380	0.51	0.69497
-0.28	0.38974	0.12	0.54776	0.52	0.69847
-0.27	0.39358	0.13	0.55172	0.53	0.70194
-0.26	0.39743	0.14	0.55567	0.54	0.70540
-0.25	0.40129	0.15	0.55962	0.55	0.70884
-0.24	0.40517	0.16	0.56356	0.56	0.71226
-0.23	0.40905	0.17	0.56749	0.57	0.71566
-0.22	0.41294	0.18	0.57142	0.58	0.71904
-0.21	0.41683	0.19	0.57535	0.59	0.72240
-0.20	0.42074	0.20	0.57926	0.60	0.72575
-0.19	0.42465	0.21	0.58317	0.61	0.72907
-0.18	0.42858	0.22	0.58706	0.62	0.73237
-0.17	0.43251	0.23	0.59095	0.63	0.73565
-0.16	0.43644	0.24	0.59483	0.64	0.73891
-0.15	0.44038	0.25	0.59871	0.65	0.74215
-0.14	0.44433	0.26	0.60257	0.66	0.74537
-0.13	0.44828	0.27	0.60642	0.67	0.74857
-0.12	0.45224	0.28	0.61026	0.68	0.75175
-0.11	0.45620	0.29	0.61409	0.69	0.75490
-0.10	0.46017	0.30	0.61791	0.70	0.75804
-0.09	0.46414	0.31	0.62172	0.71	0.76115
-0.08	0.46812	0.32	0.62552	0.72	0.76424
-0.07	0.47210	0.33	0.62930	0.73	0.76730
-0.06	0.47608	0.34	0.63307	0.74	0.77035
-0.05	0.48006	0.35	0.63683	0.75	0.77337
-0.04	0.48405	0.36	0.64058	0.76	0.77637
-0.03	0.48803	0.37	0.64431	0.77	0.77935
-0.02	0.49202	0.38	0.64803	0.78	0.78230
-0.01	0.49601	0.39	0.65173	0.79	0.78524

STANDARD NORMAL CUMULATIVE DISTRIBUTION

z	F(z)	z	F(z)	z	F(z)
0.80	0.78814	1.20	0.88493	1.60	0.94520
0.81	0.79103	1.21	0.88686	1.61	0.94630
0.82	0.79389	1.22	0.88877	1.62	0.94738
0.83	0.79673	1.23	0.89065	1.63	0.94845
0.84	0.79955	1.24	0.89251	1.64	0.94950
0.85	0.80234	1.25	0.89435	1.65	0.95053
0.86	0.80511	1.26	0.89617	1.66	0.95154
0.87	0.80785	1.27	0.89796	1.67	0.95254
0.88	0.81057	1.28	0.89973	1.68	0.95352
0.89	0.81327	1.29	0.90147	1.69	0.95449
0.90	0.81594	1.30	0.90320	1.70	0.95543
0.91	0.81859	1.31	0.90490	1.71	0.95637
0.92	0.82121	1.32	0.90658	1.72	0.95728
0.93	0.82381	1.33	0.90824	1.73	0.95818
0.94	0.82639	1.34	0.90988	1.74	0.95907
0.95	0.82894	1.35	0.91149	1.75	0.95994
0.96	0.83147	1.36	0.91309	1.76	0.96080
0.97	0.83398	1.37	0.91466	1.77	0.96164
0.98	0.83646	1.38	0.91621	1.78	0.96246
0.99	0.83891	1.39	0.91774	1.79	0.96327
1.00	0.84134	1.40	0.91924	1.80	0.96407
1.01	0.84375	1.41	0.92073	1.81	0.96485
1.02	0.84614	1.42	0.92220	1.82	0.96562
1.03	0.84850	1.43	0.92364	1.83	0.96638
1.04	0.85083	1.44	0.92507	1.84	0.96712
1.05	0.85314	1.45	0.92647	1.85	0.96784
1.06	0.85543	1.46	0.92785	1.86	0.96856
1.07	0.85769	1.47	0.92922	1.87	0.96926
1.08	0.85993	1.48	0.93056	1.88	0.96995
1.09	0.86214	1.49	0.93189	1.89	0.97062
1.10	0.86433	1.50	0.93319	1.90	0.97128
1.11	0.86650	1.51	0.93448	1.91	0.97193
1.12	0.86864	1.52	0.93574	1.92	0.97257
1.13	0.87076	1.53	0.93699	1.93	0.97320
1.14	0.87286	1.54	0.93822	1.94	0.97381
1.15	0.87493	1.55	0.93943	1.95	0.97441
1.16	0.87698	1.56	0.94062	1.96	0.97500
1.17	0.87900	1.57	0.94179	1.97	0.97558
1.18	0.88100	1.58	0.94295	1.98	0.97615
1.19	0.88298	1.59	0.94408	1.99	0.97670

STANDARD NORMAL CUMULATIVE DISTRIBUTION

z	F(z)	z	F(z)	z	F(z)
2.00	0.97725	2.40	0.99180	2.80	0.99744
2.01	0.97778	2.41	0.99202	2.81	0.99752
2.02	0.97831	2.42	0.99224	2.82	0.99760
2.03	0.97882	2.43	0.99245	2.83	0.99767
2.04	0.97932	2.44	0.99266	2.84	0.99774
2.05	0.97982	2.45	0.99286	2.85	0.99781
2.06	0.98030	2.46	0.99305	2.86	0.99788
2.07	0.98077	2.47	0.99324	2.87	0.99795
2.08	0.98124	2.48	0.99343	2.88	0.99801
2.09	0.98169	2.49	0.99361	2.89	0.99807
2.10	0.98214	2.50	0.99379	2.90	0.99813
2.11	0.98257	2.51	0.99396	2.91	0.99819
2.12	0.98300	2.52	0.99413	2.92	0.99825
2.13	0.98341	2.53	0.99430	2.93	0.99831
2.14	0.98382	2.54	0.99446	2.94	0.99836
2.15	0.98422	2.55	0.99461	2.95	0.99841
2.16	0.98461	2.56	0.99477	2.96	0.99846
2.17	0.98500	2.57	0.99492	2.97	0.99851
2.18	0.98537	2.58	0.99506	2.98	0.99856
2.19	0.98574	2.59	0.99520	2.99	0.99861
2.20	0.98610	2.60	0.99534	3.00	0.99865
2.21	0.98645	2.61	0.99547	3.01	0.99869
2.22	0.98679	2.62	0.99560	3.02	0.99874
2.23	0.98713	2.63	0.99573	3.03	0.99878
2.24	0.98745	2.64	0.99585	3.04	0.99882
2.25	0.98778	2.65	0.99598	3.05	0.99886
2.26	0.98809	2.66	0.99609	3.06	0.99889
2.27	0.98840	2.67	0.99621	3.07	0.99893
2.28	0.98870	2.68	0.99632	3.08	0.99896
2.29	0.98899	2.69	0.99643	3.09	0.99900
2.30	0.98928	2.70	0.99653	3.10	0.99903
2.31	0.98956	2.71	0.99664	3.11	0.99906
2.32	0.98983	2.72	0.99674	3.12	0.99910
2.33	0.99010	2.73	0.99683	3.13	0.99913
2.34	0.99036	2.74	0.99693	3.14	0.99916
2.35	0.99061	2.75	0.99702	3.15	0.99918
2.36	0.99086	2.76	0.99711	3.16	0.99921
2.37	0.99111	2.77	0.99720	3.17	0.99924
2.38	0.99134	2.78	0.99728	3.18	0.99926
2.39	0.99158	2.79	0.99736	3.19	0.99929

STANDARD NORMAL CUMULATIVE DISTRIBUTION

z	F(z)	z	F(z)	z	F(z)
3.20	0.99931	3.50	0.99977	3.80	0.99993
3.21	0.99934	3.51	0.99978	3.81	0.99993
3.22	0.99936	3.52	0.99978	3.82	0.99993
3.23	0.99938	3.53	0.99979	3.83	0.99994
3.24	0.99940	3.54	0.99980	3.84	0.99994
3.25	0.99942	3.55	0.99981	3.85	0.99994
3.26	0.99944	3.56	0.99981	3.86	0.99994
3.27	0.99946	3.57	0.99982	3.87	0.99995
3.28	0.99948	3.58	0.99983	3.88	0.99995
3.29	0.99950	3.59	0.99983	3.89	0.99995
3.30	0.99952	3.60	0.99984	3.90	0.99995
3.31	0.99953	3.61	0.99985	3.91	0.99995
3.32	0.99955	3.62	0.99985	3.92	0.99996
3.33	0.99957	3.63	0.99986	3.93	0.99996
3.34	0.99958	3.64	0.99986	3.94	0.99996
3.35	0.99960	3.65	0.99987	3.95	0.99996
3.36	0.99961	3.66	0.99987	3.96	0.99996
3.37	0.99962	3.67	0.99988	3.97	0.99996
3.38	0.99964	3.68	0.99988	3.98	0.99997
3.39	0.99965	3.69	0.99989	3.99	0.99997
3.40	0.99966	3.70	0.99989	4.00	0.99997
3.41	0.99968	3.71	0.99990		
3.42	0.99969	3.72	0.99990		
3.43	0.99970	3.73	0.99990		
3.44	0.99971	3.74	0.99991		
3.45	0.99972	3.75	0.99991		
3.46	0.99973	3.76	0.99992		
3.47	0.99974	3.77	0.99992		
3.48	0.99975	3.78	0.99992		
3.49	0.99976	3.79	0.99992		

TABLE IV

Cumulative t-Distribution

This table gives selected critical values of student's t-distribution. Entries in the table are values of t_p where

$$p = P(T \leq t_p)$$

where T has student's t-distribution with ν degrees of freedom.

The table covers values of p = 0.9, 0.95, 0.97, 0.99, and 0.995 and $\nu = 1(1)30(10)100, 120, 200, \infty$. For $\nu = \infty$ the critical values are actually those of the standard normal distribution.

CUMULATIVE t-DISTRIBUTION

DEGREES OF FREEDOM, ν	p				
	.900	.950	.975	.990	.995
1	3.078	6.314	12.706	31.821	63.657
2	1.886	2.920	4.303	6.965	9.925
3	1.638	2.353	3.182	4.541	5.841
4	1.533	2.132	2.776	3.747	4.604
5	1.476	2.015	2.571	3.365	4.032
6	1.440	1.943	2.447	3.143	3.707
7	1.415	1.895	2.365	2.998	3.499
8	1.397	1.860	2.306	2.896	3.355
9	1.383	1.833	2.262	2.821	3.250
10	1.372	1.812	2.228	2.764	3.169
11	1.363	1.796	2.201	2.718	3.106
12	1.356	1.782	2.179	2.681	3.055
13	1.350	1.771	2.160	2.650	3.012
14	1.345	1.761	2.145	2.624	2.977
15	1.341	1.753	2.131	2.602	2.947
16	1.337	1.746	2.120	2.583	2.921
17	1.333	1.740	2.110	2.567	2.898
18	1.330	1.734	2.101	2.552	2.878
19	1.328	1.729	2.093	2.539	2.861
20	1.325	1.725	2.086	2.528	2.845
21	1.323	1.721	2.080	2.518	2.831
22	1.321	1.717	2.074	2.508	2.819
23	1.319	1.714	2.069	2.500	2.807
24	1.318	1.711	2.064	2.492	2.797
25	1.316	1.708	2.060	2.485	2.787
26	1.315	1.706	2.056	2.479	2.779
27	1.314	1.703	2.052	2.473	2.771
28	1.313	1.701	2.048	2.467	2.763
29	1.311	1.699	2.045	2.462	2.756
30	1.310	1.697	2.042	2.457	2.750
40	1.303	1.684	2.021	2.423	2.704
50	1.299	1.676	2.008	2.403	2.678
60	1.296	1.671	2.000	2.390	2.660
70	1.294	1.667	1.994	2.381	2.648
80	1.293	1.665	1.990	2.374	2.638
90	1.291	1.662	1.987	2.369	2.632
100	1.290	1.661	1.984	2.364	2.626
120	1.289	1.658	1.980	2.358	2.617
200	1.286	1.653	1.972	2.345	2.601
∞	1.282	1.645	1.960	2.326	2.576

TABLE V

Cumulative Chi-Square (χ^2) Distribution

This table gives critical values of the χ^2 distribution. Entries in the table are values of χ_p^2 where

$$p = P(\chi^2 \leq \chi_p^2)$$

and χ^2 has a chi-square distribution with ν degrees of freedom.

The table covers values of p = 0.005, 0.01, 0.25, 0.05, 0.10, 0.90, 0.95, 0.975, 0.99, and 0.995, and ν = 1(1)100.

CUMULATIVE χ^2-DISTRIBUTION

ν	p				
	.005	.010	.025	.050	.100
1	.000039	.000157	.000982	.00393	.0158
2	.0100	.0201	.0506	.103	.211
3	.0717	.115	.216	.352	.584
4	.207	.297	.484	.711	1.06
5	.412	.554	.831	1.15	1.61
6	.676	.872	1.24	1.64	2.20
7	.989	1.24	1.69	2.17	2.83
8	1.34	1.65	2.18	2.73	3.49
9	1.73	2.09	2.70	3.33	4.17
10	2.16	2.56	3.25	3.94	4.87
11	2.60	3.05	3.82	4.57	5.58
12	3.07	3.57	4.40	5.23	6.30
13	3.57	4.11	5.01	5.89	7.04
14	4.07	4.66	5.63	6.57	7.79
15	4.60	5.23	6.26	7.26	8.55
16	5.14	5.81	6.91	7.96	9.31
17	5.70	6.41	7.56	8.67	10.1
18	6.26	7.01	8.23	9.39	10.9
19	6.84	7.63	8.91	10.1	11.7
20	7.43	8.26	9.59	10.9	12.4
21	8.03	8.90	10.3	11.6	13.2
22	8.64	9.54	11.0	12.3	14.0
23	9.26	10.2	11.7	13.1	14.8
24	9.89	10.9	12.4	13.8	15.7
25	10.5	11.5	13.1	14.6	16.5
26	11.2	12.2	13.8	15.4	17.3
27	11.8	12.9	14.6	16.2	18.1
28	12.5	13.6	15.3	16.9	18.9
29	13.1	14.3	16.0	17.7	19.8
30	13.8	15.0	16.8	18.5	20.6
31	14.5	15.7	17.5	19.3	21.4
32	15.1	16.4	18.3	20.1	22.3
33	15.8	17.1	19.0	20.9	23.1
34	16.5	17.8	19.8	21.7	24.0
35	17.2	18.5	20.6	22.5	24.8
36	17.9	19.2	21.3	23.3	25.6
37	18.6	20.0	22.1	24.1	26.5
38	19.3	20.7	22.9	24.9	27.3
39	20.0	21.4	23.7	25.7	28.2
40	20.7	22.2	24.4	26.5	29.1
41	21.4	22.9	25.2	27.3	29.9
42	22.1	23.7	26.0	28.1	30.8
43	22.9	24.4	26.8	29.0	31.6
44	23.6	25.1	27.6	29.8	32.5
45	24.3	25.9	28.4	30.6	33.4
46	25.0	26.7	29.2	31.4	34.2
47	25.8	27.4	30.0	32.3	35.1
48	26.5	28.2	30.8	33.1	35.9
49	27.2	28.9	31.6	33.9	36.8
50	28.0	29.7	32.4	34.8	37.7

CUMULATIVE X^2-DISTRIBUTION

	p				
ν	.005	.010	.025	.050	.100
51	28.7	30.5	33.2	35.6	38.6
52	29.5	31.2	34.0	36.4	39.4
53	30.2	32.0	34.8	37.3	40.3
54	31.0	32.8	35.6	38.1	41.2
55	31.7	33.6	36.4	39.0	42.1
56	32.5	34.3	37.2	39.8	42.9
57	33.2	35.1	38.0	40.6	43.8
58	34.0	35.9	38.8	41.5	44.7
59	34.8	36.7	39.7	42.3	45.6
60	35.5	37.5	40.5	43.2	46.5
61	36.3	38.3	41.3	44.0	47.3
62	37.1	39.1	42.1	44.9	48.2
63	37.8	39.9	43.0	45.7	49.1
64	38.6	40.6	43.8	46.6	50.0
65	39.4	41.4	44.6	47.4	50.9
66	40.2	42.2	45.4	48.3	51.8
67	40.9	43.0	46.3	49.2	52.7
68	41.7	43.8	47.1	50.0	53.5
69	42.5	44.6	47.9	50.9	54.4
70	43.3	45.4	48.8	51.7	55.3
71	44.1	46.2	49.6	52.6	56.2
72	44.8	47.1	50.4	53.5	57.1
73	45.6	47.9	51.3	54.3	58.0
74	46.4	48.7	52.1	55.2	58.9
75	47.2	49.5	52.9	56.1	59.8
76	48.0	50.3	53.8	56.9	60.7
77	48.8	51.1	54.6	57.8	61.6
78	49.6	51.9	55.5	58.7	62.5
79	50.4	52.7	56.3	59.5	63.4
80	51.2	53.5	57.2	60.4	64.3
81	52.0	54.4	58.0	61.3	65.2
82	52.8	55.2	58.8	62.1	66.1
83	53.6	56.0	59.7	63.0	67.0
84	54.4	56.8	60.5	63.9	67.9
85	55.2	57.6	61.4	64.7	68.8
86	56.0	58.5	62.2	65.6	69.7
87	56.8	59.3	63.1	66.5	70.6
88	57.6	60.1	63.9	67.4	71.5
89	58.4	60.9	64.8	68.2	72.4
90	59.2	61.8	65.6	69.1	73.3
91	60.0	62.6	66.5	70.0	74.2
92	60.8	63.4	67.4	70.9	75.1
93	61.6	64.2	68.2	71.8	76.0
94	62.4	65.1	69.1	72.6	76.9
95	63.2	65.9	69.9	73.5	77.8
96	64.1	66.7	70.8	74.4	78.7
97	64.9	67.6	71.6	75.3	79.6
98	65.7	68.4	72.5	76.2	80.5
99	66.5	69.2	73.4	77.0	81.4
100	67.3	70.1	74.2	77.9	82.4

CUMULATIVE χ^2-DISTRIBUTION

ν	p .900	.950	.975	.990	.995
1	2.71	3.84	5.02	6.63	7.88
2	4.61	5.99	7.38	9.21	10.6
3	6.25	7.81	9.35	11.3	12.8
4	7.78	9.49	11.1	13.3	14.9
5	9.24	11.1	12.8	15.1	16.7
6	10.6	12.6	14.4	16.8	18.5
7	12.0	14.1	16.0	18.5	20.3
8	13.4	15.5	17.5	20.1	22.0
9	14.7	16.9	19.0	21.7	23.6
10	16.0	18.3	20.5	23.2	25.2
11	17.3	19.7	21.9	24.7	26.8
12	18.5	21.0	23.3	26.2	28.3
13	19.8	22.4	24.7	27.7	29.8
14	21.1	23.7	26.1	29.1	31.3
15	22.3	25.0	27.5	30.6	32.8
16	23.5	26.3	28.8	32.0	34.3
17	24.8	27.6	30.2	33.4	35.7
18	26.0	28.9	31.5	34.8	37.2
19	27.2	30.1	32.9	36.2	38.6
20	28.4	31.4	34.2	37.6	40.0
21	29.6	32.7	35.5	38.9	41.4
22	30.8	33.9	36.8	40.3	42.8
23	32.0	35.2	38.1	41.6	44.2
24	33.2	36.4	39.4	43.0	45.6
25	34.4	37.7	40.6	44.3	46.9
26	35.6	38.9	41.9	45.6	48.3
27	36.7	40.1	43.2	47.0	49.6
28	37.9	41.3	44.5	48.3	51.0
29	3.90	104.	605.	705.	605.
30	40.3	43.8	47.0	50.9	53.7
31	41.4	45.0	48.2	52.2	55.0
32	42.6	46.2	49.5	53.5	56.3
33	43.7	47.4	50.7	54.8	57.6
34	44.9	48.6	52.0	56.1	59.0
35	46.1	49.8	53.2	57.3	60.3
36	47.2	51.0	54.4	58.6	61.6
37	48.4	52.2	55.7	59.9	62.9
38	49.5	53.4	56.9	61.2	64.2
39	50.7	54.6	58.1	62.4	65.5
40	51.8	55.8	59.3	63.7	66.8
41	52.9	56.9	60.6	65.0	68.1
42	54.1	58.1	61.8	66.2	69.3
43	55.2	59.3	63.0	67.5	70.6
44	56.4	60.5	64.2	68.7	71.9
45	57.5	61.7	65.4	70.0	73.2
46	58.6	62.8	66.6	71.2	74.4
47	59.8	64.0	67.8	72.4	75.7
48	60.9	65.2	69.0	73.7	77.0
49	62.0	66.3	70.2	74.9	78.2
50	63.2	67.5	71.4	76.2	79.5

CUMULATIVE X^2-DISTRIBUTION

ν	p .900	.950	.975	.990	.995
51	64.3	68.7	72.6	77.4	80.7
52	65.4	69.8	73.8	78.6	82.0
53	66.5	71.0	75.0	79.8	83.3
54	67.7	72.2	76.2	81.1	84.5
55	68.8	73.3	77.4	82.3	85.7
56	69.9	74.5	78.6	83.5	87.0
57	71.0	75.6	79.8	84.7	88.2
58	72.2	76.8	80.9	86.0	89.5
59	73.3	77.9	82.1	87.2	90.7
60	74.4	79.1	83.3	88.4	92.0
61	75.5	80.2	84.5	89.6	93.2
62	76.6	81.4	85.7	90.8	94.4
63	77.7	82.5	86.8	92.0	95.6
64	78.9	83.7	88.0	93.2	96.9
65	80.0	84.8	89.2	94.4	98.1
66	81.1	86.0	90.3	95.6	99.3
67	82.2	87.1	91.5	96.8	100.6
68	83.3	88.3	92.7	98.0	101.8
69	84.4	89.4	93.9	99.2	103.0
70	85.5	90.5	95.0	100.4	104.2
71	86.6	91.7	96.2	101.6	105.4
72	87.7	92.8	97.4	102.8	106.6
73	88.8	93.9	98.5	104.0	107.9
74	90.0	95.1	99.7	105.2	109.1
75	91.1	96.2	100.8	106.4	110.3
76	92.2	97.4	102.0	107.6	111.5
77	93.3	98.5	103.2	108.8	112.7
78	94.4	99.6	104.3	110.0	113.9
79	95.5	100.7	105.5	111.1	115.1
80	96.6	101.9	106.6	112.3	116.3
81	97.7	103.0	107.8	113.5	117.5
82	98.8	104.1	108.9	114.7	118.7
83	99.9	105.3	110.1	115.9	119.9
84	101.0	106.4	111.2	117.1	121.1
85	102.1	107.5	112.4	118.2	122.3
86	103.2	108.6	113.5	119.4	123.5
87	104.3	109.8	114.7	120.6	124.7
88	105.4	110.9	115.8	121.8	125.9
89	106.5	112.0	117.0	122.9	127.1
90	107.6	113.1	118.1	124.1	128.3
91	108.7	114.3	119.3	125.3	129.5
92	109.8	115.4	120.4	126.5	130.7
93	110.9	116.5	121.6	127.6	131.9
94	111.9	117.6	122.7	128.8	133.1
95	113.0	118.8	123.9	130.0	134.2
96	114.1	119.9	125.0	131.1	135.4
97	115.2	121.0	126.1	132.3	136.6
98	116.3	122.1	127.3	133.5	137.8
99	117.4	123.2	128.4	134.6	139.0
100	118.5	124.3	129.6	135.8	140.2

TABLE VI

Cumulative F-Distribution

This table gives critical values of the F-distribution. Entries in the table are values of F_p where

$$p = P(F \leq F_p)$$

and F has an F-distribution with ν_1 (numerator) and ν_2 (denominator) degrees of freedom. The table covers values ν_1 (and ν_2) = 1(1)15, 20, 24, 30, 40, 50, 60, 100, 120, ∞ and p = 0.005, 0.01, 0.025, 0.05, 0.10, 0.90, 0.95, 0.975, 0.99, 0.995.

CUMULATIVE F-DISTRIBUTION

		ν_1					
ν_2	p	1	2	3	4	5	6
1	.005	.000062	.0050	.018	.032	.044	.054
	.010	.00025	.010	.029	.047	.062	.073
	.025	.00154	.026	.057	.082	.100	.113
	.050	.00619	.054	.099	.130	.151	.167
	.100	.025	.117	.181	.220	.246	.265
	.900	39.9	49.5	53.6	55.8	57.2	58.2
	.950	161.	200.	216.	225.	230.	234.
	.975	648.	799.	864.	900.	922.	937.
	.990	4052.	5000.	5403.	5625.	5764.	5859.
	.995	16211.	20000.	21615.	22500.	23056.	23437.
2	.005	.000050	.0050	.020	.038	.055	.069
	.010	.00020	.010	.032	.056	.075	.092
	.025	.00125	.026	.062	.094	.119	.138
	.050	.0050	.053	.105	.144	.173	.194
	.100	.020	.111	.183	.231	.265	.289
	.900	8.53	9.00	9.16	9.24	9.29	9.33
	.950	18.5	19.0	19.2	19.2	19.3	19.3
	.975	38.5	39.0	39.2	39.2	39.3	39.3
	.990	98.5	99.0	99.2	99.3	99.3	99.3
	.995	198.	199.	199.	199.	199.	199.
3	.005	.000046	.0050	.021	.041	.060	.077
	.010	.00019	.010	.034	.060	.083	.102
	.025	.00116	.026	.065	.100	.129	.152
	.050	.0046	.052	.108	.152	.185	.210
	.100	.019	.109	.185	.239	.276	.304
	.900	5.54	5.46	5.39	5.34	5.31	5.28
	.950	10.1	9.55	9.28	9.12	9.01	8.94
	.975	17.4	16.0	15.4	15.1	14.9	14.7
	.990	34.1	30.8	29.5	28.7	28.2	27.9
	.995	55.6	49.8	47.5	46.2	45.4	44.8
4	.005	.000044	.0050	.022	.043	.064	.083
	.010	.00018	.010	.035	.063	.088	.109
	.025	.00111	.025	.066	.104	.135	.161
	.050	.0045	.052	.110	.157	.193	.221
	.100	.018	.108	.187	.243	.284	.314
	.900	4.54	4.32	4.19	4.11	4.05	4.01
	.950	7.71	6.94	6.59	6.39	6.26	6.16
	.975	12.2	10.6	9.98	9.60	9.36	9.20
	.990	21.2	18.0	16.7	16.0	15.5	15.2
	.995	31.3	26.3	24.3	23.2	22.5	22.0

CUMULATIVE F-DISTRIBUTION

ν_2	p	1	2	3	4	5	6
5	.005	.000043	.0050	.022	.045	.067	.087
	.010	.00017	.010	.035	.064	.091	.114
	.025	.00108	.025	.067	.107	.140	.167
	.050	.0043	.052	.111	.160	.198	.228
	.100	.017	.108	.188	.247	.290	.322
	.900	4.06	3.78	3.62	3.52	3.45	3.40
	.950	6.61	5.79	5.41	5.19	5.05	4.95
	.975	10.0	8.43	7.76	7.39	7.15	6.98
	.990	16.3	13.3	12.1	11.4	11.0	10.7
	.995	22.8	18.3	16.5	15.6	14.9	14.5
6	.005	.000043	.0050	.022	.046	.069	.090
	.010	.00017	.010	.036	.066	.094	.118
	.025	.00107	.025	.068	.109	.143	.172
	.050	.0043	.052	.112	.162	.202	.233
	.100	.017	.107	.189	.249	.294	.327
	.900	3.78	3.46	3.29	3.18	3.11	3.05
	.950	5.99	5.14	4.76	4.53	4.39	4.28
	.975	8.81	7.26	6.60	6.23	5.99	5.82
	.990	13.7	10.9	9.78	9.15	8.75	8.47
	.995	18.6	14.5	12.9	12.0	11.5	11.1
7	.005	.000042	.0050	.023	.046	.070	.093
	.010	.00017	.010	.036	.067	.096	.121
	.025	.00105	.025	.068	.110	.146	.176
	.050	.0042	.052	.113	.164	.205	.238
	.100	.017	.107	.190	.251	.297	.332
	.900	3.59	3.26	3.07	2.96	2.88	2.83
	.950	5.59	4.74	4.35	4.12	3.97	3.87
	.975	8.07	6.54	5.89	5.52	5.29	5.12
	.990	12.2	9.55	8.45	7.85	7.46	7.19
	.995	16.2	12.4	10.9	10.1	9.52	9.16
8	.005	.000042	.0050	.023	.047	.072	.095
	.010	.00017	.010	.036	.068	.097	.123
	.025	.00105	.025	.069	.111	.148	.179
	.050	.0042	.052	.113	.166	.208	.241
	.100	.017	.107	.190	.253	.299	.335
	.900	3.46	3.11	2.92	2.81	2.73	2.67
	.950	5.32	4.46	4.07	3.84	3.69	3.58
	.975	7.57	6.06	5.42	5.05	4.82	4.65
	.990	11.3	8.65	7.59	7.01	6.63	6.37
	.995	14.7	11.0	9.60	8.81	8.30	7.95

CUMULATIVE F-DISTRIBUTION

		ν_1					
ν_2	p	1	2	3	4	5	6
9	.005	.000042	.0050	.023	.047	.073	.096
	.010	.00017	.010	.037	.068	.098	.125
	.025	.00104	.025	.069	.112	.150	.181
	.050	.0042	.052	.113	.167	.210	.244
	.100	.017	.107	.191	.254	.302	.338
	.900	3.36	3.01	2.81	2.69	2.61	2.55
	.950	5.12	4.26	3.86	3.63	3.48	3.37
	.975	7.21	5.71	5.08	4.72	4.48	4.32
	.990	10.6	8.02	6.99	6.42	6.06	5.80
	.995	13.6	10.1	8.72	7.96	7.47	7.13
10	.005	.000041	.0050	.023	.048	.073	.098
	.010	.00017	.010	.037	.069	.099	.127
	.025	.00103	.025	.069	.113	.151	.183
	.050	.0041	.052	.114	.168	.211	.246
	.100	.017	.106	.191	.255	.303	.340
	.900	3.28	2.92	2.73	2.61	2.52	2.46
	.950	4.96	4.10	3.71	3.48	3.33	3.22
	.975	6.94	5.46	4.83	4.47	4.24	4.07
	.990	10.0	7.56	6.55	5.99	5.64	5.39
	.995	12.8	9.43	8.08	7.34	6.87	6.54
11	.005	.000041	.0050	.023	.048	.074	.099
	.010	.00016	.010	.037	.069	.100	.128
	.025	.00103	.025	.070	.114	.152	.185
	.050	.0041	.052	.114	.168	.213	.248
	.100	.017	.106	.191	.256	.305	.343
	.900	3.23	2.86	2.66	2.54	2.45	2.39
	.950	4.84	3.98	3.59	3.36	3.20	3.09
	.975	6.72	5.26	4.63	4.28	4.04	3.88
	.990	9.65	7.21	6.22	5.67	5.32	5.07
	.995	12.2	8.91	7.60	6.88	6.42	6.10
12	.005	.000041	.0050	.023	.048	.075	.100
	.010	.00016	.010	.037	.070	.101	.130
	.025	.00102	.025	.070	.114	.153	.186
	.050	.0041	.052	.114	.169	.214	.250
	.100	.016	.106	.192	.257	.306	.344
	.900	3.18	2.81	2.61	2.48	2.39	2.33
	.950	4.75	3.89	3.49	3.26	3.11	3.00
	.975	6.55	5.10	4.47	4.12	3.89	3.73
	.990	9.33	6.93	5.95	5.41	5.06	4.82
	.995	11.8	8.51	7.23	6.52	6.07	5.76

CUMULATIVE F-DISTRIBUTION

ν_2	p	ν_1 = 1	2	3	4	5	6
13	.005	.000041	.0050	.023	.049	.075	.100
	.010	.00016	.010	.037	.070	.102	.131
	.025	.00102	.025	.070	.115	.154	.188
	.050	.0041	.051	.115	.170	.215	.251
	.100	.016	.106	.192	.257	.307	.346
	.900	3.14	2.76	2.56	2.43	2.35	2.28
	.950	4.67	3.81	3.41	3.18	3.03	2.92
	.975	6.41	4.97	4.35	4.00	3.77	3.60
	.990	9.07	6.70	5.74	5.21	4.86	4.62
	.995	11.4	8.19	6.93	6.23	5.79	5.48
14	.005	.000041	.0050	.023	.049	.076	.101
	.010	.00016	.010	.037	.070	.102	.131
	.025	.00102	.025	.070	.115	.155	.189
	.050	.0041	.051	.115	.170	.216	.253
	.100	.016	.106	.192	.258	.308	.347
	.900	3.10	2.73	2.52	2.39	2.31	2.24
	.950	4.60	3.74	3.34	3.11	2.96	2.85
	.975	6.30	4.86	4.24	3.89	3.66	3.50
	.990	8.86	6.51	5.56	5.04	4.69	4.46
	.995	11.1	7.92	6.68	6.00	5.56	5.26
15	.005	.000041	.0050	.023	.049	.076	.102
	.010	.00016	.010	.037	.070	.103	.132
	.025	.00102	.025	.070	.116	.156	.190
	.050	.0041	.051	.115	.171	.217	.254
	.100	.016	.106	.192	.258	.309	.348
	.900	3.07	2.70	2.49	2.36	2.27	2.21
	.950	4.54	3.68	3.29	3.06	2.90	2.79
	.975	6.20	4.77	4.15	3.80	3.58	3.41
	.990	8.68	6.36	5.42	4.89	4.56	4.32
	.995	10.8	7.70	6.48	5.80	5.37	5.07
20	.005	.000040	.0050	.023	.050	.077	.104
	.010	.00016	.010	.037	.071	.105	.135
	.025	.00101	.025	.071	.117	.158	.193
	.050	.0040	.051	.115	.172	.219	.258
	.100	.016	.106	.193	.260	.312	.353
	.900	2.97	2.59	2.38	2.25	2.16	2.09
	.950	4.35	3.49	3.10	2.87	2.71	2.60
	.975	5.87	4.46	3.86	3.51	3.29	3.13
	.990	8.10	5.85	4.94	4.43	4.10	3.87
	.995	9.94	6.99	5.82	5.17	4.76	4.47

CUMULATIVE F-DISTRIBUTION

ν_2	p	ν_1 = 1	2	3	4	5	6
24	.005	.000040	.0050	.023	.050	.078	.106
	.010	.00016	.010	.038	.072	.106	.137
	.025	.00100	.025	.071	.117	.159	.195
	.050	.0040	.051	.116	.173	.221	.260
	.100	.016	.106	.193	.261	.313	.355
	.900	2.93	2.54	2.33	2.19	2.10	2.04
	.950	4.26	3.40	3.01	2.78	2.62	2.51
	.975	5.72	4.32	3.72	3.38	3.15	2.99
	.990	7.82	5.61	4.72	4.22	3.90	3.67
	.995	9.55	6.66	5.52	4.89	4.49	4.20
30	.005	.000040	.0050	.024	.050	.079	.107
	.010	.00016	.010	.038	.072	.107	.138
	.025	.00100	.025	.071	.118	.161	.197
	.050	.0040	.051	.116	.174	.222	.263
	.100	.016	.106	.193	.262	.315	.357
	.900	2.88	2.49	2.28	2.14	2.05	1.98
	.950	4.17	3.32	2.92	2.69	2.53	2.42
	.975	5.57	4.18	3.59	3.25	3.03	2.87
	.990	7.56	5.39	4.51	4.02	3.70	3.47
	.995	9.18	6.35	5.24	4.62	4.23	3.95
40	.005	.000040	.0050	.024	.051	.080	.108
	.010	.00016	.010	.038	.073	.108	.140
	.025	.00099	.025	.071	.119	.162	.199
	.050	.0040	.051	.116	.175	.224	.265
	.100	.016	.106	.194	.263	.317	.360
	.900	2.84	2.44	2.23	2.09	2.00	1.93
	.950	4.08	3.23	2.84	2.61	2.45	2.34
	.975	5.42	4.05	3.46	3.13	2.90	2.74
	.990	7.31	5.18	4.31	3.83	3.51	3.29
	.995	8.83	6.07	4.98	4.37	3.99	3.71
50	.005	.000040	.0050	.024	.051	.080	.109
	.010	.00016	.010	.038	.073	.108	.141
	.025	.00099	.025	.071	.119	.163	.201
	.050	.0040	.051	.117	.175	.225	.266
	.100	.016	.106	.194	.263	.318	.361
	.900	2.81	2.41	2.20	2.06	1.97	1.90
	.950	4.03	3.18	2.79	2.56	2.40	2.29
	.975	5.34	3.97	3.39	3.05	2.83	2.67
	.990	7.17	5.06	4.20	3.72	3.41	3.19
	.995	8.63	5.90	4.83	4.23	3.85	3.58

CUMULATIVE F-DISTRIBUTION

ν_2	p	ν_1 = 1	2	3	4	5	6
60	.005	.000040	.0050	.024	.051	.081	.110
	.010	.00016	.010	.038	.073	.109	.142
	.025	.00099	.025	.071	.120	.163	.202
	.050	.0040	.051	.117	.176	.226	.267
	.100	.016	.106	.194	.264	.318	.362
	.900	2.79	2.39	2.18	2.04	1.95	1.87
	.950	4.00	3.15	2.76	2.53	2.37	2.25
	.975	5.29	3.93	3.34	3.01	2.79	2.63
	.990	7.08	4.98	4.13	3.65	3.34	3.12
	.995	8.49	5.79	4.73	4.14	3.76	3.49
100	.005	.000039	.0050	.024	.051	.081	.111
	.010	.00016	.010	.038	.074	.110	.143
	.025	.00099	.025	.072	.120	.164	.203
	.050	.0040	.051	.117	.177	.227	.269
	.100	.016	.105	.194	.265	.320	.364
	.900	2.76	2.36	2.14	2.00	1.91	1.83
	.950	3.94	3.09	2.70	2.46	2.31	2.19
	.975	5.18	3.83	3.25	2.92	2.70	2.54
	.990	6.90	4.82	3.98	3.51	3.21	2.99
	.995	8.24	5.59	4.54	3.96	3.59	3.33
120	.005	.000039	.0050	.024	.051	.081	.111
	.010	.00016	.010	.038	.074	.110	.143
	.025	.00099	.025	.072	.120	.165	.204
	.050	.0039	.051	.117	.177	.227	.270
	.100	.016	.105	.194	.265	.320	.365
	.900	2.75	2.35	2.13	1.99	1.90	1.82
	.950	3.92	3.07	2.68	2.45	2.29	2.17
	.975	5.15	3.80	3.23	2.89	2.67	2.52
	.990	6.85	4.79	3.95	3.48	3.17	2.96
	.995	8.18	5.54	4.50	3.92	3.55	3.28
∞	.005	.000039	.0050	.024	.052	.082	.113
	.010	.00016	.010	.038	.074	.111	.145
	.025	.00098	.025	.072	.121	.166	.206
	.050	.0039	.051	.117	.178	.229	.273
	.100	.016	.105	.195	.266	.322	.367
	.900	2.71	2.30	2.08	1.94	1.85	1.77
	.950	3.84	3.00	2.60	2.37	2.21	2.10
	.975	5.02	3.69	3.12	2.79	2.57	2.41
	.990	6.63	4.61	3.78	3.32	3.02	2.80
	.995	7.88	5.30	4.28	3.72	3.35	3.09

CUMULATIVE F-DISTRIBUTION

ν_2	p	ν_1 = 7	8	9	10	11	12
1	.005	.062	.068	.073	.078	.082	.085
	.010	.082	.089	.095	.100	.104	.107
	.025	.124	.132	.139	.144	.149	.153
	.050	.179	.188	.195	.201	.206	.211
	.100	.279	.289	.298	.304	.310	.315
	.900	58.9	59.4	59.9	60.2	60.5	60.7
	.950	237.	239.	241.	242.	243.	244.
	.975	948.	957.	963.	969.	973.	977.
	.990	5928.	5981.	6023.	6056.	6083.	6106.
	.995	23715.	23925.	24091.	24224.	24334.	24426.
2	.005	.081	.091	.099	.106	.112	.118
	.010	.105	.116	.125	.132	.139	.144
	.025	.153	.165	.175	.183	.190	.196
	.050	.211	.224	.235	.244	.251	.257
	.100	.307	.321	.333	.342	.350	.356
	.900	9.35	9.37	9.38	9.39	9.40	9.41
	.950	19.4	19.4	19.4	19.4	19.4	19.4
	.975	39.4	39.4	39.4	39.4	39.4	39.4
	.990	99.4	99.4	99.4	99.4	99.4	99.4
	.995	199.	199.	199.	199.	199.	199.
3	.005	.092	.104	.115	.124	.132	0.14
	.010	.118	.132	.143	.153	.161	.168
	.025	.170	.185	.197	.207	.216	.223
	.050	.230	.246	.259	.270	.279	.287
	.100	.325	.342	.356	.367	.376	.384
	.900	5.27	5.25	5.24	5.23	5.22	5.22
	.950	8.89	8.85	8.81	8.79	8.76	8.74
	.975	14.6	14.5	14.5	14.4	14.4	14.3
	.990	27.7	27.5	27.3	27.2	27.1	27.1
	.995	44.4	44.1	43.9	43.7	43.5	43.4
4	.005	.099	.114	.126	.136	.145	0.15
	.010	.127	.143	.156	.167	.176	.105
	.025	.181	.198	.212	.224	.234	.243
	.050	.243	.261	.275	.288	.298	.307
	.100	.338	.356	.371	.384	.394	.403
	.900	3.98	3.95	3.94	3.92	3.91	3.90
	.950	6.09	6.04	6.00	5.96	5.94	5.91
	.975	9.07	8.98	8.90	8.84	8.79	8.75
	.990	15.0	14.8	14.7	14.5	14.5	14.4
	.995	21.6	21.4	21.1	21.0	20.8	20.7

CUMULATIVE F-DISTRIBUTION

ν_2	p	ν_1 = 7	8	9	10	11	12
5	.005	.105	.120	.134	.146	.156	0.16
	.010	.134	.151	.165	.177	.188	.197
	.025	.189	.208	.223	.236	.247	.257
	.050	.252	.271	.287	.301	.312	.322
	.100	.347	.367	.383	.397	.408	.418
	.900	3.37	3.34	3.32	3.30	3.28	3.27
	.950	4.88	4.82	4.77	4.74	4.70	4.68
	.975	6.85	6.76	6.68	6.62	6.57	6.52
	.990	10.5	10.3	10.2	10.1	9.96	9.89
	.995	14.2	14.0	13.8	13.6	13.5	13.4
6	.005	.109	.126	.140	.153	.164	0.17
	.010	.139	.157	.172	.186	.197	.207
	.025	.195	.215	.231	.246	.258	.268
	.050	.259	.279	.296	.311	.323	.334
	.100	.354	.375	.392	.406	.419	.429
	.900	3.01	2.98	2.96	2.94	2.92	2.90
	.950	4.21	4.15	4.10	4.06	4.03	4.00
	.975	5.70	5.60	5.52	5.46	5.41	5.37
	.990	8.26	8.10	7.98	7.87	7.79	7.72
	.995	10.8	10.6	10.4	10.3	10.1	10.0
7	.005	.113	.130	.145	.159	.171	0.18
	.010	.143	.162	.178	.192	.205	.216
	.025	.200	.221	.238	.253	.266	.277
	.050	.264	.286	.304	.319	.332	.343
	.100	.359	.381	.399	.414	.427	.438
	.900	2.78	2.75	2.72	2.70	2.68	2.67
	.950	3.79	3.73	3.68	3.64	3.60	3.57
	.975	4.99	4.90	4.82	4.76	4.71	4.67
	.990	6.99	6.84	6.72	6.62	6.54	6.47
	.995	8.89	8.68	8.51	8.38	8.27	8.18
8	.005	.115	.133	.149	.164	.176	0.19
	.010	.146	.166	.183	.198	.211	.222
	.025	.204	.226	.244	.259	.273	.285
	.050	.268	.291	.310	.326	.339	.351
	.100	.363	.386	.405	.421	.434	.446
	.900	2.62	2.59	2.56	2.54	2.52	2.50
	.950	3.50	3.44	3.39	3.35	3.31	3.28
	.975	4.53	4.43	4.36	4.30	4.24	4.20
	.990	6.18	6.03	5.91	5.81	5.73	5.67
	.995	7.69	7.50	7.34	7.21	7.10	7.01

CUMULATIVE F-DISTRIBUTION

		ν_1					
ν_2	p	7	8	9	10	11	12
9	.005	.117	.136	.153	.168	.181	0.19
	.010	.149	.169	.187	.202	.216	.228
	.025	.207	.229	.248	.265	.279	.291
	.050	.272	.295	.315	.331	.345	.358
	.100	.367	.390	.410	.426	.440	.452
	.900	2.51	2.47	2.44	2.42	2.40	2.38
	.950	3.29	3.23	3.18	3.14	3.10	3.07
	.975	4.20	4.10	4.03	3.96	3.91	3.87
	.990	5.61	5.47	5.35	5.26	5.18	5.11
	.995	6.88	6.69	6.54	6.42	6.31	6.23
10	.005	.119	.139	.156	.171	.185	0.20
	.010	.151	.172	.190	.206	.220	.233
	.025	.210	.233	.252	.269	.284	.296
	.050	.275	.299	.319	.336	.350	.363
	.100	.370	.394	.414	.431	.445	.457
	.900	2.41	2.38	2.35	2.32	2.30	2.28
	.950	3.14	3.07	3.02	2.98	2.94	2.91
	.975	3.95	3.85	3.78	3.72	3.66	3.62
	.990	5.20	5.06	4.94	4.85	4.77	4.71
	.995	6.30	6.12	5.97	5.85	5.75	5.66
11	.005	.121	.141	.158	.174	.188	0.20
	.010	.153	.174	.193	.210	.224	.237
	.025	.212	.236	.256	.273	.288	.301
	.050	.278	.302	.322	.340	.355	.368
	.100	.373	.397	.417	.434	.449	.462
	.900	2.34	2.30	2.27	2.25	2.23	2.21
	.950	3.01	2.95	2.90	2.85	2.82	2.79
	.975	3.76	3.66	3.59	3.53	3.47	3.43
	.990	4.89	4.74	4.63	4.54	4.46	4.40
	.995	5.86	5.68	5.54	5.42	5.32	5.24
12	.005	.122	.143	.161	.177	.191	0.20
	.010	.155	.176	.196	.212	.227	.241
	.025	.214	.238	.259	.276	.292	.305
	.050	.280	.305	.325	.343	.359	.372
	.100	.375	.400	.420	.438	.453	.466
	.900	2.28	2.24	2.21	2.19	2.17	2.15
	.950	2.91	2.85	2.80	2.75	2.72	2.69
	.975	3.61	3.51	3.44	3.37	3.32	3.28
	.990	4.64	4.50	4.39	4.30	4.22	4.16
	.995	5.52	5.35	5.20	5.09	4.99	4.91

CUMULATIVE F-DISTRIBUTION

ν_2	p	ν_1 = 7	8	9	10	11	12
13	.005	.124	.144	.163	.179	.194	0.21
	.010	.156	.178	.198	.215	.230	.244
	.025	.216	.240	.261	.279	.295	.309
	.050	.282	.307	.328	.346	.362	.376
	.100	.377	.402	.423	.441	.456	.469
	.900	2.23	2.20	2.16	2.14	2.12	2.10
	.950	2.83	2.77	2.71	2.67	2.63	2.60
	.975	3.48	3.39	3.31	3.25	3.20	3.15
	.990	4.44	4.30	4.19	4.10	4.02	3.96
	.995	5.25	5.08	4.94	4.82	4.72	4.64
14	.005	.125	.146	.164	.181	.196	0.21
	.010	.157	.180	.200	.217	.233	.247
	.025	.218	.242	.263	.282	.298	.312
	.050	.283	.309	.331	.349	.365	.379
	.100	.378	.404	.425	.443	.459	.472
	.900	2.19	2.15	2.12	2.10	2.07	2.05
	.950	2.76	2.70	2.65	2.60	2.57	2.53
	.975	3.38	3.29	3.21	3.15	3.09	3.05
	.990	4.28	4.14	4.03	3.94	3.86	3.80
	.995	5.03	4.86	4.72	4.60	4.51	4.43
15	.005	.126	.147	.166	.183	.198	0.21
	.010	.158	.181	.202	.219	.235	.249
	.025	.219	.244	.265	.284	.300	.315
	.050	.285	.311	.333	.351	.368	.382
	.100	.380	.406	.427	.446	.461	.475
	.900	2.16	2.12	2.09	2.06	2.04	2.02
	.950	2.71	2.64	2.59	2.54	2.51	2.48
	.975	3.29	3.20	3.12	3.06	3.01	2.96
	.990	4.14	4.00	3.89	3.80	3.73	3.67
	.995	4.85	4.67	4.54	4.42	4.33	4.25
20	.005	.129	.151	.171	.190	.206	0.22
	.010	.162	.187	.208	.227	.244	.259
	.025	.224	.250	.273	.293	.310	.325
	.050	.290	.317	.341	.360	.378	.393
	.100	.385	.412	.435	.454	.471	.486
	.900	2.04	2.00	1.96	1.94	1.91	1.89
	.950	2.51	2.45	2.39	2.35	2.31	2.28
	.975	3.01	2.91	2.84	2.77	2.72	2.68
	.990	3.70	3.56	3.46	3.37	3.29	3.23
	.995	4.26	4.09	3.96	3.85	3.76	3.68

CUMULATIVE F-DISTRIBUTION

ν_2	p	ν_1 = 7	8	9	10	11	12
24	.005	.131	.154	.175	.193	.210	0.23
	.010	.165	.189	.211	.231	.249	.265
	.025	.226	.253	.277	.297	.315	.331
	.050	.293	.321	.345	.365	.383	.399
	.100	.388	.416	.439	.459	.476	.491
	.900	1.98	1.94	1.91	1.88	1.85	1.83
	.950	2.42	2.36	2.30	2.25	2.22	2.18
	.975	2.87	2.78	2.70	2.64	2.59	2.54
	.990	3.50	3.36	3.26	3.17	3.09	3.03
	.995	3.99	3.83	3.69	3.59	3.50	3.42
30	.005	.133	.156	.178	.197	.215	0.23
	.010	.167	.192	.215	.235	.254	.270
	.025	.229	.257	.281	.302	.321	.337
	.050	.296	.325	.349	.370	.389	.405
	.100	.391	.420	.444	.464	.482	.497
	.900	1.93	1.88	1.85	1.82	1.79	1.77
	.950	2.33	2.27	2.21	2.16	2.13	2.09
	.975	2.75	2.65	2.57	2.51	2.46	2.41
	.990	3.30	3.17	3.07	2.98	2.91	2.84
	.995	3.74	3.58	3.45	3.34	3.25	3.18
40	.005	.135	.159	.181	.201	.220	0.24
	.010	.169	.195	.219	.240	.259	.276
	.025	.232	.260	.285	.307	.327	.344
	.050	.299	.329	.354	.376	.395	.412
	.100	.394	.423	.448	.469	.487	.503
	.900	1.87	1.83	1.79	1.76	1.74	1.71
	.950	2.25	2.18	2.12	2.08	2.04	2.00
	.975	2.62	2.53	2.45	2.39	2.33	2.29
	.990	3.12	2.99	2.89	2.80	2.73	2.66
	.995	3.51	3.35	3.22	3.12	3.03	2.95
50	.005	.136	.161	.183	.204	.223	0.24
	.010	.171	.197	.221	.243	.262	.280
	.025	.234	.263	.288	.310	.330	.348
	.050	.301	.331	.357	.379	.399	.416
	.100	.396	.426	.451	.472	.491	.508
	.900	1.84	1.80	1.76	1.73	1.70	1.68
	.950	2.20	2.13	2.07	2.03	1.99	1.95
	.975	2.55	2.46	2.38	2.32	2.26	2.22
	.990	3.02	2.89	2.78	2.70	2.63	2.56
	.995	3.38	3.22	3.09	2.99	2.90	2.82

CUMULATIVE F-DISTRIBUTION

		ν_1					
ν_2	p	7	8	9	10	11	12
60	.005	.137	.162	.185	.206	.225	0.24
	.010	.172	.199	.223	.245	.265	.283
	.025	.235	.264	.290	.313	.333	.351
	.050	.303	.333	.359	.382	.402	.419
	.100	.398	.428	.453	.475	.494	.510
	.900	1.82	1.77	1.74	1.71	1.68	1.66
	.950	2.17	2.10	2.04	1.99	1.95	1.92
	.975	2.51	2.41	2.33	2.27	2.22	2.17
	.990	2.95	2.82	2.72	2.63	2.56	2.50
	.995	3.29	3.13	3.01	2.90	2.82	2.74
10C	.005	.139	.164	.188	.210	.229	0.25
	.010	.174	.201	.226	.249	.270	.288
	.025	.238	.267	.294	.317	.338	.357
	.050	.305	.336	.363	.386	.407	.426
	.100	.400	.431	.457	.479	.499	.516
	.900	1.78	1.73	1.69	1.66	1.64	1.61
	.950	2.10	2.03	1.97	1.93	1.89	1.85
	.975	2.42	2.32	2.24	2.18	2.12	2.08
	.990	2.82	2.69	2.59	2.50	2.43	2.37
	.995	3.13	2.97	2.85	2.74	2.66	2.58
120	.005	.139	.165	.189	.211	.231	0.25
	.010	.174	.202	.227	.250	.271	.290
	.025	.238	.268	.295	.318	.340	.359
	.050	.306	.337	.364	.388	.408	.427
	.100	.401	.432	.458	.480	.500	.518
	.900	1.77	1.72	1.68	1.65	1.63	1.60
	.950	2.09	2.02	1.96	1.91	1.87	1.83
	.975	2.39	2.30	2.22	2.16	2.10	2.05
	.990	2.79	2.66	2.56	2.47	2.40	2.34
	.995	3.09	2.93	2.81	2.71	2.62	2.54
∞	.005	.141	.168	.193	.216	.237	0.26
	.010	.177	.206	.232	.256	.278	.298
	.025	.241	.272	.300	.325	.347	.367
	.050	.310	.342	.369	.394	.416	.435
	.100	.405	.436	.463	.487	.507	.525
	.900	1.72	1.67	1.63	1.60	1.57	1.55
	.950	2.01	1.94	1.88	1.83	1.79	1.75
	.975	2.29	2.19	2.11	2.05	1.99	1.94
	.990	2.64	2.51	2.41	2.32	2.25	1.18
	.995	2.90	2.74	2.62	2.52	2.43	2.36

CUMULATIVE F-DISTRIBUTION

		ν_1					
ν_2	p	13	14	15	20	24	30
1	.005	.088	.090	.093	.101	.105	0.11
	.010	.110	.113	.115	.124	.128	.132
	.025	.156	.159	.161	.170	.175	.180
	.050	.214	.217	.220	.230	.235	.240
	.100	.319	.322	.325	.336	.342	.347
	.900	60.9	61.1	61.2	61.7	62.0	62.3
	.950	245.	245.	246.	248.	249.	250.
	.975	980.	983.	985.	993.	997.	1001.
	.990	6126.	6143.	6157.	6209.	6235.	6261.
	.995	24504.	24572.	24630.	24836.	24940.	25044.
2	.005	.122	.126	.130	.143	.150	0.16
	.010	.149	.153	.157	.171	.178	.186
	.025	.201	.206	.210	.224	.232	.239
	.050	.263	.267	.272	.286	.294	.302
	.100	.362	.367	.371	.386	.394	.402
	.900	9.41	9.42	9.42	9.44	9.45	9.46
	.950	19.4	19.4	19.4	19.4	19.5	19.5
	.975	39.4	39.4	39.4	39.4	39.5	39.5
	.990	99.4	99.4	99.4	99.4	99.5	99.5
	.995	199.	199.	199.	199.	199.	199.
3	.005	.144	.150	.154	.172	.181	0.19
	.010	.174	.180	.185	.202	.212	.222
	.025	.230	.236	.241	.259	.269	.279
	.050	.293	.299	.304	.323	.332	.342
	.100	.391	.396	.402	.420	.430	.439
	.900	5.21	5.20	5.20	5.18	5.18	5.17
	.950	8.73	8.71	8.70	8.66	8.64	8.62
	.975	14.3	14.3	14.3	14.2	14.1	14.1
	.990	27.0	26.9	26.9	26.7	26.6	26.5
	.995	43.3	43.2	43.1	42.8	42.6	42.5
4	.005	.160	.167	.172	.193	.205	0.22
	.010	.192	.199	.204	.226	.237	.249
	.025	.250	.257	.263	.285	.296	.308
	.050	.315	.321	.327	.349	.360	.372
	.100	.411	.418	.423	.445	.456	.467
	.900	3.89	3.88	3.87	3.84	3.83	3.82
	.950	5.89	5.87	5.86	5.80	5.77	5.75
	.975	8.71	8.68	8.66	8.56	8.51	8.46
	.990	14.3	14.2	14.2	14.0	13.9	13.8
	.995	20.6	20.5	20.4	20.2	20.0	19.9

CUMULATIVE F-DISTRIBUTION

ν_2	p	ν_1 13	14	15	20	24	30
5	.005	.173	.180	.186	.210	.223	0.24
	.010	.206	.213	.220	.244	.257	.270
	.025	.265	.273	.280	.304	.317	.330
	.050	.331	.338	.345	.369	.382	.395
	.100	.426	.433	.440	.463	.475	.488
	.900	3.26	3.25	3.24	3.21	3.19	3.17
	.950	4.66	4.64	4.62	4.56	4.53	4.50
	.975	6.49	6.46	6.43	6.33	6.28	6.23
	.990	9.82	9.77	9.72	9.55	9.47	9.38
	.995	13.3	13.2	13.1	12.9	12.8	12.7
6	.005	.182	.190	.197	.224	.238	0.25
	.010	.216	.224	.232	.258	.273	.288
	.025	.277	.286	.293	.320	.334	.349
	.050	.343	.351	.358	.385	.399	.413
	.100	.438	.446	.453	.478	.491	.505
	.900	2.89	2.88	2.87	2.84	2.82	2.80
	.950	3.98	3.96	3.94	3.87	3.84	3.81
	.975	5.33	5.30	5.27	5.17	5.12	5.07
	.990	7.66	7.60	7.56	7.40	7.31	7.23
	.995	9.95	9.88	9.81	9.59	9.47	9.36
7	.005	.190	.199	.206	.235	.251	0.27
	.010	.225	.234	.241	.270	.286	.303
	.025	.287	.296	.304	.333	.348	.364
	.050	.353	.362	.369	.398	.413	.428
	.100	.448	.456	.463	.490	.504	.519
	.900	2.65	2.64	2.63	2.59	2.58	2.56
	.950	3.55	3.53	3.51	3.44	3.41	3.38
	.975	4.63	4.60	4.57	4.47	4.41	4.36
	.990	6.41	6.36	6.31	6.16	6.07	5.99
	.995	8.10	8.03	7.97	7.75	7.64	7.53
8	.005	.197	.206	.214	.244	.261	0.28
	.010	.232	.242	.250	.281	.297	.315
	.025	.295	.304	.313	.343	.360	.377
	.050	.361	.371	.379	.409	.425	.441
	.100	.456	.464	.472	.500	.515	.531
	.900	2.49	2.48	2.46	2.42	2.40	2.38
	.950	3.26	3.24	3.22	3.15	3.12	3.08
	.975	4.16	4.13	4.10	4.00	3.95	3.89
	.990	5.61	5.56	5.52	5.36	5.28	5.20
	.995	6.94	6.87	6.81	6.61	6.50	6.40

CUMULATIVE F-DISTRIBUTION

ν_2	p	ν_1 13	14	15	20	24	30
9	.005	.203	.212	.220	.253	.271	0.29
	.010	.239	.248	.257	.289	.307	.326
	.025	.302	.312	.320	.353	.370	.388
	.050	.368	.378	.386	.418	.435	.452
	.100	.462	.471	.479	.509	.525	.541
	.900	2.36	2.35	2.34	2.30	2.28	2.25
	.950	3.05	3.03	3.01	2.94	2.90	2.86
	.975	3.83	3.80	3.77	3.67	3.61	3.56
	.990	5.05	5.01	4.96	4.81	4.73	4.65
	.995	6.15	6.09	6.03	5.83	5.73	5.62
10	.005	.207	.217	.226	.260	.279	0.30
	.010	.244	.254	.263	.297	.316	.336
	.025	.308	.318	.327	.361	.379	.398
	.050	.374	.384	.393	.426	.444	.462
	.100	.468	.477	.486	.516	.533	.550
	.900	2.27	2.26	2.24	2.20	2.18	2.16
	.950	2.89	2.86	2.84	2.77	2.74	2.70
	.975	3.58	3.55	3.52	3.42	3.37	3.31
	.990	4.65	4.60	4.56	4.41	4.33	4.25
	.995	5.59	5.53	5.47	5.27	5.17	5.07
11	.005	.212	.222	.231	.266	.286	0.31
	.010	.248	.259	.268	.304	.323	.344
	.025	.313	.323	.332	.368	.387	.407
	.050	.380	.390	.399	.433	.451	.470
	.100	.473	.482	.491	.523	.540	.557
	.900	2.19	2.18	2.17	2.12	2.10	2.08
	.950	2.76	2.74	2.72	2.65	2.61	2.57
	.975	3.39	3.36	3.33	3.23	3.17	3.12
	.990	4.34	4.29	4.25	4.10	4.02	3.94
	.995	5.16	5.10	5.05	4.86	4.76	4.65
12	.005	.215	.226	.235	.272	.292	0.31
	.010	.252	.263	.273	.309	.330	.352
	.025	.317	.328	.337	.374	.394	.415
	.050	.384	.395	.404	.439	.458	.478
	.100	.477	.487	.496	.528	.546	.564
	.900	2.13	2.12	2.10	2.06	2.04	2.01
	.950	2.66	2.64	2.62	2.54	2.51	2.47
	.975	3.24	3.21	3.18	3.07	3.02	2.96
	.990	4.10	4.05	4.01	3.86	3.78	3.70
	.995	4.84	4.77	4.72	4.53	4.43	4.33

CUMULATIVE F-DISTRIBUTION

ν_2	p	ν_1 13	14	15	20	24	30
13	.005	.219	.229	.239	.277	.298	0.32
	.010	.256	.267	.277	.315	.336	.359
	.025	.321	.332	.342	.379	.400	.422
	.050	.388	.399	.408	.445	.464	.485
	.100	.481	.491	.500	.533	.551	.570
	.900	2.08	2.07	2.05	2.01	1.98	1.96
	.950	2.58	2.55	2.53	2.46	2.42	2.38
	.975	3.11	3.08	3.05	2.95	2.89	2.84
	.990	3.91	3.86	3.82	3.66	3.59	3.51
	.995	4.57	4.51	4.46	4.27	4.17	4.07
14	.005	.222	.233	.243	.281	.303	0.33
	.010	.259	.270	.281	.320	.341	.365
	.025	.324	.336	.346	.384	.405	.428
	.050	.392	.403	.412	.449	.470	.491
	.100	.484	.494	.504	.538	.556	.576
	.900	2.04	2.02	2.01	1.96	1.94	1.91
	.950	2.51	2.48	2.46	2.39	2.35	2.31
	.975	3.01	2.98	2.95	2.84	2.79	2.73
	.990	3.75	3.70	3.66	3.51	3.43	3.35
	.995	4.36	4.30	4.25	4.06	3.96	3.86
15	.005	.224	.235	.246	.286	.308	0.33
	.010	.262	.274	.284	.324	.346	.370
	.025	.328	.339	.349	.389	.410	.433
	.050	.395	.406	.416	.454	.474	.496
	.100	.487	.498	.507	.542	.561	.581
	.900	2.00	1.99	1.97	1.92	1.90	1.87
	.950	2.45	2.42	2.40	2.33	2.29	2.25
	.975	2.92	2.89	2.86	2.76	2.70	2.64
	.990	3.61	3.56	3.52	3.37	3.29	3.21
	.995	4.18	4.12	4.07	3.88	3.79	3.69
20	.005	.234	.246	.258	.301	.327	0.35
	.010	.273	.285	.297	.340	.365	.392
	.025	.339	.352	.363	.406	.430	.456
	.050	.407	.419	.430	.471	.493	.518
	.100	.498	.510	.520	.557	.578	.600
	.900	1.87	1.86	1.84	1.79	1.77	1.74
	.950	2.25	2.22	2.20	2.12	2.08	2.04
	.975	2.64	2.60	2.57	2.46	2.41	2.35
	.990	3.18	3.13	3.09	2.94	2.86	2.78
	.995	3.61	3.55	3.50	3.32	3.22	3.12

CUMULATIVE F-DISTRIBUTION

ν_2	p	ν_1 = 13	14	15	20	24	30
24	.005	.240	.252	.264	.310	.337	0.37
	.010	.279	.292	.304	.350	.376	.405
	.025	.346	.359	.370	.415	.441	.468
	.050	.413	.426	.437	.480	.504	.530
	.100	.504	.516	.527	.566	.588	.611
	.900	1.81	1.80	1.78	1.73	1.70	1.67
	.950	2.15	2.13	2.11	2.03	1.98	1.94
	.975	2.50	2.47	2.44	2.33	2.27	2.21
	.990	2.98	2.93	2.89	2.74	2.66	2.58
	.995	3.35	3.30	3.25	3.06	2.97	2.87
30	.005	.246	.259	.271	.320	.349	0.38
	.010	.285	.299	.311	.360	.388	.419
	.025	.352	.366	.378	.426	.453	.482
	.050	.420	.433	.445	.490	.516	.543
	.100	.511	.523	.534	.575	.598	.622
	.900	1.75	1.74	1.72	1.67	1.64	1.61
	.950	2.06	2.04	2.01	1.93	1.89	1.84
	.975	2.37	2.34	2.31	2.20	2.14	2.07
	.990	2.79	2.74	2.70	2.55	2.47	2.39
	.995	3.11	3.06	3.01	2.82	2.73	2.63
40	.005	.252	.266	.279	.331	.362	0.40
	.010	.292	.306	.319	.371	.401	.435
	.025	.360	.374	.387	.437	.466	.498
	.050	.427	.441	.454	.502	.529	.558
	.100	.518	.530	.542	.585	.610	.636
	.900	1.69	1.68	1.66	1.61	1.57	1.54
	.950	1.97	1.95	1.92	1.84	1.79	1.74
	.975	2.25	2.21	2.18	2.07	2.01	1.94
	.990	2.61	2.56	2.52	2.37	2.29	2.20
	.995	2.89	2.83	2.78	2.60	2.50	2.40
5C	.005	.256	.270	.284	.338	.370	0.41
	.010	.296	.311	.325	.378	.410	.445
	.025	.364	.379	.392	.445	.475	.508
	.050	.432	.446	.459	.509	.537	.568
	.100	.522	.535	.547	.592	.617	.644
	.900	1.66	1.64	1.63	1.57	1.54	1.50
	.950	1.92	1.89	1.87	1.78	1.74	1.69
	.975	2.18	2.14	2.11	1.99	1.93	1.87
	.990	2.51	2.46	2.42	2.27	2.18	2.10
	.995	2.76	2.70	2.65	2.47	2.37	2.27

CUMULATIVE F-DISTRIBUTION

		ν_1					
ν_2	p	13	14	15	20	24	30
60	.005	.259	.274	.287	.343	.376	0.41
	.010	.299	.314	.328	.383	.416	.453
	.025	.368	.383	.396	.450	.481	.515
	.050	.435	.450	.463	.514	.543	.575
	.100	.525	.538	.550	.596	.622	.650
	.900	1.64	1.62	1.60	1.54	1.51	1.48
	.950	1.89	1.86	1.84	1.75	1.70	1.65
	.975	2.13	2.09	2.06	1.94	1.88	1.82
	.990	2.44	2.39	2.35	2.20	2.12	2.03
	.995	2.68	2.62	2.57	2.39	2.29	2.19
100	.005	.265	.280	.295	.354	.389	0.43
	.010	.306	.321	.336	.394	.429	.469
	.025	.374	.390	.404	.461	.494	.531
	.050	.442	.457	.471	.525	.555	.590
	.100	.531	.545	.558	.606	.633	.664
	.900	1.59	1.57	1.56	1.49	1.46	1.42
	.950	1.82	1.79	1.77	1.68	1.63	1.57
	.975	2.04	2.00	1.97	1.85	1.78	1.71
	.990	2.31	2.27	2.22	2.07	1.98	1.89
	.995	2.52	2.46	2.41	2.23	2.13	2.02
120	.005	.266	.282	.297	.356	.393	0.43
	.010	.307	.323	.338	.397	.433	.474
	.025	.376	.392	.406	.464	.498	.536
	.050	.444	.459	.473	.527	.559	.594
	.100	.533	.547	.560	.609	.636	.667
	.900	1.58	1.56	1.54	1.48	1.45	1.41
	.950	1.80	1.77	1.75	1.66	1.61	1.55
	.975	2.01	1.98	1.94	1.82	1.76	1.69
	.990	2.28	2.23	2.19	2.03	1.95	1.86
	.995	2.48	2.42	2.37	2.19	2.09	1.98
∞	.005	.274	.291	.307	.372	.412	0.46
	.010	.316	.333	.349	.413	.452	.498
	.025	.385	.402	.417	.480	.517	.560
	.050	.453	.469	.484	.543	.577	.616
	.100	.542	.556	.570	.622	.652	.687
	.900	1.52	1.50	1.49	1.42	1.38	1.34
	.950	1.72	1.69	1.67	1.57	1.52	1.46
	.975	1.90	1.86	1.83	1.71	1.64	1.57
	.990	2.13	2.08	2.04	1.88	1.79	1.70
	.995	2.29	2.24	2.19	2.00	1.90	1.79

CUMULATIVE F-DISTRIBUTION

ν_2	p	ν_1 = 40	50	60	100	120	∞
1	.005	.113	.116	.118	.121	.122	.127
	.010	.137	.139	.141	.145	.146	.151
	.025	.184	.187	.189	.193	.194	.199
	.050	.245	.248	.250	.254	.255	.260
	.100	.353	.356	.358	.363	.364	.370
	.900	62.5	62.7	62.8	63.0	63.1	63.3
	.950	251.	252.	252.	253.	253.	254.
	.975	1006.	1008.	1010.	1013.	1014.	1018.
	.990	6287.	6303.	6313.	6334.	6339.	6366.
	.995	25148.	25211.	25253.	25337.	25359.	25465.
2	.005	.165	.169	.173	.179	.181	0.19
	.010	.193	.198	.201	.207	.209	.217
	.025	.247	.252	.255	.261	.263	.271
	.050	.309	.314	.317	.324	.326	.334
	.100	.410	.415	.418	.424	.426	.434
	.900	9.47	9.47	9.47	9.48	9.48	9.49
	.950	19.5	19.5	19.5	19.5	19.5	19.5
	.975	39.5	39.5	39.5	39.5	39.5	39.5
	.990	99.5	99.5	99.5	99.5	99.5	99.5
	.995	199.	199.	199.	199.	199.	200.
3	.005	.201	.207	.211	.220	.222	0.23
	.010	.232	.238	.242	.251	.253	.264
	.025	.289	.295	.299	.308	.310	.321
	.050	.352	.358	.363	.371	.373	.384
	.100	.449	.455	.459	.467	.469	.480
	.900	5.16	5.15	5.15	5.14	5.14	5.13
	.950	8.59	8.58	8.57	8.55	8.55	8.53
	.975	14.0	14.0	14.0	14.0	13.9	13.9
	.990	26.4	26.4	26.3	26.2	26.2	26.1
	.995	42.3	42.2	42.1	42.0	42.0	41.8
4	.005	.229	.236	.242	.252	.255	0.27
	.010	.261	.269	.274	.285	.287	.301
	.025	.320	.327	.332	.343	.346	.359
	.050	.384	.391	.396	.406	.409	.422
	.100	.478	.485	.490	.500	.502	.514
	.900	3.80	3.80	3.79	3.78	3.78	3.76
	.950	5.72	5.70	5.69	5.66	5.66	5.63
	.975	8.41	8.38	8.36	8.32	8.31	8.26
	.990	13.7	13.7	13.7	13.6	13.6	13.5
	.995	19.8	19.7	19.6	19.5	19.5	19.3

CUMULATIVE F-DISTRIBUTION

ν_2	p	ν_1 = 40	50	60	100	120	∞
5	.005	.251	.260	.266	.279	.282	0.30
	.010	.285	.293	.299	.312	.315	.331
	.025	.344	.353	.359	.371	.374	.390
	.050	.408	.417	.422	.434	.437	.452
	.100	.501	.509	.514	.525	.527	.541
	.900	3.16	3.15	3.14	3.13	3.12	3.10
	.950	4.46	4.44	4.43	4.41	4.40	4.36
	.975	6.18	6.14	6.12	6.08	6.07	6.02
	.990	9.29	9.24	9.20	9.13	9.11	9.02
	.995	12.5	12.5	12.4	12.3	12.3	12.1
6	.005	.269	.279	.286	.301	.304	0.32
	.010	.304	.314	.321	.335	.338	.357
	.025	.364	.374	.381	.394	.398	.415
	.050	.428	.437	.444	.456	.460	.477
	.100	.519	.528	.533	.545	.548	.564
	.900	2.78	2.77	2.76	2.75	2.74	2.72
	.950	3.77	3.75	3.74	3.71	3.70	3.67
	.975	5.01	4.98	4.96	4.92	4.90	4.85
	.990	7.14	7.09	7.06	6.99	6.97	6.88
	.995	9.24	9.17	9.12	9.03	9.00	8.88
7	.005	.285	.296	.304	.320	.324	0.35
	.010	.320	.331	.339	.354	.358	.379
	.025	.381	.392	.399	.414	.418	.437
	.050	.445	.455	.462	.476	.479	.498
	.100	.534	.543	.550	.563	.566	.583
	.900	2.54	2.52	2.51	2.50	2.49	2.47
	.950	3.34	3.32	3.30	3.27	3.27	3.23
	.975	4.31	4.28	4.25	4.21	4.20	4.14
	.990	5.91	5.86	5.82	5.75	5.74	5.65
	.995	7.42	7.35	7.31	7.22	7.19	7.08
8	.005	.299	.311	.319	.336	.341	0.36
	.010	.334	.346	.354	.371	.376	.398
	.025	.395	.407	.415	.431	.435	.456
	.050	.459	.469	.477	.492	.496	.516
	.100	.547	.557	.563	.577	.581	.599
	.900	2.36	2.35	2.34	2.32	2.32	2.29
	.950	3.04	3.02	3.01	2.97	2.97	2.93
	.975	3.84	3.81	3.78	3.74	3.73	3.67
	.990	5.12	5.07	5.03	4.96	4.95	4.86
	.995	6.29	6.22	6.18	6.09	6.06	5.95

CUMULATIVE F-DISTRIBUTION

ν_2	p	ν_1 = 40	50	60	100	120	∞
9	.005	.310	.323	.332	.351	.356	0.38
	.010	.346	.359	.368	.386	.391	.415
	.025	.408	.420	.428	.446	.450	.473
	.050	.471	.482	.490	.506	.511	.532
	.100	.558	.568	.575	.590	.594	.613
	.900	2.23	2.22	2.21	2.19	2.18	2.16
	.950	2.83	2.80	2.79	2.76	2.75	2.71
	.975	3.51	3.47	3.45	3.40	3.39	3.33
	.990	4.57	4.52	4.48	4.41	4.40	4.31
	.995	5.52	5.45	5.41	5.32	5.30	5.19
10	.005	.321	.335	.344	.364	.370	0.40
	.010	.357	.371	.380	.399	.405	.431
	.025	.419	.432	.440	.459	.464	.488
	.050	.481	.494	.502	.519	.523	.546
	.100	.567	.578	.586	.601	.605	.626
	.900	2.13	2.12	2.11	2.09	2.08	2.06
	.950	2.66	2.64	2.62	2.59	2.58	2.54
	.975	3.26	3.22	3.20	3.15	3.14	3.08
	.990	4.17	4.12	4.08	4.01	4.00	3.91
	.995	4.97	4.90	4.86	4.77	4.75	4.64
11	.005	.330	.345	.355	.376	.382	0.41
	.010	.367	.381	.391	.411	.417	.445
	.025	.428	.442	.451	.471	.476	.502
	.050	.491	.504	.512	.530	.535	.559
	.100	.576	.587	.595	.611	.615	.637
	.900	2.05	2.04	2.03	2.00	2.00	1.97
	.950	2.53	2.51	2.49	2.46	2.45	2.40
	.975	3.06	3.03	3.00	2.96	2.94	2.88
	.990	3.86	3.81	3.78	3.71	3.69	3.60
	.995	4.55	4.49	4.44	4.36	4.34	4.23
12	.005	.339	.354	.365	.387	.393	0.42
	.010	.375	.390	.401	.422	.420	.044
	.025	.437	.451	.461	.481	.487	.514
	.050	.499	.512	.522	.540	.545	.571
	.100	.583	.595	.603	.620	.625	.647
	.900	1.99	1.97	1.96	1.94	1.93	1.90
	.950	2.43	2.40	2.38	2.35	2.34	2.30
	.975	2.91	2.87	2.85	2.80	2.79	2.72
	.990	3.62	3.57	3.54	3.47	3.45	3.36
	.995	4.23	4.17	4.12	4.04	4.01	3.90

CUMULATIVE F-DISTRIBUTION

ν_2	p	ν_1 = 40	50	60	100	120	∞
13	.005	.346	.362	.374	.397	.403	0.44
	.010	.383	.399	.410	.432	.438	.470
	.025	.445	.460	.470	.491	.497	.526
	.050	.507	.520	.530	.550	.555	.582
	.100	.590	.602	.611	.628	.633	.657
	.900	1.93	1.92	1.90	1.88	1.88	1.85
	.950	2.34	2.31	2.30	2.26	2.25	2.21
	.975	2.78	2.74	2.72	2.67	2.66	2.60
	.990	3.43	3.38	3.34	3.27	3.25	3.17
	.995	3.97	3.91	3.87	3.78	3.76	3.65
14	.005	.353	.370	.382	.406	.413	0.45
	.010	.390	.406	.418	.441	.448	.481
	.025	.452	.467	.478	.500	.506	.536
	.050	.513	.528	.538	.558	.563	.591
	.100	.596	.609	.618	.636	.640	.665
	.900	1.89	1.87	1.86	1.83	1.83	1.80
	.950	2.27	2.24	2.22	2.19	2.18	2.13
	.975	2.67	2.64	2.61	2.56	2.55	2.49
	.990	3.27	3.22	3.18	3.11	3.09	3.00
	.995	3.76	3.70	3.66	3.57	3.55	3.44
15	.005	.360	.377	.389	.415	.421	0.46
	.010	.397	.413	.425	.450	.456	.491
	.025	.458	.474	.485	.508	.514	.546
	.050	.520	.534	.545	.566	.571	.600
	.100	.602	.615	.624	.642	.647	.672
	.900	1.85	1.83	1.82	1.79	1.79	1.76
	.950	2.20	2.18	2.16	2.12	2.11	2.07
	.975	2.58	2.55	2.52	2.47	2.46	2.40
	.990	3.13	3.08	3.05	2.98	2.96	2.87
	.995	3.58	3.52	3.48	3.39	3.37	3.26
20	.005	.385	.405	.419	.449	.457	0.50
	.010	.422	.441	.455	.484	.491	.532
	.025	.484	.502	.514	.541	.548	.585
	.050	.544	.560	.572	.596	.603	.637
	.100	.623	.638	.648	.669	.675	.704
	.900	1.71	1.69	1.68	1.65	1.64	1.61
	.950	1.99	1.97	1.95	1.91	1.90	1.84
	.975	2.29	2.25	2.22	2.17	2.16	2.09
	.990	2.69	2.64	2.61	2.54	2.52	2.42
	.995	3.02	2.96	2.92	2.83	2.81	2.69

CUMULATIVE F-DISTRIBUTION

ν_2	p	ν_1					
		40	50	60	100	120	∞
24	.005	.400	.421	.437	.470	.479	0.53
	.010	.437	.458	.473	.504	.513	.558
	.025	.498	.518	.531	.561	.568	.610
	.050	.558	.576	.588	.615	.622	.659
	.100	.635	.651	.662	.685	.691	.723
	.900	1.64	1.62	1.61	1.58	1.57	1.53
	.950	1.89	1.86	1.84	1.80	1.79	1.73
	.975	2.15	2.11	2.08	2.02	2.01	1.94
	.990	2.49	2.44	2.40	2.33	2.31	2.21
	.995	2.77	2.70	2.66	2.57	2.55	2.43
30	.005	.416	.440	.457	.494	.504	0.56
	.010	.454	.477	.493	.528	.538	.589
	.025	.515	.536	.551	.583	.592	.639
	.050	.573	.593	.606	.636	.643	.685
	.100	.649	.666	.678	.703	.710	.745
	.900	1.57	1.55	1.54	1.51	1.50	1.46
	.950	1.79	1.76	1.74	1.69	1.68	1.62
	.975	2.01	1.97	1.94	1.88	1.87	1.79
	.990	2.30	2.24	2.21	2.13	2.11	2.01
	.995	2.52	2.46	2.42	2.32	2.30	2.18
40	.005	.436	.462	.481	.523	.534	0.60
	.010	.473	.498	.517	.556	.567	.628
	.025	.533	.557	.573	.610	.620	.674
	.050	.591	.612	.627	.660	.669	.717
	.100	.664	.683	.696	.724	.731	.772
	.900	1.51	1.48	1.47	1.43	1.42	1.38
	.950	1.69	1.66	1.64	1.59	1.58	1.51
	.975	1.88	1.83	1.80	1.74	1.72	1.64
	.990	2.11	2.06	2.02	1.94	1.92	1.80
	.995	2.30	2.23	2.18	2.09	2.06	1.93
50	.005	.449	.477	.498	.543	.556	0.63
	.010	.486	.513	.533	.576	.588	.657
	.025	.546	.571	.589	.628	.639	.700
	.050	.602	.625	.641	.677	.687	.741
	.100	.674	.694	.708	.738	.746	.792
	.900	1.46	1.44	1.42	1.39	1.38	1.33
	.950	1.63	1.60	1.58	1.52	1.51	1.44
	.975	1.80	1.75	1.72	1.66	1.64	1.55
	.990	2.01	1.95	1.91	1.82	1.80	1.68
	.995	2.16	2.10	2.05	1.95	1.93	1.79

CUMULATIVE F-DISTRIBUTION

ν_2	p	ν_1 40	50	60	100	120	∞
60	.005	.458	.488	.510	.559	.572	0.65
	.010	.495	.524	.545	.591	.604	.679
	.025	.555	.581	.600	.642	.654	.720
	.050	.611	.635	.652	.689	.700	.759
	.100	.682	.702	.717	.749	.757	.806
	.900	1.44	1.41	1.40	1.36	1.35	1.29
	.950	1.59	1.56	1.53	1.48	1.47	1.39
	.975	1.74	1.70	1.67	1.60	1.58	1.48
	.990	1.94	1.88	1.84	1.75	1.73	1.60
	.995	2.08	2.01	1.96	1.86	1.83	1.69
100	.005	.479	.512	.537	.595	.611	0.71
	.010	.516	.548	.572	.626	.641	.736
	.025	.575	.604	.625	.674	.688	.772
	.050	.629	.656	.675	.719	.731	.804
	.100	.698	.720	.737	.773	.783	.844
	.900	1.38	1.35	1.34	1.29	1.28	1.21
	.950	1.52	1.48	1.45	1.39	1.38	1.28
	.975	1.64	1.59	1.56	1.48	1.46	1.35
	.990	1.80	1.74	1.69	1.60	1.57	1.43
	.995	1.91	1.84	1.79	1.68	1.65	1.49
120	.005	.485	.519	.545	.605	.623	0.73
	.010	.522	.555	.579	.636	.652	.755
	.025	.580	.610	.632	.683	.698	.788
	.050	.634	.662	.682	.727	.740	.819
	.100	.702	.725	.742	.780	.791	.856
	.900	1.37	1.34	1.32	1.28	1.26	1.19
	.950	1.50	1.46	1.43	1.37	1.35	1.25
	.975	1.61	1.56	1.53	1.45	1.43	1.31
	.990	1.76	1.70	1.66	1.56	1.53	1.38
	.995	1.87	1.80	1.75	1.64	1.61	1.43
∞	.005	.518	.560	.592	.673	.699	1.00
	.010	.554	.594	.625	.701	.724	1.00
	.025	.611	.647	.675	.742	.763	1.00
	.050	.663	.695	.720	.779	.798	1.00
	.100	.726	.754	.774	.824	.838	1.00
	.900	1.30	1.26	1.24	1.18	1.17	1.00
	.950	1.39	1.35	1.32	1.24	1.22	1.00
	.975	1.48	1.43	1.39	1.30	1.27	1.00
	.990	1.59	1.52	1.47	1.36	1.32	1.00
	.995	1.67	1.59	1.53	1.40	1.36	1.00

Appendix D
ANSWERS TO SELECTED PROBLEMS

CHAPTER 2

2.1: 35,910; 2.2: 165; 2.3: 56; 2.4: 56; 2.5: 5040; 2.6: 1/3;
2.7: 2/3; 2.8: 11/36; 2.9: 1/36; 2.10:a: $P(A) = 1/2$, $P(B) = 1/2$,
$P(E) = 1/6$; b: A and B; d: $P(C \text{ or } D) = 5/6$; 2.11: 1/69,300;
2.12: 7/12 (assuming no "curve" on the grades);
2.13:a: $(1/12)^4$; b: $(1/3)^4$; c: $(7/12)^4$;
2.14: $P(1) = 12!\ (0.05)(0.95)^{11}/11!\ 1!$, $P(5) = 12!(0.05)^5(0.95)^7/7!\ 5!$;
2.15: P(one needs service call) = 0.64; 2.16: 0.0182;
2.17: 0.0144; 2.18: b: 0.9182; 2.19: 0.9817;
2.20: P(accepting) = 0.7659; 2.22: 0.657;
2.23:a: 18/35; b: 1/7; c: 0.414; 2.24: P(2 alike) = 10/36;
2.25:a: 0.43046; b: 0.18689; c: 3; d: P(30 good) = 0.04239;
2.26:b: 0.6678; c: 4420; 2.27:a: 0.09693; b: 0.0494, c: 1.48 ¢/lb.;
2.28:a: 0.2589; b: 0.3150; c: 0.3665. $P(\geq 12 \text{ points}) = 0.2013$.

CHAPTER 3

3.15:a: 0.88, b: 0.128, c: 0.829;

3.18: $s_X = 1.27$, $s_{\overline{X}} = 0.366$; 3.19: $P(r \geq 1) = \dfrac{12!\ (0.84)^{11}(0.16)}{11!\ 1!}$

$+\ 12!\ (0.84)^{12}(0.16)^0/12!\ 0!$;

3.20: $m = 10.08$, $P(11 \leq r \leq 12) = e^{-10.08}\left[\dfrac{(10.08)^{11}}{11!} + \dfrac{(10.08)^{12}}{12!}\right]$

3.21: 0.168; 3.22: 0.000519;

3.23: $P(\text{failure in 9th year}) = \sum_{i=1}^{9} P(\text{failure in } i^{th} \text{ year}) = 0.33514$;

3.24: $P(1) = 0.23423$, $P(2) = 0.001909$; 3.25: $P(6 \text{ or } 7 \text{ or } 8) = 0.0064$;

3.27: $m = 0.48$, $P(0) = 0.619$, $P(1) = 0.297$, $P(2) = 0.071$, $P(3) = 0.011$;

3.28: $P(0) = 0.1377$; 3.29: 4 broken at 95% CI, 6 broken at 99% CI;

3.30: $m = 0.85$.

CHAPTER 4

4.1:a: $\bar{x} = \bar{z}(250)$; b: $\bar{x} = \bar{w} + 25$; 4.2: $\bar{x} = 44.9833$, $s^2 = 3518.7896$;

4.3: $\bar{x} = 144.5$, $s^2 = 2788.4$; 4.4: $\bar{x} = 82.782$, $s^2 = 56.648$;

4.5: $\bar{x} = 2.79$, $s^2 = 8.4706$; 4.6: $\bar{x} = 0.92085$, $s^2 = 0.50503$;

4.7: $\bar{x} = 79.04$, $s^2 = 2.7944$; 4.8: $\bar{x} = 14.63$, $s^2 = 16.8037$;

4.9: $\bar{x} = 0.9358$, $s^2 = 0.0026$; 4.10: $\bar{x} = 1449.3$, $s^2 = 42{,}053.07$;

4.11: $\bar{x} = 51.19$, $s^2 = 193.854$.

CHAPTER 5

5.1: $k = \sqrt{2\theta}; f(x) = x/\theta$; 5.2: $a = 3$; 5.3:a: $A = 1/18$; b: $\mu = 3.07$, $\sigma^2 = 0.328$; c: $P(3 \leq x \leq 4) = 0.556$; d: $P(x<3) = 0.444$;

5.6: $\mu_1' = 1/2$, $\mu_2 = 1/4$; 5.7: $\mu_1' = m$, $\mu_2 = m$; 5.8: 0.0256;

5.13: $Var(x) = 276$.

CHAPTER 6

6.1: $\bar{x} = 2.501 \times 10^{11}$ ft./lb., $s_x^2 = 1.11 \times 10^{20}$, range on α is 2.446×10^{11} to 2.556×10^{11} ft./lb.;

6.3: $\bar{x} = 11.411$, $s_X^2 = 3.5551$, $z_{0.025} = -1.96$, range on $SO_4^=$ is 10.24 to 12.58 $\mu g/m^3$;

6.4: $\bar{x} = \$516.62$, $s_x = \$8.551$, $z_{0.025} = -1.96$; range on cost = \$500.06 to \$533.58. Note: disregard all units if either cost and/or volume specification are not met;

6.5: $\bar{x} = 1.09375$, $\sigma_X^2 = 0.00071257$, $z = 2.7178$ so confidence level $= 99.3\%$;

6.6: $\bar{x} = 50.46$ psig, $s_{\bar{X}} = 0.45318$ psig/sample, $z_{0.005} = -2.575$, range = 49.29 to 51.63 psig;

6.7: $\bar{x} = 12.75$, $s_{\bar{X}} = 0.98107$, $z_{calc} = 1.4131$, confidence level $= 84.2\%$;

6.8: $\bar{x} = 0.91$, $s_x = 2.9277 \times 10^{-3}$, $F(z) = 0.833$, instructor was not fair as 61% is passing and he failed at 66.6%;

6.9: $s_{\bar{X}} = 1.1686$, 95% C.I. is $58.18 < \mu < 63.32$;

6.10: $\bar{x} = 17.8$, $s_{\bar{X}} = 2.190892$;

6.11: $s_X^2 = 0.033928$, 99% C.I. is 0.745 % Si $< \mu < 0.931\%$ Si;

6.17: $f = 12.55$, $t_f = \pm 3.0356$, $s_{\bar{X}_A - \bar{X}_B} = 2.073$;

6.18: $s_{\bar{D}} = 1.823$, $\bar{d} = 5.5$, 95% C.I. is $1.191 < \mu < 9.809$;

6.22: $\sigma_{\bar{X}} = 1.55\%$; 6.23: 0.1915; 6.24: $-0.0496 < P_1 - P_2 < 0.0208$;

6.25: $s_{\hat{P}_1 - \hat{P}_2} = 0.047191$.

CHAPTER 7

7.1: $z = 1.725$, accept H_0: $\mu_1 = \mu_2$;

7.2: $\bar{x} = 60.75$, $s_X^2 = 16.386$, $t = -3.6368$; at $\alpha = 0.05$, accept H_0: $\mu < \mu_0 = 65$;

7.4: $\bar{x} = 17.8$, $s_X^2 = 480$; 95% CI is $13.46 < \mu < 21.14$;

7.5: $\bar{x} = 39$, $s_X = 2.4$, $t = -1.8633$, accept H_0: $\mu = \mu_0 = 40$; CI is $37.474 < \mu < 40.526$;

7.6: $\bar{x} = 0.838$, $s_{\bar{X}} = 0.0336$, $t = -0.357$; accept H_0: $\mu_{new} \geq \mu = 0.85$;

7.7: $\bar{x} = 0.197$ in., $s_X = 0.005$ in.; $t_{99,0.025} = 1.9824$; reject H_0: $\mu_1 = \mu_2 = 0.1875$ at both levels of α;

7.8: $s_A^2 = 17.95$, $s_B^2 = 15.61$, $T = -1.615$; $s_{\bar{X}_A - \bar{X}_B} = 2.0542$; accept H_0: $\mu_A = \mu_B$ at $\alpha = 0.02$;

7.10: $\bar{\bar{x}} = 69.07$, $s_{\bar{\bar{X}}} = 0.0789$, $T = 2.914$; $\alpha \leqq 0.072$ to accept H_0;

7.14: $s_{\Delta P}^2 = 2.06105$, $s_F^2 = 0.060475$; to test H_0: $\mu_{\Delta P_{Expt}} = \mu_{\Delta P_{Theory}}$, $t = -2.815$ which is outside the acceptance region for $\alpha = 0.05$;

7.16: $\bar{x}_1 = 4.69$, $\bar{x}_2 = 4.63$, $s_{X_1} = 29.15 \times 10^{-3}$, $s_{X_2} = 5.612 \times 10^{-2}$

$F_{4,4,0.025} = 0.104$; $F_{4,4,0.975} = 9.60$. The 95% CI on σ_1/σ_2 is $0.2326 < \sigma_1/\sigma_2 < 2.233$. From F-test using s_1/s_2 and $\sqrt{}$ of above F values, accept H_0: $\sigma_1 = \sigma_2$.

Alternate solution: $t_f = 2.122$, $f \cong 6.01$, since $t_{6.01,0.025} = -2.447$, accept H_0: $\mu_1 = \mu_2$ by t-test.

7.17:a: $s_{\bar{X}} = 37.4$, $\bar{X}_2 = 2365$; $t = 7.674$ so reject H_0 and conclude that the modification is helpful; b: $2260 < \mu < 2470$ lb./day wastage;

7.18:a: $SS_T = 14,754.$; $SS_M = 14,045.$; $SS_E = 563.2$. Reject H_0: $\alpha_i = 0$ at 5% level; b: $18.67 < \mu_1 < 28.93$; $25.41 < \mu_2 < 32.99$;

7.19: $s_p^2 = 13.9194$, $s_X^2 = 67.71$. As $F_{1,18,0.05} = 4.41$ we reject H_0: cleaning had no effect. H_0 would be accepted at 99% level.

7.20: $\bar{d} = 0$, $s_{\bar{D}} = 33.28$, $t = 0$; $t_{3,0.975} = 4.176$ so accept H_0: $\mu_{\bar{D}} = 0$.

Alternate solution: calculate $\chi^2 = 27.459$; $\chi^2_{3,0.95} = 7.815$. As $\chi^2_{calc} > \chi^2_{tab}$, reject H_0: $0 = E$;

7.21: $t_f = -3.0593$, $f = 12.17$, accept H_0: $\mu_{140} < \mu_{120}$,

7.22: $\bar{x}_A = 76.2$, $\bar{x}_B = 78.286$, $s_A^2 = 1.075$, $s_B^2 = 3.5714$; a: $t_{f,0.05} = -1.822$, $f = 9.57$. Accept H_0: $\mu_B > \mu_A$ by 1-tailed test as $T = 2.449$. Alternate solution: $\chi^2_{calc} = 19.9468$; $\chi^2_{6,0.05} = 12.592$ so accept H_0: $\sigma_B > \sigma_A$; b: 0.953 chance of being wrong by t-test; 0.5% chance of being wrong according to χ^2 test;

7.23: $s^2_{x_1} = 8.75$; $s^2_{x_2} = 6.625$; $F_{9,9,0.95} = 3.18$ so we accept H_0: $\sigma_1^2 = \sigma_2^2$. As the variances are thus shown to be statistically equal, pool them to get $s_p^2 = 7.6875$ and $t_p = 2.21$ which does fall in the 99% acceptance region for the difference in means;

7.25: $s_X^2 = 0.027461$, $s_Y^2 = 0.0048073$;

7.26: $\chi^2 = 12.32$ $\chi^2_{3,0.95} = 7.81$ so reject hypothesis of uniform population;

7.27: $\chi^2 = 6.881$ $\chi^2_{6,0.95} = 11.1$ so accept hypothesis;

7.28: $\chi^2 = 3.408$ $\chi^2_{2,0.95} = 5.99$ so accept hypothesis of normality;

7.29: Method I: F-test, $s_p^2 = 1.36$; $s_X^2 = 3.97$; $F_{1,10,0.05} = 4.96$ so accept H_0: wetting agent had no effect;

Method II: t-test. $\bar{d} = 1.15$, $s_D^2 = 1.05$; $t_{calc} = 2.74$ $t_{5,0.95} = 2.015$ for 1-tailed test so we reject H_0: wetting agent had no significant effect;

7.30: $\chi^2_{calc} = 6.1667$, $\chi^2_{3\times1,0.95} = 7.815$ so accept H_0: no regional differences;

7.31: χ^2 32.6048 vs. $\chi^2_{17,0.95} = 27.6$ so reject H_0: predictive method is accurate;

7.32: χ^2 0.00772; compare to $\chi^2_{6,0.025}$ and $\chi^2_{6,0.975}$ and reject the null hypothesis;

7.34: $\chi^2 = 9$ vs. $\chi^2_{9,0.99} = 6.635$ so coin is probably biased;

7.36: $s_{\bar{d}} = 6.05923$, $-14.8269 < \mu_D < 14.8269$, $\bar{d} = 0$, dye probably has no effect; $\chi^2 = 14.46$ which is not in the 95% CI for χ^2_6.

7.37: $\chi^2 = 17.38 > \chi^2_{5,0.95} = 11.1$ so must conclude that the system was not installed properly.

7.38: Method I: F-test. $s^2_p = 0.4815$; $s^2_X = 0.9025$ so we accept H_0: no difference in predicted and actual values at the 95% level; Method II: χ^2-test. $\chi^2_{calc} = 0.37723$; $\chi^2_{5,0.95} = 1.145$. Again accept H_0 at 95% level.

CHAPTER 8

8.1:a: $s^2_p = 1.808$, $s^2_Y = 24.596$, reject H_0: $\alpha_i = 0$ at 5% level;

b: $t_{20,0.95} = 2.086$; $36.522 < \mu_{III} < 38.812$ for highest yield;

8.2:a: (coke yield): $s^2_p = 13.7923$, $s^2_Y = 5.6205$; accept H_0: hydrogenation of feed does not affect coke yield; b: (conversion): $s^2_p = 96.248$, $s^2_Y = 54.908$; accept H_0: no significant differenc in conversion;

8.3:a: $s^2_p = 0.89391$, $s^2_Y = 5.1566$; accept H_0: there is a significant difference between populations; to compare (A + B) vs. (C + I $s^2_p = 0.85286$, $s^2_Y = 15.2420$ and we accept H_0: $\mu_{A+B} \neq \mu_{C+D}$;

8.6: $s^2_p = 0.075$, $s^2_Y = 0.08988$; $F_{5,12,0.99} = 5.06$ so accept H_0: $\alpha_i = 0$;

8.7: $s_Y^2 = 0.0014167$, $s_p^2 = 0.0007458$, $s_{\bar{p}} = 0.01362$, $t_{12,0.995} = 3.055$; accept H_0: $\alpha_i = 0$ at both 95% and 99% levels; $\mu_{.1} = 0.385$ 3.055(0.01362);

8.9: $s_p^2 = 0.1678$, $s_Y^2 = 0.357$, accept H_0: different gases have no effect;

8.10: $s_p^2 = 0.008085$, $s_Y^2 = 0.81334$; reject H_0 and conclude that flow rate does affect CO_2 leakage;

8.11: $s_p^2 = 66.083$, $f = 1.411$; accept H_0: no significant difference in lab techniques;

8.12: $SS_T = 2.3236$, $SS_M = 2.3104$, $SS_C = 0.00425$, $SS_E = 0.0084$; accept H_0: $\alpha_i = 0$, $\beta_j = 0$ and conclude no difference between batches or analysts;

8.13: $SS_T = 119{,}047.54$; $SS_C = 15{,}449.66$; $SS_R = 1.6366$; $SS_E = 31.27334$; at $\alpha = 0.05$, inlet concentration affects outlet gas composition but lab groups do not;

8.14: $SS_T = 760.430$; $SS_C = 189.393$; $SS_R = 0.433333$; $SS_E = 0.186667$; both flow rate and lab group affect the HTU (at $\alpha = 0.05$);

8.15: $SS_T = 310{,}933{,}275$; $SS_{col} = 5{,}653{,}437.5$; $SS_R = 170{,}343{,}939.6$; $SS_E = 7{,}306{,}879.2$; accept H_{0_1}: supply pressure does not affect output; reject H_{0_2}: model has a significant effect on output;

8.16: a: $SS_{col} = 95.7905$, $SS_{row} = 556.402$, $SS_{AC} = 959.305$; $SS_{EE} = 307.1125$, $SS_{SE} = 272.225$. By F-test, we find sampling errors to be insignificant so we pool the error sources to get $MS_{pooled\ error} = 19.9772$. As the resulting $f > F$ at $\alpha = 0.05$, we accept H_{0_1}: $\alpha_i = 0$ and H_{0_2}: $\beta_j = 0$. b: $SS_{level} = 17.2925$, $SS_E = 562.045$, column and row effects are unchanged. $f_{col} = 4.772$, $f_{row} = 3.080$, $f_{level} = 0.861$.

Reject insignificance of row and column effects; accept insignificance of level effect. All tests at $\alpha = 0.05$;

8.18: $SS_M = 7.196$, $SS_{flow} = 2.148$, $SS_{conc} = 1.492$, $MS_E = 0.007662$. By 5% F-tests, both null hypotheses are rejected and we conclude that flow rate and amine concentration affect performance;

8.19: $SS_M = 2{,}601.907$; $SS_{col} = 685.815$; $SS_{row} = 33.163$; $SS_E = 9.525$. Reject both null hypotheses and conclude that both temperature and pipe diameter influence flow rate, the former possibly more strongly than the latter;

8.20: $f_{feed\ rate} = 165.707$, $f_{conc} = 7.607$, reject both null hypotheses at $\alpha = 0.05$ and conclude that feed rate and amount of catalyst affect the % conversion. $MS_E = 1.90276$;

8.21: $SS_M = 2{,}596{,}126.5625$; $SS_{col} = 39{,}934.1875$; $SS_{row} = 324{,}082.1875$; $SS_E = 9{,}232.0625$. Reject the null hypotheses that liquid and gas rates have no effect at $\alpha = 0.01$;

8.22: $SS_{col} = 24.97$; $SS_{row} = 1.62$; $SS_E = 0.16$; reject H_{0_1}: blue colorant is insignificant; reject H_{0_2}: yellow colorant is insignificant;

8.23: $SS_T = 2164.06$; $SS_{col} = 12.8833$; $SS_{row} = 1.8067$; $SS_E = 0.0067$. Reject the null hypotheses of no effects on greenness of the yellow and blue colorants;

8.24: $SS_T = 572.07$; $SS_M = 460.041$; $SS_{col} = 47.33$; $SS_{row} = 27.96$. Accept $H_{0_1}: \alpha_i = 0$ and $H_{0_2}: \beta_j = 0$;

8.25: $SS_{col} = 246.966$; $SS_{row} = 58.157$; $SS_E = 91.006$. Accept H_{0_1}: pull angle does not affect force. Reject H_{0_2}: different connectors do not affect force. $\alpha = 0.05$ in both tests;

8.26: $SS_M = 101,798.13$; $SS_{col} = 6274.307$; $SS_{row} = 4.380$; $SS_{AC} = 7083.911$; $f_{SE} = 0.636$ so pool errors. Based on pooled errors, $f_{cols} = 11.771$, $f_{rows} = 0.024$. Surface pretreatment effects are found insignificant; adhesive effect is significant;

8.27: Method I: no interaction. $SS_{TC} = 14,481.344$; $SS_{row} = 9.276$; $SS_{col} = 13,761.23$. At $\alpha = 0.05$, accept H_{0_1}: primer has no effect; reject H_{0_2}: adhesive system does not contribute significantly to variation in data;
Method II: interactions exist as measured by repetition effect. $SS_M = 169,183.126$; $SS_{rep} = 19.2777$; $SS_{primer \times adhesive} = 14,036.7154$. At $\alpha = 0.05$, replication and primer effects are insignificant but adhesive and adhesive × primer interaction are significant with f values of 232.15 and 236.8, respectively;

8.28: $SS_T = 105,788$; $SS_C = 105.25$; $SS_R = 326.25$; at $\alpha = 0.05$, both students and sleep time affect test-taking ability;

8.30: $SS_T = 156.5279$; $SS_M = 133.0061$; $SS_R = 10.92024$; $SS_C = 11.6725$; at $\alpha = 0.05$, conclude that catalyst effectiveness is affected by both space velocity and surface area;

8.31: $s_p^2 = 19.7599$, $s_Y^2 = 18.344$, accept H_0: there is no significant difference in areas.

CHAPTER 9

9.1: $\Sigma xy = 14.626$, $\Sigma y^2 = 0.005331$, $\Sigma d^2 = 0.000442$, $s_{\hat{\beta}_1} = 5.025 \times 10^{-5}$, $s_{\hat{\beta}_0} = 0.9785 \times 10^{-2}$, $C_p = 0.353332 + 3.34292 \times 10^{-4}\, t$;

9.2: $\hat{Y} = 6.00 + 0.8X$, $s_{\hat{\beta}_1} = 0.2$; 9.3: $\Sigma\,\theta^2 = 279.9997$,

$\Sigma F = 872.0988$, $(\Sigma F)^2 = 760{,}556.4$, $F = 41.4239 + 0.727023\,\theta$;

9.4: at 1.1% yellow, $s_{\overline{Y}} = 0.0516$, $L = 81.49709 - 1.93353$ (% blue);

at 1.9% yellow, $s_{\overline{Y}} = 0.12187$, $L = 81.64635 - 2.18323$ (% blue);

9.5: $\Sigma x_1^2 = 46.8125$, $\Sigma\, y^2 = 61 \times 10^{-6}$, $\Sigma x_1 y = 5.561005$,

$\hat{C} = -89.5017 \times 10^{-3} + 11.2732 \times 10^{-4}$ (% T); 9.6: $X = 3.53327$

$+ 2.06668Y$, $r^2 = 0.9929$;

9.6: for loading zone, $12.64 < L/G < 21.7$: $-\Delta P = 0.3344 - 0.01111(L/G)$; for safe operating region, $12.64 \leq L/G < 5.557$: $-\Delta P = 1.186 - 0.07853(L/G)$; for flooding zone, $5.557 \leq L/G < 7.6$, $-\Delta P = 2.391 - 0.2953(L/G)$; loading point is at intersection of first two equations: $(L/G) = 12.637$, $-\Delta P = 0.1940$ in H_2O; flooding point is at intersection of second and third equations: $(L/G) = 5.5571$, $-\Delta P = 0.7500$ in H_2O;

9.7: $\hat{Y} = 3.2666 + 1.98789X$, $r^2 = 0.99516$, $\Sigma x^2 = 82.49951$, $\Sigma xy = 164$;

9.8: $s^2_{E_1} = 1.86222 \times 10^{-5}$, $\Sigma x_1^2 = 0.717779$, $\overline{x}_1 = 0.509559$,

$t_{\beta_1} = -25.8626$, $s^2_{E_2} = 1.30333 \times 10^{-5}$, $s_{\hat{\beta}_0} = 0.0024289$, t_{β_0}

$= 617.085$; for testing H_0: $\beta_{0_1} = \beta_{0_2}$, $t_f = 0.6132$, $f = 21.41$;

9.9: a: $\hat{Y} = -0.26672 + 3.73417X$, $s_{\hat{\beta}_1} = 0.40765$, $t_{calc} = 9.1602$,

b: $\hat{Y} = 3.48945 + 2.44597X_w + 0.82089X_a$, $SS_E = 48{,}380.21$

$F_{calc} = 49.138$, Reject H_0: regression is not significant;

9.10: $\hat{Y} = 45.1972 - 2.68408X_1 + 4.20910X_2$, $R^2 = 0.94808$;

9.11:a: $X = \beta_0 + \beta_1 Z_1 + \beta_2 Z_2 + \varepsilon$, $\hat{X} = 14.7075 - 1.2042Z_1 - 0.46288Z_2$

c: $R^2 = 0.82523$;

9.16: $\hat{Y} = -0.8868835 + 0.76595168X$, $s_{\hat{\beta}_0} = 0.02767$, $s_{\hat{\beta}_1} = 0.03725$,

$r^2 = 0.97246$;

9.18: $\hat{Y}_1 = 3.09204 + 0.193694X_1 + 0.0362872X_1^2$, $R_1^2 = 0.99956$,

$\hat{Y}_2 = 2.41588 + 0.180067X_2 + 0.034122X_2^2$, $R_2^2 = 0.998707$.

$F_{calc,1} = 398.37$, $F_{calc,2} = 68.567$;

9.20: (100% gasoline): $\hat{Y} = -72.6072 + 0.807145X - 0.944949 \times 10^{-3}X^2$,

$R^2 = 0.996678$; (100% acetone, $142^{\circ}F$ endpoint): $\hat{Y} = -2507.72 + 27.5217X - 0.0641013X^2$, $R^2 = 0.89897$; (50% gasoline, 50% acetone): $\hat{Y} = 148.651 - 3.0168X + 0.05172X^2$, $R^2 = 0.94478$; (70% gasoline, 30% acetone): $\hat{Y} = 152.229 - 2.1634X + 0.0442202X^2$, $R^2 = 0.98786$;

9.21: $MV = 43.123445 - 40.784467 \log_{10}(NH_3 \text{ concentration})$;

$T_{\beta_0} = 38.68$; $T_{\beta_1} = -72.48$;

9.24: $\widehat{\log N_{St}N_{Pr}^{0.2}} = -0.0166505 - 0.625178 \log N_{Pr}$, $\Sigma y^2 = 0.552948$,

$\Sigma x^2 = 1.384811$, $\Sigma xy = -0.865753$, $r^2 = 0.978844$;

9.25: $\widehat{\log U} = -0.0853233 + 0.390123 \log v_{max}$, $\Sigma x^2 = 0.821564$,

$\Sigma xy = 0.320511$, $\Sigma y^2 = 0.139633$;

9.26: (Run 12-18-1): $\widehat{\log Y} = -2.54230 + 0.690076 \log X$ where Y = fractional weight gain, X = time; $s_{\hat{\beta}_0} = 0.02129$, $s_{\hat{\beta}_1} = 0.006197$, $t_{\hat{\beta}_1} = 111.47$ so reject H_0: $\beta_1 = 0$. For Run 12-18-2, $\widehat{\log Y} = -2.51246 + 0.687049 \log X$, $s_{\hat{\beta}_1} = 0.001632$, $s_{\hat{\beta}_0} = 0.005671$, $t_{\beta_0} = 442$ so reject H_0: $\beta_0 = 0$. To compare intercepts, H_{0_1} : $\beta_0 = \beta_0'$ for which $t_f = -8.011383$, f = 39.078. For slopes, H_{0_2} : $\beta_1 = \beta_1'$ for which $t_f = 3.441$ and f = 38.966. Both hypotheses are rejected;

9.27: (Run 1-13-3): $\hat{Y} = 0.003151X^{0.70405}$ where Y = fractional weight gain, X = time. For Run 1-13-6, $\hat{Y} = 2.116X^{0.70842}$. To compare runs, calculate the sum of squares due to regression, $SS_R = b_1\Sigma xy$ and compare runs by $F = SS_{R_1}/SS_{R_2}$, etc.

$SS_R(1\text{-}13\text{-}3) = -58.35$; $SS_R(1\text{-}13\text{-}6) = -59.18$, $SS_R(12\text{-}18\text{-}2) = -59.98$;

9.28: $\widehat{\log k} = -8.96025 + 0.0483922\ T$ where T = ^oC; $s_E^2 = 0.0381656$,

$t_{\beta_1} = 13.107$, $t_{\beta_0} = 14.772$; 95% CI on β_0 is -10.9626 to -6.9579,

95% CI on β_1 is 0.03890 to 0.05788;

9.29: 0.916938; 9.31: 0.13341; 9.32: 1.1% yellow, 0.99925; 1.5% yellow, 0.99655; 1.9% yellow, 0.99879; 9.33: 0.97535;

9.34: 0.99287; 9.35: 0.99516; 9.36: Run 1, 0.98653; Run 2, 0.99052;

9.38: 0.94808; 9.39: 0.82525; 9.40: 0.993818;

9.45: $\hat{Y} = 9.5663 + 0.452X$, $r^2 = 0.9708$;

9.52: 0.995; 9.56: 0.99902; 9.59: 0.995474; 9.60: 0.971;

9.61: 0.97582; 9.62: 0.94898; 9.63: 0.99034;

9.62: $U = \hat{\beta}_0 + \hat{\beta}_1 \log_{10} \Delta T = -1022.4986 + 626.0556 \log_{10} \Delta T$;

$F = 398$ vs. $F_{1,3,0.95} = 10.1$ so accept H_A: regression is significant; $R^2 = 0.9925$; $T_{\beta_1} = 19.962$ so reject H_0: slope = 0.

CHAPTER 10

10.2: $\beta_0' = 10.432$, $\beta_1' = 3.450$, $\hat{Y} = 0.082 + 0.8625X$,

$r^2 = SS_{\beta_1'}/(SS_T - SS_{\beta_0'}) = 0.996$;

10.3: $SS_{\beta_0'} = 9159.2$, $SS_{\beta_1'} = 4656.964$, $SS_{\beta_2'} = 6.113$, $SS_{\beta_3'} = 190.096$,

$SS_{Res} = 4.836$, $\hat{Y} = -21.84 + 2.158X$, $r^2 = 0.963$;

10.4: $SS_T = 210.14$, $SS_{\beta_0'}$ 202.248, $SS_{\beta_1'}$ 7.569, $SS_{\beta_2'} = 0.315$,

$SS_{\beta_3'} = 0.001$, $SS_{Res} = 0.007$; $\hat{Y} = 7.1985 - 1.20X + 0.15X^2$,

$R^2 = 0.999$;

$SS_{'3} = 0.001$, $SS_{Res} = 0.007$; $Y = 7.1985 - 1.20X + 0.15X^2$,

$R^2 = 0.999$;

10.5: $\hat{z} = -0.0172 + 1.0286R^2$ from F-tests at $\alpha = 0.1$ level;

10.8: $SS_{\beta_0'} = 2016.4$, $SS_{\beta_1'} = 326.03$, $SS_{\beta_2'} = 0.371$, $SS_{Res} = 1.20$

$\hat{Y} = 3.27 + 1.988X$, $r^2 = 0.995$;

10.13: $\hat{Y} = 95.9750 + 1.41171X + 4.8072\ (10^{-2})X^2 - 1.0366(10^{-3})X^3$
$+ 7.07(10^{-6})X^4$, $R^2 = 0.99902$;

10.14: β_2' term not significant, $\hat{Y} = 133.625 + 0.03667X' + 0.286713(X')^2$

$+ 0.059828(X')^3 - 0.013986(X')^4$ where $X' = (X/10 - 4.5)$, $R^2 = 0.990968$;

10.15: $\hat{Y} = 131.68 + 0.083X$, $r^2 = 0.971$;

10.16: From a quartic orthogonal model, all terms significant but the last. $\hat{Y} = 128.493 + 2.4624X - 0.09027X^2 + 0.0009703X^3$,

$R^2 = 0.97582$;

10.17: In a cubic orthogonal model, the quadratic term was insignificant at $\alpha = 0.05$; $\ddot{Y} = 124.723 - 0.1058X - 9.34335(10^{-3})X^2$
$+ 6.2289(10^{-5})X^3$, $R^2 = 0.94898$;

10.18: All terms but the last significant at $\alpha = 0.05$ in the cubic orthogonal model. $\hat{Y} = 162.119 - 2.5548X + 0.0476X^2$,

$R^2 = 0.99034$;

CHAPTER 11

11.1: 4; 11.2: 12 for $\alpha = 0.05$; 11.3: $\hat{\sigma} = 1.9925$, n is about 50;

11.4: $n = 8$ for $\alpha = 0.05$, $n = 11$ for $\alpha = 0.01$;

11.5: $SS_{Tr} = 21{,}851.1$, $SS_{EE} = 1.74$, $SS_{SE} = 0.08$, reject H_0: experimental error is indistinguishable from sampling error; reject H_0: sample source is not significant for this Model I situation;

11.6: $MS_{Tr} = 22{,}699.4$, $MS_{EE} = 133.18$, $MS_{SE} = 7328.266$, $f_{EE} = 0.018$ so accept H_0: experimental error and sampling error measure the same thing and pool their SS to calculate a pooled error term: $MS_{PE} = 165.8099$, $f_{Tr} = 273.8$ so reject H_0: $\Sigma\alpha_j^2 = 0$;

11.7: $MS_{EE} = 3.9333$, $MS_{SE} = 12.6055$, $f_{EE} = MS_{EE}/MS_{PE} = 0.312$ but no need to pool the error terms as $\nu_{SE} = 45$ is sufficient. $f_{Tr} = 356.67$ so reject H_0: $\Sigma\alpha_j^2 = 0$;

11.9: $MS_{EE} = 39.95$, $MS_{SE} = 0.1083$, $f_1 = 368.8$ so we reject H_0: experimental error is insignificant. Comparing MS_{Tr} to MS_{EE} gives $f_2 = 6.326$ so reject H_0: age does not affect invertase activity;

11.10: $SS_{Tr} = 2510.96$, $SS_B = 1.936$, $SS_E = 8.384$, $f_{Tr} = 299.5$ so reject H_0: age does not affect invertase activity.

11.11: $f_{shift} = 1.2644$, $f_{site} = 0.3083$, $f_{shift \times site} = 0.0758$ so we accept all Model I hypotheses;

11.14: $f_H = 1.785$, $f_M = 5.193$, $f_{H \times M} = 1 \times 10^{-5}$ so accept all Model I hypotheses;

11.17: $f_D = 2.877$, $f_S = 2.230$, $f_{DS} = 0.113$;

11.22: $s_{\hat{\beta}_0} = 0.990$, $s_{\hat{\beta}_1} = 0.443$, $1.78 \leq \beta_1 \leq 3.822$;

11.23: Method I: no interaction, no sampling error. $SS_M = 266{,}725.25$; $SS_{col} = 17{,}721.175$; $SS_{row} = 13.202$; $SS_{level} = 4256.007$; $SS_E = 4386.9$. Reject hypotheses of no effect of thickness and adhesive type. Accept null hypothesis of no effect due to primer;

Method II: no interaction, sampling error exists. SS_{AC} = 25,381.63; SS_{SE} = 1000.57; SS_{EE} = 3386.16. f_{calc} = $MS_{SE}/MS_{EE} << 1$, the error sources are pooled to give MS_p = 48.73. Results of F-tests at $\alpha = 0.05$ are the same as in Method I;

Method III: all interactions exist. Calculated F-values are: $f_P < 1$; $f_T = 71.38$; $f_A = 99.08$; $f_{PT} < 1$; $f_{PA} = 5.32$, $f_{TA} = 12.93$; $f_{PTA} = 18.93$. Degrees of freedom for error = 75;

11.25: c: f = 5; d: 2; e: 133%; 11.29: b: 1.1;

11.30: Each main factor has 2 degrees of freedom. The second-order interaction has 8 d.f. and there are 27 d.f. for error.

11.33: $f_{movement} = 41.67$, $f_{depth} = 15.00$, $f_{stock} = 6.67$, $f_{movement \times feed\ rate} = 6.67$, $f_{movement \times depth} = 3.75$;

11.38: f(adjusted) = 34.05 so reject H_0: ammonia concentration has no effect on performance, T = -0.51 so accept H_0: phase rate ratio is not important;

11.39: f(adjusted) = 44.37, T = -4.89 so reject both null hypotheses.

INDEX

A

α, probability of type 1 error, 180
Accuracy, 462
Additive law of probability, 15
Additive model, 243-245
Additivity of effects, 248-252, 259-261
Adjoint of a matrix, 590
Alternate hypothesis, 180
Among cells sum of squares, 508
Analysis of correlation, 363-373
Analysis of covariance, 555-562
  assumptions, 556
  model, 556
  tests of hypotheses, 558
Analysis of regression, 301-394
Analysis of variance, 241-282
  assumptions, 245, 248 249, 462, 474, 484, 524, 534
  comparing individual means, 252-255
  completely randomized design, 462-499
  components of variance model, 248
computational procedure, 245-249, 261-264
degrees of freedom, 252, 261
description of use, 241
efficiency, 553-555
equal subclass numbers within groups, 261
factorials, 474-476, 482
fixed effects model; see Model I
Graeco-Latin square design, 533-534
group comparison, 247-249
  among groups, 247-253
  between and within groups, 245-253
homogeneity of variances, test for, 278-282
hypothesis; see analysis of variance models
incomplete block design, 545-549
interaction in, 448, 473, 481
Latin square design, 522-533
Model I, fixed treatment effects, 474
Model II, random treatment effects, 475
Model III, mixed model, 476

Model IV, mixed model, 476
multiple linear regression, 334
one-way, Model I, 475
one-way, Model II, 475
one-way classification, 241, 245-249
one-way hypothesis tests, 250, 252-254
one-way model, 244, 249
one-way table, 253
pooling errors, 149, 197, 490, 511
random effects model, 475
randomized complete block design, 500-504, 508-521
in regression, 321-328, 335, 434-442
relative efficiency, 553
simple linear regression, 321-328
split-plot design, 541-545
subsampling, 468-472, 508-521
three-way classification, 481-484
model, 481
table, 482
two-way, Model I, 474
two-way classification, 244, 249, 259-270
two-way hypothesis tests, 260, 262, 266
two-way layout, one observation per cell, 500
two-way model, 244, 249
two-way table, 262
unequal subclass numbers within groups, 247, 257
Anderson, R. L. and Houseman, E. E., 444
Answers to problems, 655-669
Antagonistic effect, 473
AOV; see Analysis of variance
Approximations, Poisson to binomial, 62
Arithmetic mean, 103
Arrangement, order of, 11-13
Array, 242
Attribute of data, 31
Average; see Mean, Median, Mode, etc.

B

β, population regression coefficient, 302, 424
β, probability of type 2 error, 180
$\hat{\beta}$, sample regression coefficient, 303-304
β', regression coefficients in orthogonal polynomials, 425
Balanced blocks, 447
Balanced incomplete blocks, 546-549
Bartlett's test for equality (homogeneity) of variance, 278-282
Bayes' theorem, 22-24

Bernoulli distribution, 57
Best estimator, 129-131
Between groups sum of squares, 247
Bias, 505, 530
Binomial, expansion, 56
Binomial, expected value, 56-59, 123
Binomial, mean, 56
Binomial, Poisson approximation, 62
  mean, 56
  probabilities, 56
  probability function, 56
  variance, 57
Binomial distribution, 55-59
  approximation by Poisson, 60, 62
  cumulative, 56
    table of values, 600
  expectation, 123
  mean and variance of, 56, 57, 123
  negative, 67-68
  point, 57
Blocking, 447, 448
Blocks
  incomplete, 545-549
  randomized complete, 499-522
Boundary, class, 37
Box-Wilson composite rotatable design, 549-553

C

Carmer, S. G. and Seif, R. D., 444
Central limit theorem, 73
Central moments, definition, 121
Central tendency, measures of, 103
Chi-square
  definition, 81, 134-136
  degrees of freedom, 134
  density function, 82, 134
  distribution, 82, 134
  expected value, 135
  goodness of fit, 215-216
  parameters, 135
  table of values, 625
  tests, 208-210, 213-220, 278-282
    Bartlett's, 278-282
    contingency, 217-222
    goodness of fit, 215-216
    homogeneity of variances, 278-282
    parameters of, 135
    of a proportion, 213, 215
    variance of normal population, 208-210
    variance, 135
Class
  boundary, 37
  definition of, 37, 39
  frequency, 37
  length, 37, 39, 43
  mark, 37

Coding, 103, 107
Coefficient(s)
  correlation, 363
  for orthogonal polynomials, 426, 433-437
  regression, 302, 333, 344, 353
  for selected treatment comparisons, 512-519
  of variation, 106
Cofactor in matrix operations, 588
Combinations
  of n things taken r at a time, 13, 14
  treatment, 446
Comparison
  multiple, 271-278
    Duncan's multiple range test, 276
    Scheffé method, 271-274, 277
  of difference of two means, 146-153
  of means, 146-153
  of paired data, 153-156, 202-204
  of two proportions, 167-169
  of regression lines, 388-392
  of variances, 159-162, 210-212
Complement of an event (set), 14
Completely randomized design, 462-499
  advantages, 462-463
  analysis of variance, 463-468
  assumptions, 463, 469, 474-476, 484
  calculations, 465, 484
  comparison of selected treatments, 463
  components of variance, 463, 468, 473, 481
  definition of, 462
  degrees of freedom, 464, 468, 469, 474, 475, 482
  difference between treatment means, 467
  equal numbers of observations, 464-467
  expected mean squares, 464, 470, 475, 482, 494
  factorials, 473-481
  interaction, 473, 478
  missing data, 467
  models, 463, 468, 473
  subsampling, 468-472
  table, 464, 469, 474-476, 482
  tests of hypotheses, 463, 470, 476, 477
  unequal numbers of observations, 467-468
  variance of a treatment mean, 465, 470
  when to use, 463
Component of variance model, 249, 259

Composite hypothesis, 179
Compound event, 14-16
Computerized solutions for:
  analysis of covariance, 559-562
  analysis of variance
    one-way, equal subclass numbers, 255-257
    one-way, unequal subclass numbers, 257-259
    two-way, 264-266
    two-way with interactions, 477-481
  comparison of two means, 204
  experimental distributions, 46-51
  frequency polygons, 50, 51
  histogram, 47-49
  measures of location and variation, 110, 204
  multiple comparisons
    Duncan's multiple range test, 274-276
    Scheffé method, 274-278
  regression analysis
    multiple linear, 339-343
    non-linear, 357-362
    quadratic, 346-352
    simple linear, 322-328
  subsampling in:
    one-way analysis of variance, 470-472
    two-way analysis of variance, 508-511
Conditional mean, 302
Conditional probability, 17-20
Confidence interval, 133
  definition of, 133
  difference between means of two normal populations, 146-149
  difference of two proportions, 167-169
  for contrasts, 266-270
  for paired observations, 154
  for proportions, 165
  for variance ratios, 159-162
  mean of a normal population, 141, 142
  variance of a normal population, 157-159
Confidence interval statement, 133, 135, 138, 140
Confidence limits
  for contrasts, 266-270
  definition, 133
  difference between two means, 146, 149, 152, 154
  intercept of straight line, 313
  mean, univariate, 141-146
  proportion, 165, 166
  ratio of two standard deviations, 162
  ratio of two variances, 160-162
  regression coefficient, 313
  slope of straight line, 313
  standard deviation, 162

two-sided, 133
in two-way analysis of variance, 266-270
variance, 157-159
Confounded designs, 447
Confounding, 537-543
Conjugate of a matrix, 587
Consistent estimator, 130, 131
Contingency test, 217-220
Continuous
definition of, 35, 36
distribution, 35, 115
variance, 54
distribution function, 35, 36
probability density function; see specific distribution desired
probability distribution, 54
random variable, 36, 115
variable, 32, 36
Contrast, 266-270
multiple, 268-269
orthogonal, 512-519
simple, 266-268, 270
Control
lack of, 447
local, 447
Correction for bias
Latin square, 530
randomized complete block, 506
Correlation
analysis, 363-373
coefficient, 363
definition, 363
estimator of, 366
index, 364, 368
methods, 364-373
multiple linear, 368-373
partial, 370
sample covariance, 366
simple linear, 364-368
testing for, 367
Counting rule, 11
Covariance, sample, 366
Critical region, 183
Cumulative distribution, 33, 35
Cumulative distribution function
binomial, 56
continuous, 34-36
definition of, 34, 35
discrete, 33-35
normal, 76
Poisson, 59-67
properties, 34-36, 52-55
Cumulative frequency
definition of, 32, 35
distribution; see specific distribution desired
Cumulative frequency polygon, 44, 46, 48
Curve fitting; see regression
Curvilinear regression, 343-352

D

D, difference between paired observations, 154, 202-204

Data, missing
  Latin-square, 529-530
  randomized complete block, 504-507
  transformation of, 392-394
Degrees of freedom (d.f.)
  chi-square, 81, 135
  F-variate, 82, 138
  partitioning, 248, 261
  regression, 281
  t distribution, 82, 136
Density function
  Bernoulli, 57
  binomial, 56
  chi-square, 82, 135
  continuous probability, 35-36
  discrete, 34
  exponential, 59-67, 72
  F, 82, 138-140
  gamma, 73
  hypergeometric, 69
  joint, 117
  negative binomial, 67
  normal, 74
  probability, 33, 36
  standard normal, 76
  t, 82, 136-138
  Weibull, 71
Dependent variable, 249
Derivative of a matrix, 592
Design
  Box-Wilson, 549-553
  completely randomized, 462-499
  efficiency, 553-555
  experimental, 445-562
    advantages, 445
    check list of pertinent points, 445-446
    examples of approach to, 446-448
    nature and value of, 446-448
    principles of, 446
    steps in, 445
  of experiments, 445-562
  factorial, 473-488, 534-542
  Graeco-Latin square, 533-534
  incomplete block, 545-549
  Latin square, 522-533
  nested, 493-499
  randomized complete block, 499-522
  split plot, 541-545
Descriptive statistics, 102-110
Determinants, 588
Deviation
  from regression, 310
  standard, 103, 106
    confidence limit, 158, 162
Difference(s)
  among means, 245-259
  between paired observations (D), 202-204
  between two means, 193-202
  between two proportions, 206-208
  among variances, 210-212
Differentiation of a matrix, 592

Discrete distribution, 33-35, 113-115
  binomial, 55-59
  hypergeometric, 68-71
  negative binomial, 67
  Poisson, 59-67
    variance of, 60
  Weibull, 71, 72
Discrete probability function, 33
Discrete random variable, 32, 113-115
Dispersion, measures of; see Variability
Distribution
  approximation by Poisson, 60, 62
  Bernoulli, 57
  binomial, 55-59
  chi-square, 81, 134
  continuous, 35, 71-83, 115
  cumulative, 34
  discrete, 33-35, 113-115
  expected value of, 112
  experimental, 36-51
  exponential, 59-67, 72
  F, 82, 138
  frequency, 33, 39, 44, 46
  function, 34-36
  gamma, 73
  Gaussian; see Normal distribution
  hypergeometric, 68-71
  joint, 116-120
  of linear combinations of random variables, 113, 114, 116
  of mean, 74
  negative binomial, 67
  negative exponential, 67
  normal, 73-81
  Poisson, 59-67
  probability; see desired distribution
  sampling, 118-120
  standard normal, 76-81
  of sum of squares, 81
  t, 82, 136
  Weibull, 71
Duncan's multiple range test, 276

E

Effects
  additivity of, 462
  antagonistic, 473
  fixed, 463, 474
  interaction, 473
  main, 535
  random, 474, 475
  simple, 536
  synergistic, 473
Efficiency
  of estimator, 129
  of Latin square, 554
  of randomized complete block, 554
  relative, 553-554

Empirical distribution, 36-46
Empty set (event), 7
Equality
  of means, 193-202
  of variances, 210-212
Equally likely events, 10
Equations
  normal, 305, 333, 344, 426
  solution by matrices, 591
Error
  of estimate, 311-314
  experimental, 447
    partitioning of, 447, 469
    reduction, 447
  kind
    type 1, 180
    type 2, 180
  pooled, 148, 197, 490, 511
  sampling, 244
  sources of, 461
  standard, of regression coefficient, 311-314
  sum of squares, 246
Estimate(s)
  best, 129-131
  interval, 128, 132-134
  minimum variance, 129, 130
  point, 128, 132
  pooled variance, 245
  regression, 303-304
  unbiased, 129
Estimation
  of confidence interval
    for difference between means of two normal populations, 146-149
    for difference of two proportions, 167-169
    for mean of a normal population, 140-146
    for paired observations, 153-156
    for a proportion, 162-166
    for a slope and intercept of a straight line, 305, 308, 309
    statement, 131
    for variance of a normal population, 157-159
    for variance ratios, 159-162
  point, 128-131
  statistical, 127-133
Estimator(s)
  best, 129-131
  consistent, 130, 131
  correlation, 366
  definition, 128
  efficient, 129
  interval, 132-134
    of regression coefficients, 313
    unbiased, 132-134
  minimum variance, 129
  point, 127-131
  pooled, 245-246

properties, 128-131
regression, 303, 329, 333
unbiased, 120, 128, 246, 261
Events, 7
complementary, 14
exhaustive, 14
mutually exclusive, 7, 10, 14, 16
simple, 7
Expectation
of binomial distribution, 123
of chi-square distribution, 135
of a function, 112
mathematical, 112
of mean squares in analysis of variance, 246, 261, 464, 468, 469, 474, 482, 494, 500, 509, 526
of a multivariate function, 113, 118
of random variable, 113, 115
properties of, 113-116
regression, 302-303
Expected mean squares, 246, 251, 261
Expected value(s) of
binomial distribution, 123
chi-square distribution, 135
constant, 113
constant times a random variable, 113
continuous random variable, 54, 115, 116
definition, 110-116
discrete random variable, 53, 113-115
linear combinations of random variables, 113, 118
notation, 53, 54
population variance, 112
proportion, 162
sample mean, 119
sample variance, 119, 120
Experiment
design of, 445-562
error defined, 446
random, 7
Experimental design, nature of, 1, 445, 448
Experimental unit, 445
Experimental value
distribution, 36ff.
partition of; <u>see</u> Analysis of variance, models
Exponential density function, 59-67, 72

F

F-test
equality of variances, 210-212
variance, 210
F-variate
definition of, 82, 138
degrees of freedom, 82, 138
distribution, 82, 138-140
table of, 630
f, observed frequency, 33

Factor
  definition, 241, 446
  level, 446
Factorial, 473-488, 534-542
  advantages, 477, 535
  calculations, 474-488
  completely randomized design, 473-484
  component of variance model, 473, 481
  confounding, 537-542
  definition of, 473
  degrees of freedom, 474, 475
  disadvantages, 537
  effects, 473
  expected mean squares, 474, 475, 482
  fixed effects model, 474
  interaction
    definition of, 473
    effects, 474
  main effects, 477, 535
  meaning of, 473
  mixed model, 476
  Model I (fixed), 474
  Model II (random), 475
  Model III (mixed), 476
  Model IV (mixed), 476
  notation, 535
  random effects model, 475
  randomized complete block design, 521, 522
  subsampling, 468-472, 508-521
  tests of hypotheses, 476
  treatment combinations, 473-481, 521
Fisher, R. A., and Yates, F., 444
Fitting
  of constants, 304-308
  of curves, 343-352, 354-355
  in multiple regression, 332-343
  in simple linear regression, 301-330
Fixed effects, 474
Four-way analysis of variance, 488-493
  pooling errors, 490
Freedom, degrees of
  chi-square, 134
  F-variate, 82, 138, 210
  t-distribution, 82, 136, 137
Frequency
  class, 37
  cumulative, 34
  distribution, 34
  histogram, 39, 43
  observed (f), 33, 42, 44
  polygon, 44-46, 50, 51
    cumulative, 44, 46, 48
  relative, 8, 42, 44
  relative cumulative, 42, 44, 46
Function
  chi-square, 81, 82, 134

cumulative distribution, 34
density; see distribution desired
gamma, 81
probability
binomial, 55-59
chi-square, 81, 134
exponential, 59-67, 72
F, 82, 138
hypergeometric, 68
negative binomial, 67
negative exponential, 67
normal, 73-81
point, 57
Poisson, 59-67
t, 82, 136
Weibull, 71-72

G

Gamma distribution, 73
Gamma function, 81
General linear model, 243-245
Goodness of fit tests, 215-217, 369-370
Graeco-Latin squares, 533-534
advantages, 533
arrangements, 533
assumptions, 534
model, 534
Grand population mean, 250
Grandage, A. H. E., 444
Grouped data, 245-248
sample mean of, 247
sample variance of, 247
Grouping, 448

H

H_0 hypothesis (or null hypothesis), 179
Heterogeneity
chi-square, 213-217
of variances, 208-212
Hierarchical designs, 495
Histogram, frequency, 42, 43, 45, 49
Homogeneity of variances
assumption of, 280-282
test for, 278-282
Hypergeometric distribution, 68-71
Hypotheses
alternative, 180
composite, 179
definition, 3, 179
null, 179
regression, 308, 318, 321, 334
simple, 179
tests of, 181-220, 308, 317-319

I

Incomplete block design, 545-549
Independent variable, 241, 302
Individual comparisons, 193-202, 252-255
Inference, statistical, 127
Inferences about populations, 127

Interaction
  definitions of, 448, 473
  in three-way analysis, 481-484
  in two-way analysis, 473-477
Intersection, 7
Interval
  class, 37
  confidence
    definition of, 133
    for contrasts, 266-270
    statement, 133
  estimate, 133
  estimator, 132-134
Interval estimators
  for difference of means of two normal populations, 146-149
  for difference of two proportions, 167-169
  for mean and variance of normal populations, 141-146, 158, 159
  for ratio of variances of two normal populations, 159-162
Intervals for, confidence
  difference of two means, 146-153
  mean, 141-146
  multiple comparisons, 271-278
  proportion, 165, 166
  ratio of two variances, 159-162
  variance, 157-159
Inverse of a matrix, 590
Inverse prediction in simple linear regression, 319-321

J

Joint distribution, 116-120
Joint probability function, 117
Joint probability density function, 117

L

λ, parameter of Poisson distribution, 60
Lack of fit, 328
  test for (in regression), 329, 435
Latin square, 522-533
Latin square design, 522-523
  advantages, 522
  analysis of variance, 524-529
  assumptions, 524
  calculations, 525-529
  components of variance, 529
  degrees of freedom, 526
  disadvantages, 523, 525
  examples of, 524
  expected mean squares, 526
  individual degrees of freedom, 526
  missing observations, 529-533
  model, 524
  relative efficiency, 554
  test of hypotheses, 524
Law of large numbers, 130-131

Laws of probability, 14-21
Least squares
  method of, 3, 304, 333
  non-linear, 354-355
  regression, 304, 333, 354
Length, class, 37, 39, 43
Level of factor, 446
Level of significance, 180
Limits, confidence, 132-134
  difference between two means, 146-153, 154-156
  mean, univariate, 141-146
  proportion, 165-169
  ratio of two standard deviations, 162
  ratio of two variances, 160-162
  in two-way analysis of variance, 266-268
  variance, 157-159, 266-270
Linear combination of random variables, 244, 249, 259
Linear correlation, coefficient of, 364-368
Linear equations, 364-368
Linear model, 301-303
Linear regression, 301-330
  in one independent variable, 301-330
  in two or more independent variables, 332-343
Linearity
  assumption of, 301
  of regression, 301-303
  test for, 308
Linearizing transformations in regression, 344, 353-354, 392-394
Location, measure of, 103-105
Logarithmic transformation, 353-354

M

Main effect, 535
Main plot; see whole plot
Mark, class, 37
Mathematical concepts, 7
Mathematical expectation, 112
Mathematical model; see specific regression models and experimental designs
Matrices
  algebraic manipulations, 585-592
  applications to regression, 330-332
Mean(s)
  arithmetic, 103
  binomial distribution, 56, 123
  chi-square distribution, 135
  comparison of two means, 146-153
  conditional, 302
  confidence interval for, 141-146
  confidence limits for, 141-146, 187

correction for, 261
definition of, 103
deviation from, 306
differences among several, 245-258, 271-278
differences between two, 193-202
distribution of, 74
estimation of, 140-146
expected value of, 119
Poisson distribution, 60
population (μ), 108
of probability distribution function, 52; also see Expected value
random variables, 32
sample ($\overline{X}$), 55, 103, 119
square
expected, 251, 252, 261
between groups, 247
within groups, 246
standard error, 106
tests concerning, 186-204
weighted, 104
Measure
of location, 103-105
of precision, 108
Measurement, definition of, 32
Median
definition, 105
sample, 103, 105
Method of least squares, 3, 304, 333
Minimum variance estimator, 130
Missing data or plot
Latin square, 529-533
randomized complete block, 504-507
Mixed model, 476
Mode
definition, 105
sample, 103, 105
Model
additive, 249, 259
analysis of variance, 249, 259
for completely randomized design, 463
F-test, 476
for factorials, 474-476
for Latin square, 524
general linear, 243-245
linear, 301, 303, 332, 344, 345
mixed, 476
multiple linear, 332
one-way, 244, 249
orthogonal, 425
for randomized complete block, 499
for regression, 301, 332, 344, 345, 351, 353-355
three-way, 481
two-way, 244, 259

Moments
definition, 121
mean, 121, 122
origin, 121
Multiple comparisons, 271-278
Duncan's multiple range test, 276
Scheffé method, 271-274, 277
Multiple linear correlation, 368-373
Multiple linear regression, 332-343
Multiplication
of matrices, 586
of probabilities, 15
Multiplicative law of probability, 15
Multivariate functions, expectation of, 113, 118
Mutually exclusive events, 7, 10, 14, 16

N

N, size of population, 33, 108
N (μ,σ), normally distributed with mean μ and standard deviation σ, 449
n, sample size, 33
n_o, average group size in ANOVA, 467
ν, degrees of freedom, 211
ν, parameter in χ^2-distribution, 82, 134
ν, parameter in F-distribution, 82, 138
ν, parameter in t-distribution, 82, 136
Negative binomial distribution, 67
Negative exponential distribution, 67
Nested designs, 493-499
model, 494
tests of hypotheses, 495
Nonlinear regression, 351, 354-355
Nonlinear transformations, 353-355
Normal approximation to distribution of sample means, 74
Normal distribution, 73-81
approximation to distribution of sample means, 74
areas, table of, 615
central limit theorem, 73
cumulative distribution, 76
density function of, 74
equation of, 74
mean of, 74
parameters, 76
percentage points of; see Table of areas, 615
probability density function, 74
properties of standarized, 76
sampling from, 74
standard, 76-81
variance of, 74, 157-159

Normal equations
  orthogonal, 426
  regression, 305, 333, 344
Normality, assumption of, 467
Normalizing transformations, 353-354, 394
Notation, 242, 243
Null hypothesis, 179
Null set; see Empty set
Numbers
  in classes, 37-43
    equal, 467, 469
    unequal, 468
  of replicates, 448-462

O

Observation(s)
  definition of, 32
  paired, 153-156, 202-204
One-sided (tailed) test, 182, 189, 190, 194, 195, 205, 207, 209
One-way analysis of variance, 244, 245-259
  table, 253
    equal observations within groups, 469
    unequal observations within groups, 468
Origin, moments about, 121
Orthogonal comparisons, 502-504
Orthogonal contrasts, 267-270 512-519
Orthogonal normal equations, 426
Orthogonal polynomials, 424-442
  model, 425
  partition of treatment sum of squares, 502
  quadratic form, 433
  recursive formula, 427
  regression, 425-428, 433, 434
  tests of significance, 434-442, 502-503, 517-519
Orthogonal polynomial coefficients, 429-432
Outcome of an experiment, 6
  sample points, 7

P

P, probability, 10-11
Π, product sign, 15
Paired comparisons, 153-156, 202-204
Paired observations, 153-156, 202-204
Parameters
  of binomial distribution, 56, 57
  of chi-square distribution, 81, 135
  definition, 52
  of F distribution, 82, 138
  of gamma distribution, 73
  of negative binomial distribution, 67
  of normal distribution, 75
  of Poisson distribution, 60
  of regression, 302

of standard normal distribution, 76
of t distribution, 82, 136
of Weibull distribution, 71
Partial correlation, 370
Partition
of experimental values; see Analysis of variance, models
of degrees of freedom, 248, 261
of sum of squares in regression, 307, 334
of variance, 248, 261
Permutations, 11-13
Planning of experiments, 445-446
Plot
missing, 504-507
split or sub, 541-545
whole, 542-544
Point binomial distribution, 57
Point estimate, 128-131
Point probability function, 33
Poisson approximation to binomial, 60, 62
Poisson distribution, 59-67
cumulative, 60
equation for, 60
Poisson probability function, 60
Poisson table of cumulative probabilities, 609
Polygon, frequency, 44-46, 50, 51
cumulative, 44-46, 50, 51
Polynomial
orthogonal, 424-442
regression, 343-352
second degree, 344-351, 439-442
Pooled estimate of variance, 149, 245, 246
Population
binomial, 55
correlation index, 366
covariance, 366
definition, 2, 31
finite, 32
grand mean, 250
mean, 108
interval estimator in normal case, 140-142, 146-148
sample mean as estimator of, 108, 142-146
median, 105
mode, 105
moments, 121, 122
variance, 108
interval estimator, in normal case, 157-159
pooled, 149, 245-246
sample variance as estimator of, 108
Power of a test, 190
Precision
definition of, 462
measure of, 108
(predicted (or estimated) value of regression coefficient,

305, 329, 333, 344, 353-354
Prediction (or regression) equation; *see* desired model
Prediction interval, 303-304
Principle of least squares, 3, 304, 333
Principles of experimental design, 445-446
Probabilities, binomial, 55-59
Probability
- a posterori, 21-24
- a priori, 21
- addition law, 15
- Bayes' theorem, 21, 22
- classical definition, 8
- compound, 15
- computation rules, 14-21
- concepts, 6-11
- conditional, 14, 17-20
- continuous distribution, 35, 54
- cumulative distribution function, 34
- definition of, 8, 9, 11
- density function, 33, 36
- discrete distribution, 33
- distribution, 36
- estimates of, 36, 44, 78
- event, 10
- function, 10
  - binomial, 55-59
  - chi-square, 81, 135
  - discrete, 33
  - exponential, 59-67, 72
  - F, 82, 138-140
  - gamma, 73
  - hypergeometric, 68-71
  - negative binomial, 67
  - normal, 73-81
  - point, 33
  - Poisson, 59-67
  - properties, 33, 54
  - t, 82, 136
  - Weibull, 71
- independence, definition, 14
- joint, 15, 117
- laws of, general, 14-21
- multiplicative law, 15
- notation, 7
- point function, 33
- properties, general, 7, 14-21
- relative frequency definition, 8, 36
- type 1 and 2 errors, 180

Probability distribution associated with a random variable, 54
- continuous, 54, 115
  - cumulative distribution, 36
  - expected value of, 54
  - mean of, 54
  - moments about mean, 121
  - moments about origin, 121

variance of, 54
definition of, 34, 35, 115
discrete, 54
cumulative distribution, 34
expected value of, 53
mean of, 52, 53
moments about mean, 121
moments about origin, 121
variance of, 53
repetitive events, 20
simple events, 8
Product sign (Π), 15
Properties
of cumulative distribution, 34, 54
of estimators, 128-131
of expectations, 113-115
Proportion, estimation and tests of, 162-169, 205-208
comparison of two, 167-169, 206-208, 213-215
confidence interval for difference between two, 165
variance of a, 163

Q

Quadratic form of regression, 344, 433
Quadratic regression, 344-351, 439-442

R

R, coefficient of correlation, 363
R^2, correlation index, 364
r, linear correlation coefficient, 364-368
r, sample correlation coefficient, 363-367
$r_{X_1X_2}$, r_{X_1Y}, partial correlation coefficients, 370
ρ, population correlation coefficient, 366
Random effects, 467, 475
Random experiment, 7
Random interval, 133
Random sample, 2, 102
Random sampling distributions, 119-120
chi-square, 81, 134
difference between two means, 193-202
F, 82, 138
mean, 103
normal populations, 73-81
sample proportion, 205
standard deviation, 106
t, 82, 136
variance, 105-107
Random variable
continuous, 36
definition of, 32
discrete, 32
expectation of, 115
Randomization
concept, 446
use in experimental design, 446

Randomized blocks; see Randomized complete block design
Randomized complete block design, 499-522
  advantages, 499
  analysis of variance, 500-504, 509
  assumptions, 499-500
  bias correction, 506
  calculations, 501-503
  components of variance, 499
  definition, 499
  efficiency, 554
  expected mean squares, 500, 509
  factorial treatments, 521-522
  method of use, 499
  missing observations, 504-507
  model, 499
  orthogonal polynomials with, 502, 512
  paired observations, 507-508
  partition of treatment sum of squares, 512-519
  purpose of, 499
  subdivision of experimental error, 508-509
  subsampling, 508-521
  t-test (paired observations), 508
  tables
    one observation, 500
    regression, 515-517
    subsampling, 509, 519
    tests of hypotheses, 501
Range
  definition of, 108
  sample, 108
Ratio of variances, 160-162, 210-212
Reduction in sums of squares, 306
Region
  acceptance, 181-184
  critical, 181, 183, 184
  rejection, 183, 184
Regression
  analysis, 301-394
    transformations, 344, 353-355
  analysis of variance
    multiple linear, 334-335
    simple linear, 321-328
    tables, 322, 335
  assumptions; see regression models
  cautions about, 354, 394, 425
  coefficient, 302, 333, 344, 353
    variances, 310, 312
  in comparisons, 388-392
  confidence ellipse, 315
  correlation analysis, 363-373
    linear regression, 364-368

multiple linear regression, 368-373
sample covariance, 366
tests of hypotheses, 367-368
correlation index, 363, 365
cubic, 377-382, 436-439
curvilinear, 343-352
definition, 301-302
degrees of freedom, 311
deviations from, 310
in each group, 388-392
error sum of squares, 304, 333, 354, 435
estimates, 303, 304
estimators, 304-306, 314
expectations, 302,
exponential, 351, 353-355
general remarks, 301
graphical interpretation, 304
of group means, 388-392
hypothesis, simple linear, 308
interval estimates, simple linear, 313, 314, 320
inverse prediction, 319-321
lack of fit, 328, 435
least squares, 304, 333, 354
line of Y on X, 302
linear correlation index, 364
linear model, 301, 303
by matrix techniques, 330-332
methods, 301
model
curvilinear, 343
multiple linear, 332
nth degree, 343
nonlinear, 351, 353-355
orthogonal, 425
polynomial, 343
quadratic, 344
simple linear, 301
multiple correlation index, 368
multiple linear, 332-343
nonlinear, 351, 354-355
normal equations, 305, 333, 344
orthogonal polynomials, 424-442
analysis of variance, 437-438, 440-441
lack of fit sum of squares, 435
model, 425
normal equations, 426
procedure, 435
quadratic form, 433
reduction in sum of squares, 434
table of values, 429-432
parameters, 302
partitioning sum of squares, 307, 334
polynomial, 343-352
pooled coefficient, 389-392

predicted values, 303, 319-321, 324, 333, 344, 353-355
quadratic, 344-351, 439-442
quadratic form, 304, 433
reduction in sum of squares, 307
relation to analysis of covariance, 366
relation to analysis of variance, 321-328, 335 434-442
several samples, 388, 389
simple linear, 301-330
sum of squares
deviation about regression, 307, 322, 334
due to regression, 307, 322, 334
error, 307, 322, 334
partitioning in simple linear regression, 307
tests of hypothesis
equality of slopes, 388-392
multiple linear, 334, 335
simple linear, 317-319
simultaneous linear, 318
through a point, 329-330
through origin, 317, 329
transformation of data, 392-394
error structure, 393
propagation of error, 392
reason for, 394
uses, 301
using matrices, 330-332
Rejection region, 183-184
Relative efficiency, 553-554
Relative frequency, 37
definition of probability, 8
Repeatability, 462
Replicates, number of, 448-462
tables of, 451, 456
Replication, 446

S

S, sample standard deviation, 106
S^2, sample variance, 105
$S^2_{\hat{\beta}_0}$, variance of intercept, 313
$S^2_{\hat{\beta}_1}$, variance of regression coefficient, 310
SS_E, error sum of squares, 306
$S_{\overline{X}}$, standard error of sample mean, 106
$S_{\overline{X}_1-\overline{X}_2}$, standard error of difference between two sample means, 150
$S^2_{\hat{Y}}$, standard error of predicted value of dependent variable in a regression analysis, 313
σ^2, population variance, 112
Sample
correlation coefficient, 363
covariance, 366
definition of, 2, 31
mean, 55, 103-105, 119, 141

expectation, 119
measurements, 103
median, 105
mode, 105
moments, 121, 122
obtaining, 445, 450
points, 7
random, 2, 102
range, 108
regression coefficient, 302, 333, 344, 353
size, 448-462
space, 7, 10, 32
finite, 11
standard deviation, 103, 106
standard error, 106
transformations, 103, 107
variance, 55, 105
computation of, for grouped and ungrouped data, 245-249
distribution of, for normal samples, 208-212
expectation, 119, 120
weighted mean, 104
Sampling
error, 244
without replacement, 68
Scheffé method for multiple comparisons, 271-274, 277
Scientific method, relation to statistics, 1
Set
of all possible outcomes of an experiment, 7
definition of, 7
empty, 7
number of elements in, 7
theory, 7, 10
union, 7
Significance level, 180
Significance test; see Testing of hypotheses
Simple effect, 536
Simple hypothesis, 179
Simple linear regression, 301-330
Simple random sample, 102
Size of experiment, 461-462
Size of sample, 448-462
Split plot (subplot)
analysis of variance, 544
definition of, 541-545
design, 542
model, 543
Squares, least, 4, 304, 333, 354
Standard deviation
confidence limit, 162
definition, 55
of mean, 106
Standard error
of difference between two means, 194, 197
of estimate, 308
of mean, 106

of predicted value in regression, 312, 313
of sample, 106
$\hat{Y}$, 312-314
Standard normal distribution
definition, 76-81
table of areas, 615
Statistical Analysis System, SAS®, 593-599
multiple treatment comparisons, 274-277
PROC ANOVA, 255, 265
PROC CHART, 47
PROC FREQ, 47
PROC GLM, 258, 323, 358, 471, 510, 560
PROC MEANS, 110, 204
PROC PLOT, 50
PROC PRINT, 47
PROC RSQUARE, 374, 376
PROC STEPWISE, 373-378
PROC TTEST, 201-202
Statistical control, 447
Statistical design of experiments, 445-448
Statistical estimation, 127-133
Statistical inference, 127
Statistical tests of hypotheses, 181-184
Statistics
definition (as a science), 1
descriptive, 2
relation to probability, 6
relation to research, 1
relation to scientific method, 1
scope of, 1
Steps in designing experiments, 445-446
"Student's" t, 82, 136; see also t distribution
Subplot, 542-544
Subsampling, 468-472, 508-521
Subset, 31
Sum
of matrices, 586
of squares
among cells, 508
corrected, 248, 306, 334
due to regression, 306, 334
errors, 246, 306
between groups, 247
within groups, 246
Synergistic effect, 473

T

t distribution, 82, 136
degrees of freedom, 82, 136, 137
density function, 82, 136
relation to normal distribution, 137
table of values, 623
t-test
concerning differences of means of two normal populations, 193
concerning means of normal populations, unknown vari-

ances, 194-202
t, random variable
definition, 82
degrees of freedom, 82, 136, 137
density function, 82, 136
Tables
analysis of covariance, 557
binomial distribution, 600
chi-square, 625
chi-square test of variance, 208-210
F distribution, 630
F test of ratio of two variances, 210-212
factorial treatment combinations, 474-476, 482
Latin square AOV, 526
number of replications
one-class data, 451
two-class data, 456
one-way AOV, 253
equal subclass numbers, 469
unequal subclass numbers, 468
orthogonal polynomial coefficients, 429-432
Poisson, 609
randomized complete block AOV
1 observation per cell, 500
subsampling, 509, 519
size of sample, 448-462
standard normal distribution, 615
subsampling, 469
t distribution, 623
t-test of difference between two means, 195
t-test of mean, 190
t-test for paired observations, 203
testing a proportion, 205
testing the difference between two proportions, 207
three-way AOV, 482
two-way AOV, 262
Test
power of, 180
significance level, 180
Test procedures, 182
Test statistic, 182
Tests of hypotheses
analysis of covariance, 558
analysis of variance
one-way, 250, 252-254
three-way, 484
two-way, 260, 262, 266
Bartletts', 278-282
chi-square, 208-210, 213-220
contingency, 217-220
for contrasts, 266-270
critical region, 183
definition, 179
difference between J means, 245-259
difference between two means, 193-202
difference between two proportions, 206-208

equality of variances, 210-212
general procedure, 182
goodness of fit, 215, 369
homogeneity of variances, 278-282
lack of fit, 435
mean, 140-153, 181-193
notation for
one-tailed, 184
two-tailed, 183
paired observations, 153-156, 202-204
proportion, 162-169, 205-208, 213-215
regression
equality of slopes, 388-392
multiple linear, 334-338
orthogonal polynomials, 435
quadratic, 348, 351
simple linear, 317-319, 322
simultaneous linear, 388-392
two-sided, 260, 262, 266
variance, 208-212

Tests of significance for orthogonal polynomials, 434-442

Theoretical distributions, 50, 52-83

Three-way analysis of variance, 481-484
assumptions, 484
model, 481
table, 482

Transformations
arithmetic, 103, 107, 344
exponential, 353
general discussion, 103, 394
logarithmic, 344, 353
nonlinear, 353-355
in regression analysis, 344, 353-355
sample, 103, 107
z, 76-81

Transpose of a matrix, 588

Treatment
combination, 446
comparison, 461-462
contrast; see Orthogonal contrasts
definition of, 241, 446

Two-sided (tailed) confidence limits, 132-138

Two-sided (tailed) tests, 182-184, 189, 190, 194, 195, 205, 207, 209

Two-way analysis of variance, 244, 249, 259-270
table, 262

Type 1 error, 180

Type 2 error, 180

U

Unbiased estimator, 129, 246, 261

Union, 7

Unequal observations
per cell, 467-468
per group, 247, 257
subclass numbers, 467-468
Unit, experimental, 446

V

Value
chi-square, 134-136
expected, 111-120
Variability, measures of, 105-110
Variable
continuous, 32, 35
dependent, 249
discrete, 32
independent, 241
random, 32
Variance
analysis of, 241-282; see also Analysis of variance
multiple linear, 334-335
simple linear, 321-328
three-way, 481-484
two-way confidence limit, 266-268
of binomial distribution, 57, 123
of chi-square distribution, 135
comparison of two, 159-162
confidence limits for, 158, 159
of continuous distributions, 52-54
definition of, 52
of discrete distributions, 52-53
equality, 210, 278
Bartletts' test, 278-282
F-test, 210-212
estimation of, 105-109
of group means, 247
homogeneity of regression models, 354
homogeneous, 280-282
normal distribution, 73-81
partition, 248, 261
pooled estimate, 149, 245
population, 108
of a proportion, 163
random variables, 113, 115
ratio, 160-162
regression coefficient, 310, 312
sample, 55, 105
of sample mean, 106
significance tests concerning; see Chi-square tests and F-tests
of standard normal distributions, 76
of t distribution, 82
tests about, 208-212, 278-282
of treatment mean, 465, 470, 553
unbiased estimator of, 129
Variation
coefficient of, 106

measures, 105-109
partition, 248, 261
Venn diagram, 16, 18

W

Weibull distribution, 71, 72
Weighted mean, w, 104
Whole plot, 542-544
Within groups sum of squares, 247-248

X

X, independent variable in regression, 302
$\overline{X}$, sample mean, 55, 103

Y

Y, dependent variable in regression, 302
$\hat{Y}$, estimated (or predicted) value, 303
$\hat{Y}$, value of Y estimated from regression, 303, 304
$Y-\hat{Y}$, deviation from regression, 307, 310

Z

z transformation, 76